FPGA PROTOTYPING
BY VHDL EXAMPLES

FPGA PROTOTYPING BY VHDL EXAMPLES
Xilinx MicroBlaze MCS SoC
Second Edition

Pong P. Chu
Cleveland State University

Registered Office
John Wiley & Sons, Inc., 111 River Street, Hoboken, NJ 07030, USA

Editorial Office
111 River Street, Hoboken, NJ 07030, USA

For details of our global editorial offices, customer services, and more information about Wiley products visit us at www.wiley.com.

Wiley also publishes its books in a variety of electronic formats and by print-on-demand. Some content that appears in standard print versions of this book may not be available in other formats.

Library of Congress Cataloging-in-Publication Data
Names: Chu, Pong P., 1959- author.
Title: FPGA prototyping by VHDL examples : Xilinx MicroBlaze MCS SoC / by Pong P. Chu, Cleveland State University.
Description: Second edition. | Hoboken, NJ, USA : Wiley, 2017. | Includes bibliographical references and index. |
Identifiers: LCCN 2017015720 (print) | LCCN 2017016756 (ebook) | ISBN 9781119282754 (pdf) | ISBN 9781119282761 (epub) | ISBN 9781119282747 (cloth)
Subjects: LCSH: Field programmable gate arrays--Design and construction. | Prototypes, Engineering. | VHDL (Computer hardware description language)
Classification: LCC TK7895.G36 (ebook) | LCC TK7895.G36 C485 2017 (print) | DDC 621.39/5--dc23
LC record available at https://lccn.loc.gov/2017015720

Cover image: Olympic National Park, Courtesy of Pong P. Chu
Cover design by Wiley

10 9 8 7 6 5 4 3 2 1

To my mother, Chi-Te, my wife, Lee, and my daughter, Patricia

CONTENTS

PART II   EMBEDDED SOC I:
VANILLA FPRO SYSTEM

PART III EMBEDDED SOC II: BASIC I/O CORES

PART IV   EMBEDDED SOC III: VIDEO CORES

20 Introduction to the Video System   439

PART V   EPILOGUE

PREFACE

HDL (hardware description language) and *FPGA* (field-programmable gate array) devices allow designers to quickly develop and simulate a sophisticated digital circuit, realize it on a prototyping device, and verify operation of the physical implementation. As the capacity of FPGA devices continues to grow, a device can accommodate an SoC (system on a chip) design, which integrates a processor, memory modules, I/O peripherals, and custom hardware accelerators into a single chip. This book uses a "learning by doing" approach and illustrates the FPGA and HDL development and design process by a series of examples in the SoC context.

The examples start with simple gate-level circuits, progress gradually through the RT (register-transfer) level modules, and lead to a functional embedded system with custom I/O peripherals and hardware accelerators. A simple SoC framework, *FPro* (abbreviated from the book title "FPGA Prototyping"), is introduced as a platform to integrate all the design examples together. An FPro system contains a Xilinx MicroBlaze MCS soft-core processor, a video subsystem, and the MMIO (memory-mapped I/O) subsystem that can incorporate custom I/O cores. Except for the processor, all components are designed and coded from scratch. All the hardware and software examples can be synthesized, compiled, and physically tested on the prototyping board.

Focus and audience

Focus The primary focus of this book is on developing efficient and reliable digital systems and effectively using HDL as a tool to describe the intended hardware. The HDL language itself is not the main subject and its coverage is limited to a small synthesizable subset. The book uses about a dozen proven code templates to provide the skeletal structures of various types of circuits. These templates are general and can easily be integrated to construct a large, complex system. Although this approach limits the "freedom" of syntactic expression, it helps us steer our effort to develop an innovative and efficient hardware architecture.

After discussing the fundamentals in Part I, the book illustrates more complicated and sophisticated designs in the SoC context. Along the way, readers will learn many system-level concepts, including the derivation of a soft-core processor and *IP* (*intellectual property*) core based system, the partition and integration of software and hardware, and the development of custom I/O peripherals and hardware accelerators.

Although the book is intended for beginning designers, the examples follow strict design guidelines and prepare readers for future endeavors. The coding and design practice is "forward compatible," which means that:

- The same practice can be applied to large design in the future.
- The same practice can aid other system development tasks, including simulation, timing analysis, verification, and testing.
- The same practice can be applied to ASIC technology and different types of FPGA devices.
- The code can be accepted by synthesis software from different vendors.

Audience and prerequisites   The intended audience is students in an advanced digital design course as well as practicing engineers who wish to learn about FPGA- and HDL-based development. Readers need to have a basic knowledge of digital systems, usually a required course in electrical engineering and computer engineering curricula, and a working knowledge of the C/C++ language. Prior exposure to computer architecture, embedded system, and operating system is not necessary but will be helpful.

Changes for the MicroBlaze MCS SoC Edition

This book is the successor edition of *FPGA Prototyping by VHDL Examples: Xilinx Spartan 3 Version*. The most significant change is that the new edition presents the hardware in the SoC context and covers many system-level concepts. Instead of treating each module as an isolated entity, the book integrates them into a single coherent SoC platform that allows readers to explore both hardware and software "programmability" and develop complex and interesting embedded system projects. The major revisions in this edition are:

- Add four general-purpose peripheral modules: multi-channel PWM (pulse width modulation), I^2C controller, SPI controller, and XADC (Xilinx analog-to-digital converter) controller.
- Introduce a music synthesizer constructed with a DDFS (direct digital frequency synthesis) module and an ADSR (attack-decay-sustain-release) envelope generator.
- Expand the original video controller into a complete stream-based video subsystem that incorporates a video synchronization circuit, a test-pattern generator, an OSD (on-screen-display) controller, a sprite generator, and a frame buffer.

- Introduce basic concepts of software-hardware co-design with Xilinx Micro-Blaze MCS soft-core processor.
- Provide an overview of the bus interconnect and interface circuit.
- Introduce basic embedded system software development.
- Suggest additional modules and peripherals for interesting and challenging projects.

Logistics

FPGA prototyping board   This book is prepared to be used with the *Nexys 4 DDR* FPGA prototyping board manufactured by Digilent Inc. It contains an Artix FPGA device and the needed I/O peripherals. All HDL codes and discussions in the book can be applied to this board directly. The less expensive *Basys 3* board can be used as well. This board incorporates fewer I/O peripherals and contains a smaller FPGA device.

Most peripherals discussed in the book are de facto industrial standards and the corresponding HDL codes can be used for other FPGA boards as long as they provide adequate analog interface circuits and connectors. Another option is to use stand-alone I/O peripheral modules or to construct the circuits on a breadboard.

Software   The book uses the Xilinx *Vivado WebPack edition* for hardware development and the Xilinx *SDK* for software development. Both software packages are free and can be downloaded from Xilinx's website.

PC accessories   The design examples involve interfaces to several PC peripheral devices, including a USB keyboard, a USB mouse, a VGA compatible monitor, and a powered speaker. These accessories are widely available and probably can be obtained from an old PC.

Book organization

The book is divided into four parts. Part I introduces the elementary HDL constructs and their hardware counterparts, and demonstrates the construction of a basic digital circuit with these constructs. It consists of six chapters:

- Chapter 1 describes the skeleton of an HDL program, the basic language syntax, and the logical operators. Gate-level combinational circuits are derived with these language constructs.
- Chapter 2 provides an overview of an FPGA device, prototyping board, and development flow.
- Chapter 3 introduces HDL's relational and arithmetic operators and routing constructs. These correspond to medium-sized components, such as comparators, adders, and multiplexers. Module-level combinational circuits are derived with these language constructs.
- Chapter 4 presents the codes for memory elements and the construction of "regular" sequential circuits, such as counters and shift registers, in which the state transitions exhibit a regular pattern.
- Chapter 5 discusses the construction of a finite state machine (FSM), which is a sequential circuit whose state transitions do not exhibit a simple, regular pattern.

- Chapter 6 presents the construction of an FSM with data path (FSMD). The FSMD is used to implement the register-transfer (RT) methodology, in which the system operation is described by data transfers and manipulations among registers.
- Chapter 7 covers the methods to infer FPGA's internal memory modules, which can then be used to construct buffers and lookup tables.

Part II introduces the hardware construction of an FPro system and the development of embedded software. A basic "vanilla" FPro system, which contains a timer core, a UART (universal asynchronous receiver and transmitter) core, a GPI (general-purpose input) core, and a GPO (general-purpose output) core, is used to illustrate the key concepts of the process. It consists of four chapters:

- Chapter 8 introduces the SoC development and provides an overview of the hardware organization and software structure of the FPro platform.
- Chapter 9 discusses the software development for an embedded system and the basic coding techniques to access low-level I/O cores.
- Chapter 10 covers the FPro bus protocol and the bus interface circuit and demonstrates the construction of basic GPI, GPO, and timer cores.
- Chapter 11 presents the construction of a more sophisticated UART core and the derivation of software device drivers.

Part III applies the techniques from Parts I and II to develop an array of I/O cores for the peripherals on the Nexys 4 DDR prototyping board. The I/O cores are constructed from scratch with custom hardware and device driver. Part III consists of nine chapters:

- Chapter 12 discusses the Xilinx device's internal analog-to-digital converter (XADC) and derives an interface circuits to retrieve the analog readings.
- Chapter 13 presents the design of a multi-channel PWM core and demonstrates its application for LED brightness adjustment and servo motor control.
- Chapter 14 converts the seven-segment LED control circuit and the switch debouncing circuit of Part I into I/O cores and integrates them into an FPro system.
- Chapter 15 provides an overview of the SPI protocol, covers the design of an SPI controller core, and shows its operation with Nexys 4 DDR board's ADXL362 three-axis accelerometer.
- Chapter 16 provides an overview of the I^2C protocol, discusses the design of an I^2C controller core, and demonstrates its operation with Nexys 4 DDR board's ADT7420 temperature sensor.
- Chapter 17 covers the design of a PS2 controller core, which can be connected to a PS2 mouse or a PS2 keyboard, and discusses the device driver routines to read and decode keyboard's scan codes and to obtain and process mouse's movement information and button activities.
- Chapter 18 discusses the construction of a DDFS (direct digital frequency synthesis) controller core with amplitude and frequency modulation and demonstrates its application as a music synthesizer.
- Chapter 19 augments the music synthesizer with an ADSR (attack-decay-sustain-release) envelope generator core, which can produce sound mimicking various music instruments.

Part IV discusses the development of a stream-based video subsystem. The subsystem provides a framework to generate and mix multiple video sources into a single video data stream for display. It consists of four chapters:

- Chapter 20 introduces the concept of stream data processing and constructs a basic video system with a test-pattern generator, a color-to-grayscale conversion circuit, and a frame synchronization circuit.
- Chapter 21 provides an overview of the FPro video subsystem framework and the FPro video core structure and demonstrates the stream interface with a line buffer.
- Chapter 22 presents the design of a sprite circuit, which adds an overlay of small animated objects on the screen, and applies the technique for a mouse pointer core and a "Pac-Man ghost character" core.
- Chapter 23 discusses the design of an OSD (on-screen-display) controller core, which produces an overlay of text similar to the subtitles on a TV screen.
- Chapter 24 covers the design of a frame buffer, which maintains a bit map for one screen.

The book also includes an Appendix with four tutorials. The tutorials consist of the following:

- Develop, synthesize, and implement a digital circuit on the Nexys 4 DDR board with Vivado.
- Perform simulation of an HDL program with Vivado's built-in simulator.
- Configure and instantiate Xilinx IP cores.
- Construct a basic FPro system with a Xilinx microBlaze MCS IP core and develop software with the Xilinx SDK platform.

Companion Website

On an accompanying website (`http://academic.csuohio.edu/chu_p`), additional information is available, including the following materials:

- Errata
- HDL and C/C++ code listings and relevant files
- Links to synthesis and simulation software
- Links to reference materials

The book contains a number of color figures. They are shown as grayscale in the printed version. These figures can be found on the website as well.

Errata   The book is self-prepared, which means that the author has produced all aspects of the text, including illustrations, tables, code listings, indexing, and formatting. As errors are always bound to happen, the accompanying website provides an updated errata sheet and a place to report errors.

P. P. CHU

Cleveland, Ohio
May, 2017

ACKNOWLEDGMENTS

The author would like to thank Dr. R. James Duckworth and Dr. Jackie F. Woldering for their suggestions and feedback. Part of this material is based upon work supported by the National Science Foundation under Grant No. 1504030. Any opinions, findings, and conclusions or recommendations expressed in this material are those of the author and do not necessarily reflect the views of the National Science Foundation.

All trademarks used or referred to in this book are the property of their respective owners.

P. P. Chu

PART I

BASIC DIGITAL CIRCUITS DEVELOPMENT

CHAPTER 1

GATE-LEVEL COMBINATIONAL CIRCUIT

HDL (hardware description language) is used to describe and model digital systems. VHDL is one of the two major HDLs. In this chapter, we use a simple comparator to illustrate the skeleton of a VHDL program. The description uses only logical operators and represents a gate-level combinational circuit, which is composed of simple logic gates.

1.1 OVERVIEW OF VHDL

VHDL stands for "VHSIC (very high-speed integrated circuit) hardware description language." It was originally sponsored by the U.S. Department of Defense and later transferred to the IEEE (Institute of Electrical and Electronics Engineers). The language is formally defined by IEEE Standard 1076. The standard was first ratified in 1987 and revised several times. The main versions are referred to as VHDL-87, VHDL-93, VHDL-2000, and VHDL-2008. The synthesis subset does not change significantly after VHDL-93 and this book mainly follows VHDL-93.

VHDL is intended for describing and modeling a digital system at various levels and is an extremely complex language. The focus of this book is on hardware design rather than the language. Instead of covering every aspect of VHDL, we introduce the key VHDL synthesis constructs by examining a collection of examples. Detailed VHDL coverage may be explored through the sources listed in the Bibliographic Notes section.

Table 1.1   Truth table of a 1-bit equality comparator

input	output
i0 i1	*eq*
0 0	1
0 1	0
1 0	0
1 1	1

In this chapter, we introduce the HDL concepts, basic VHDL language constructs, logical operators, and program structure. A simple gate-level combinational circuit is used for demonstration. In Chapter 3, we cover the more sophisticated VHDL operators and constructs and examine module-level combinational circuits, which are composed of intermediate-sized components, such as adders, comparators, and multiplexers.

1.2 GENERAL DESCRIPTION

Consider a 1-bit equality comparator with two inputs, i0 and i1, and an output, eq. The eq signal is asserted when i0 and i1 are equal. The truth table of this circuit is shown in Table 1.1.

Assume that we want to use basic logic gates, which include *not*, *and*, *or*, and *xor* cells, to implement the circuit. One way to describe the circuit is to use a sum-of-products format. The logic expression is

$$eq = i0 \cdot i1 + i0' \cdot i1'$$

One possible corresponding VHDL code is shown in Listing 1.1. We examine the language constructs and statements of this code in the following subsections.

Listing 1.1   Gate-level implementation of a 1-bit comparator

```
library ieee;
use ieee.std_logic_1164.all;
entity eq1 is
   port(
      i0, i1 : in std_logic;
      eq     : out std_logic
   );
end eq1;

architecture sop_arch of eq1 is
   signal p0, p1: std_logic;
begin
   -- sum of two product terms
   eq <= p0 or p1;
   -- product terms
   p0 <= (not i0) and (not i1);
   p1 <= i0 and i1;
end sop_arch;
```

1.2.1 Basic lexical rules

VHDL is case insensitive, which means that upper- and lowercase letters can be used interchangeably, and free formatting, which means that spaces and blank lines can be inserted freely. It is good practice to add proper spaces to make the code clear and to associate special meaning with cases. In this book, we reserve uppercase letters for constants.

An *identifier* is the name of an object and is composed of 26 letters, digits, and the underscore (_), as in i0, i1, and data_bus1_enable. The identifier must start with a letter.

The comments start with -- and the text after it is ignored. In this book, the VHDL keywords are shown in boldface type, as in **entity**, and the comments are shown in italics type, as in

> — *this is a comment*

1.2.2 Library and package

The first two lines,

```
library ieee;
use ieee.std_logic_1164.all;
```

invoke the std_logic_1164 package from the ieee library. The package and library allow us to add additional types, operators, functions, etc., to VHDL. The two statements are needed because a special data type is used in the code.

1.2.3 Entity declaration

The entity declaration

```
entity eq1 is
   port(
      i0, i1: in std_logic;
      eq: out std_logic
   );
end eq1;
```

essentially outlines the I/O signals of the circuit. The first line indicates that the name of the circuit is eq1, and the port section specifies the I/O signals. The basic format for an I/O port declaration is

```
signal_name1, signal_name2, ... : mode data_type;
```

The mode term can be **in** or **out**, which indicates that the corresponding signals flow "into" or "out of" of the circuit. It can also be **inout**, for bidirectional signals.

1.2.4 Data type and operators

VHDL is a *strongly typed language*, which means that an object must have a data type and only the defined values and operations can be applied to the object. Although VHDL is rich in data types, our discussion is limited to a small set of predefined types that are suitable for synthesis, mainly the std_logic type and its variants.

std_logic type   The `std_logic` type is defined in the `std_logic_1164` package and consists of nine values. Three of the values, `'0'`, `'1'`, and `'Z'`, which stand for logical 0, logical 1, and high impedance, can be synthesized. Two values, `'U'` and `'X'`, which stand for "uninitialized" and "unknown" (e.g., when signals with `'0'` and `'1'` values are tied together), may be encountered in simulation. The other four values, `'-'`, `'H'`, `'L'`, and `'W'`, are not used in this book.

A signal in a digital circuit frequently contains multiple bits, such as a data bus. The `std_logic_vector` data type, which is defined as an array with elements of `std_logic`, can be used for this purpose. For example, let `a` be an 8-bit input port. It can be declared as

```
a: in std_logic_vector(7 downto 0);
```

We can use a term like `a(7 downto 4)` to specify a desired range and a term like `a(1)` to access a single element of the array. The array can also be declared in ascending order:

```
a: in std_logic_vector(0 to 7);
```

We generally avoid this format since it is more natural to associate the MSB with the leftmost position.

Logical operators   Several logical operators, including **not**, **and**, **or**, and **xor**, are defined over the `std_logic_vector` and `std_logic` data type. Bitwise operation is used when an operator is applied to an object with the `std_logic_vector` data type. Note that the **and**, **or**, and **xor** operators have the same precedence and we need to use parentheses to specify the desired order of evaluation, as in

```
(a and b) or (c and d)
```

1.2.5   Architecture body

The architecture body,

```
architecture sop_arch of eq1 is
   signal p0, p1: std_logic;
begin
   -- sum of two product terms
   eq <= p0 or p1;
   -- product terms
   p0 <= (not i0) and (not i1);
   p1 <= i0 and i1;
end sop_arch;
```

describes operation of the circuit. VHDL allows multiple bodies associated with an entity, and thus the body is identified by the name `sop_arch` ("sum-of-products architecture").

The architecture body may include an optional declaration section, which specifies constants, internal signals, and so on. Two internal signals are declared in this program:

```
signal p0, p1: std_logic;
```

The main description, encompassed between **begin** and **end**, contains three *concurrent statements*. Unlike a program in a conventional language, such as C,

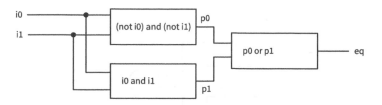

Figure 1.1 Graphical representation of a comparator program.

in which the statements are executed sequentially, concurrent statements are like circuit parts that operate in parallel. The signal on the left-hand side of a statement can be considered as the output of that part, and the expression specifies the circuit function and corresponding input signals. For example, consider the statement

```
eq <= p0 or p1;
```

It is a circuit that performs the or operation. When p0 or p1 changes its value, this statement is activated and the expression is evaluated. The new value is assigned to eq after the default propagation delay.

The graphical representation of this program is shown in Figure 1.1. The three circuit parts represent the three concurrent statements. The connections among these parts are implicitly specified by the signal and port names. The order of the concurrent statements is clearly irrelevant and the statements can be rearranged arbitrarily.

1.2.6 Code of a 2-bit comparator

We can expand the comparator to 2-bit inputs. Let the input be a and b and the output be aeqb. The aeqb signal is asserted when both bits of a and b are equal. The code is shown in Listing 1.2.

Listing 1.2 Gate-level implementation of a 2-bit comparator

```
library ieee;
use ieee.std_logic_1164.all;
entity eq2 is
    port(
        a, b : in std_logic_vector(1 downto 0);
        aeqb : out std_logic
    );
end eq2;

architecture sop_arch of eq2 is
    signal p0, p1, p2, p3: std_logic;
begin
    -- sum of product terms
    aeqb <= p0 or p1 or p2 or p3;
    -- product terms
    p0 <= ((not a(1)) and (not b(1))) and ((not a(0)) and (not b(0)));
    p1 <= ((not a(1)) and (not b(1))) and (a(0) and b(0));
    p2 <= (a(1) and b(1)) and ((not a(0)) and (not b(0)));
    p3 <= (a(1) and b(1)) and (a(0) and b(0));
end sop_arch;
```

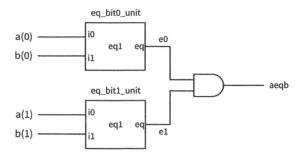

Figure 1.2 Construction of a 2-bit comparator from 1-bit comparators.

The a and b ports are now declared as a two-element std_logic_vector. Derivation of the architecture body is similar to that of a 1-bit comparator. The p0, p1, p2, and p3 signals represent the results of the four product terms, and the final result, aeqb, is the logic expression in sum-of-products format.

1.3 STRUCTURAL DESCRIPTION

A digital system is frequently composed of several smaller subsystems. This allows us to build a large system from simpler or predesigned components. VHDL provides a mechanism, known as *component instantiation*, to perform this task. This type of code is called a *structural description*.

An alternative to the design of the 2-bit comparator of Section 1.2.6 is to utilize the previously constructed 1-bit comparators as the building blocks. The diagram is shown in Figure 1.2, in which two 1-bit comparators are used to check the two individual bits and their results are fed to an and cell. The aeqb signal is asserted only when the two bits are equal.

The corresponding code is shown in Listing 1.3. Note that the entity declaration is the same and thus is not included.

Listing 1.3 Structural description of a 2-bit comparator

```
architecture struc_arch of eq2 is
   signal e0, e1 : std_logic;
begin
   -- instantiate two 1-bit comparators
   eq_bit0_unit : entity work.eq1(sop_arch)
      port map(
         i0 => a(0),
         i1 => b(0),
         eq => e0
      );
   eq_bit1_unit : entity work.eq1(sop_arch)
      port map(
         i0 => a(1),
         i1 => b(1),
         eq => e1
      );
   -- a and b are equal if individual bits are equal
   aeqb <= e0 and e1;
end struc_arch;
```

The code includes two component instantiation statements, whose syntax is

```
unit_label: entity lib_name.entity_name(arch_name)
    port map(
        formal_signal=>actual_signal,
        formal_signal=>actual_signal,
             . . .
    );
```

The first portion of the statement specifies which component is used. The `unit_label` term gives a unique id for an instance, the `lib_name` term indicates where (i.e., in which library) the component resides, and the `entity_name` and `arch_name` terms indicate the names of the entity and architecture. The `arch_name` term is optional. If it is omitted, the last compiled architecture body will be used. The second portion is port mapping, which indicates the connection between *formal signals*, which are I/O ports declared in a component's entity declaration, and *actual signals*, which are the signals used in the architecture body.

The first component instantiation statement is

```
eq_bit0_unit : entity work.eq1(sop_arch)
    port map(
        i0 => a(0),
        i1 -> b(0),
        eq => e0
    );
```

The `work` library is the default library in which the compiled entity and architecture units are stored, and `eq1` and `sop_arch` are the names of the entity and architecture defined in Listing 1.1. The port mapping reflects the connections shown in Figure 1.2. The component instantiation statement is also a concurrent statement and represents a circuit that is encompassed in a "black box" whose function is defined in another module.

This example demonstrates the close relationship between a block diagram and code. The code is essentially a textual description of a schematic. Although it is a clumsy way for humans to comprehend a diagram, it puts all representations in a single HDL framework.

The component instantiation statement is added in VHDL 93. Older codes may use the mechanism in VHDL 87, in which a component must first be declared (i.e., made known) and then used. The code in this format is shown in Listing 1.4.

Listing 1.4 Structural description with VHDL-87

```
architecture vhd_87_arch of eq2 is
    -- component declaration
    component eq1
        port(
            i0, i1 : in std_logic;
            eq     : out std_logic
        );
    end component;
    signal e0, e1: std_logic;
begin
    -- instantiate two 1-bit comparators
    eq_bit0_unit: eq1   -- use the declared name, eq1
        port map(
            i0=>a(0),
```

```
        i1=>b(0),
        eq=>e0
    );
  eq_bit1_unit: eq1    — use the declared name, eq1
    port map(
        i0=>a(1),
        i1=>b(1),
        eq=>e1
    );
  — a and b are equal if individual bits are equal
  aeqb <= e0 and e1;
end vhd_87_arch;
```

Note that the original clause

```
eq_bit0_unit: entity work.eq1(sop_arch)
```

is replaced by a clause with the declared component name

```
eq_bit0_unit: eq1
```

1.4 TOP-LEVEL SIGNAL MAPPING

When an HDL program is targeted to a physical device of a prototyping board, the design is subject to a variety of *constraints*. One constraint is the locations of the I/O pins. For example, the switches and LEDs of the board are "pre-wired" to specific I/O pins of the FPGA device and they cannot be altered. The pin assignment is defined in a *constraint file*, which is processed in conjunction with HDL files.

The designs of this book use a constraint file that specifies the pin assignment for all the I/O signals on the Nexys 4 DDR prototyping board. To use this file, the top-level HDL module must have the same predefined I/O signal names. This can be achieved by creating an HDL file to "wrap" the original design and map its original I/O signals to the prototyping board's I/O signals. For example, we name the I/O pins connected to the slide switches and LEDs as sw and led and specify their pin assignment in the constraint file. For a physical implementation, the a and b signals of the previous comparator circuit can be connected to the four switches and the output, aeqb, can be connected to an LED. The corresponding wrapping code is shown in Listing 1.5.

Listing 1.5   Top-level wrapping circuit

```
library ieee;
use ieee.std_logic_1164.all;
entity eq_top is
    port(
        sw  : in  std_logic_vector(3 downto 0); — 4 switches
        led : out std_logic_vector(0 downto 0)  — 1 red LED
    );
end eq_top;

architecture struc_arch of eq_top is
begin
    — instantiate 2-bit comparator
    eq2_unit : entity work.eq2(struc_arch)
        port map(
```

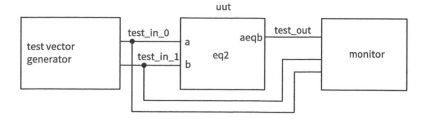

Figure 1.3   Testbench for a 2-bit comparator.

```
       a    => sw(3 downto 2),
       b    => sw(1 downto 0),
       aeqb => led(0)
   );
end struc_arch;
```

The code essentially maps the "logical" port names of the comparator to the physical signals on the prototyping board. Note that the output `led` signal is defined as a one-element vector to accommodate future expansion. The procedure to include the constraint file is demonstrated in Appendix A.2.

1.5   TESTBENCH

After code is developed, it can be *simulated* in a host computer to verify the correctness of the circuit operation and can be *synthesized* to a physical device. Simulation is usually performed within the same HDL framework. We create a special program, known as a *testbench*, to mimic a physical lab bench. The sketch of a 2-bit comparator testbench program is shown in Figure 1.3. The `uut` block is the unit under test, the `test vector generator` block generates testing input patterns, and the `monitor` block examines the output responses.

A simple testbench for the 2-bit comparator is shown in Listing 1.6.

Listing 1.6   Testbench for a 2-bit comparator

```
library ieee;
use ieee.std_logic_1164.all;
entity eq2_tb is
end eq2_tb;

architecture tb_arch of eq2_tb is
   signal test_in0, test_in1: std_logic_vector(1 downto 0);
   signal test_out: std_logic;
begin
   -- instantiate the circuit under test
   uut: entity work.eq2(struc_arch)
      port map(
         a=>test_in0,
         b=>test_in1,
         aeqb=>test_out
      );
   -- test vector generator
   process
   begin
```

```
        -- test vector 1
        test_in0 <= "00";
        test_in1 <= "00";
        wait for 200 ns;
        -- test vector 2
        test_in0 <= "01";
        test_in1 <= "00";
        wait for 200 ns;
        -- test vector 3
        test_in0 <= "01";
        test_in1 <= "11";
        wait for 200 ns;
        -- test vector 4
        test_in0 <= "10";
        test_in1 <= "10";
        wait for 200 ns;
        -- test vector 5
        test_in0 <= "10";
        test_in1 <= "00";
        wait for 200 ns;
        -- test vector 6
        test_in0 <= "11";
        test_in1 <= "11";
        wait for 200 ns;
        -- test vector 7
        test_in0 <= "11";
        test_in1 <= "01";
        wait for 200 ns;
        -- terminate simulation
        assert false
            report "Simulation Completed"
          severity failure;
    end process;
end tb_arch;
```

The code consists of a component instantiation statement, which creates an instance of a 2-bit comparator, and a process statement, which generates a sequence of test patterns.

The process statement is a special VHDL construct in which the operations are performed sequentially. Each test pattern is generated by three statements. For example,

```
        -- test vector 2
        test_in0 <= "01";
        test_in1 <= "00";
        wait for 200 ns;
```

The first two statements specify the values for the test_in0 and test_in1 signals, and the third indicates that the two values will last for 200 ns.

The code has no monitor. We can observe the input and output waveforms on a simulator's display, which can be treated as a "virtual logic analyzer." The simulated timing diagram of this testbench is shown in Figure 1.4. Writing code for a comprehensive test vector generator and a monitor requires detailed knowledge of VHDL and is beyond the scope of this book. This listing can serve as a testbench template for other combinational circuits. We can substitute the **uut** instance and modify the test patterns according to the new circuit.

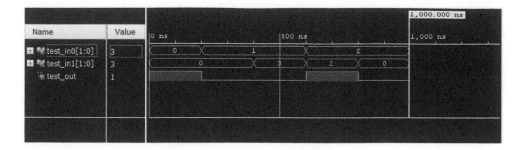

Figure 1.4   Simulated waveforms.

1.6   BIBLIOGRAPHIC NOTES

A short bibliographic section appears at the end of each chapter to provide some of
the most relevant references for further exploration. A comprehensive bibliography
is included at the end of the book.

VHDL is a complex language. *The Designer's Guide to VHDL* by P. J. Ashenden
provides detailed coverage of the language's syntax and constructs. The author's
RTL Hardware Design Using VHDL: Coding for Efficiency, Portability, and Scalability
provides a comprehensive discussion on developing effective, synthesizable
codes. The derivation of the testbench for a large digital system is a difficult
task. *Writing Testbenches: Functional Verification of HDL Models, 2nd edition*, by
J. Bergeron focuses on this topic.

1.7   SUGGESTED EXPERIMENTS

At the end of each chapter, some experiments are suggested as exercises. The experiments help us better understand the concepts and provide a hands-on opportunity
to design and debug actual circuits.

1.7.1   Code for gate-level greater-than circuit

Develop the HDL codes in Experiment 2.6.1. The code can be simulated and
synthesized after we complete Chapter 2.

1.7.2   Code for gate-level binary decoder

Develop the HDL codes in Experiment 2.6.2. The code can be simulated and
synthesized after we complete Chapter 2.

CHAPTER 2

OVERVIEW OF FPGA AND EDA SOFTWARE

An FPGA (field-programmable gate array) prototyping board is used to implement the design examples and projects of this book. We provide an overview of FPGA devices, the Nexys 4 DDR prototyping board, and the development process in this chapter.

2.1 FPGA

2.1.1 Overview of a general FPGA device

An *FPGA (field-programmable gate array)* is a logic device that contains a two-dimensional array of generic *logic cells* and *programmable switches*. The conceptual structure of an FPGA device is shown in Figure 2.1. A logic cell can be configured (i.e., *programmed*) to perform a simple function, and a programmable switch can be customized to provide interconnections among the logic cells. A custom design can be implemented by specifying the function of each logic cell and selectively setting the connection of each programmable switch. Once the design and synthesis are completed, we can use a simple adaptor cable to download the desired logic cell and switch configuration to the FPGA device and obtain the custom circuit. Since this process can be done "in the field" rather than "in a fabrication facility (fab)," the device is known as *field programmable*.

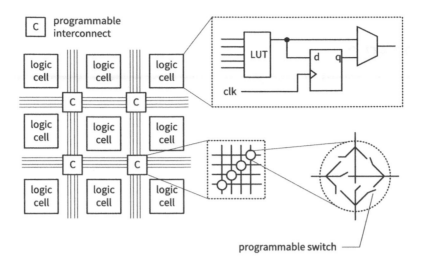

Figure 2.1   Conceptual structure of an FPGA device.

LUT-based logic cell   A logic cell usually contains a small configurable combinational circuit with a D FF (D-type flip-flop). The most common method to implement a configurable combinational circuit is an *LUT* (*lookup table*). An n-input LUT can be considered as a small 2^n-by-1 memory. By properly writing the memory content, we can use the LUT to implement any n-input combinational function. The conceptual diagram of a five-input LUT-based logic cell is shown in the top right of Figure 2.1. Note that the output of the LUT can be used directly or stored to the D FF. The latter can be used to implement sequential circuits.

Macro cell   Most FPGA devices also embed certain *macro cells* or *macro blocks*. These are designed and fabricated at the transistor level, and their functionalities complement the general logic cells. Commonly used macro cells include memory blocks, combinational multipliers, clock management circuits, and I/O interface circuits. Advanced FPGA devices may even contain one or more prefabricated processor cores.

2.1.2   Overview of the Xilinx Artix-7 devices

This book uses Xilinx Artix-7 family FPGA devices and we provide a brief overview of this device in this section.

Logic cell, slice, and CLB   The most basic element of an Artix-7 device is the *logic cell* (*LC*). A logic cell contains one LUT, which can be configured either as one 6-input LUT or as two 5-input LUTs, and two FFs. In addition, a logic cell contains a carry circuit, which is used to implement arithmetic functions, and a multiplexing circuit, which is used to implement wide multiplexers. Some LUTs can also be configured as a small *distributed SRAM* (static random access memory) module or a *shift register*. To increase flexibility and improve performance, eight logic cells are combined together with a special internal routing structure. In Xilinx terms, four logic cells are grouped to form a *slice*, and two slices are grouped to form a *configurable logic block* (CLB).

Table 2.1   Devices in the Artix-7 family

Device	Num. of LCs	Num. of 36-Kb BRAMs	BRAM bits	Num. of DSP slices	Num. of MMCMs
XC7A15T	16,640	25	900K	45	5
XC7A35T	33,280	50	1,800K	90	5
XC7A50T	52,160	75	2,700K	120	5
XC7A75T	75,520	105	3,780K	180	6
XC7A100T	101,440	135	4,860K	240	6
XC7A200T	215,360	365	13,140K	740	10

Macro cell   The Artix-7 device contains several types of macro cells. The *MMCM* (*mixed-mode clock manager*) macro cell is a clock management core that can produce a wide range of frequencies from a single oscillator input, reduce clock skew, and adjust the phase shift of a clock signal. The *BRAM* (*block random access memory*) macro cell is a 36K-bit dual-port synchronous SRAM that can be arranged in various types of configurations. The *DSP* (*digital signal processing*) macro cell is composed of a 25-by-18 binary multiplier and a 48-bit accumulator and is intended to support computation intensive DSP algorithms. An *IOB* (*input/output block*) macro cell is associated with a physical I/O pin of the FPGA device. It can be configured to support a wide variety of I/O signaling standards and high-speed serial data links. The *XADC* (*Xilinx analog-to-digital converter*) contains two 12-bit analog-to-digital converters. In addition to these, the device may also include special blocks for the gigabit ethernet transceivers and the PCI express bus.

Devices in the Artix-7 family   Although Artix-7 FPGA devices have similar types of logic cells and macro cells, their densities differ. The family contains an array of devices of various densities. The numbers of logic cells, 36K-bit BRAMs, DSP slices, and MMCMs of the devices are summarized in Table 2.1. The *Nexys 4 DDR* prototyping board used in the book contains an Artix XC7A100T device. The simpler *Basys 3* board contains a smaller Artix XC7A35T device

2.2   OVERVIEW OF THE DIGILENT NEXYS 4 DDR BOARD

The Digilent Nexys 4 DDR board is designed around an Artix XC7A100T device and has an array of built-in peripherals. The layouts of the board are shown in Figure 2.2.

The main components and connectors are as follows:

1. Power jack for optional external power supply
2. Shared USB JTAG and UART port
3. Artix XC7A100T FPGA device
4. Pmod port (JD)
5. Pmod port (JC)
6. Sixteen discrete LEDs
7. Sixteen slide switches
8. Temperature sensor
9. Eight-digit seven-segment LED display

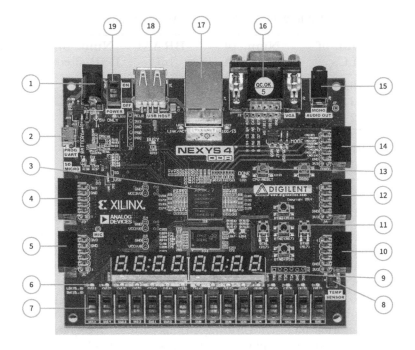

Figure 2.2   Nexys 4 DDR board.

10. Pmod port (JB)
11. Five pushbutton switches
12. Pmod port (JA)
13. Soft-core processor reset button
14. Pmod port with analog input (connected to XADC)
15. Audio jack
16. VGA port
17. Ethernet connector
18. USB host port (connected to USB mouse/keyboard)
19. Power-on switch

2.3   DEVELOPMENT FLOW

The simplified development flow of an FPGA-based system is shown in Figure 2.3. To facilitate further reading, we follow the terms used in the Xilinx documentation. The left portion of the flow is the *refinement and programming process*, in which a system is transformed from an abstract textual HDL description to a device cell-level configuration and then downloaded to the FPGA device. The right portion is the *validation process*, which checks whether the system meets the functional specification and performance goals. The major steps in the flow are:

1. Design the system and derive the HDL file(s). We may need to add a separate constraint file to specify certain implementation constraints, such as the pin assignment and the clock frequency.

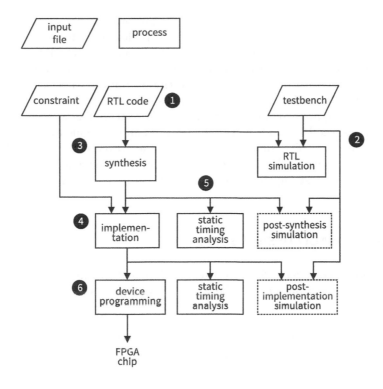

Figure 2.3   Development flow.

2. Develop the testbench in HDL and perform *RTL simulation*. The RTL term reflects the fact that the HDL code is done at the *register transfer level*. The simulation is performed to verify code syntax and to confirm that the HDL description meets the intended specification and functions.

3. Perform *synthesis*. The synthesis process is generally known as *logic synthesis*, in which the software transforms the HDL constructs to generic gate-level components, such as simple logic gates and FFs.

4. Perform *implementation*. The *implementation* process consists of three smaller subprocesses: translate, technology mapping, and placement and routing. The *translate* subprocess merges multiple design files to a single netlist. The *technology mapping* subprocess maps the generic gates in the netlist to FPGA's logic cells. The *placement and routing* subprocess derives the physical layout inside the FPGA chip. It places the cells in physical locations and determines the routes to connect various signals.

5. Examine the timing report. In the Xilinx flow, *static timing analysis*, which determines various timing parameters, such as the setup time slack, is performed at the end of the synthesis process and at the end of the implementation process. The latter is more accurate since the wire delays (correlated to path's length) are known and can be used for calculation.

6. Generate and download the programming file. In this process, a configuration file, which is also known as a *bit file*, is generated according to the final netlist. This file is downloaded to an FPGA device serially to configure the logic cells and programmable switches. The physical circuit can be verified accordingly.

The optional *post-synthesis simulation* can be performed after synthesis, and the optional *post-implementation simulation* can be performed after implementation. Post-synthesis simulation uses a synthesized netlist to replace the RTL description and checks the correctness of the synthesis process. Post-implementation uses the final netlist, along with detailed wire delay timing data, to perform simulation. Because of the complexity of the netlist, post-synthesis and post-implementation simulation may require a significant amount of time. If we follow good design and coding practices, the HDL code will be synthesized and implemented correctly. We only need to use the RTL simulation to check the correctness of the HDL code and use static timing analysis to verify the relevant timing information. Both post-synthesis and post-implementation simulations may be omitted from the development flow.

2.4 XILINX VIVADO DESIGN SUITE

We use *Vivado Design Suite* for hardware development in this book. Vivado is an integrated design environment for the Xilinx FPGA product and incorporates all the software tools discussed in Figure 2.3. A typical Vivado window is shown in Figure 2.4. A "watered-down" version, *Vivado WebPack edition*, can be downloaded for free and is adequate for the design and projects in this book.

As FPGA's capability and capacity continue to grow, the EDA (electronic design automation) tools evolve in a similar pace. The software is updated and patched almost on a quarterly basis. While the detailed description of the vendor's tools is beyond the scope of this book, several short tutorials are provided in the Appendix. Appendix A.1 provides an overview of the Vivado Design Suite environment, Appendix A.2 illustrates the hardware development flow, and Appendix A.3 demonstrates the RTL simulation.

2.5 BIBLIOGRAPHIC NOTES

Relevant information for the Artix-7 device can be found in its data sheets. Xilinx's *7 Series FPGAs Overview* provides a high-level overview and its user guide, *UG474 7 Series FPGAs Configurable Logic Block User Guide* gives a detailed explanation of the logic cells. *The Design Warrior's Guide to FPGAs* by Clive Maxfield provides a comprehensive review of FPGA-related issues.

The detailed layout and relevant information of the Nexys 4 DDR board can be found in *Nexys 4 DDR FPGA Board Reference Manual*.

2.6 SUGGESTED EXPERIMENTS

2.6.1 Gate-level greater-than circuit

The greater-than circuit compares two inputs, a and b, and asserts an output when a is greater than b. We want to create a 4-bit greater-than circuit from the bottom up and use only gate-level logical operators. Design the circuit as follows:

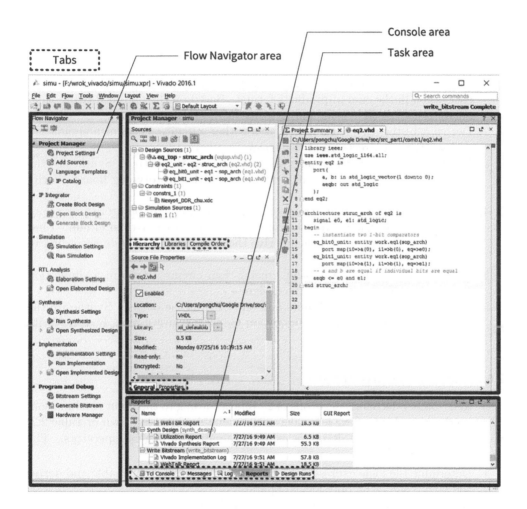

Figure 2.4   Vivado window.

Table 2.2   Truth table of a 2-to-4 decoder with enable

	input		output
en	$a(1)$	$a(0)$	$bcode$
0	–	–	0000
1	0	0	0001
1	0	1	0010
1	1	0	0100
1	1	1	1000

1. Derive the truth table for a 2-bit greater-than circuit and obtain the logic expression in the sum-of-products format. Based on the obtained expression, derive the HDL code using only logical operators.
2. Derive a testbench for the 2-bit greater-than circuit. Perform a simulation and verify the correctness of the design.
3. Use four switches as the inputs and one LED as the output. Synthesize the circuit and download the configuration file to the prototyping board. Verify its operation.
4. Use the 2-bit greater-than circuits and 2-bit equality comparators and a minimal number of "glue gates" to construct a 4-bit greater-than circuit. First draw a block diagram and then derive the structural HDL code according to the diagram.
5. Derive a testbench for the 4-bit greater-than circuit. Perform a simulation and verify the correctness of the design.
6. Use eight switches as the inputs and one LED as the output. Synthesize the circuit and download the configuration file to the prototyping board. Verify its operation.

2.6.2   Gate-level binary decoder

An n-to-2^n binary decoder asserts one of 2^n bits according to the input combination. The functional table of a 2-to-4 decoder with an enable signal is shown in Table 2.2. We want to create several decoders using only gate-level logical operators. The procedure is as follows:

1. Determine the logic expressions for the 2-to-4 decoder with enable and derive the HDL code using only logical operators.
2. Derive a testbench for the decoder. Perform a simulation and verify the correctness of the design.
3. Use two switches as the inputs and four LEDs as the outputs. Synthesize the circuit and download the configuration file to the prototyping board. Verify its operation.
4. Use the 2-to-4 decoders to derive a 3-to-8 decoder. First draw a block diagram and then derive the structural HDL code according to the diagram.
5. Derive a testbench for the 3-to-8 decoder. Perform a simulation and verify the correctness of the design.

6. Use three switches as the inputs and eight LEDs as the outputs. Synthesize the circuit and download the configuration file to the prototyping board. Verify its operation.
7. Use the 2-to-4 decoders to derive a 4-to-16 decoder. First draw a block diagram and then derive the structural HDL code according to the diagram.
8. Derive a testbench for the 4-to-16 decoder. Perform a simulation and verify the correctness of the design.

CHAPTER 3

RT-LEVEL COMBINATIONAL CIRCUIT

The gate-level circuits discussed in Chapter 1 utilize simple logical operators to describe gate-level design, which is composed of simple logic cells. In this chapter, we examine the HDL description of module-level circuits, which are composed of intermediate-sized components, such as adders, comparators, and multiplexers. Since these components are the basic building blocks used in *register-transfer methodology*, it is sometimes referred to as RT-level (or RTL) design. We first discuss more sophisticated VHDL operators and routing constructs and then demonstrate the RT-level combinational circuit design through a series of examples.

3.1 RT-LEVEL COMPONENTS

In addition to the logical operators, relational operators and several arithmetic operators can also be synthesized automatically. These operators correspond to intermediate-sized module-level components, such as comparators and adders. We examine these operators in this section and also cover miscellaneous synthesis-related VHDL constructs. Tables 3.1 and 3.2 summarize the operators and their applicable data types used in this book.

Table 3.1   Operators and data types of VHDL-93 and IEEE `std_logic_1164` package

Operator	Description	Data type of operands	Data type of result
a ** b	exponentiation	integer	integer
a * b	multiplication		
a / b	division	*integer type for constants and*	
a + b	addition	*array boundaries, not synthesis*	
a − b	subtraction		
a & b	concatenation	1-D array, element	1-D array
a = b	equal to	any	boolean
a /= b	not equal to		
a < b	less than	scalar or 1-D array	boolean
a <= b	less than or equal to		
a > b	greater than		
a >= b	greater than or equal to		
not a	negation	boolean, std_logic,	same as operand
a **and** b	and	std_logic_vector	
a **or** b	or		
a **xor** b	xor		

Table 3.2   Overloaded operators and data types in the IEEE `numeric_std` package

Overloaded operator	Description	Data type of operands	Data type of result
a * b	arithmetic	unsigned, natural	unsigned
a + b	operation	signed, integer	signed
a − b			
a = b			
a /= b			
a < b	relational	unsigned, natural	boolean
a <= b	operation	signed, integer	boolean
a > b			
a >= b			

Table 3.3   Type conversions between `std_logic_vector` and numeric data types

Data type of a	To data type	Conversion function/type casting
unsigned, signed	std_logic_vector	std_logic_vector(a)
signed, std_logic_vector	unsigned	unsigned(a)
unsigned, std_logic_vector	signed	signed(a)
unsigned, signed	integer	to_integer(a)
natural	unsigned	to_unsigned(a, size)
integer	signed	to_signed(a, size)

3.1.1   Relational operators

Six relational operators are defined in the VHDL standard: = (equal to), /= (not equal to), < (less than), <= (less than or equal to), > (greater than), and >= (greater than or equal to). These operators compare operands of the same data type and return a value of the boolean data type. In this book, we don't use the boolean data type directly but embed it in routing constructs. This is discussed in Sections 3.2 and 3.4. During synthesis, comparators are inferred for these operators.

3.1.2   Arithmetic operators

In the VHDL standard, arithmetic operations are defined for the integer data type and for the natural data type, which is a subtype of integer containing zero and positive integers. We usually prefer to have more control in synthesis and define the exact number of bits and format (i.e., signed or unsigned). The IEEE numeric_std package is developed for this purpose. In this book, we use the integer and natural data types for constants and array boundaries but not for synthesis.

IEEE numeric_std package   The IEEE numeric_std package adds two new data types, unsigned and signed, and defines the relational and arithmetic operators over the new data types (known as *operator overloading*). The unsigned and signed data types are defined as an array with elements of the std_logic data type. The array is interpreted as the binary representation of unsigned or signed integers. We have to add an additional use statement to invoke the package:

```
library ieee;
use ieee.std_logic_1164.all;
use ieee.numeric_std.all;    -- invoke numeric_std package
```

The synthesizable overloaded operators are summarized in Table 3.2.

Multiplication is a complicated operation and synthesizing the multiplication operator * from normal logic cells consumes a lot of resources. Modern FPGA devices provide prefabricated multiplier macro cells. When the * operator is encountered, the synthesis software infers a macro cell rather than synthesizing the circuit from logic cells. Xilinx Artix-7 FPGA family embeds the prefabricated 25-by-18 binary multipliers in DSP slices and Vivado software can infer these for the * operator. While the multiplication operator is supported, we need to be aware of the limitation on the number and input width of the multiplier macro cells and use them with care.

Type conversion   Because VHDL is a strongly typed language, std_logic_vector, unsigned, and signed are treated as different data types even when all of them are defined as an array with elements of the std_logic data type. A *conversion function* or *type casting* is needed to convert signals of different data types. The conversion is summarized in Table 3.3. Note that the std_logic_vector data type is not interpreted as a number and thus cannot be converted directly to an integer, and vice versa.

The following examples illustrate the common mistakes and remedies for type conversion. Assume that some signals are declared as follows:

```
library ieee;
use ieee.std_logic_1164.all;
use ieee.numeric_std.all;
. . .
signal s1, s2, s3, s4, s5, s6 : std_logic_vector(3 downto 0);
signal u1, u2, u3, u4, u5, u6, u7 : unsigned(3 downto 0);
. . .
```

Let us first consider the following assignment statements:

```
u1 <= s1;    -- not ok, type mismatch
u2 <= 5;     -- not ok, type mismatch
s2 <= u3;    -- not ok, type mismatch
s3 <= 5;     -- not ok, type mismatch
```

They are all invalid because of type mismatch. The righthand-side expression must be converted to the data type of the lefthand-side signal:

```
u1 <= unsigned(s1);      -- ok, type casting
u2 <= to_unsigned(5,4);  -- ok, conversion function
s2 <= std_logic_vector(u3);  -- ok, type casting
s3 <= std_logic_vector(to_unsigned(5,4));  -- ok
```

Note that two type conversions are needed for the last statement.

Let us consider statements that involve arithmetic operations. The following statements are valid since the + operator is defined with the **unsigned** and **natural** types in the IEEE **numeric_std** package.

```
u4 <= u2 + u1;   -- ok, both operands unsigned
u5 <= u2 + 1;    -- ok, operands unsigned and natural
```

In contrast, the following statements are invalid since no overloaded arithmetic operation is defined for the **std_logic_vector** data type:

```
s5 <= s2 + s1;  -- not ok, + undefined over the types
s6 <= s2 + 1;   -- not ok, + undefined over the types
```

To fix the problem, we must convert the operands to the **unsigned** (or **signed**) data type, perform addition, and then convert the result back to the **std_logic_vector** data type. The revised code becomes

```
s5 <= std_logic_vector(unsigned(s2) + unsigned(s1));  -- ok
s6 <= std_logic_vector(unsigned(s2) + 1);             -- ok
```

Other arithmetic packages   There are several non-IEEE arithmetic packages, which are the **std_logic_arith**, **std_logic_unsigned**, and **std_logic_signed** packages. The **std_logic_arith** package is similar to the **numeric_std** package. The other two packages do not introduce any new data type but define overloaded arithmetic operators over the **std_logic_vector** data type. This approach eliminates the need for data conversion.

VHDL-2008 introduces two new standard arithmetic packages, which are the **numeric_std_unsigned** and **numeric_std_signed** packages. They define overloaded arithmetic operators over the **std_logic_vector** data type and are similar to the **std_logic_unsigned**, and **std_logic_signed** packages.

Although using these packages seems to be less cumbersome initially, it is not a good practice. A large digital system is composed of many subsystems. These subsystems may be developed independently and use different types of number formats. It will be difficult to integrate and interface these subsystems since std_logic_vector is used to represent both signed and unsigned numbers. We do not use these packages in this book.

3.1.3   Other synthesis-related VHDL constructs

Concatenation operator   The concatenation operator, &, combines segments of elements and small arrays to form a large array. The following example illustrates its use:

```
signal a1 : std_logic;
signal a4 : std_logic_vector(3 downto 0);
signal b8, c8, d8 : std_logic_vector(7 downto 0);
. . .
b8 <= a4 & a4;
c8 <= a1 & a1 & a4 & "00";
d8 <= b8(3 downto 0) & c8(3 downto 0);
```

Implementation of the concatenation operator involves reconnection of the input and output signals and only requires "wiring."

One major application of the & operator is to perform shifting operations. Although both VHDL standard and numeric_std package define shift functions, they sometimes cannot be synthesized automatically. The & operator can be used for shifting a signal for a fixed amount, as shown in the following example:

```
signal a : std_logic_vector(7 downto 0);
signal rot, shl, sha : std_logic_vector(7 downto 0);
. . .
-- rotate a to right 3 bits
rot <= a(2 downto 0) & a(7 downto 3);
-- shift a to right 3 bits and insert 0 (logic shift)
shl <= "000" & a(7 downto 3);
-- shift a to right 3 bits and insert MSB
-- (arithmetic shift)
sha <= a(7) & a(7) & a(7) & a(7 downto 3);
```

An additional routing circuit is needed if the amount of shifting is not fixed. The design of a barrel shifter is discussed in Section 3.7.3.

'Z' value of std_logic   The std_logic data type has a value of 'Z', which implies *high impedance* or an open circuit. It is not a normal logic value and can only be synthesized by a *tristate buffer*. The symbol and function table of a tristate buffer are shown in Figure 3.1. Operation of the buffer is controlled by an enable signal, oe ("output enable"). When it is '1', the input is passed to output. On the other hand, when it is '0', the y output appears to be an open circuit. The code of the tristate buffer is

```
y <= a_in when oe='1' else 'Z';
```

The most common application for a tristate buffer is to implement a *bidirectional port* to better utilize a physical I/O pin. A simple example is shown in Figure 3.2. The dir signal controls the direction of signal flow of the bi pin. When it is '0', the tristate buffer is in the high-impedance state and the sig_out signal is blocked.

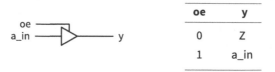

oe	y
0	Z
1	a_in

Figure 3.1 Symbol and functional table of a tristate buffer.

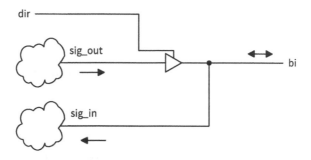

Figure 3.2 Single-buffer bidirectional I/O port.

The pin is used as an input port and the input signal is routed to the `sig_in` signal. When the `dir` signal is '1', the pin is used as an output port and the `sig_out` signal is routed to an external circuit. The HDL code can be derived according to the diagram:

```
entity bi_demo is
    port(
        bi: inout std_logic;
        . . .
    )
begin
    sig_out <= output_expression;
    . . .
    some_signal <= expression_with_sig_in;
    . . .
    bi      <= sig_out when dir='1' else 'Z';
    sig_in <= bi;
    . . .
```

Note that the mode of the `bi` port must be declared as **inout** for bidirectional operation.

In today's FPGA devices, a tristate buffer exists only in the I/O macro cell of a physical pin. Thus, the tristate buffer cannot be used for internal logic. It can only be used for I/O ports that are mapped to the physical pins of an FPGA device.

3.1.4 Summary

Because of the nature of a strongly typed language, the data type frequently confuses a new VHDL user. Since this book is focused on synthesis, only a small set of data types and operators are needed. Their uses can be summarized as follows:

- Use the `std_logic` and `std_logic_vector` data types in entity port declaration and for the internal signals that involve no arithmetic operations.
- Use the 'Z' value only to infer a tristate buffer.
- Use the IEEE `numeric_std` package and its `unsigned` or `signed` data types for the internal signals that involve arithmetic operation.
- Use the data type casting or conversion functions in Table 3.3 to convert signals and expressions among the `std_logic_vector` and various numerical data types.
- Use VHDL's built-in `integer` data type and arithmetic operators for constant and array boundary expressions, but not for synthesis (i.e., not used as a data type for a signal).
- Embed the result of a relational operation, which is in the `boolean` data type, in routing constructs (discussed in Section 3.2).
- Use a user-defined two-dimensional data type for two-dimensional storage array (discussed in Chapter 7) and bus controller slot interface (discussed in Chapter 10).
- Use a user-defined *enumerate data type* for the symbolic states of a finite state machine (discussed in Chapter 5).

3.2   ROUTING CIRCUIT WITH CONCURRENT ASSIGNMENT STATEMENTS

The *conditional signal assignment* and *selected signal assignment* statements are concurrent statements. Their behaviors are somewhat like the if and case statements of a conventional programming language. Instead of being executed sequentially, these statements are mapped to a routing network during synthesis.

3.2.1   Conditional signal assignment statement

Syntax and conceptual implementation   The simplified syntax of a conditional signal assignment statement is

```
signal_name <= value_expr_1 when boolean_expr_1 else
               value_expr_2 when boolean_expr_2 else
                    . . .
               value_expr_n;
```

The Boolean expressions are evaluated successively in turn until one is found to be `true` and the corresponding value expression is assigned to the signal. The `value_expr_n` is assigned if all Boolean expressions are evaluated to be `false`.

The conditional signal assignment statement implies a cascading priority routing network. Consider the following statement:

```
r <= a + b + c when m = n else
     a - b      when m > n else
     c + 1;
```

The routing is done by a sequence of 2-to-1 multiplexers. The diagram and truth table of a 2-to-1 multiplexer are shown in Figure 3.3(a), and the conceptual diagram of the statement is shown in Figure 3.3(b). If the first Boolean condition (i.e., m=n) is `true`, the result of a+b+c is routed to r. Otherwise, the data connected to the

(a) Diagram of a 2-to-1 multiplexer

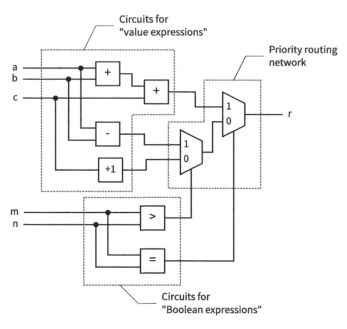

(b) Diagram of a conditional signal assignment statement

Figure 3.3   Implementation of a conditional signal assignment statement.

Table 3.4   Function table of a four-request priority encoder

input r	output pcode
1 – – –	100
0 1 – –	011
0 0 1 –	010
0 0 0 1	001
0 0 0 0	000

0 port is passed to r. We need to trace the path along the 0 port and check the next Boolean condition (i.e., m>n) to determine whether the result of a−b or c+1 is routed to the output.

Note that all the Boolean expressions and value expressions are evaluated concurrently. The values from the Boolean circuits set the selection signals of the multiplexers to route the desired value to the output. The number of cascading stages increases proportionally to the number of when-else clauses. A large number of when-else clauses will lead to a long cascading chain and introduce a large propagation delay.

Examples   We use two simple examples to demonstrate use of the conditional signal assignment statement. The first example is a priority encoder. The priority encoder has four requests, r(4), r(3), r(2), and r(1), which are grouped as a single 4-bit r input, and r(4) has the highest priority. The output is the binary code of the highest order request. The function table is shown in Table 3.4. The HDL code is shown in Listing 3.1.

Listing 3.1   Priority encoder using a conditional signal assignment statement

```
library ieee;
use ieee.std_logic_1164.all;
entity prio_encoder is
   port(
       r     : in  std_logic_vector(4 downto 1);
       pcode : out std_logic_vector(2 downto 0)
   );
end prio_encoder;

architecture cond_arch of prio_encoder is
begin
   pcode <= "100" when (r(4) = '1') else
            "011" when (r(3) = '1') else
            "010" when (r(2) = '1') else
            "001" when (r(1) = '1') else
            "000";
end cond_arch;
```

The code first checks the r(4) request and assigns "100" to pcode if it is asserted. It continues to check the r(3) request if r(4) is not asserted and repeats the process until all requests are examined.

The second example is a binary decoder. An n-to-2^n binary decoder asserts 1 bit of the 2^n-bit output according to the input combination. The functional table of

Table 3.5   Truth table of a 2-to-4 decoder with enable

	input		output
en	a(1)	a(0)	y
0	–	–	0000
1	0	0	0001
1	0	1	0010
1	1	0	0100
1	1	1	1000

a 2-to-4 decoder is shown in Table 3.5. The circuit also has a control signal, en, which enables the decoding function when asserted. The HDL code is shown in Listing 3.2.

Listing 3.2   Binary decoder using a conditional signal assignment statement

```
library ieee;
use ieee.std_logic_1164.all;
entity decoder_2_4 is
   port(
      a  : in   std_logic_vector(1 downto 0);
      en : in   std_logic;
      y  : out  std_logic_vector(3 downto 0)
   );
end decoder_2_4;

architecture cond_arch of decoder_2_4 is
begin
   y <= "0000" when (en = '0') else
        "0001" when (a = "00") else
        "0010" when (a = "01") else
        "0100" when (a = "10") else
        "1000";    —— a="11"
end cond_arch;
```

The code first checks whether en is not asserted. If the condition is false (i.e., en is '1'), it tests the four binary combinations in sequence.

3.2.2   Selected signal assignment statement

Syntax and conceptual implementation   The simplified syntax of a selected signal assignment statement is

```
with sel select
   sig <= value_expr_1 when choice_1,
          value_expr_2 when choice_2,
          value_expr_3 when choice_3,
          . . .
          value_expr_n when others;
```

The selected signal assignment statement is somewhat like a case statement in a traditional programming language. It assigns an expression to a signal according to the value of the sel signal. A choice (i.e., choice_i) must be a valid value or a set of valid values of sel. The choices have to be *mutually exclusive* (i.e., no value

(a) Diagram and functional table of a 4-to-1 multiplexer

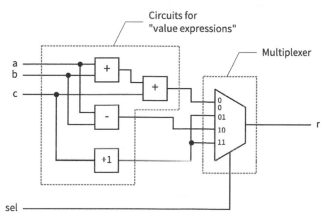

(b) Diagram of a selected signal assignment statement

Figure 3.4   Implementation of a selected signal assignment statement.

can be used more than once) and *all inclusive* (i.e., all values must be used). In other words, all possible values of sel must be covered by one and only one choice. The reserved word, **others**, is used in the end to cover unused values. Since the sel signal usually has the std_logic_vector data type, the **others** term is always needed to cover the unsynthesizable values ('X', 'U', etc.).

The selected signal assignment statement implies a multiplexing structure. Consider the following statement:

```
signal sel: std_logic_vector(1 downto 0);
 . . .
with sel select
   r <= a + b + c    when "00",
        a - b         when "10",
        c + 1         when others;
```

For synthesis purposes, the sel signal can assume four possible values: "00", "01", "10", and "11". It implies a 2^2-to-1 multiplexer with sel as the selection signal. The diagram and functional table of the 2^2-to-1 multiplexer are shown in Figure 3.4(a), and the conceptual diagram of the statement is shown in Figure 3.4(b). The evaluated result of a+b+c is routed to r when sel is "00", the result of a-b is routed when sel is "10", and the result of c+1 is routed when sel is "01" or "11".

Again, note that all value expressions are evaluated concurrently. The sel signal is used as the selection signal to route the desired value to the output. The width (i.e., number of input ports) of the multiplexer increases geometrically with the number of bits of the `sel` signal.

Example   We use the same encoder and decoder circuits to illustrate use of the selected signal assignment statement. The code for the priority encoder is shown in Listing 3.3. The entity declaration is identical to that in Listing 3.1 and is omitted.

Listing 3.3   Priority encoder using a selected signal assignment statement

```
architecture sel_arch of prio_encoder is
begin
   with r select
      pcode <= "100" when "1000"|"1001"|"1010"|"1011"|
                           "1100"|"1101"|"1110"|"1111",
               "011" when "0100"|"0101"|"0110"|"0111",
               "010" when "0010"|"0011",
               "001" when "0001",
               "000" when others;   -- r="0000"
end sel_arch;
```

The code exhaustively lists all possible combinations of the `r` signal and the corresponding output values. Note that the | symbol is used if the choice is more than one value.

The code for the 2-to-4 decoder is shown in Listing 3.4.

Listing 3.4   Binary decoder using a selected signal assignment statement

```
architecture sel_arch of decoder_2_4 is
   signal s: std_logic_vector(2 downto 0);
begin
   s <= en & a;
   with s select
      y <= "0000" when "000"|"001"|"010"|"011",
           "0001" when "100",
           "0010" when "101",
           "0100" when "110",
           "1000" when others;   -- s="111"
end sel_arch;
```

We concatenate `en` and `a` to form a 3-bit signal, `s`, and use it as the selection signal. The remaining code again exhaustively lists all possible combinations and the corresponding output values.

3.3   MODELING WITH A PROCESS

3.3.1   Process

To facilitate system modeling, VHDL contains a number of *sequential statements*, which are executed in sequence. Since their behavior is different from that of a normal concurrent circuit model, these statements are encapsulated inside a *process*. A process itself is a concurrent statement. It can be thought of as a black box whose behavior is described by sequential statements.

Sequential statements include a rich variety of constructs, but many of them don't have clear hardware counterparts. A poorly coded process frequently leads to unnecessarily complex implementation or cannot be synthesized at all. Detailed discussion of sequential statements and processes is beyond the scope of this book. For synthesis, we restrict the use of the process to two purposes:

- Describe routing structures with *if* and *case* statements.
- Construct templates for memory elements (discussed in Chapter 4).

The simplified syntax of a process with a sensitivity list is

```
process(sensitivity_list)
begin
   sequential statement;
   sequential statement;
   . . .
end process;
```

The `sensitivity_list` is a list of signals to which the process responds (i.e., is "sensitive to"). For a combinational circuit, all the input signals should be included in this list. The body of a process is composed of any number of sequential statements.

3.3.2   Sequential signal assignment statement

The simplest sequential statement is a *sequential* signal assignment statement. The simplified syntax is

```
   sig <= value_expression;
```

The statement must be encapsulated inside a process.

Although its syntax is similar to that of a simple *concurrent* signal assignment statement, the semantics are different. When a signal is assigned multiple times inside a process, only the last assignment takes effect. For example, the code segment

```
process(a,b)
begin
   c <= a and b;
   c <= a or b;
end process;
```

is the same as

```
process(a,b)
begin
   c <= a or b;
end process;
```

On the other hand, if they are concurrent signal assignment statements, as in

```
   -- not within a process
   c <= a and b;
   c <= a or b;
```

the code infers an and cell and an or cell, whose outputs are tied together. It is not allowed in most device technology and thus is a design error.

The semantics of assigning a signal multiple times inside a process is subtle and can sometimes be error prone. Detailed explanations can be found in the references cited in the bibliographic section. We use multiple assignments only to avoid unintended memory, as discussed in Section 3.4.4.

3.4   ROUTING CIRCUIT WITH IF AND CASE STATEMENTS

If and *case* statements are two other commonly used sequential statements. In synthesis, they can be used to describe routing structures.

3.4.1   If statement

Syntax and conceptual implementation   The simplified syntax of an if statement is

```
if boolean_expr_1 then
    sequential_statements;
elsif boolean_expr_2 then
    sequential_statements;
elsif boolean_expr_3 then
    sequential_statements;
   . . .
else
    sequential_statements;
end if;
```

It has one *then branch*, one or more optional *elsif branches*, and one optional *else branch*. The Boolean expressions are evaluated sequentially until an expression is evaluated as **true** or the else branch is reached, and the statements in the corresponding branch will be executed.

An if statement and a concurrent conditional signal assignment statement are somewhat similar. The two statements are equivalent if each branch of the if statement contains only a single sequential signal assignment statement. For example, the previous statement

```
r <= a + b + c when m = n else
        a - b       when m > 0 else
        c + 1;
```

can be rewritten as

```
process(a,b,c,m,n)
begin
    if m = n then
        r <= a + b + c;
    elsif m > 0 then
        r <= a - b;
    else
        r <= c + 1;
    end if;
end;
```

As in a conditional signal assignment statement, the if statement infers a similar priority routing structure during synthesis.

Example   The codes of the same priority encoder and binary decoder written with an if statement are shown in Listings 3.5 and 3.6. They are similar to those in Listings 3.1 and 3.2. Note that the if statement must be encapsulated inside a process.

Listing 3.5 Priority encoder using an if statement

```
architecture if_arch of prio_encoder is
begin
    process(r)
    begin
        if (r(4) = '1') then
            pcode <= "100";
        elsif (r(3) = '1') then
            pcode <= "011";
        elsif (r(2) = '1') then
            pcode <= "010";
        elsif (r(1) = '1') then
            pcode <= "001";
        else
            pcode <= "000";
        end if;
    end process;
end if_arch;
```

Listing 3.6 Binary decoder using an if statement

```
architecture if_arch of decoder_2_4 is begin
    process(en,a)
    begin
        if (en='0') then
            y <= "0000";
        elsif (a="00") then
            y <= "0001";
        elsif (a="01")then
            y <= "0010";
        elsif (a="10")then
            y <= "0100";
        else
            y <= "1000";
        end if;
    end process;
end if_arch;
```

3.4.2 Case statement

Syntax and conceptual implementation   The simplified syntax of a case statement is

```
case sel is
    when choice_1 =>
        sequential statements;
    when choice_2 =>
        sequential statements;
    . . .
    when others =>
        sequential statements;
end case;
```

A case statement uses the `sel` signal to select a set of sequential statements for execution. As in a selected signal assignment statement, a choice (i.e., `choice_i`) must be a valid value or a set of valid values of `sel`, and the choices have to be mutually exclusive and all inclusive. Note that the **others** term at the end covers the unused values.

A case statement and a concurrent selected signal assignment statement are somewhat similar. The two statements are equivalent if each branch of the case statement contains only a single sequential signal assignment statement. For example, the previous statement

```
with sel select
   r <= a + b + c    when "00",
        a - b         when "10",
        c + 1         when others;
```

can be rewritten as

```
process(a,b,c,sel)
begin
   case sel is
      when "00" =>
         r <= a + b + c;
      when "10" =>
         r <= a - b;
      when others =>
         r <= c + 1;
   end case;
end;
```

As in a selected signal assignment statement, the case statement infers a similar multiplexing structure during synthesis.

Example   The codes of the same priority encoder and decoder written with a case statement are shown in Listings 3.7 and 3.8. As in Listings 3.3 and 3.4, the codes exhaustively lists all possible input combinations and the corresponding output values.

Listing 3.7   Priority encoder using a case statement

```
architecture case_arch of prio_encoder is
begin
   process(r)
   begin
      case r is
         when "1000"|"1001"|"1010"|"1011"|
              "1100"|"1101"|"1110"|"1111" =>
            pcode <= "100";
         when "0100"|"0101"|"0110"|"0111" =>
            pcode <= "011";
         when "0010"|"0011" =>
            pcode <= "010";
         when "0001" =>
            pcode <= "001";
         when others =>
            pcode <= "000";
      end case;
   end process;
end case_arch;
```

Listing 3.8   Binary decoder using a case statement

```vhdl
architecture case_arch of decoder_2_4 is
   signal s: std_logic_vector(2 downto 0);
begin
   s <= en & a;
   process(s)
   begin
      case s is
         when "000"|"001"|"010"|"011" =>
            y <= "0000";
         when "100" =>
            y <= "0001";
         when "101" =>
            y <= "0010";
         when "110" =>
            y <= "0100";
         when others =>
            y <= "1000";
      end case;
   end process;
end case_arch;
```

3.4.3   Comparison to concurrent statements

The preceding subsections show that the simple if and case statements are equivalent to the conditional and selected signal assignment statements. However, an if or case statement allows *any number* and *any type* of sequential statements in their branches and thus is more flexible and versatile. Disciplined use can make the code more descriptive and even make a circuit more efficient.

This can be illustrated by two code segments. First, consider a circuit that sorts the values of two input signals and routes them to the `large` and `small` outputs. This can be done by using two conditional signal assignment statements:

```vhdl
large <= a when a > b else b;
small <= b when a > b else a;
```

Since there are two relational operators (i.e., two >) in code, synthesis software may infer two greater-than comparators. The same function can be coded by a single if statement:

```vhdl
process(a,b)
begin
   if a > b then
      large <= a;
      small <= b;
   else
      large <= b;
      small <= a;
   end if;
end;
```

The code consists of only a single relational operator.

Second, let us consider a circuit that routes the maximum value of three input signals to the output. This can be clearly described by nested two-level if statements:

```
process(a,b,c)
begin
    if (a > b) then
        if (a > c) then
            max <= a;
        else
            max <= c;
        end if;
    else
        if (b > c) then
            max <= b;
        else
            max <= c;
        end if;
    end if;
end process;
```

We can translate the if statement to a "single-level" conditional signal assignment statement:

```
max <= a when ((a > b) and (a > c)) else
       c when (a > b) else
       b when (b > c) else
       c;
```

Since no nesting is allowed, the code is less intuitive. If concurrent statements must be used, a better alternative is to describe the circuit with three conditional signal assignment statements:

```
signal ac_max, bc_max: std_logic;
. . .
ac_max <= a when (a > c) else c;
bc_max <= b when (b > c) else c;
max    <= ac_max when (a > b) else bc_max;
```

3.4.4   Unintended memory

Although a process is flexible, a subtle error in code may infer incorrect implementation. One common problem is the inclusion of intended memory in a combinational circuit. The VHDL standard specifies that a signal will *keep its previous value* if it is not assigned in a process. During synthesis, this infers an internal state (via a closed feedback loop) or a memory element (such as a latch).

To prevent unintended memory, we should observe the following rules while developing code for a combinational circuit:

- Include all input signals in the sensitivity list.
- Include the else branch in an if statement.
- Assign a value to every signal in every branch.

For example, the following code segment tries to generate a greater-than (i.e., gt) and an equal-to (i.e., eq) output signal:

```
process(a)                 — b missing from sensitivity list
begin
   if (a > b) then    — eq not assigned in this branch
      gt <= '1';
   elsif (a = b) then — gt not assigned in this branch
      eq <= '1';
   end if;                  — else branch is omitted
end process;
```

Although the syntax is correct, it violates all three rules. For example, gt will keep its previous value when the a>b expression is `false` and a latch will be inferred accordingly. The correct code should be

```
process(a,b)
begin
   if (a > b) then
      gt <= '1';
      eq <= '0';
   elsif (a = b) then
      gt <= '0';
      eq <= '1';
   else
      gt <= '0';
      eq <= '0';
   end if;
end process;
```

Since multiple sequential signal assignment statements arc allowed inside a process, we can correct the problem by assigning a default value in the beginning:

```
process(a,b)
begin
   gt <= '0';                — assign default value
   eq <= '0';
   if (a > b) thcn
      gt <= '1';
   elsif (a = b) then
      eq <= '1';
   end if;
end process;
```

The gt and eq signals assume '0' if they are not assigned a value later. As discussed earlier, assigning a signal multiple times inside a process can be error prone. For synthesis, this should not be used in other context and should be considered as shorthand to satisfy the "assigning all signals in all branches" rule.

3.5 CONSTANTS AND GENERICS

3.5.1 Constants

HDL code frequently uses constant values in expressions and array boundaries. One good design practice is to replace the "hard literals" with symbolic constants. It makes code clear and helps future maintenance and revision. The constant definition is included in the architecture's declaration section and its syntax is

```
constant const_name: data_type := value_expression;
```

For example, we can declare two constants as

```
constant DATA_WIDTH: integer := 8;
constant DATA_RANGE: integer := 2**DATA_WIDTH - 1;
```

The constant expression is evaluated during preprocessing and thus requires no physical circuit. In this book, we use capital letters for constants.

The use of a constant can best be explained by an example. Assume that we want to design an adder with the carry-out bit. One way to do it is to extend the input by 1 bit and then perform regular addition. The MSB of the summation becomes the carry-out bit. The code is shown in Listing 3.9.

Listing 3.9   Adder using a hard literal

```
library ieee;
use ieee.std_logic_1164.all;
use ieee.numeric_std.all;
entity add_w_carry is
   port(
      a, b : in  std_logic_vector(3 downto 0);
      cout : out std_logic;
      sum  : out std_logic_vector(3 downto 0)
   );
end add_w_carry;

architecture hard_arch of add_w_carry is
   signal a_ext, b_ext, sum_ext : unsigned(4 downto 0);
begin
   a_ext   <= unsigned('0' & a);
   b_ext   <= unsigned('0' & b);
   sum_ext <= a_ext + b_ext;
   sum     <= std_logic_vector(sum_ext(3 downto 0));
   cout    <= sum_ext(4);
end hard_arch;
```

The code is for a 4-bit adder. Hard literals, such as 3 and 4, are used for the ranges, as in unsigned(4 downto 0) and sum_ext(3 downto 0), and the MSB, as in sum_ext(4). If we want to revise the code for an 8-bit adder, these literals have to be modified manually. This will be a tedious and error prone process if the code is complex and the literals are referred to in many places.

To improve the readability, we can use a symbolic constant, N, to represent the number of bits of the adder. The revised architecture body is shown in Listing 3.10.

Listing 3.10   Adder using a constant

```
architecture const_arch of add_w_carry is
   constant N : integer := 4;
   signal a_ext, b_ext, sum_ext : unsigned(N downto 0);
begin
   a_ext   <= unsigned('0' & a);
   b_ext   <= unsigned('0' & b);
   sum_ext <= a_ext + b_ext;
   sum     <= std_logic_vector(sum_ext(N - 1 downto 0));
   cout    <= sum_ext(N);
end const_arch;
```

The constant makes the code easier to understand and maintain.

3.5.2   Generics

VHDL provides a construct, known as a *generic*, to pass information into an entity and component. Since a generic cannot be modified inside the architecture, it functions somewhat like a constant. A generic is declared inside an entity declaration, just before the port declaration:

```
entity entity_name is
   generic(
      generic_name: data_type := default_value;
      generic_name: data_type := default_value;
      . . .
      generic_name: data_type := default_value
   )
   port(
      port_name: mode data_type;
      . . .
   );
end entity_name;
```

For example, the previous adder code can be modified to use the adder width as a generic, as shown in Listing 3.11.

Listing 3.11   Adder using a generic

```
library ieee;
use ieee.std_logic_1164.all;
use ieee.numeric_std.all;
entity gen_add_w_carry is
   generic(N : integer := 4);
   port(
      a, b : in  std_logic_vector(N - 1 downto 0);
      cout : out std_logic;
      sum  : out std_logic_vector(N - 1 downto 0)
   );
end gen_add_w_carry;

architecture arch of gen_add_w_carry is
   signal a_ext, b_ext, sum_ext : unsigned(N downto 0);
begin
   a_ext   <= unsigned('0' & a);
   b_ext   <= unsigned('0' & b);
   sum_ext <= a_ext + b_ext;
   sum     <= std_logic_vector(sum_ext(N - 1 downto 0));
   cout    <= sum_ext(N);
end arch
```

The N generic is declared in line 5 with a default value of 4. After N is declared, it can be used in the port declaration and architecture body, just like a constant.

If the adder is later used as a component in other code, we can assign the desired value to the generic in component instantiation. This is known as *generic mapping*. The default value will be used if generic mapping is omitted. Use of the generic in component instantiation is shown below.

```
signal a4, b4, sum4   : unsigned(3 downto 0);
signal a8, b8, sum8   : unsigned(7 downto 0);
signal a16, b16, sum16: unsigned(15 downto 0);
signal c4, c8, c16    : std_logic;
```

```
    . . .
  -- instantiate 8-bit adder
  adder_8_unit: work.gen_add_w_carry(arch)
     generic map(N=>8)
     port map(a=>a8, b=>b8, cout=>c8, sum=>sum8));
  -- instantiate 16-bit adder
  adder_16_unit: work.gen_add_w_carry(arch)
     generic map(N=>16)
     port map(a=>a16, b=>b16, cout=>c16, sum=>sum16));
  -- instantiate 4-bit adder
  -- (generic mapping omitted, default value 4 used)
  adder_4_unit: work.gen_add_w_carry(arch)
     port map(a=>a4, b=>b4, cout=>c4, sum=>sum4));
```

A generic provides a mechanism to create *scalable code*, in which the "width" of a circuit can be adjusted to meet a specific need. This makes code more portable and encourages design reuse.

3.6   REPLICATED STRUCTURE

Many digital circuits exhibit a well-patterned structure, such as a one-dimensional cascading chain or a two-dimensional mesh, and can be implemented as a repetitive composition of basic building blocks.

3.6.1   Loop statements

A replicated structure can be described by the VHDL *for-generate* and *for-loop* statements. The former is a concurrent statement and the latter is a sequential statement.

The simplified syntax of the for-generate statement is

```
gen_label:
for loop_index in loop_range generate
   concurrent statement;
   concurrent statement;
   . . .
end generate;
```

The for-generate statement repeats the loop body of concurrent statements for a fixed number of iterations. The `loop_range` term specifies a range of values between the left and right bounds. The range has to be *static*, which means that it has to be determined before the time of execution (synthesis). It is normally specified by the width parameters. The `loop_index` term is used to keep track of the iteration and takes a successive value from `loop_range` in each iteration, starting from the leftmost value. The index automatically takes the data type of `loop_range`'s element and does not need to be declared. The loop body contains a collection of concurrent statements and represents a stage of the iterative circuit. During synthesis, the loop is "unrolled" and flattened. The for-generate statement is frequently used in conjunction with generics to create scalable and reusable codes.

The for-loop statement is similar to the for-generate statement but is a sequential statement and can only be used within a process. The simplified syntax of the for-loop statement is

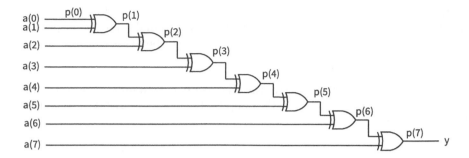

Figure 3.5   Reduced-xor circuit.

```
for loop_index in loop_range loop
    sequential statement;
    sequential statement;
        . . .
end loop;
```

3.6.2   Example

We use a reduced-xor circuit to demonstrate the use of the for-generate statement. A reduced-xor function is to apply xor operations over all bits of an input signal. For example, let $a_7a_6a_5a_4a_3a_2a_1a_0$ be an 8-bit signal. The reduced-xor operation of this signal is

$$a_7 \oplus a_6 \oplus a_5 \oplus a_4 \oplus a_3 \oplus a_2 \oplus a_1 \oplus a_0$$

Since this function returns '1' if there are odd number of 1's in its input, it can be used to determine the odd parity of the input signal.

The HDL code of the direct implementation of an eight-bit reduced-xor circuit is shown in Listing 3.12.

Listing 3.12   Initial description of a reduced-xor circuit

```
library ieee;
use ieee.std_logic_1164.all;
entity reduced_xor8 is
    port(
        a : in std_logic_vector(7 downto 0);
        y : out std_logic
    );
end reduced_xor8;

architecture cascade_arch of reduced_xor8 is
begin
    y <= a(0) xor a(1) xor a(2) xor a(3) xor
        a(4) xor a(5) xor a(6) xor a(7);
end cascade_arch;
```

While the code is simple, it cannot be "parameterized" or reused with other input widths.

Closer observation shows that the HDL code constitutes a linear cascading chain, as shown in Figure 3.5. The basic building block in a stage is the xor gate. We divide the chain into stages and number the stages from left to right, starting at the 0th stage, and name the intermediate results the p signal. The diagram show that there is a clear relationship among the two input signals and the output signal of an xor gate. For the ith stage, the three signals can be expressed as

```
p(i) <= a(i) xor p(i-1);
```

Based on this equation, we can derive the HDL code using a for-generate statement. The code is shown in Listing 3.13. The loop body iterates WIDTH-1 times and thus infers WIDTH-1 xor gates.

Listing 3.13   Parameterized reduced-xor circuit

```
library ieee;
use ieee.std_logic_1164.all;
entity reduced_xor is
   generic(WIDTH: integer := 8);
   port(
      a : in std_logic_vector(WIDTH-1 downto 0);
      y : out std_logic
   );
end reduced_xor;

architecture gen_linear_arch of reduced_xor is
   signal p: std_logic_vector(WIDTH-1 downto 0);
begin
   p(0) <= a(0);
   xor_gen:
   for i in 1 to (WIDTH-1) generate
      p(i) <= a(i) xor p(i-1);
   end generate;
   y <= p(WIDTH-1);
end gen_linear_arch;
```

The revised code infers the same circuit but is parameterized. It can be instantiated and reused with any input width.

3.7 DESIGN EXAMPLES

3.7.1 Hexadecimal digit to seven-segment LED decoder

The sketch of a seven-segment LED, display is shown in Figure 3.6(a). It consists of seven LED bars and a single round LED decimal point. On the prototyping board, the seven-segment LED is configured as active low, which means that an LED segment is lit if the corresponding control signal is '0'.

A hexadecimal digit to seven-segment LED decoder treats a 4-bit input as a hexadecimal digit and generates appropriate LED patterns, as shown in Figure 3.6(b). For completeness, we assume that there is also a 1-bit input, dp, which is connected directly to the decimal point LED. The LED control signals, dp, g, f, e, d, c, b, and a, are grouped together as a single 8-bit signal, sseg. The code is shown in Listing 3.14. It uses one selected signal assignment statement to list all the desired patterns for the seven LSBs of the sseg signal. The MSB is connected to dp.

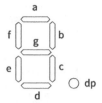

(a) Diagram of a seven-segment LED display

(b) Hexadecimal digit patterns

Figure 3.6   Seven-segment LED display and hexadecimal patterns.

Listing 3.14   Hexadecimal digit to seven-segment LED decoder

```
library ieee;
use ieee.std_logic_1164.all;
entity hex_to_sseg is
   port(
      hex  : in   std_logic_vector(3 downto 0);
      dp   : in   std_logic;
      sseg : out  std_logic_vector(7 downto 0)
   );
end hex_to_sseg;

architecture arch of hex_to_sseg is
begin
   with hex select
      sseg(6 downto 0) <=
         "1000000" when "0000",
         "1111001" when "0001",
         "0100100" when "0010",
         "0110000" when "0011",
         "0011001" when "0100",
         "0010010" when "0101",
         "0000010" when "0110",
         "1111000" when "0111",
         "0000000" when "1000",
         "0010000" when "1001",
         "0001000" when "1010", --a
         "0000011" when "1011", --b
         "1000110" when "1100", --c
         "0100001" when "1101", --d
         "0000110" when "1110", --e
         "0001110" when others; --f
   sseg(7) <= dp;
end arch;
```

The Nexys 4 DDR board contains an eight-digit seven-segment LED display and the Basys 3 board contains a four-digit seven-segment LED display. To facilitate both boards, our discussion focuses on a four-digit display in Part I of the book. We extend the design to an eight-digit display in Chapter 14 of Part III.

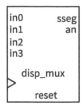

(a) Block diagram of an LED time-multiplexing module

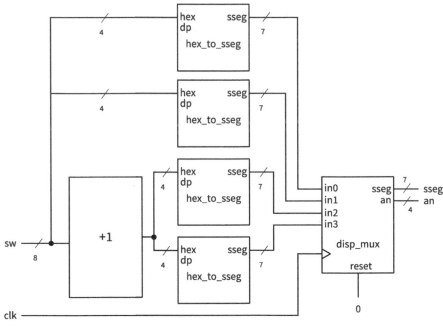

(b) Block diagram of a decoder testing circuit

Figure 3.7   LED time-multiplexing module and decoder testing circuit.

To save the number of I/O pins, a time-multiplexing scheme is sometimes used in a multi-digit seven-segment LED display. The block diagram of a four-digit time-multiplexing module, disp_mux, is shown in Figure 3.7(a). The inputs are in0, in1, in2, and in3, which correspond to four 8-bit seven-segment LED patterns, and the outputs are an, which is a 4-bit signal that enables the four displays individually, and sseg, which is the shared 8-bit signal that controls the eight LED segments. The circuit generates a properly timed enable signal and routes the four input patterns to the output alternatively. The design of this module is discussed in Chapter 4. For now, we just treat it as a black box that takes four seven-segment LED patterns, and instantiate it in the code.

Testing circuit   We use a simple 8-bit increment circuit to verify operation of the decoder. The sketch is shown in Figure 3.7(b). The sw input is connected to eight slide switches of the prototyping board. It is fed to an incrementor to obtain sw+1. The original and incremented sw signals are then passed to four decoders to

display the four hexadecimal digits on the right four digits of the seven-segment
LED display. The code is shown in Listing 3.15.

Listing 3.15   Hex-to-LED decoder testing circuit

```vhdl
library ieee;
use ieee.std_logic_1164.all;
use ieee.numeric_std.all;
entity hex_to_sseg_test is
   port(
      clk  : in  std_logic;
      sw   : in  std_logic_vector(7 downto 0);
      an   : out std_logic_vector(3 downto 0);
      sseg : out std_logic_vector(7 downto 0)
   );
end hex_to_sseg_test;

architecture arch of hex_to_sseg_test is
   signal inc        : std_logic_vector(7 downto 0);
   signal led1, led0 : std_logic_vector(7 downto 0);
   signal led3, led2 : std_logic_vector(7 downto 0);
begin
   -- increment input
   inc <= std_logic_vector(unsigned(sw) + 1);
   -- instance for 4 LSBs of input
   sseg_unit_0 : entity work.hex_to_sseg
      port map(
         hex  => sw(3 downto 0),
         dp   => '0',
         sseg => led0
      );
   -- instance for 4 MSBs of input
   sseg_unit_1 : entity work.hex_to_sseg
      port map(
         hex  => sw(7 downto 4),
         dp   => '0',
         sseg => led1
      );
   -- instance for 4 LSBs of incremented value
   sseg_unit_2 : entity work.hex_to_sseg
      port map(
         hex  => inc(3 downto 0),
         dp   => '1',
         sseg => led2
      );
   -- instance for 4 MSBs of incremented value
   sseg_unit_3 : entity work.hex_to_sseg
      port map(
         hex  => inc(7 downto 4),
         dp   => '1',
         sseg => led3
      );
   disp_unit : entity work.disp_mux
      port map(
         clk   => clk,
         reset => '0',
         in0   => led0,
         in1   => led1,
         in2   => led2,
         in3   => led3,
         an    => an,
         sseg  => sseg
```

```
      );
end arch;
```

We can follow the procedure in Appendix A.2 to synthesize and implement the circuit on the prototyping board. The disp_mux.vhd file, which contains the code for the time-multiplexing module, and the constraint file must be included in the Xilinx Vivado project during synthesis.

3.7.2   Sign-magnitude adder

An integer can be represented in *sign-magnitude* format, in which the MSB is the sign and the remaining bits form the magnitude. For example, 3 and −3 become "0011" and "1011" in 4-bit sign-magnitude format.

A sign-magnitude adder performs an addition operation in this format. The operation can be summarized as follows:

- If the two operands have the same sign, add the magnitudes and keep the sign.
- If the two operands have different signs, subtract the smaller magnitude from the larger one and keep the sign of the number that has the larger magnitude.

One possible implementation is to divide the circuit into two stages. The first stage sorts the two input numbers according to their magnitudes and routes them to the max and min signals. The second stage examines the signs and performs addition or subtraction on the magnitude accordingly. Note that since the two numbers have been sorted, the magnitude of max is always larger than that of min and the final sign is the sign of max.

The code is shown in Listing 3.16, which realizes the two-stage implementation scheme. For clarity, we split the input number internally and use separate sign and magnitude signals. A generic, N, is used to represent the width of the adder. Note that the relevant magnitude signals are declared as unsigned to facilitate the arithmetic operation, and type conversions are performed at the beginning and end of the code.

Listing 3.16   Sign-magnitude adder

```
library ieee;
use ieee.std_logic_1164.all;
use ieee.numeric_std.all;
entity sign_mag_add is
   generic(N : integer := 4);              -- default 4 bits
   port(
      a, b : in  std_logic_vector(N - 1 downto 0);
      sum  : out std_logic_vector(N - 1 downto 0)
   );
end sign_mag_add;

architecture arch of sign_mag_add is
   signal mag_a, mag_b             : unsigned(N - 2 downto 0);
   signal mag_sum, max, min        : unsigned(N - 2 downto 0);
   signal sign_a, sign_b, sign_sum : std_logic;
begin
   mag_a  <= unsigned(a(N - 2 downto 0));
   mag_b  <= unsigned(b(N - 2 downto 0));
   sign_a <= a(N - 1);
   sign_b <= b(N - 1);
```

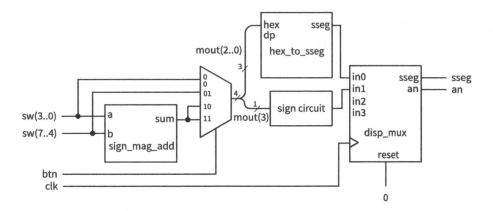

Figure 3.8   Sign-magnitude adder testing circuit.

```
—  sort according to magnitude
process(mag_a, mag_b, sign_a, sign_b)
begin
    if mag_a > mag_b then
        max        <= mag_a;
        min        <= mag_b;
        sign_sum <= sign_a;
    else
        max        <= mag_b;
        min        <= mag_a;
        sign_sum <= sign_b;
    end if;
end process;
—  add/sub magnitude
mag_sum <= max + min when (sign_a = sign_b) else max - min;
—form output
    sum      <= std_logic_vector(sign_sum & mag_sum);
end arch;
```

Testing circuit   We use a 4-bit sign-magnitude adder to verify the circuit operation. The sketch of the testing circuit is shown in Figure 3.8. The two input numbers are connected to eight slide switches, and the sign and magnitude are shown on two right digits of the seven-segment LED display. The rightmost LED digit shows the 3-bit magnitude, which is appended with a '0' in front and fed to the hexadecimal to seven-segment LED decoder. The next LED digit displays the sign bit, which is blank for the plus sign and is lit with a middle LED segment for the minus sign. The two LED patterns are then fed to the time-multiplexing module, disp_mux, as explained in Section 3.7.1.

Two pushbuttons are used as the selection signal of a multiplexer to route an operand or the sum to the display circuit. Five general-purpose pushbutton switches are available on the Nexys 4 DDR and Basys 3 prototyping boards and are labeled btnc (center), btnl (left), btnd (down), btnr (right), and btnu (up). To reduce the number of input ports in entity declaration, we group all pushbutton inputs into one multi-bit signal, btn, in which the LSB (i.e., btn(0)) corresponds to btnu and the remaining indexes are increased clockwise.

The code follows the block diagram and is shown in Listing 3.17.

Listing 3.17   Sign-magnitude adder testing circuit

```vhdl
library ieee;
use ieee.std_logic_1164.all;
use ieee.numeric_std.all;
entity sm_add_test is
   port(
      clk  : in  std_logic;
      btn  : in  std_logic_vector(1 downto 0);
      sw   : in  std_logic_vector(7 downto 0);
      an   : out std_logic_vector(3 downto 0);
      sseg : out std_logic_vector(7 downto 0)
   );
end sm_add_test;

architecture arch of sm_add_test is
   signal sum, mout, oct       : std_logic_vector(3 downto 0);
   signal led3, led2, led1, led0 : std_logic_vector(7 downto 0);
begin
   -- instantiate adder
   sm_adder_unit : entity work.sign_mag_add
      generic map(N => 4)
      port map(
         a   => sw(3 downto 0),
         b   => sw(7 downto 4),
         sum => sum
      );

   -- 3-to-1 mux to select a number to display
   with btn select
      mout <= sw(3 downto 0) when "00",   -- a
              sw(7 downto 4) when "01",   -- b
              sum when others;            -- sum

   -- magnitude displayed on rightmost 7-seg LED
   oct <= '0' & mout(2 downto 0);
   sseg_unit : entity work.hex_to_sseg
      port map(hex => oct, dp => '1', sseg => led0);

   -- sign displayed on 2nd 7-seg LED
   led1 <= "10111111" when mout(3) = '1' else -- middle bar
           "11111111";                        -- blank
   -- other two 7-seg LEDs blank
   led2 <= "11111111";
   led3 <= "11111111";

   -- instantiate display multiplexer
   disp_unit : entity work.disp_mux
      port map(
         clk => clk,
         reset => '0',
         in0 => led0,
         in1 => led1,
         in2 => led2,
         in3 => led3,
         an  => an,
         sseg => sseg
      );
end arch;
```

3.7.3 Barrel shifter

Although VHDL has built-in shift functions, they sometimes cannot be synthesized automatically. In this subsection, we examine an 8-bit barrel shifter that rotates an arbitrary number of bits to right. The circuit has an 8-bit data input, a, and a 3-bit control signal, amt, which specifies the amount to be rotated. The first design uses a selected signal assignment statement to exhaustively list all combinations of the amt signal and the corresponding rotated results. The code is shown in Listing 3.18.

Listing 3.18   Barrel shifter using a selected signal assignment statement

```
library ieee;
use ieee.std_logic_1164.all;
entity barrel_shifter is
   port(
      a    : in std_logic_vector(7 downto 0);
      amt  : in std_logic_vector(2 downto 0);
      y    : out std_logic_vector(7 downto 0)
   );
end barrel_shifter ;

architecture sel_arch of barrel_shifter is
begin
   with amt select
      y <= a                                  when "000",
           a(0)           & a(7 downto 1) when "001",
           a(1 downto 0) & a(7 downto 2) when "010",
           a(2 downto 0) & a(7 downto 3) when "011",
           a(3 downto 0) & a(7 downto 4) when "100",
           a(4 downto 0) & a(7 downto 5) when "101",
           a(5 downto 0) & a(7 downto 6) when "110",
           a(6 downto 0) & a(7) when others; -- "111"
end sel_arch;
```

While the code is straightforward, it will become cumbersome when the number of input bits increases. Furthermore, a large number of choices implies a wide multiplexer, which makes synthesis difficult and leads to a large propagation delay. Alternatively, we can construct the circuit by stages. In the nth stage, the input signal is either passed directly to output or rotated right by 2^n positions. The nth stage is controlled by the nth bit of the amt signal. Assume that the 3 bits of amt are $m_2 m_1 m_0$. The total rotated amount after three stages is $m_2 2^2 + m_1 2^1 + m_0 2^0$, which is the desired rotating amount. The code for this scheme is shown in Listing 3.19.

Listing 3.19   Barrel shifter using multi-stage shifts

```
architecture multi_stage_arch of barrel_shifter is
   signal s0, s1: std_logic_vector(7 downto 0);
begin
   -- stage 0, shift 0 or 1 bit
   s0 <= a(0) & a(7 downto 1) when amt(0)='1' else a;
   -- stage 1, shift 0 or 2 bits
   s1 <= s0(1 downto 0) & s0(7 downto 2) when amt(1)='1' else s0;
   -- stage 2, shift 0 or 4 bits
   y <= s1(3 downto 0) & s1(7 downto 4) when amt(2)='1' else s1;
end multi_stage_arch ;
```

Testing circuit   To test the circuit, we can use eight slide switches for the **a** signal, three pushbutton switches for the `amt` signal, and the eight discrete LEDs for the output.

A top-level HDL file is created to wrap the barrel shifter circuit and maps its signals to the prototyping board's signals. The code is shown in Listing 3.20.

Listing 3.20   Barrel shifter testing circuit

```
library ieee;
use ieee.std_logic_1164.all;
use ieee.numeric_std.all;
entity shifter_test is
   port(
      sw  : in  std_logic_vector(7 downto 0);
      btn : in  std_logic_vector(2 downto 0);
      led : out std_logic_vector(7 downto 0)
   );
end shifter_test;

architecture arch of shifter_test is
begin
   shift_unit : entity work.barrel_shifter(multi_stage_arch)
      port map(
         a   => sw,
         amt => btn,
         y   => led
      );
end arch;
```

3.7.4   Simplified floating-point adder

Floating point is another format to represent a number. With the same number of bits, the range in floating-point format is much larger than that in signed integer format. Although VHDL has a built-in floating-point data type, it is too complex to be synthesized automatically.

Detailed discussion of floating-point representation is beyond the scope of this book. We use a simplified 13-bit format in this example and ignore the round-off error. The representation consists of a sign bit, s, which indicates the sign of the number (1 for negative); a 4-bit exponent field, e, which represents the exponent; and an 8-bit significand field, f, which represents the significand or the fraction. In this format, the value of a floating-point number is $(-1)^s * .f * 2^e$. The $.f * 2^e$ is the magnitude of the number and $(-1)^s$ is just a formal way to state that "s equal to 1 implies a negative number." Since the sign bit is separated from the rest of the number, floating-point representation can be considered as a variation of the sign-magnitude format.

We also make the following assumptions:

- Both exponent and significand fields are in unsigned format.
- The representation has to be either normalized or zero. *Normalized representation* means that the MSB of the significand field must be '1'. If the magnitude of the computation result is smaller than the smallest normalized nonzero magnitude, $0.10000000 * 2^{0000}$, it must be converted to zero.

Under these assumptions, the largest and smallest nonzero magnitudes are $0.11111111 * 2^{1111}$ and $0.10000000 * 2^{0000}$, and the range is about 2^{16} (i.e., $\frac{0.11111111*2^{1111}}{0.10000000*2^{0000}}$).

		sort	align	add/sub	normalize
eg. 1	+0.54E3	-0.87E4	-0.87E4	-0.87E4	-0.87E4
	-0.87E4	+0.54E3	+0.05E4	+0.05E4	+0.05E4
				-0.82E4	-0.82E4
eg. 2	+0.54E3	-0.55E3	-0.55E3	-0.55E3	-0.55E3
	-0.55E3	+0.54E3	+0.54E3	+0.54E3	+0.54E3
				-0.01E3	-0.10E2
eg. 3	+0.54E0	-0.55E0	-0.55E0	-0.55E0	-0.55E0
	-0.55E0	+0.54E0	+0.54E0	+0.54E0	+0.54E0
				-0.01E0	-0.00E0
eg. 4	+0.56E3	+0.56E3	+0.56E3	+0.56E3	+0.56E3
	+0.52E3	+0.52E3	+0.52E3	+0.52E3	+0.52E3
				+1.08E3	+0.10E4

Figure 3.9   Floating-point addition examples.

Our floating-point adder design follows the process of adding numbers manually in scientific notation. This process can best be explained by examples. We assume that the widths of the exponent and significand are 1 and 2 digits, respectively. Decimal format is used for clarity. The computations of several representative examples are shown in Figure 3.9. The computation is done in four major steps:

1. *Sorting*: puts the number with the larger magnitude on the top and the number with the smaller magnitude on the bottom (we call the sorted numbers "big number" and "small number").

2. *Alignment*: aligns the two numbers so they have the same exponent. This can be done by adjusting the exponent of the small number to match the exponent of the big number. The significand of the small number has to shift to the right according to the difference in exponents.

3. *Addition/subtraction*: adds or subtracts the significands of two aligned numbers.

4. *Normalization*: adjusts the result to normalized format. Three types of normalization procedures may be needed:

   - After a subtraction, the result may contain leading zeros in front, as in example 2.
   - After a subtraction, the result may be too small to be normalized and thus needs to be converted to zero, as in example 3.
   - After an addition, the result may generate a carry-out bit, as in example 4.

Our binary floating-point adder design uses a similar algorithm. To simplify the implementation, we ignore the rounding. During alignment and normalization, the lower bits of the significand will be discarded when shifted out. The design is divided into four stages, each corresponding to a step in the foregoing algorithm. The suffixes, b, s, a, r, and n, used in signal names are for "big number," "small number," "aligned number," "result of addition/subtraction," and "normalized number," respectively. The code is developed according to these stages, as shown in Listing 3.21.

Listing 3.21   Simplified floating-point adder

```vhdl
library ieee;
use ieee.std_logic_1164.all;
use ieee.numeric_std.all;
entity fp_adder is
   port (
      sign1, sign2 : in std_logic;
      exp1, exp2   : in std_logic_vector(3 downto 0);
      frac1, frac2 : in std_logic_vector(7 downto 0);
      sign_out     : out std_logic;
      exp_out      : out std_logic_vector(3 downto 0);
      frac_out     : out std_logic_vector(7 downto 0)
   );
end fp_adder ;

architecture arch of fp_adder is
   -- suffix b, s, a, n for big, small, aligned, normalized number
   signal signb, signs      : std_logic;
   signal expb, exps, expn  : unsigned(3 downto 0);
   signal fracb, fracs      : unsigned(7 downto 0);
   signal fraca, fracn      : unsigned(7 downto 0);
   signal sum_norm          : unsigned(7 downto 0);
   signal exp_diff          : unsigned(3 downto 0);
   signal sum               : unsigned(8 downto 0);
   signal lead0             : unsigned(2 downto 0);
begin
   -- 1st stage: sort to find the larger number
   process (sign1, sign2, exp1, exp2, frac1, frac2)
   begin
      if (exp1 & frac1) > (exp2 & frac2) then
         signb <= sign1;
         signs <= sign2;
         expb <= unsigned(exp1);
         exps <= unsigned(exp2);
         fracb <= unsigned(frac1);
         fracs <= unsigned(frac2);
      else
         signb <= sign2;
         signs <= sign1;
         expb <= unsigned(exp2);
         exps <= unsigned(exp1);
         fracb <= unsigned(frac2);
         fracs <= unsigned(frac1);
      end if;
   end process;

   -- 2nd stage: align smaller number
   exp_diff <= expb - exps;
   with exp_diff select
      fraca <=
         fracs                            when "0000",
         "0"      & fracs(7 downto 1) when "0001",
         "00"     & fracs(7 downto 2) when "0010",
         "000"    & fracs(7 downto 3) when "0011",
         "0000"   & fracs(7 downto 4) when "0100",
         "00000"  & fracs(7 downto 5) when "0101",
         "000000" & fracs(7 downto 6) when "0110",
```

```vhdl
            "0000000" & fracs(7)              when "0111",
            "00000000"                        when others;

    -- 3rd stage: add/substract
    sum <= ('0' & fracb) + ('0' & fraca) when signb=signs else
           ('0' & fracb) - ('0' & fraca);

    -- 4th stage: normalize
    -- count leading 0s
    lead0 <= "000" when (sum(7)='1') else
             "001" when (sum(6)='1') else
             "010" when (sum(5)='1') else
             "011" when (sum(4)='1') else
             "100" when (sum(3)='1') else
             "101" when (sum(2)='1') else
             "110" when (sum(1)='1') else
             "111";
    -- shift significand according to leading 0
    with lead0 select
       sum_norm <=
          sum(7 downto 0)                 when "000",
          sum(6 downto 0) & '0'           when "001",
          sum(5 downto 0) & "00"          when "010",
          sum(4 downto 0) & "000"         when "011",
          sum(3 downto 0) & "0000"        when "100",
          sum(2 downto 0) & "00000"       when "101",
          sum(1 downto 0) & "000000"      when "110",
          sum(0) &          "0000000" when others;

    -- normalize with special conditions
    process(sum, sum_norm, expb, lead0)
    begin
       if sum(8)='1' then -- w/ carry out; shift frac to right
          expn <= expb + 1;
          fracn <= sum(8 downto 1);
       elsif (lead0 > expb) then  -- too small to normalize;
          expn <= (others=>'0');  -- set to 0
          fracn <= (others=>'0');
       else
          expn <= expb - lead0;
          fracn <= sum_norm;
       end if;
    end process;

    -- form output
    sign_out <= signb;
    exp_out  <= std_logic_vector(expn);
    frac_out <= std_logic_vector(fracn);
end arch;
```

The circuit in the first stage compares the magnitudes and routes the big number to the signb, expb, and fracb signals and the smaller number to the signs, exps, and fracs signals. The comparison is done between exp1&frac1 and exp2&frac2. It implies that the exponents are compared first, and if they are the same, the significands are compared.

The circuit in the second stage performs alignment. It first calculates the difference between the two exponents, which is expb-exps, and then shifts the significand, fracs, to the right by this amount. The aligned significand is labeled fraca. The circuit in the third stage performs sign-magnitude addition, similar to that in

Section 3.7.2. Note that the operands are extended by 1 bit to accommodate the carry-out bit.

The circuit in the fourth stage performs normalization, which adjusts the result to make the final output conform to the normalized format. The normalization circuit is constructed in three segments. The first segment counts the number of leading zeros. It is somewhat like a priority encoder. The second segment shifts the significands to the left by the amount specified by the leading-zero counting circuit. The last segment checks the carry-out and zero conditions and generates the final normalized number.

The floating-point adder has two 13-bit input operands and one 13-bit output sum. It is difficult to test the circuit on the prototyping board. A testbench can be developed to verify its operation with input patterns similar to those discussed in Figure 3.9.

3.8 BIBLIOGRAPHIC NOTES

The Designer's Guide to VHDL by P. J. Ashenden provides detailed coverage on the VHDL constructs discussed in this chapter, and the author's *RTL Hardware Design Using VHDL: Coding for Efficiency, Portability, and Scalability* discusses the coding and optimization schemes and gives additional design examples.

3.9 SUGGESTED EXPERIMENTS

3.9.1 Multi-function barrel shifter

Consider an 8-bit shifting circuit that can perform rotating right or rotating left. An additional 1-bit control signal, `lr`, specifies the desired direction.

1. Design the circuit using one rotate-right circuit, one rotate-left circuit, and one 2-to-1 multiplexer to select the desired result. Derive the code.
2. Derive a testbench and use simulation to verify operation of the code.
3. Synthesize the circuit, program the FPGA, and verify its operation.
4. This circuit can also be implemented by one rotate-right shifter with pre- and post-reversing circuits. The reversing circuit either passes the original input or reverses the input bitwise (e.g., if an 8-bit input is $a_7a_6a_5a_4a_3a_2a_1a_0$, the reversed result becomes $a_0a_1a_2a_3a_5a_5a_6a_7$). Repeat steps 2 and 3.
5. Check the report files and compare the number of logic cells and propagation delays of the two designs.
6. Expand the code for a 16-bit circuit and synthesize the code. Repeat steps 1 to 5.
7. Expand the code for a 32-bit circuit and synthesize the code. Repeat steps 1 to 5.

3.9.2 Parameterized barrel shifter

The barrel shifter discussed in Section 3.7.3 only accepts an eight-bit input signal. We want to construct a parameterized module in which the input width can be specified by a generic. To simplified the design, we assume that width of the input is 2^N and the N is used as the generic.

1. Design the circuit and derive the code.
2. Derive a testbench and use simulation to verify operation of the code.
3. Design a testing circuit for a 16-bit input and derive the code.
4. Synthesize the circuit, program the FPGA, and verify its operation.

3.9.3 Dual-priority encoder

A dual-priority encoder returns the codes of the highest or second-highest priority requests. The input is a 12-bit `req` signal and the outputs are `first` and `second`, which are the 4-bit binary codes of the highest and second-highest priority requests, respectively.

1. Design the circuit and derive the code.
2. Derive a testbench and use simulation to verify operation of the code.
3. Design a testing circuit that displays the two output codes on the seven-segment LED display of the prototyping board, and derive the code.
4. Synthesize the circuit, program the FPGA, and verify its operation.

3.9.4 BCD incrementor

The binary-coded-decimal (BCD) format uses 4 bits to represent 10 decimal digits. For example, 259_{10} is represented as "0010 0101 1001" in BCD format. A BCD incrementor adds 1 to a number in BCD format. For example, after incrementing, "0010 0101 1001" (i.e., 259_{10}) becomes "0010 0110 0000" (i.e., 260_{10}).

1. Design a three-digit 12-bit incrementor and derive the code.
2. Derive a testbench and use simulation to verify operation of the code.
3. Design a testing circuit that displays three digits on the seven-segment LED display and derive the code.
4. Synthesize the circuit, program the FPGA, and verify its operation.

3.9.5 Floating-point greater-than circuit

A floating-point greater-than circuit compares two floating-point numbers and asserts output, `gt`, when the first number is larger than the second number. Assume that the two numbers are represented in the format discussed in Section 3.7.4.

1. Design the circuit and derive the code.
2. Derive a testbench and use simulation to verify operation of the code.
3. Design a testing circuit and derive the code.
4. Synthesize the circuit, program the FPGA, and verify its operation.

3.9.6 Floating-point and signed integer conversion circuit

A number may need to be converted to different formats in a large system. Assume that we use the 13-bit format in Section 3.7.4 for the floating-point representation and the 8-bit `signed` data type for the integer representation. An integer-to-floating-point conversion circuit converts an 8-bit integer input to a normalized, 13-bit floating-point output. A floating-point-to-integer conversion circuit reverses the operation. Since the range of a floating-point number is much larger, conversion may lead to the underflow condition (i.e., the magnitude of the converted number is smaller than "00000001") or the overflow condition (i.e., the magnitude of the converted number is larger than "01111111").

1. Design an integer-to-floating-point conversion circuit and derive the code.
2. Derive a testbench and use simulation to verify operation of the code.
3. Design a testing circuit and derive the code.
4. Synthesize the circuit, program the FPGA, and verify its operation.
5. Design a floating-point-to-integer conversion circuit. In addition to the 8-bit integer output, the design should include two status signals, `uf` and `of`, for the underflow and overflow conditions. Derive the code and repeat steps 2 to 4.

3.9.7   Enhanced floating-point adder

The floating-point adder in Section 3.7.4 discards the lower bits when they are shifted out (it is known as *round to zero*). A more accurate method is to *round to the nearest even*, as defined in the *IEEE Standard for Binary Floating-Point Arithmetic* (IEEE Std 754). Three extra bits, known as the *guard*, *round*, and *sticky bits*, are required to implement this method. If you learned floating-point arithmetic before, modify the floating-point adder in Section 3.7.4 to accommodate the round-to-the-nearest-even method.

CHAPTER 4

REGULAR SEQUENTIAL CIRCUIT

A sequential circuit is a circuit with memory. Modern development follows *synchronous design methodology* and uses a common clock signal to control storage elements. In this chapter, we describe the HDL codes for basic storage elements, introduce the design and coding of "regular sequential circuits," in which the state transitions in the circuit exhibit a "regular" pattern, as in a counter or shift register.

4.1 INTRODUCTION

A sequential circuit is a circuit with *memory*, which forms the *internal state* of the circuit. Unlike a combinational circuit, in which the output is a function of input only, the output of a sequential circuit is a function of the input and the internal state. The *synchronous design methodology* is the most commonly used practice in designing a sequential circuit. In this methodology, all storage elements are controlled (i.e., synchronized) by a global clock signal and the data is sampled and stored at the rising or falling edge of the clock signal. It allows designers to separate the storage components from the circuit and greatly simplifies the development process. This methodology is the most important principle in developing a large, complex digital system and is the foundation of most synthesis, verification, and testing algorithms. All of the designs in the book follow this methodology.

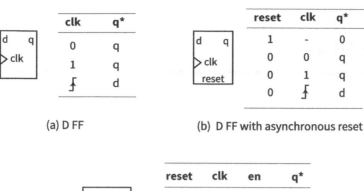

(a) D FF   (b) D FF with asynchronous reset

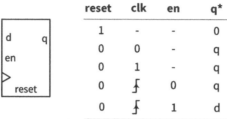

(c) D FF with synchronous enable

Figure 4.1   Block diagram and functional table of a D FF.

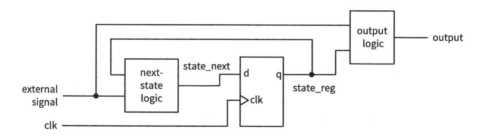

Figure 4.2   Block diagram of a synchronous system.

4.1.1   D FF and register

The most basic storage component in a sequential circuit is a D-type flip-flop (D FF). The symbol and function table of a positive edge-triggered D FF are shown in Figure 4.1(a). The value of the d signal is sampled at the rising edge of the clk signal and stored to the FF. A D FF may contain an asynchronous reset signal to clear the FF to '0'. Its symbol and function table are shown in Figure 4.1(b). Note that the reset operation is independent of the clock signal.

A D FF provides 1-bit storage. A collection of D FFs can be grouped together to store multiple bits and is known as a *register*.

4.1.2   Basic block system

The block diagram of a synchronous system is shown in Figure 4.2. It consists of the following parts:

- *State register*: a collection of D FFs controlled by the same clock signal
- *Next-state logic*: combinational logic that uses the external input and internal state (i.e., the output of register) to determine the new value of the register
- *Output logic*: combinational logic that generates the output signal

4.1.3   Code development

Our code development follows the basic block diagram in Figure 4.2. The key is to separate the memory component (i.e., the register) from the system. Once the register is isolated, the remaining portion is a pure combinational circuit, and the coding and analysis schemes discussed in previous chapters can be applied accordingly. While this approach may make the code somewhat more cumbersome at times, it helps us better visualize the circuit architecture and avoid unintended memory and subtle mistakes.

Based on the characteristics of the next-state logic, we divide sequential circuits into three categories:

- *Regular sequential circuit*. The state transitions in the circuit exhibit a "regular" pattern, as in a counter or shift register. The next-state logic is constructed primarily by a predesigned, "regular" component, such as an incrementor or shifter.
- *FSM*. The state transitions in the circuit do not exhibit a simple, repetitive pattern. The next-state logic is constructed by "random logic" and synthesized from scratch. It should be called a random sequential circuit, but is commonly known as an FSM (*finite state machine*).
- *FSMD*. The circuit consists of a regular sequential circuit and an FSM. The two parts are known as a *data path* and a *control path*, and the complete circuit is known as an FSMD (*FSM with data path*). This type of circuit is used to implement an algorithm represented by *register-transfer* (RT) methodology, which describes system operation by a sequence of data transfers and manipulations among registers.

The three types of circuits are discussed in this and two subsequent chapters.

4.1.4   Sequential circuit coding style

Some literatures present different "coding styles" of a sequential circuit. The term refres to the way to group the three main blocks in Figure 4.2. The main difference among different styles is whether to combine the register and combinational logic (i.e., the next-state logic and output logic). The first style is to separate the register and the combinational logic and code the register as an isolated process segment. The second style is to combine the register and next-state logic into a single process.

The first coding style mirrors the block diagram and the description can be easily mapped to a physical implementation. Because of the multiple segments and additional internal signals, the resulting code tends to be longer. The second coding style is more compact but harder to identify the next-state logic.

One common problem encountered by new HDL users is the confusion about a registered and a combinational output and the inference of unintended latches and buffers. The first code style can help us think more closely to "hardware" and have a better "mental picture" of the intended physical organization. Thus, despite its cumbersomeness, this style is used throughout the book. After we have a good comprehension of the hardware design, other styles can be easily learned and adopted.

4.2 HDL CODE OF THE FF AND REGISTER

The coding practice in this book separates the memory components into an individual code segment, as discussed in the previous section. We use three types of components:

- D FF
- Register
- Register file and synchronous SRAM (static RAM)

The coding of FFs and registers is described in the following subsections and the last item is discussed in Chapter 7.

4.2.1 D FF

We consider three types of D FFs:

- D FF without asynchronous reset
- D FF with asynchronous reset
- D FF with synchronous enable

The first two are the most basic memory components and can be found in the library of any device technology. The third can be constructed from a simple D FF. We include the code since it is a frequently used memory component and can be mapped to the FF of the FPGA device's logic cell.

D FF without asynchronous reset   The function table of a D FF is shown in Figure 4.1(a) and the code is shown in Listing 4.1.

Listing 4.1   D FF without asynchronous reset

```
library ieee;
use ieee.std_logic_1164.all;
entity d_ff is
   port(
      clk : in std_logic;
      d   : in std_logic;
      q   : out std_logic
   );
end d_ff;

architecture arch of d_ff is
begin
   process(clk)
   begin
      if (clk'event and clk='1') then
         q <= d;
      end if;
   end process;
end arch;
```

The rising edge is checked by the `clk'event and clk='1'` expression, which represents that there is a change in the `clk` signal (i.e., an "event") and the new value is '1'. If this condition is true, the value of d is stored to q, and if this condition is false, q keeps its previous value (i.e., memorizes the value sampled earlier). Note that only the clk signal is included in the sensitivity list. This is consistent with the fact that the d signal is sampled only at the rising edge of the clk signal, and change in its value does not trigger any immediate response.

An alternative for the Boolean condition is to use the rising_edge() function from the std_logic_1164 package. The if statement becomes

```
if (rising_edge(clk)) then
    . . .
```

D FF with asynchronous reset A D FF may contain an asynchronous reset signal, as shown in the function table of Figure 4.1(b). The signal clears the D FF to '0' any time and is not controlled by the clock signal. It actually has a higher priority than the regularly sampled input. Using an asynchronous reset signal violates the synchronous design methodology and thus should be avoided in normal operation. Its major application is to perform system initialization. For example, we can generate a short reset pulse to force a system to an initial state after turning on the power. The code for a D FF with asynchronous reset is shown in Listing 4.2.

Listing 4.2   D FF with asynchronous reset

```
library ieee;
use ieee.std_logic_1164.all;
entity d_ff_reset is
   port(
       clk   : in  std_logic;
       reset : in  std_logic;
       d     : in  std_logic;
       q     : out std_logic
   );
end d_ff_reset;

architecture arch of d_ff_reset is
begin
   process(clk, reset)
   begin
      if (reset = '1') then
         q <= '0';
      elsif (clk'event and clk = '1') then
         q <= d;
      end if;
   end process;
end arch;
```

Note that the reset signal is included in the sensitivity list, and its condition is checked before the rising-edge condition.

D FF with synchronous enable A D FF may include an additional control signal, en, to enable the FF to sample the input value. Its symbol and functional table are shown in Figure 4.1(c). Note that the en signal is examined only at the rising edge of the clock and thus is synchronous. If it is not asserted, the FF keeps its previous value. The code is shown in Listing 4.3.

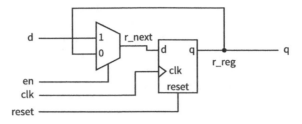

Figure 4.3   D FF with synchronous enable.

Listing 4.3   One-process coding style for a D FF with synchronous enable

```
library ieee;
use ieee.std_logic_1164.all;
entity d_ff_en is
   port(
      clk   : in  std_logic;
      reset : in  std_logic;
      en    : in  std_logic;
      d     : in  std_logic;
      q     : out std_logic
   );
end d_ff_en;

architecture arch of d_ff_en is
begin
   process(clk, reset)
   begin
      if (reset = '1') then
         q <= '0';
      elsif (clk'event and clk = '1') then
         if (en = '1') then
            q <= d;
         end if;
      end if;
   end process;
end arch;
```

The enabling feature of this D FF is useful in maintaining synchronism between a fast subsystem and a slow subsystem. For example, assume that the operation rates of a fast and a slow subsystem are 100M Hz and 1M Hz. Instead of using a derived 1M-Hz clock to drive the slow subsystem, we can generate a periodic enable tick that is asserted one clock cycle every 100 clock cycles. The slow subsystem is disabled (i.e., keeps the previous state) for the remaining 99 clock cycles. The same scheme can also be applied to eliminate a gated clock signal.

Since the enable signal is synchronous, this circuit can be constructed by a regular D FF and simple next-state logic. The code is shown in Listing 4.4, and its block diagram is shown in Figure 4.3.

Listing 4.4   Two-segment coding style for a D FF with synchronous enable

```
architecture two_seg_arch of d_ff_en is
   signal r_reg, r_next : std_logic;
begin
   — D FF
   process(clk, reset)
```

```
   begin
      if (reset = '1') then
         r_reg <= '0';
      elsif (clk'event and clk = '1') then
         r_reg <= r_next;
      end if;
   end process;
   -- next-state logic
   r_next <= d when en = '1' else r_reg;
   -- output logic
   q <= r_reg;
end two_seg_arch;
```

For clarity, we use suffixes _next and _reg to emphasize the next input value and the registered output of an FF. They are connected to the d and q signals of a D FF. The earlier one-process code can be considered as shorthand for this more explicit description.

4.2.2   Register

A register is a collection of D FFs that are controlled by the same clock and reset signals. Like a D FF, a register can have an optional asynchronous reset signal and a synchronous enable signal. The code is identical to that of a D FF except that the array data type, std_logic_vector, is needed for the relevant input and output signals. For example, an 8-bit register with asynchronous reset is shown in Listing 4.5.

Listing 4.5   Register

```
library ieee;
use ieee.std_logic_1164.all;
entity reg_reset is
   port(
      clk    : in   std_logic;
      reset  : in   std_logic;
      d      : in   std_logic_vector(7 downto 0);
      q      : out  std_logic_vector(7 downto 0)
   );
end reg_reset;

architecture arch of reg_reset is
begin
   process(clk, reset)
   begin
      if (reset = '1') then
         q <= (others => '0');
      elsif (clk'event and clk = '1') then
         q <= d;
      end if;
   end process;
end arch;
```

Note that the expression (others=>'0') means that all elements are assigned to '0' and is equivalent to "00000000" in this case.

4.3   SIMPLE DESIGN EXAMPLES

We illustrate the construction of several simple, representative sequential circuits in this section.

4.3.1   Shift register

Free-running shift register   A free-running shift register shifts its content to the left or right by one position in each clock cycle. There is no other control signal. The code for an N-bit free-running shift-right register is shown in Listing 4.6.

Listing 4.6   Free-running shift register

```vhdl
library ieee;
use ieee.std_logic_1164.all;
entity free_run_shift_reg is
   generic(N : integer := 8);
   port(
      clk   : in  std_logic;
      reset : in  std_logic;
      s_in  : in  std_logic;
      s_out : out std_logic
   );
end free_run_shift_reg;

architecture arch of free_run_shift_reg is
   signal r_reg  : std_logic_vector(N-1 downto 0);
   signal r_next : std_logic_vector(N-1 downto 0);
begin
   -- register
   process(clk, reset)
   begin
      if (reset = '1') then
         r_reg <= (others => '0');
      elsif (clk'event and clk = '1') then
         r_reg <= r_next;
      end if;
   end process;
   -- next-state logic (shift right 1 bit)
   r_next <= s_in & r_reg(N-1 downto 1);
   -- output
   s_out  <= r_reg(0);
end arch;
```

The next-state logic is a 1-bit shifter, which shifts **r_reg** right one position and inserts the serial input, **s_in**, to the MSB. Since the 1-bit shifter involves only reconnection of the input and output signals, no real logic is needed.

Universal shift register   A universal shift register can load parallel data, shift its content left or right, or remain in the same state. It can perform parallel-to-serial operation (first loading parallel input and then shifting) or serial-to-parallel operation (first shifting and then retrieving parallel output). The desired operation is specified by a 2-bit control signal, **ctrl**. The code is shown in Listing 4.7.

Listing 4.7 Universal shift register

```vhdl
library ieee;
use ieee.std_logic_1164.all;
entity univ_shift_reg is
   generic(N: integer := 8);
   port(
      clk   : in  std_logic;
      reset : in  std_logic;
      ctrl  : in std_logic_vector(1 downto 0);
      d     : in std_logic_vector(N-1 downto 0);
      q     : out std_logic_vector(N-1 downto 0)
   );
end univ_shift_reg;

architecture arch of univ_shift_reg is
   signal r_reg: std_logic_vector(N-1 downto 0);
   signal r_next: std_logic_vector(N-1 downto 0);
begin
   -- register
   process(clk,reset)
   begin
      if (reset='1') then
         r_reg <= (others=>'0');
      elsif (clk'event and clk='1') then
         r_reg <= r_next;
      end if;
   end process;
   -- next-state logic
   with ctrl select
     r_next <=
       r_reg                        when "00",  --no op
       r_reg(N-2 downto 0) & d(0)   when "01",  --shift left;
       d(N-1) & r_reg(N-1 downto 1) when "10",  --shift right;
       d                            when others; -- load
   -- output
   q <= r_reg;
end arch;
```

The next-state logic uses a 4-to-1 multiplexer to select the desired next value of the register. Note that the LSB and MSB of d (i.e., d(0) and d(N-1)) are used as serial input for the shift-left and shift-right operations.

4.3.2 Binary counter and variant

Free-running binary counter A free-running binary counter circulates through a binary sequence repeatedly. For example, a 4-bit binary counter counts from "0000", "0001", ..., to "1111" and wraps around. The code for a parameterized N-bit free-running binary counter is shown in Listing 4.8.

Listing 4.8 Free-running binary counter

```vhdl
library ieee;
use ieee.std_logic_1164.all;
use ieee.numeric_std.all;
entity free_run_bin_counter is
   generic(N: integer := 8);
   port(
      clk      : in  std_logic;
      reset    : in  std_logic;
```

Table 4.1   Function table of a universal binary counter

syn_clr	load	en	up	q*	Operation
1	–	–	–	$00\cdots00$	synchronous clear
0	1	–	–	d	parallel load
0	0	1	1	q+1	count up
0	0	1	0	q-1	count down
0	0	0	–	q	pause

```
    max_tick : out std_logic;
    q        : out std_logic_vector(N-1 downto 0)
  );
end free_run_bin_counter;

architecture arch of free_run_bin_counter is
   signal r_reg  : unsigned(N-1 downto 0);
   signal r_next : unsigned(N-1 downto 0);
begin
   -- register
   process(clk,reset)
   begin
      if (reset='1') then
         r_reg <= (others=>'0');
      elsif (clk'event and clk='1') then
         r_reg <= r_next;
      end if;
   end process;
   -- next-state logic
   r_next <= r_reg + 1;
   -- output logic
   q <= std_logic_vector(r_reg);
   max_tick <= '1' when r_reg=(2**N - 1) else '0';
end arch;
```

The next-state logic is an incrementor, which adds 1 to the register's current value. By definition of the + operator in the IEEE numeric_std package, the operation implicitly wraps around after the r_reg reaches "1...1". The circuit also consists of an output status signal, max_tick, which is asserted when the counter reaches the maximum value, "1...1" (which is equal to $2^N - 1$).

The max_tick signal represents a special type of signal that is asserted for a single clock cycle. In this book, we call this type of signal a *tick* and use the suffix _tick to indicate a signal with this property. It is commonly used to interface with the enable signal of other sequential circuits.

Universal binary counter   A universal binary counter is more versatile. It can count up or down, pause, be loaded with a specific value, or be synchronously cleared. Its functions are summarized in Table 4.1. Note the difference between the reset and syn_clr signals. The former is asynchronous and should only be used for system initialization. The latter is sampled at the rising edge of the clock and can be used in normal synchronous design. The code for this counter is shown in Listing 4.9.

Listing 4.9   Universal binary counter

```vhdl
library ieee;
use ieee.std_logic_1164.all;
use ieee.numeric_std.all;
entity univ_bin_counter is
    generic(N : integer := 8);
    port(
        clk      : in  std_logic;
        reset    : in  std_logic;
        syn_clr  : in  std_logic;
        load     : in  std_logic;
        en       : in  std_logic;
        up       : in  std_logic;
        d        : in  std_logic_vector(N-1 downto 0);
        max_tick : out std_logic;
        min_tick : out std_logic;
        q        : out std_logic_vector(N-1 downto 0)
    );
 end univ_bin_counter;

architecture arch of univ_bin_counter is
    signal r_reg  : unsigned(N-1 downto 0);
    signal r_next : unsigned(N-1 downto 0);
begin
    -- register
    process(clk,reset)
    begin
        if (reset='1') then
            r_reg <= (others=>'0');
        elsif (clk'event and clk='1') then
            r_reg <= r_next;
        end if;
    end process;
    -- next-state logic
    r_next <= (others=>'0') when syn_clr='1' else
              unsigned(d)   when load='1' else
              r_reg + 1     when en ='1' and up='1' else
              r_reg - 1     when en ='1' and up='0' else
              r_reg;
    -- output logic
    q <= std_logic_vector(r_reg);
    max_tick <= '1' when r_reg=(2**N - 1) else '0';
    min_tick <= '1' when r_reg=0 else '0';
end arch;
```

The next-state logic follows the function table and uses a conditional signal assignment to prioritize the desired operations.

Mod-m counter   A mod-m counter counts from 0 to $m - 1$ and wraps around. A parameterized mod-m counter is shown in Listing 4.10. It has two generics. One is M, which specifies the limit, m, and the other is N, which specifies the number of bits needed and should be equal to $\lceil \log_2 M \rceil$. The code is shown in Listing 4.10, and the default value is for a mod-10 counter.

Listing 4.10   Mod-m counter

```vhdl
library ieee;
use ieee.std_logic_1164.all;
use ieee.numeric_std.all;
entity mod_m_counter is
```

```vhdl
   generic(
      N : integer := 4;        -- number of bits
      M : integer := 10        -- mod-M
   );
   port(
      clk      : in  std_logic;
      reset    : in  std_logic;
      max_tick : out std_logic;
      q        : out std_logic_vector(N-1 downto 0)
   );
end mod_m_counter;

architecture arch of mod_m_counter is
   signal r_reg: unsigned(N-1 downto 0);
   signal r_next: unsigned(N-1 downto 0);
begin
   -- register
   process(clk,reset)
   begin
      if (reset='1') then
         r_reg <= (others=>'0');
      elsif (clk'event and clk='1') then
         r_reg <= r_next;
      end if;
   end process;
   -- next-state logic
   r_next <= (others=>'0') when (r_reg=(M-1)) else r_reg + 1;
   -- output logic
   q <= std_logic_vector(r_reg);
   max_tick <= '1' when r_reg=(M-1) else '0';
end arch;
```

The next-state logic is constructed by a conditional signal assignment statement. If the counter reaches M-1, the new value is cleared to 0. Otherwise, it is incremented by 1.

Inclusion of the N parameter in the code is somewhat redundant since its value depends on M. A more elegant way is to define a function that calculates N from M automatically. In VHDL, this can be done by creating a user-defined *function* in a *package* and invoking the package before the entity declaration. This is beyond the scope of this book and the details may be found in the references cited in the bibliographic section.

4.4   TESTBENCH FOR SEQUENTIAL CIRCUITS

A testbench is a program that mimics a physical lab bench, as discussed in Section 1.5. Developing a comprehensive testbench is beyond the scope of this book. We discuss a simple testbench for the previous universal binary counter in this section. It can serve as a template for other sequential circuits. The code for the testbench is shown in Listing 4.11.

Listing 4.11   Testbench for a universal binary counter

```vhdl
library ieee;
use ieee.std_logic_1164.all;

entity bin_counter_tb is
end bin_counter_tb;
```

```vhdl
architecture arch of bin_counter_tb is
   constant THREE      : integer := 3;
   constant T          : time    := 10 ns; -- clk period
   signal clk          : std_logic;
   signal reset        : std_logic;
   signal syn_clr      : std_logic;
   signal load, en, up : std_logic;
   signal d            : std_logic_vector(THREE-1 downto 0);
   signal max_tick     : std_logic;
   signal min_tick     : std_logic;
   signal q            : std_logic_vector(THREE-1 downto 0);
begin
   --****************************************************************
   -- instantiation
   --****************************************************************
   counter_unit : entity work.univ_bin_counter(arch)
      generic map(N => THREE)
      port map(
         clk      => clk,
         reset    => reset,
         syn_clr  => syn_clr,
         load     => load,
         en       => en,
         up       => up,
         d        => d,
         max_tick => max_tick,
         min_tick => min_tick,
         q        => q
      );
   --****************************************************************
   -- clock
   --****************************************************************
   -- 20 ns clock running forever
   process
   begin
      clk <= '0';
      wait for T / 2;
      clk <= '1';
      wait for T / 2;
   end process;
   --****************************************************************
   -- reset
   --****************************************************************
   -- reset asserted for T/2
   reset <= '1', '0' after T / 2;
   --****************************************************************
   -- other stimulus
   --****************************************************************
   process
   begin
      --****************************************************************
      -- initial input
      --****************************************************************
      syn_clr <= '0';
      load    <= '0';
      en      <= '0';
      up      <= '1';                        -- count up
      d       <= (others => '0');
      wait until falling_edge(clk);
      wait until falling_edge(clk);
```

```vhdl
   --*******************************************************
   -- test load
   --*******************************************************
   load <= '1';
   d    <= "011";
   wait until falling_edge(clk);
   load <= '0';
   -- pause 2 clocks
   wait until falling_edge(clk);
   wait until falling_edge(clk);
   --*******************************************************
   -- test syn_clear
   --*******************************************************
   syn_clr <= '1';                     -- clear
   wait until falling_edge(clk);
   syn_clr <= '0';
   --*******************************************************
   -- test up counter and pause
   --*******************************************************
   en      <= '1';                     -- count
   up      <= '1';
   for i in 1 to 10 loop               -- count 10 clocks
      wait until falling_edge(clk);
   end loop;
   en <= '0';
   wait until falling_edge(clk);
   wait until falling_edge(clk);
   en <= '1';
   wait until falling_edge(clk);
   wait until falling_edge(clk);
   --*******************************************************
   -- test down counter
   --*******************************************************
   up <= '0';
   for i in 1 to 10 loop               -- run 10 clocks
      wait until falling_edge(clk);
   end loop;
   --*******************************************************
   -- other wait conditions
   --*******************************************************
   -- continue until q=2
   wait until q = "010";
   wait until falling_edge(clk);
   up <= '1';
   -- continue until min_tick changes value
   wait on min_tick;
   wait until falling_edge(clk);
   up <= '0';
   wait for 4 * T;                     -- wait for 80 ns
   en <= '0';
   wait for 4 * T;
   --*******************************************************
   -- terminate simulation
   --*******************************************************
   assert false
      report "Simulation Completed"
      severity failure;
   end process;
end arch;
```

The `clk` signal is assigned between '0' and '1' alternately, and each value lasts for half a period. Note that the process has no sensitivity list and repeats itself forever.

The reset stimulus involves one statement:

```
reset <= '1', '0' after T/2;
```

It indicates that the `reset` signal is set to '1' initially and changed to '0' after half a period. The statement represents the "power-on" condition, in which the `reset` signal is asserted momentarily to clear the system to the initial state. Note that, by default, the '`U`' value (for uninitialized), not '`0`', is assigned to a signal with the `std_logic` type. Using a short reset pulse is a good mechanism to perform system initialization.

The last process statement generates a stimulus for other input signals. We first test the load and clear operations and then exercise counting in both directions. The final **assert false** statement forces the simulator to terminate simulation.

For a synchronous system with positive edge-triggered FFs, an input signal must be stable around the rising edge of the clock signal to satisfy the setup and hold time constraints. One easy way to achieve this is to change an input signal's value during the '1'-to-'0' transition of the `clk` signal. The `falling_edge` function of the `std_logic_1164` package checks this condition, and we can use it in a wait statement:

```
wait until falling_edge(clk);
```

Note that each statement represents a new falling edge, which corresponds to the advancement of one clock cycle. In our template, we generally use this statement to specify the progress of time. For multiple clock cycles, we can use a loop statement:

```
for i in 1 to 10 loop  -- count 10 clocks
   wait until falling_edge(clk);
end loop;
```

There are other useful forms of wait statements, as shown at the end of the process. We can wait until a special condition, such as "when q is equal to 2",

```
wait until q="010";
```

or wait until a signal changes, such as

```
wait on min_tick;
```

or wait for an absolute time, such as

```
wait for 4*T;  -- wait for 4 clock periods
```

If an input signal is modified after these statements, we need to make sure that the input change does not occur at the rising edge of the clock. An additional

```
wait until falling_edge(clk);
```

statement should be added when needed.

We can compile the code and perform simulation. Part of the simulated waveform is shown in Figure 4.4.

4.5 CASE STUDY

After examining several simple circuits, we discuss the design of more sophisticated examples in this section.

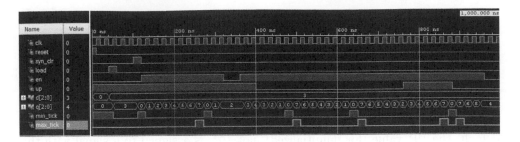

Figure 4.4   Testbench waveform.

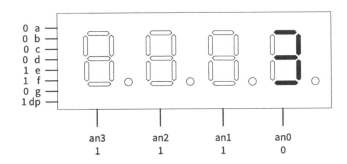

Figure 4.5   Time-multiplexed seven-segment LED display.

4.5.1   LED time-multiplexing circuit

The Nexys 4 DDR board contains an eight-digit seven-segment LED display and the Basys 3 board contains a four-digit seven-segment LED display. To facilitate both boards, our discussion focuses on a four-digit display in Part I of the book.

To save the number of I/O pins, a time-multiplexing sharing scheme is used in some multi-digit seven-segment LED displays. In this scheme, each individual seven-segment LED has its own enable signal but shares eight common signals to light the segments. All signals are active-low (i.e., enabled when a signal is '0'). The schematic of displaying '3' on the rightmost LED is shown in Figure 4.5. Note that the enable signal (i.e., **an**) is "1110". This configuration clearly can enable only one seven-segment LED at a time. We can *time-multiplex* the four LED patterns by enabling the four individual seven-segment LEDs in turn, as shown in the simplified timing diagram in Figure 4.6. If the refreshing rate of the enable signal is fast enough, the human eye cannot distinguish the on and off intervals of the LEDs and perceives that all four seven-segment LEDs are lit simultaneously. This scheme reduces the number of I/O pins from 32 to 12 (i.e., eight LED segments

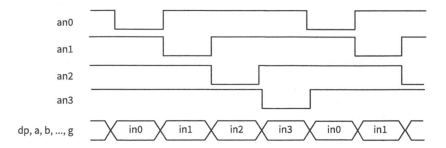

Figure 4.6   Timing diagram of a time-multiplexed seven-segment LED display.

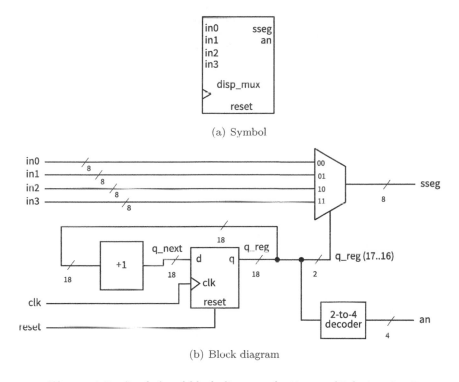

(a) Symbol

(b) Block diagram

Figure 4.7   Symbol and block diagram of a time-multiplexing circuit.

plus four enable signals) but requires a time-multiplexing circuit. Two variations of the circuit are discussed in the following subsections.

Time multiplexing with LED patterns   The symbol and block diagram of the four-digit time-multiplexing circuit are shown in Figure 4.7. It takes four seven-segment LED patterns, in3, in2, in1, and in0, and passes them to the output, sseg, in accordance with the enable signal.

The refresh rate of the enable signal has to be fast enough to fool our eyes but should be slow enough so that the LEDs can be turned on and off completely. The rate around the range 1000 Hz should work properly. In our design, we use an 18-bit binary counter for this purpose. The two MSBs are decoded to generate the enable signal and are used as the selection signal for multiplexing. The refreshing rate of an individual bit, such as an(0), becomes $\frac{100M}{2^{16}}$ Hz, which is about 1600 Hz. The code is shown in Listing 4.12.

Listing 4.12 LED time-multiplexing circuit with LED patterns

```vhdl
library ieee;
use ieee.std_logic_1164.all;
use ieee.numeric_std.all;
entity disp_mux is
   port(
      clk   : in  std_logic;
      reset : in  std_logic;
      in0   : in  std_logic_vector(7 downto 0);
      in1   : in  std_logic_vector(7 downto 0);
      in2   : in  std_logic_vector(7 downto 0);
      in3   : in  std_logic_vector(7 downto 0);
      an    : out std_logic_vector(3 downto 0);
      sseg  : out std_logic_vector(7 downto 0)
   );
end disp_mux;

architecture arch of disp_mux is
   -- refreshing rate around 1600 Hz (100MHz/2^16)
   constant N    : integer := 18;
   signal q_reg  : unsigned(N-1 downto 0);
   signal q_next : unsigned(N-1 downto 0);
   signal sel    : std_logic_vector(1 downto 0);
begin
   -- register
   process(clk, reset)
   begin
      if reset = '1' then
         q_reg <= (others => '0');
      elsif (clk'event and clk = '1') then
         q_reg <= q_next;
      end if;
   end process;
   -- next-state logic for the counter
   q_next <= q_reg + 1;
   -- 2 MSBs of counter to control 4-to-1 multiplexing
   -- and to generate active-low enable signal
   sel <= std_logic_vector(q_reg(N-1 downto N-2));
   process(sel, in0, in1, in2, in3)
   begin
      case sel is
         when "00" =>
            an(3 downto 0) <= "1110";
            sseg           <= in0;
         when "01" =>
            an(3 downto 0) <= "1101";
            sseg           <= in1;
         when "10" =>
            an(3 downto 0) <= "1011";
            sseg           <= in2;
         when others =>
            an(3 downto 0) <= "0111";
            sseg           <= in3;
      end case;
   end process;
end arch;
```

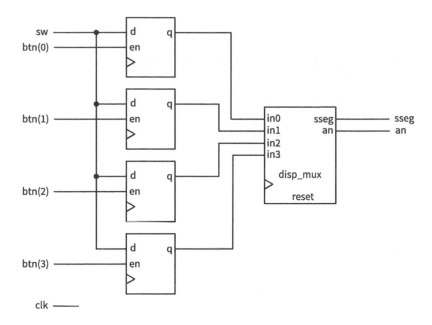

Figure 4.8   LED time multiplexing testing circuit.

Since the circuit is designed for a four-digit seven-segment LED display, a four-bit **an** signal is used. For the Nexys 4 DDR board, four MSBs of the eight-bit **an** signal are not connected and all LED segments of the leftmost four digits are lit. If desired, we can turn off these digits by declaring an eight-bit **an** signal and connecting the four MSBs with 1's, as in

```
  . . .
an : out std_logic_vector(7 downto 0);
  . . .
an(7 downto 4) <= "1111";
  . . .
```

We use the testing circuit in Figure 4.8 to verify operation of the LED time-multiplexing circuit. It uses four 8-bit registers to store the LED patterns. The registers use the same eight switches as input but are controlled by individual enable signals. When we press a button, the corresponding register is enabled and the switch pattern is loaded to that register. The code is shown in Listing 4.13.

Listing 4.13   Testing circuit for time multiplexing with LED patterns

```vhdl
library ieee;
use ieee.std_logic_1164.all;
use ieee.numeric_std.all;
entity disp_mux_test is
   port(
      clk  : in  std_logic;
      btn  : in  std_logic_vector(3 downto 0);
      sw   : in  std_logic_vector(7 downto 0);
      an   : out std_logic_vector(3 downto 0);
      sseg : out std_logic_vector(7 downto 0)
   );
```

```vhdl
end disp_mux_test;

architecture arch of disp_mux_test is
   signal d3_reg, d2_reg : std_logic_vector(7 downto 0);
   signal d1_reg, d0_reg : std_logic_vector(7 downto 0);
begin
   disp_unit : entity work.disp_mux
      port map(
         clk => clk, reset => '0',
         in3 => d3_reg, in2 => d2_reg, in1 => d1_reg,
         in0 => d0_reg, an => an, sseg => sseg);
   -- registers for 4 led patterns
   process(clk)
   begin
      if (clk'event and clk = '1') then
         if (btn(3) = '1') then
            d3_reg <= sw;
         end if;
         if (btn(2) = '1') then
            d2_reg <= sw;
         end if;
         if (btn(1) = '1') then
            d1_reg <= sw;
         end if;
         if (btn(0) = '1') then
            d0_reg <= sw;
         end if;
      end if;
   end process;
end arch;
```

Time multiplexing with hexadecimal digits   The most common application of a seven-segment LED is to display a hexadecimal digit. The decoding circuit is discussed in Section 3.7.1. To display four hexadecimal digits with the previous time-multiplexing circuit, four decoding circuits are needed. A better alternative is first to multiplex the hexadecimal digits and then to decode the result, as shown in Figure 4.9.

This scheme requires only one decoding circuit and reduces the width of the 4-to-1 multiplexer from 8 bits to 5 bits (i.e., 4 bits for the hexadecimal digit and 1 bit for the decimal point). The code is shown in Listing 4.14. In addition to clock and reset, the input consists of four 4-bit hexadecimal digits, hex3, hex2, hex1, and hex0, and four decimal points, which are grouped as one signal, dp_in.

Listing 4.14   LED time-multiplexing circuit with hexadecimal digits

```vhdl
library ieee;
use ieee.std_logic_1164.all;
use ieee.numeric_std.all;
entity disp_hex_mux is
   port(
      clk, reset : in   std_logic;
      hex3, hex2 : in   std_logic_vector(3 downto 0);
      hex1, hex0 : in   std_logic_vector(3 downto 0);
      dp_in      : in   std_logic_vector(3 downto 0);
      an         : out  std_logic_vector(3 downto 0);
      sseg       : out  std_logic_vector(7 downto 0)
   );
end disp_hex_mux;
```

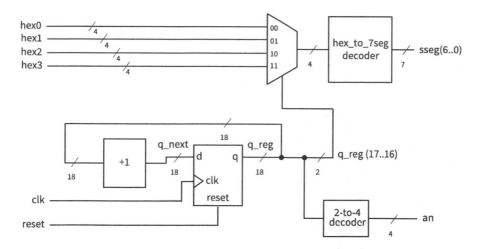

Figure 4.9 Block diagram of a hexadecimal time-multiplexing circuit.

```
architecture arch of disp_hex_mux is
   -- each 7-seg led enabled (2 18/4)*10 ns (16 ms)
   constant N      : integer := 18;
   signal q_reg    : unsigned(N - 1 downto 0);
   signal q_next   : unsigned(N - 1 downto 0);
   signal sel      : std_logic_vector(1 downto 0);
   signal hex      : std_logic_vector(3 downto 0);
   signal dp       : std_logic;
begin
   -- register
   process(clk, reset)
   begin
      if reset = '1' then
         q_reg <= (others => '0');
      elsif (clk'event and clk = '1') then
         q_reg <= q_next;
      end if;
   end process;
   -- next-state logic for the counter
   q_next <= q_reg + 1;
   -- 2 MSBs of counter to control 4-to-1 multiplexing
   sel <= std_logic_vector(q_reg(N - 1 downto N - 2));
   process(sel, hex0, hex1, hex2, hex3, dp_in)
   begin
      case sel is
         when "00" =>
            an(3 downto 0) <= "1110";
            hex            <= hex0;
            dp             <= dp_in(0);
         when "01" =>
            an(3 downto 0) <= "1101";
            hex            <= hex1;
            dp             <= dp_in(1);
         when "10" =>
            an(3 downto 0) <= "1011";
            hex            <= hex2;
            dp             <= dp_in(2);
         when others =>
```

```
                an(3 downto 0)  <= "0111";
                hex             <= hex3;
                dp              <= dp_in(3);
      end case;
   end process;
   -- hex-to-7-segment led decoding
   with hex select
      sseg(6 downto 0) <=
         "1000000" when "0000",
         "1111001" when "0001",
         "0100100" when "0010",
         "0110000" when "0011",
         "0011001" when "0100",
         "0010010" when "0101",
         "0000010" when "0110",
         "1111000" when "0111",
         "0000000" when "1000",
         "0010000" when "1001",
         "0001000" when "1010",    --a
         "0000011" when "1011",    --b
         "1000110" when "1100",    --c
         "0100001" when "1101",    --d
         "0000110" when "1110",    --e
         "0001110" when others;    --f
   -- decimal point
   sseg(7) <= dp;
end arch;
```

To verify operation of this circuit, we define eight slide switches as two 4-bit unsigned numbers, add the two numbers, and show the two numbers and their sum on the four-digit seven-segment LED display. The code is shown in Listing 4.15.

Listing 4.15   Testing circuit for time multiplexing with hexadecimal digits

```
library ieee;
use ieee.std_logic_1164.all;
use ieee.numeric_std.all;
entity hex_mux_test is
   port(
      clk  : in  std_logic;
      sw   : in  std_logic_vector(7 downto 0);
      an   : out std_logic_vector(3 downto 0);
      sseg : out std_logic_vector(7 downto 0)
   );
end hex_mux_test;

architecture arch of hex_mux_test is
   signal a, b : unsigned(7 downto 0);
   signal sum  : std_logic_vector(7 downto 0);
begin
   disp_unit : entity work.disp_hex_mux
      port map(
         clk   => clk,
         reset => '0',
         hex3  => sum(7 downto 4),
         hex2  => sum(3 downto 0),
         hex1  => sw(7 downto 4),
         hex0  => sw(3 downto 0),
         dp_in => "1011",
         an    => an,
         sseg  => sseg
      );
```

```
   a    <= "0000" & unsigned(sw(3 downto 0));
   b    <= "0000" & unsigned(sw(7 downto 4));
   sum <= std_logic_vector(a + b);
end arch;
```

Simulation consideration   Many sequential circuit examples in the book operate at a relatively slow rate, as does the enable pulse of the LED time-multiplexing circuit. This can be done by generating a single-clock enable tick from a counter. An 18-bit counter is used in this circuit:

```
   constant N: integer:=18;
   signal q_reg, q_next: unsigned(N-1 downto 0);
    .  .  .
   q_next <= q_reg + 1;
```

Because of the counter's size, simulating this type of circuit consumes a significant amount of computation time (i.e., 2^{18} clock cycles for one iteration). Since our main interest is in the multiplexing part of the code, most simulation time is wasted. It is more efficient to use a smaller counter in simulation. We can do this by modifying the constant statement

```
   constant N: integer:=4;
```

when constructing the testbench. This requires only 2^4 clock cycles for one iteration and allows us to better exercise and observe the key operations.

Instead of using a constant statement and modifying code between simulation and synthesis, an alternative is to define a generic for the relevant parameter. During instantiation, we can assign different values for simulation and synthesis.

4.5.2   Stopwatch

We consider the design of a stopwatch in this subsection. The watch displays the time in three decimal digits, and counts from 00.0 to 99.9 seconds and wraps around. It contains a synchronous clear signal, clr, which returns the count to 00.0, and an enable signal, go, which enables and suspends the counting. This design is basically a BCD (binary-coded decimal) counter, which counts in BCD format. In this format, a decimal number is represented by a sequence of 4-bit BCD digits. For example, 139_{10} is represented as "0001 0011 1001" and the next number in sequence is 140_{10}, which is represented as "0001 0100 0000".

Since the FPGA board has a 100M-Hz clock, we first need a mod-10,000,000 counter that generates a one-clock-cycle tick every 0.1 second. The tick is then used to enable counting of the three-digit BCD counter.

Design I   Our first design of the BCD counter uses a cascading structure of three decade (i.e., mod-10) counters, representing counts of 0.1, 1, and 10 seconds, respectively. The decade counter has an enable signal and generates a one-clock-cycle tick when it reaches 9. We can use these signals to "hook" the three counters. For example, the 10-second counter is enabled only when the enable tick of the mod-10,000,000 counter is asserted and both the 0.1- and 1-second counters are 9. The code is shown in Listing 4.16.

Listing 4.16 Cascading description for a stopwatch

```vhdl
library ieee;
use ieee.std_logic_1164.all;
use ieee.numeric_std.all;
entity stop_watch is
   port(
      clk        : in std_logic;
      go, clr    : in std_logic;
      d2, d1, d0 : out std_logic_vector(3 downto 0)
   );
end stop_watch;

architecture cascade_arch of stop_watch is
   constant DVSR: integer := 10000000;
   signal ms_reg, ms_next : unsigned(23 downto 0);
   signal d2_reg, d2_next : unsigned(3 downto 0);
   signal d1_reg, d1_next : unsigned(3 downto 0);
   signal d0_reg, d0_next : unsigned(3 downto 0);
   signal d1_en, d2_en, d0_en : std_logic;
   signal ms_tick, d0_tick, d1_tick : std_logic;
begin
   -- register
   process(clk)
   begin
      if (clk'event and clk='1') then
         ms_reg <= ms_next;
         d2_reg <= d2_next;
         d1_reg <= d1_next;
         d0_reg <= d0_next;
      end if;
   end process;
   -- next-state logic
   -- 0.1 sec tick generator: mod-10000000
   ms_next <=
      (others=>'0') when clr='1' or (ms_reg=DVSR and go='1') else
      ms_reg + 1    when go='1' else
      ms_reg;
   ms_tick <= '1' when ms_reg=DVSR else '0';
   -- 0.1 sec counter
   d0_en <= '1' when ms_tick='1' else '0';
   d0_next <=
      "0000"      when (clr='1') or (d0_en='1' and d0_reg=9) else
      d0_reg + 1 when d0_en='1' else
      d0_reg;
   d0_tick <= '1' when d0_reg=9 else '0';
   -- 1 sec counter
   d1_en <= '1' when ms_tick='1' and d0_tick='1' else '0';
   d1_next <=
      "0000"      when (clr='1') or (d1_en='1' and d1_reg=9) else
      d1_reg + 1 when d1_en='1' else
      d1_reg;
   d1_tick <= '1' when d1_reg=9 else '0';
   -- 10 sec counter
   d2_en <=
      '1' when ms_tick='1' and d0_tick='1' and d1_tick='1' else '0';
   d2_next <=
      "0000"      when (clr='1') or (d2_en='1' and d2_reg=9) else
      d2_reg + 1 when d2_en='1' else
      d2_reg;
   -- output logic
   d0 <= std_logic_vector(d0_reg);
```

```
    d1 <= std_logic_vector(d1_reg);
    d2 <= std_logic_vector(d2_reg);
end cascade_arch;
```

Note that all registers are controlled by the same clock signal. This example illustrates how to use a one-clock-cycle enable tick to maintain synchronicity. An inferior approach is to use the output of the lower counter as the clock signal for the next stage. Although it may appear to be simpler, it violates the synchronous design principle and is a very poor practice.

Design II An alternative for the three-digit BCD counter is to describe the entire structure in a nested if statement. The nested conditions indicate that the counter reaches .9, 9.9, and 99.9 seconds. The code is shown in Listing 4.17.

Listing 4.17 Nested if-statement description for a stopwatch

```
architecture if_arch of stop_watch is
    constant DVSR: integer := 10000000;
    signal ms_reg, ms_next : unsigned(23 downto 0);
    signal d2_reg, d2_next : unsigned(3 downto 0);
    signal d1_reg, d1_next : unsigned(3 downto 0);
    signal d0_reg, d0_next : unsigned(3 downto 0);
    signal ms_tick         : std_logic;
begin
    -- register
    process(clk)
    begin
        if (clk'event and clk='1') then
            ms_reg <= ms_next;
            d2_reg <= d2_next;
            d1_reg <= d1_next;
            d0_reg <= d0_next;
        end if;
    end process;
    -- next-state logic
    -- 0.1 sec tick generator: mod-10000000
    ms_next <=
        (others=>'0') when clr='1' or (ms_reg=DVSR and go='1') else
        ms_reg + 1 when go='1' else
        ms_reg;
    ms_tick <= '1' when ms_reg=DVSR else '0';
    -- 0.1 sec counter
    process(d0_reg,d1_reg,d2_reg,ms_tick,clr)
    begin
        -- default
        d0_next <= d0_reg;
        d1_next <= d1_reg;
        d2_next <= d2_reg;
        if clr='1' then
            d0_next <= "0000";
            d1_next <= "0000";
            d2_next <= "0000";
        elsif ms_tick='1' then
            if (d0_reg/=9) then
                d0_next <= d0_reg + 1;
            else            -- reach XX9
                d0_next <= "0000";
                if (d1_reg/=9) then
                    d1_next <= d1_reg + 1;
                else        -- reach X99
```

```
                  d1_next <= "0000";
                  if (d2_reg/=9) then
                     d2_next <= d2_reg + 1;
                  else -- reach 999
                     d2_next <= "0000";
                  end if;
               end if;
            end if;
        end if;
     end process;
     -- output logic
     d0 <= std_logic_vector(d0_reg);
     d1 <= std_logic_vector(d1_reg);
     d2 <= std_logic_vector(d2_reg);
end if_arch;
```

Verification circuit To verify operation of the stopwatch, we can combine it with the previous hexadecimal LED time-multiplexing circuit to display the output of the watch. The code is shown in Listing 4.18. Note that the first digit of the LED is assigned to 0 and the go and clr signals are mapped to two buttons of the prototyping board.

Listing 4.18 Testing circuit for a stopwatch

```
library ieee;
use ieee.std_logic_1164.all;
entity stop_watch_test is
   port(
      clk  : in  std_logic;
      btn  : in  std_logic_vector(3 downto 0);
      an   : out std_logic_vector(3 downto 0);
      sseg : out std_logic_vector(7 downto 0)
   );
end stop_watch_test;

architecture arch of stop_watch_test is
   signal d2, d1, d0 : std_logic_vector(3 downto 0);
begin
   -- instantiate 7-seg display
   disp_unit : entity work.disp_hex_mux
      port map(
         clk    => clk,
         reset  => '0',
         hex3   => "0000",
         hex2   => d2,
         hex1   => d1,
         hex0   => d0,
         dp_in  => "1101",
         an     => an,
         sseg   => sseg
      );
   -- instantiate stop watch
   watch_unit : entity work.stop_watch(cascade_arch)
      port map(
         clk => clk,
         go  => btn(1),
         clr => btn(0),
         d2  => d2,
         d1  => d1,
         d0  => d0
```

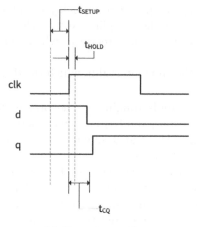

(a) Normal operation

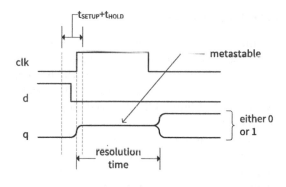

(b) Timing violation

Figure 4.10   Timing diagram of a D FF.

```
        );
end arch;
```

4.6   TIMING AND CLOCKING

The clock signal synchronizes and coordinates the activities of a sequential circuit and plays the key role of sequential circuit timing. Detailed discussion of timing and clocking is beyond the scope of this book. This section provides an overview of basic concepts.

4.6.1   Timing of FF

The basic timing diagram of a D FF is shown in Figure 4.10(a). It includes three main timing parameters:

- t_{CQ}: *clock-to-q delay*, the propagation delay required for the d input to show up at the q output after the sampling edge of the clock signal.
- t_{SETUP}: *setup time*, the time interval in which the d signal must be stable *before* the clock edge.
- t_{HOLD}: *hold time*, the time interval in which the d signal must be stable *after* the clock edge.

The t_{CQ} parameter corresponds roughly to the propagation delay of a combinational component. The t_{SETUP} and t_{HOLD} parameters, in contrast, are *timing constraints*. They specify that the d signal must be stable in a small window around the sampling edge of the clock.

If the d signal changes within the setup or hold time window, which is known as *setup time violation* or *hold time violation*, the D FF may enter a *metastable state*, in which the q signal becomes neither 0 nor 1. The condition resolves after a random amount of *resolution time* and the q signal eventually reaches a stable 0 or 1, as shown in Figure 4.10(b). If not properly controlled, the metastable state can propagate to subsequent logic and make the entire system unstable.

4.6.2   Maximum operating frequency

One of the most difficult design aspects of a sequential circuit is to ensure that the system timing does not violate the setup and hold time constraints. The key idea behind the *synchronous design methodology*, in which a common clock signal is used to control all storage elements, is to group all FFs together and treat them as a single register, as shown in Figure 4.2. We need to perform timing analysis on only one memory component.

The timing of a sequential circuit is characterized by f_{max}. The reciprocal of f_{max} specifies t_{CLOCK}, the minimum clock period, which can be interpreted as the interval between two sampling edges of the clock signal. To ensure correct operation, the next value (i.e., state_next in Figure 4.2) must be generated and stabilized within this interval. Assume that the maximum propagation delay of next-state logic is t_{COMB}. The minimum clock period can be obtained by adding the propagation delays and setup time constraint of the closed loop in Figure 4.2:

$$t_{CLOCK} = t_{CQ} + t_{COMB} + t_{SETUP}$$

and the maximum clock rate is the reciprocal:

$$f_{max} = \frac{1}{t_{CLOCK}} = \frac{1}{t_{CQ} + t_{COMB} + t_{SETUP}}$$

For a given FPGA device, t_{CQ} and t_{SETUP} are fixed. The only way to increase f_{max} is to use a faster combinational logic to reduce t_{COMB}. Synthesis software sometimes can identify the slowest path and reduce its delay by adding extra logic (i.e., larger area). A typical area–delay curve is shown in Figure 4.11, in which each point is a possible implementation. The software usually starts with the minimum-area implementation and traverses through the curve to reach a point that satisfies the designated clock rate. Of course, the trade-off can be achieved only in a limited range. We cannot increase the performance indefinitely.

In Vivado Design Suite, the desired clock rate can be specified in the constraint file and the synthesis and placement and routing software try to achieve

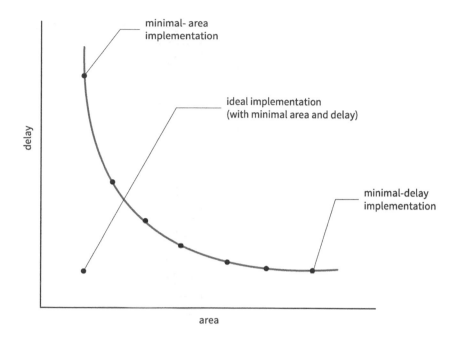

Figure 4.11   Area-delay trade-off curve.

the desired goal. For example, the 100M-Hz (i.e., 10.00-ns period) oscillator of the Nexys 4 DDR board is used as the clock signal in our design and connected to the clk port. The corresponding statement in the constraint file is

```
create_clock -add -name sys_clk_pin -period 10.00
            -waveform {0 5} [get_ports {clk}];
```

4.6.3   Clock tree

In a digital circuit, each input port of a gate and each wire introduce a small resistance and a small capacitance. The output port of a gate has to charge or discharge (i.e., "drive") all capacitors when a signal switches state. A typical gate can drive up to half a dozen gates. A complex synchronous design may contain hundreds of thousands of FFs and all the FFs are controlled by the same clock signal. A *clock tree*, also referred to as a *clock distribution network*, is the circuit that distributes the clock signal to all FFs in the system.

The clock tree uses multiple levels of buffers to increase the driving capability and applies a special routing algorithm to balance the distribution network and minimize the differences in propagation delays. A conceptual three-level clock distribution network is shown in Figure 4.12, in which each buffer can drive four input ports.

Modern FPGA devices contain one or more prerouted and prefabricated clock trees. In the Xilinx literature, a tree is sometimes referred to as a *clock buffer*. It can be thought as the root buffer of a clock tree. The Artix devices has about a dozen distribution networks.

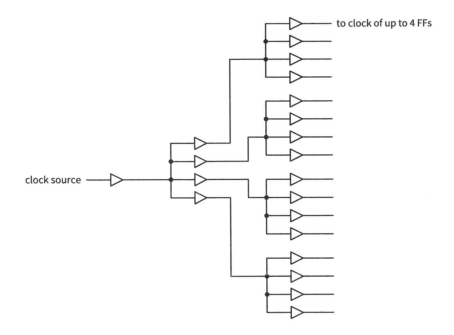

Figure 4.12   Conceptual clock tree.

4.6.4   GALS system and CDC

In a complex digital system, multiple clocks may exist or become necessary in certain situations. For example, a system may interact with external devices or exchange information through communication links. These external devices and links can be driven by external clock sources or impose different clock rates. As discussed earlier, the synchronous methodology is the fundamental principle in today's digital system development, and most design and analysis schemes are based on this methodology. Thus, even with multiple clocks, we still want to apply this methodology as much as possible. The basic approach is to divide a system into multiple synchronous subsystems and design special interfaces between the subsystems. This scheme is known as a *globally asynchronous locally synchronous* (*GALS*) system and it allows us to continuously apply the synchronous methodology to design a much larger system.

In a GALS system, a subsystem driven by the same clock system is known as a *clock domain*. A major task in designing a GALS system is the interface between the clock domains; i.e., how to exchange information and transfer data between two clock domains, which is known as *clock domain crossing* (*CDC*). Since the circuit in one domain has no clock information about another domain, a signal may switch at the clock's sampling edge of another domain, which leads to a setup or hold time violation. The main advantage of the synchronous design methodology is that it provides a systematic way to *avoid a timing violation*. Since a timing violation in the domain crossing is not avoidable, the design must focus on what to do *after a timing violation occurs*. Special *synchronization circuits, handshake schemes*, and *dual-clock FIFO buffers* are needed to manage the CDC issues.

Modern FPGA devices contain *PLL (phase-locked loop)* or *clock management* macro cells, which can generate clock signals of different frequencies and phases. They also include pre-designed dual-clock FIFO buffer cores. The video synchronization core discussed in Chapter 21 uses these components to interface the 100M-Hz main system domain with the VGA device driven by a 25M-Hz video pixel clock.

4.7 BIBLIOGRAPHIC NOTES

The bibliographic information for this chapter is similar to that for Chapter 3.

4.8 SUGGESTED EXPERIMENTS

4.8.1 Programmable square wave generator

A programmable square wave generator is a circuit that can generate a square wave with variable on (i.e., logic '1') and off (i.e., logic '0') intervals. The durations of the intervals are specified by two 4-bit control signals, m and n, which are interpreted as unsigned integers. The on and off intervals are $m*100$ ns and $n*100$ ns, respectively (recall that the period of the onboard oscillator is 10 ns). Design a programmable square wave generator circuit. The circuit should be completely synchronous. We need a logic analyzer or oscilloscope to verify its operation.

4.8.2 PWM and LED dimmer

The duty cycle of a square wave is defined as the percentage of the on interval (i.e., logic '1') in a period. A PWM (pulse width modulation) circuit can generate an output with variable duty cycles. For a PWM with 4-bit resolution, a 4-bit control signal, w, specifies the duty cycle. The w signal is interpreted as an unsigned integer and the duty cycle is $\frac{w}{16}$.

1. Design a PWM circuit with 4-bit resolution and verify its operation using a logic analyzer or oscilloscope.
2. Modify the LED time-multiplexing circuit to include the PWM circuit for the an signal. The PWM circuit specifies the percentage of time that the LED display is on. We can control the perceived brightness by changing the duty cycle. Verify the circuit's operation by observing 1 bit of an on a logic analyzer or oscilloscope.
3. Replace the LED time-multiplexing circuit of Listing 4.18 with the new design and use four slide switches to control the duty cycle. Verify operation of the circuit. It may be necessary to go to a dark area to see the effect of dimming.

4.8.3 Rotating square circuit

In a seven-segment LED display, a square pattern can be created by enabling the a, b, f, and g segments or the c, d, e, and g segments. We want to design a circuit that circulates the square patterns in the four-digit seven-segment LED display. The clockwise circulating pattern is shown in Figure 4.13. The circuit should have

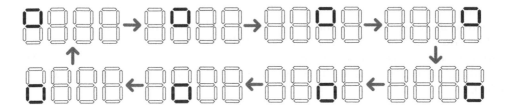

Figure 4.13 Pattern for Experiment 4.8.3.

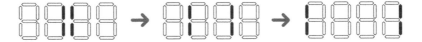

Figure 4.14 Pattern for Experiment 4.8.4.

an input, en, which enables or pauses the circulation, and an input, cw, which specifies the direction (i.e., clockwise or counterclockwise) of the circulation.

Design the circuit and verify its operation on the prototyping board. Make sure that the circulation rate is slow enough for visual inspection.

4.8.4 Heartbeat circuit

We want to create a "heartbeat" for the prototyping board. It repeats the simple pattern in the four-digit seven-segment display, as shown in Figure 4.14, at a rate of 72 Hz. Design the circuit and verify its operation on the prototyping board.

4.8.5 Rotating LED banner circuit

Four symbols can be shown on a four-digit seven-segment LED display at a time. We can show more information if the data is rotated and moved continuously. For example, assume that the message is 10 digits (i.e., "0123456789"). The display can show the message as "0123", "1234", "2345", ..., "6789", "7890", ..., "0123". The circuit should have an input, en, which enables or pauses the rotation, and an input, dir, which specifies the direction (i.e., rotate left or right).

Design the circuit and verify its operation on the prototyping board. Make sure that the rotation rate is slow enough for visual inspection.

4.8.6 Enhanced stopwatch

Modify the stopwatch with the following extensions:
- Add an additional signal, up, to control the direction of counting. The stopwatch counts up when the up signal is asserted and counts down otherwise.
- Add a minute digit to the display. The LED display format should be like M.SS.D, where D represents 0.1 second and its range is between 0 and 9, SS represents seconds and its range is between 00 and 59, and M represents minutes and its range is between 0 and 9.

Design the new stopwatch and verify its operation with a testing circuit.

CHAPTER 5

FSM

An FSM (finite state machine) is a sequential circuit that transits among a finite number of internal states. The transitions depend on the current state and external input and do exhibit a simple, "regular" pattern. In this chapter, we provide an overview of the basic characteristics and representation of FSMs and discuss the derivation of HDL codes.

5.1 INTRODUCTION

An FSM (finite state machine) is used to model a system that transits among a finite number of internal states. The transitions depend on the current state and external input. Unlike a regular sequential circuit, the state transitions of an FSM do not exhibit a simple, repetitive pattern. Its next-state logic is usually constructed from scratch and is sometimes known as "random" logic. This is different from the next-state logic of a regular sequential circuit, which is composed mostly of "structured" components, such as incrementors and shifters.

In practice, the main application of an FSM is to act as the controller of a large digital system, which examines the external commands and status and activates proper control signals to control operation of a *data path*, which is usually composed of regular sequential components. This is known as an FSMD (finite state machine with data path) and is discussed in Chapter 6.

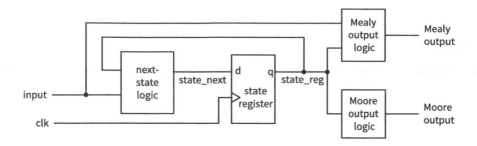

Figure 5.1 Block diagram of a synchronous FSM.

5.1.1 Mealy and Moore outputs

The basic block diagram of an FSM is the same as that of a regular sequential circuit and is repeated in Figure 5.1. It consists of a state register, next-state logic, and output logic. An FSM is known as a *Moore machine* if the output is only a function of state, and is known as a *Mealy machine* if the output is a function of state and external input. Both types of output may exist in a complex FSM, and we simply refer to it as containing a Moore output and a Mealy output. The Moore and Mealy outputs are similar but not identical. Understanding their subtle differences is the key for controller design. The example in Section 5.3.1 illustrates the behaviors and constructions of the two types of outputs.

5.1.2 FSM representation

An FSM is usually specified by an abstract *state diagram* or *ASM chart* (algorithmic state machine chart), both capturing the FSM's input, output, states, and transitions in a graphical representation. The two representations provide the same information. The FSM representation is more compact and better for simple applications. The ASM chart representation is somewhat like a flowchart and is more descriptive for applications with complex transition conditions and actions.

State diagram A state diagram is composed of *nodes*, which represent states and are drawn as circles, and annotated *transitional arcs*. A single node and its transition arcs are shown in Figure 5.2(a). A logic expression expressed in terms of input signals is associated with each transition arc and represents a specific condition. The arc is taken when the corresponding expression is evaluated as true.

The Moore output values are placed inside the circle since they depend only on the current state. The Mealy output values are associated with the conditions of transition arcs since they depend on the current state and external input. To reduce clutter in the diagram, only asserted output values are listed. The output signal takes the default (i.e., unasserted) value otherwise.

A representative state diagram is shown in Figure 5.3(a). The FSM has three states, two external input signals (i.e., a and b), one Moore output signal (i.e., y1), and one Mealy output signal (i.e., y0). The y1 signal is asserted when the FSM is in the s0 or s1 state. The y0 signal is asserted when the FSM is in the s0 state and the a and b signals are "11".

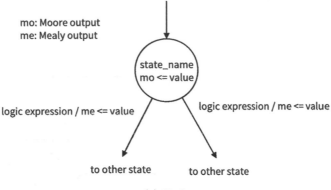

mo: Moore output
me: Mealy output

state_name
mo <= value

logic expression / me <= value   logic expression / me <= value

to other state   to other state

(a) Node

mo: Moore output
me: Mealy output

state entry

state
name

mo <= value   state box

decision box

T   Boolean   F
condition

conditional
output box

me <= value

exit to other ASM   exit to other ASM
block   block

(b) ASM block

Figure 5.2   Symbol of a state.

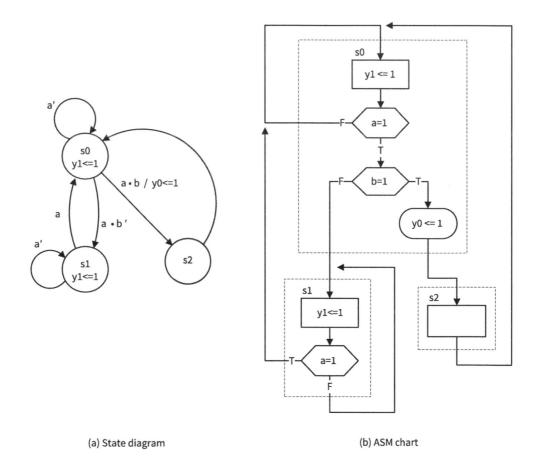

(a) State diagram   (b) ASM chart

Figure 5.3   Example of an FSM.

ASM chart   An ASM chart is composed of a network of ASM blocks. An *ASM block* consists of one *state box* and an optional network of *decision boxes* and *conditional output boxes*. A representative ASM block is shown in Figure 5.2(b).

A state box represents a state in an FSM, and the asserted Moore output values are listed inside the box. Note that it has only one exit path. A decision box tests the input condition and determines which exit path to take. It has two exit paths, labeled T and F, which correspond to the **true** and **false** values of the condition. A conditional output box lists asserted Mealy output values and is usually placed after a decision box. It indicates that the listed output signal can be activated only when the corresponding condition in the decision box is met.

A state diagram can easily be converted to an ASM chart, and vice versa. The corresponding ASM chart of the previous FSM state diagram is shown in Figure 5.3(b).

5.2   FSM CODE DEVELOPMENT

The procedure of developing code for an FSM is similar to that of a regular sequential circuit. We first separate the state register and then derive the code for the combinational next-state logic and output logic. The main difference is the next-state logic. For an FSM, the code for the next-state logic follows the flow of a state diagram or ASM chart.

For clarity and flexibility, we use the VHDL's *enumerated data type* to represent the FSM's states. The enumerated data type can best be explained by an example. Consider the FSM of Section 5.1.2, which has three states: s0, s1, and s2. We can introduce a user-defined enumerated data type as follows:

```
type eg_state_type is (s0, s1, s2);
```

The data type simply lists (i.e., *enumerates*) all symbolic values. Once the data type is defined, it can be used for the signals, as in

```
signal state_reg, state_next: eg_state_type;
```

During synthesis, software automatically maps the values in an enumerated data type to binary representations, a process known as *state assignment*. Although there is a mechanism to perform this manually, it is rarely needed.

The complete code of the FSM is shown in Listing 5.1. It consists of segments for the state register, next-state logic, Moore output logic, and Mealy output logic.

Listing 5.1   FSM example

```
library ieee;
use ieee.std_logic_1164.all;
entity fsm_eg is
   port(
      clk     : in std_logic;
      reset   : in std_logic;
      a, b    : in std_logic;
      y0, y1 : out std_logic
   );
end fsm_eg;

architecture mult_seg_arch of fsm_eg is
   type eg_state_type is (s0, s1, s2);
```

```vhdl
    signal state_reg, state_next: eg_state_type;
begin
   -- state register
   process(clk,reset)
   begin
      if (reset='1') then
         state_reg <= s0;
      elsif (clk'event and clk='1') then
         state_reg <= state_next;
      end if;
   end process;
   -- next-state logic
   process(state_reg,a,b)
   begin
      case state_reg is
         when s0 =>
            if a='1' then
               if b='1' then
                  state_next <= s2;
               else
                  state_next <= s1;
               end if;
            else
               state_next <= s0;
            end if;
         when s1 =>
            if (a='1') then
               state_next <= s0;
            else
               state_next <= s1;
            end if;
         when s2 =>
            state_next <= s0;
      end case;
   end process;
   -- Moore output logic
   process(state_reg)
   begin
      case state_reg is
         when s0|s1 =>
            y1 <= '1';
         when s2 =>
            y1 <= '0';
      end case;
   end process;
   -- Mealy output logic
   process(state_reg,a,b)
   begin
      case state_reg is
         when s0 =>
            if (a='1') and (b='1') then
               y0 <= '1';
            else
               y0 <= '0';
            end if;
         when s1 | s2 =>
            y0 <= '0';
      end case;
   end process;
end mult_seg_arch;
```

The key part is the next-state logic. It uses a case statement with the `state_reg` signal as the selection expression. The next state (i.e., `state_next` signal) is determined by the current state (i.e., `state_reg`) and external input. The code for each state basically follows the activities inside each ASM block of Figure 5.3(b).

An alternative code is to merge next-state logic and output logic into a single combinational block, as shown in Listing 5.2.

Listing 5.2   FSM with merged combinational logic

```
architecture two_seg_arch of fsm_eg is
   type eg_state_type is (s0, s1, s2);
   signal state_reg, state_next: eg_state_type;
begin
   -- state register
   process(clk,reset)
   begin
      if (reset='1') then
         state_reg <= s0;
      elsif (clk'event and clk='1') then
         state_reg <= state_next;
      end if;
   end process;
   -- next-state/output logic
   process(state_reg,a,b)
   begin
      state_next <= state_reg;  -- default back to same state
      y0 <= '0';   -- default 0
      y1 <= '0';   -- default 0
      case state_reg is
         when s0 =>
            y1 <= '1';
            if a='1' then
               if b='1' then
                  state_next <= s2;
                  y0 <= '1';
               else
                  state_next <= s1;
               end if;
            -- no else branch
            end if;
         when s1 =>
            y1 <= '1';
            if (a='1') then
               state_next <= s0;
            -- no else branch
            end if;
         when s2 =>
            state_next <= s0;
      end case;
   end process;
end two_seg_arch;
```

Note that the default output values are listed at the beginning of the code.

The code for the next-state logic and output logic follows the ASM chart closely. Once a detailed state diagram or ASM chart is derived, converting an FSM to HDL code is almost a mechanical procedure. Listings 5.1 and 5.2 can serve as templates for this purpose. Note that both versions of the codes use a separate register segment, following the coding style suggested in Section 4.1.4.

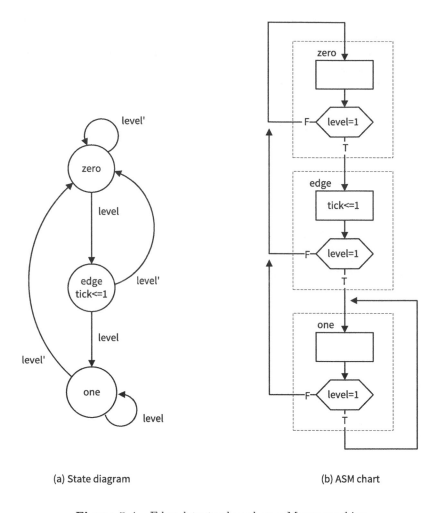

(a) State diagram   (b) ASM chart

Figure 5.4   Edge detector based on a Moore machine.

5.3   DESIGN EXAMPLES

5.3.1   Rising-edge detector

The rising-edge detector is a circuit that generates a short, one-clock-cycle pulse (we call it a *tick*) when the input signal changes from '0' to '1'. It is usually used to indicate the onset of a slow time-varying input signal. We design the circuit using both Moore and Mealy machines, and compare their differences.

Moore-based design   The state diagram and ASM chart of a Moore machine–based edge detector are shown in Figure 5.4. The `zero` and `one` states indicate that the input signal has been '0' and '1' for a while. The rising edge occurs when the input changes to '1' in the `zero` state. The FSM moves to the `edge` state and the output, `tick`, is asserted in this state. A representative timing diagram is shown at the middle of Figure 5.5. The code is shown in Listing 5.3.

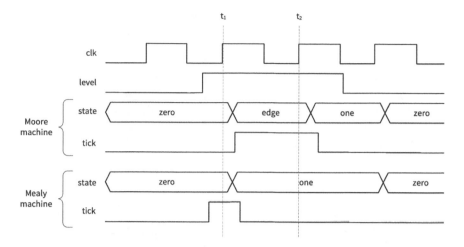

Figure 5.5   Timing diagram of two edge detectors.

Listing 5.3   Moore machine–based edge detector

```
library ieee;
use ieee.std_logic_1164.all;
entity edge_detect is
   port(
      clk    : in std_logic;
      reset  : in std_logic;
      level  : in std_logic;
      tick   : out std_logic
   );
end edge_detect;

architecture moore_arch of edge_detect is
   type state_type is (zero, edge, one);
   signal state_reg, state_next: state_type;
begin
   -- state register
   process(clk,reset)
   begin
      if (reset='1') then
         state_reg <= zero;
      elsif (clk'event and clk='1') then
         state_reg <= state_next;
      end if;
   end process;
   -- next-state/output logic
   process(state_reg,level)
   begin
      state_next <= state_reg;
      tick <= '0';
      case state_reg is
         when zero=>
            if level= '1' then
               state_next <= edge;
            end if;
         when edge =>
            tick <= '1';
            if level= '1' then
```

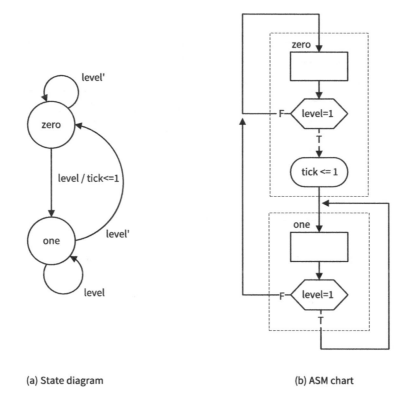

(a) State diagram   (b) ASM chart

Figure 5.6   Edge detector based on a Mealy machine.

```
            state_next <= one;
        else
            state_next <= zero;
        end if;
    when one =>
        if level= '0' then
            state_next <= zero;
        end if;
    end case;
  end process;
end moore_arch;
```

Mealy-based design   The state diagram and ASM chart of a Mealy machine–based edge detector are shown in Figure 5.6. The zero and one states have similar meaning. When the FSM is in the zero state and the input changes to '1', the output is asserted immediately. The FSM moves to the one state at the rising edge of the next clock and the output is deasserted. A representative timing diagram is shown at the bottom of Figure 5.5. Note that due to the propagation delay, the output signal is still asserted at the rising edge of the next clock (i.e., at t_1). The code is shown in Listing 5.4.

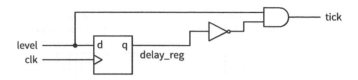

Figure 5.7   Gate-level implementation of an edge detector.

Listing 5.4   Mealy machine–based edge detector

```
architecture mealy_arch of edge_detect is
   type state_type is (zero, one);
   signal state_reg, state_next: state_type;
begin
   -- state register
   process(clk,reset)
   begin
      if (reset='1') then
         state_reg <= zero;
      elsif (clk'event and clk='1') then
         state_reg <= state_next;
      end if;
   end process;
   -- next-state/output logic
   process(state_reg,level)
   begin
      state_next <= state_reg;
      tick <= '0';
      case state_reg is
         when zero=>
            if level= '1' then
               state_next <= one;
               tick <= '1';
            end if;
         when one =>
            if level= '0' then
               state_next <= zero;
            end if;
      end case;
   end process;
end mealy_arch;
```

Direct implementation   Since the transitions of the edge detector circuit are very simple, it can be implemented without using an FSM. We include this implementation for comparison purposes. The circuit diagram is shown in Figure 5.7. It can be interpreted that the output is asserted only when the current input is '1' and the previous input, which is stored in the register, is '0'. The corresponding code is shown in Listing 5.5.

Listing 5.5   Gate-level implementation of an edge detector

```
architecture gate_level_arch of edge_detect is
   signal delay_reg: std_logic;
begin
   -- delay register
   process(clk,reset)
   begin
```

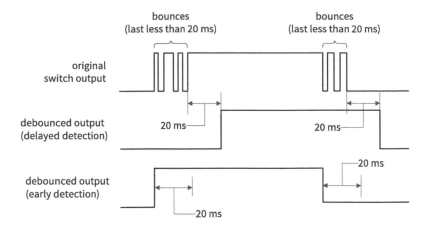

Figure 5.8   Original and debounced waveforms.

```
   if (reset='1') then
      delay_reg <= '0';
   elsif (clk'event and clk='1') then
      delay_reg <= level;
   end if;
 end process;
 -- decoding logic
   tick <= (not delay_reg) and level;
end gate_level_arch;
```

Although the descriptions in Listings 5.4 and 5.5 appear to be very different, they describe the same circuit. The circuit diagram can be derived from the FSM if we assign '0' and '1' to the **zero** and **one** states.

Comparison   Whereas both Moore and Mealy machine–based designs can generate a short tick at the rising edge of the input signal, there are several subtle differences. The Mealy machine–based design requires fewer states and responds faster, but the width of its output may vary and input glitches may be passed to the output.

The choice between the two designs depends on the subsystem that uses the output signal. Most of the time the subsystem is a synchronous system that shares the same clock signal. Since the FSM's output is sampled only at the rising edge of the clock, the width and glitches do not matter as long as the output signal is stable around the edge. Note that the Mealy output signal is available for sampling at t_1, which is one clock cycle faster than the Moore output, which is available at t_2. Therefore, the Mealy machine–based circuit is preferred for this type of application.

5.3.2 Debouncing circuit

A prototyping board usually contains a collection of pushbutton switches and slide switches. When we move a switch, its contact may bounce back and forth a few times before settling down. The bounces lead to glitches in the signal, as shown at the top of Figure 5.8. The bounces usually settle within 20 ms. The purpose of a *debouncing circuit* is to filter out the glitches associated with switch transitions.

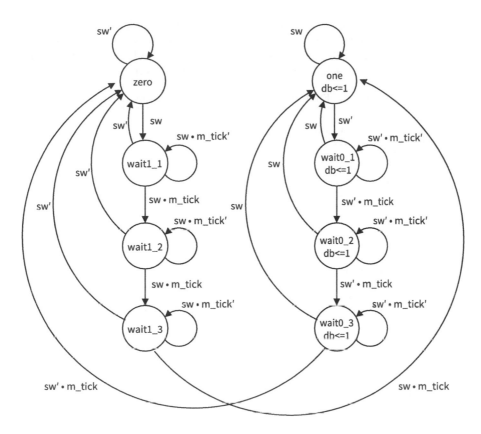

Figure 5.9   State diagram of a debouncing circuit.

We can design a debouncing circuit inside the FPGA device for the switches. The two possible debounced output signals are shown in the two bottom parts of Figure 5.8. In the *delayed detection* scheme, the circuit generates an output until the input waveform is stable for 20 ms. In the *early detection* scheme, the circuit responds immediately after a transition but ignores the subsequent changes for the next 20 ms. The delayed detection scheme is discussed in this subsection and the early detection scheme is left as an exercise in Experiment 5.5.2. An alternative FSMD-based scheme is discussed in Section 6.2.1.

An FSM-based design uses a free-running 10-ms timer and an FSM. The timer generates a one-clock-cycle enable tick (the m_tick signal) every 10 ms and the FSM uses this information to keep track of whether the input value is stabilized. The state diagram of this FSM is shown in Figure 5.9. The zero and one states indicate that the switch input signal, sw, has been stabilized with '0' and '1' values. Assume that the FSM is initially in the zero state. It moves to the wait1_1 state when sw changes to '1'. At the wait1_1 state, the FSM waits for the assertion of m_tick. If sw becomes '0' in this state, it implies that the width of the '1' value does not last long enough and the FSM returns to the zero state. This action repeats two more times for the wait1_2 and wait1_3 states. The operation from the one state is similar except that the sw signal must be '0'.

Since the 10-ms timer is free-running and the `m_tick` tick can be asserted at any time, the FSM checks the assertion three times to ensure that the `sw` signal is stabilized for at least 20 ms (it is actually between 20 and 30 ms). The code is shown in Listing 5.6. It includes a 10-ms timer and the FSM.

Listing 5.6   FSM implementation of a debouncing circuit

```vhdl
library ieee;
use ieee.std_logic_1164.all;
use ieee.numeric_std.all;
entity db_fsm is
   port(
      clk   : in std_logic;
      reset : in std_logic;
      sw    : in std_logic;
      db    : out std_logic
   );
end db_fsm;

architecture arch of db_fsm is
   constant N: integer := 20;   -- 2^N * 10ns = 10ms tick
    type db_state_type is
         (zero,wait1_1,wait1_2,wait1_3,one,wait0_1,wait0_2,wait0_3);
   signal q_reg, q_next : unsigned(N-1 downto 0);
   signal m_tick        : std_logic;
   signal state_reg     : db_state_type;
   signal state_next    : db_state_type;
begin
   --***************************************************************
   -- counter to generate 10 ms tick
   --***************************************************************
   process(clk,reset)
   begin
      if (clk'event and clk='1') then
         q_reg <= q_next;
      end if;
   end process;
   -- next state logic
   q_next <= q_reg + 1;
   --output tick
   m_tick <= '1' when q_reg=0 else '0';
   --***************************************************************
   -- debouncing FSM
   --***************************************************************
   -- state register
   process(clk,reset)
   begin
      if (reset='1') then
         state_reg <= zero;
      elsif (clk'event and clk='1') then
         state_reg <= state_next;
      end if;
   end process;
   -- next-state/output logic
   process(state_reg,sw,m_tick)
   begin
      state_next <= state_reg; --default: back to same state
      db <= '0';   -- default 0
      case state_reg is
         when zero =>
            if sw='1' then
```

```
                    state_next <= wait1_1;
                 end if;
              when wait1_1 =>
                 if sw='0' then
                    state_next <= zero;
                 else
                    if m_tick='1' then
                       state_next <= wait1_2;
                    end if;
                 end if;
              when wait1_2 =>
                 if sw='0' then
                    state_next <= zero;
                 else
                    if m_tick='1' then
                       state_next <= wait1_3;
                    end if;
                 end if;
              when wait1_3 =>
                 if sw='0' then
                    state_next <= zero;
                 else
                    if m_tick='1' then
                       state_next <= one;
                    end if;
                 end if;
              when one =>
                 db <='1';
                 if sw='0' then
                    state_next <= wait0_1;
                 end if;
              when wait0_1 =>
                 db <='1';
                 if sw='1' then
                    state_next <= one;
                 else
                    if m_tick='1' then
                       state_next <= wait0_2;
                    end if;
                 end if;
              when wait0_2 =>
                 db <='1';
                 if sw='1' then
                    state_next <= one;
                 else
                    if m_tick='1' then
                       state_next <= wait0_3;
                    end if;
                 end if;
              when wait0_3 =>
                 db <='1';
                 if sw='1' then
                    state_next <= one;
                 else
                    if m_tick='1' then
                       state_next <= zero;
                    end if;
                 end if;
           end case;
        end process;
   end arch;
```

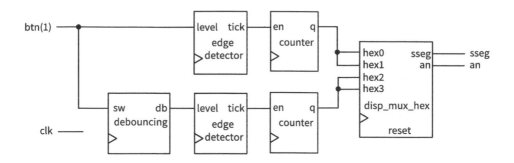

Figure 5.10   Debouncing testing circuit.

5.3.3   Testing circuit

We use a bounce counting circuit to verify the operation of the rising-edge detector and the debouncing circuit. The block diagram is shown in Figure 5.10. The input of the verification circuit is from a pushbutton switch. In the lower part, the signal is first fed to the debouncing circuit and then to the rising-edge detector. Therefore, a one-clock-cycle tick is generated each time the pushbutton switch is pressed and released. The tick in turn controls the enable input of an 8-bit counter, whose content is passed to the LED time-multiplexing circuit and shown on the left two digits of the prototyping board's seven-segment LED display. In the upper part, the input signal is fed directly to the edge detector without the debouncing circuit, and the number is shown on the right two digits of the prototyping board's seven-segment LED display. The bottom counter thus counts one desired 0-to-1 transition as well as the bounces.

The code uses component instantiation to realize the block diagram and is shown in Listing 5.7. It uses two pushbutton switches, in which `btn(1)` is served as the test input and `btn(0)` clears the counters to 0's.

Listing 5.7   Verification circuit for a debouncing circuit and rising-edge detector

```
library ieee;
use ieee.std_logic_1164.all;
use ieee.numeric_std.all;
entity debounce_test is
   port(
      clk  : in  std_logic;
      btn  : in  std_logic_vector(1 downto 0);
      an   : out std_logic_vector(3 downto 0);
      sseg : out std_logic_vector(7 downto 0)
   );
end debounce_test;

architecture arch of debounce_test is
   signal q1_reg, q1_next   : unsigned(7 downto 0);
   signal q0_reg, q0_next   : unsigned(7 downto 0);
   signal b_count, d_count  : std_logic_vector(7 downto 0);
   signal btn_reg, db_reg   : std_logic;
   signal db_level, db_tick : std_logic;
   signal btn_tick, clr     : std_logic;
begin
```

```
--******************************************************************
-- component instantiation
--******************************************************************
-- instantiate hex display time-multiplexing circuit
disp_unit : entity work.disp_hex_mux
    port map(
        clk   => clk,
        reset => '0',
        hex3  => b_count(7 downto 4),
        hex2  => b_count(3 downto 0),
        hex1  => d_count(7 downto 4),
        hex0  => d_count(3 downto 0),
        dp_in => "1011",
        an    => an,
        sseg  => sseg
    );
-- instantiate debouncing circuit
db_unit : entity work.db_fsm(arch)
    port map(
        clk   => clk,
        reset => '0',
        sw    => btn(1),
        db    => db_level
    );

--******************************************************************
-- edge detection circuits
--******************************************************************
process(clk)
begin
    if (clk'event and clk = '1') then
        btn_reg <= btn(1);
        db_reg  <= db_level;
    end if;
end process;
btn_tick <= (not btn_reg) and btn(1);
db_tick  <= (not db_reg) and db_level;

--******************************************************************
-- two counters
--******************************************************************
clr <= btn(0);
process(clk)
begin
    if (clk'event and clk='1') then
        q1_reg <= q1_next;
        q0_reg <= q0_next;
    end if;
end process;
-- next-state logic for the counter
q1_next <= (others=>'0') when clr='1' else
            q1_reg + 1     when btn_tick='1' else
            q1_reg;
q0_next <= (others=>'0') when clr='1' else
            q0_reg + 1     when db_tick='1' else
            q0_reg;
--output
b_count <= std_logic_vector(q1_reg);
d_count <= std_logic_vector(q0_reg);
end arch;
```

The seven-segment display shows the accumulated numbers of 0-to-1 edges of bounced and debounced pushbutton switch inputs. After pressing the switch several times, we can determine the average number of bounces for each transition. Note that some pushbutton switches may exhibit very few bounces.

5.4   BIBLIOGRAPHIC NOTES

The bibliographic information for this chapter is similar to that for Chapter 3.

5.5   SUGGESTED EXPERIMENTS

5.5.1   Dual-edge detector

A dual-edge detector is similar to a rising-edge detector except that the output is asserted for one clock cycle when the input changes from 0 to 1 (i.e., rising edge) and 1 to 0 (i.e., falling edge).

1. Design the circuit based on the Moore machine and draw the state diagram and ASM chart.
2. Derive the HDL code based on the state diagram of the ASM chart.
3. Derive a testbench and use simulation to verify operation of the code.
4. Replace the rising detectors in Section 5.3.3 with dual-edge detectors and verify their operations.
5. Repeat steps 1 to 4 for a Mealy machine–based design.

5.5.2   Early detection debouncing circuit

The early detection debouncing scheme is discussed in Section 5.3.2. The output timing diagram is shown at the bottom of Figure 5.8. When the input changes from 0 to 1, the FSM responds immediately. The FSM then ignores the input for about 20 ms to avoid glitches. After this amount of time, the FSM starts to check the input for the falling edge. Follow the design procedure in Section 5.3.2 to design the alternative circuit.

1. Derive the state diagram and ASM chart for the circuit.
2. Derive the HDL code.
3. Derive the HDL code based on the state diagram and ASM chart.
4. Derive a testbench and use simulation to verify operation of the code.
5. Replace the debouncing circuit in Section 5.3.3 with the alternative design and verify its operation.

5.5.3   Parking lot occupancy counter

Consider a parking lot with a single entry and exit gate. Two pairs of photo sensors are used to monitor the activity of cars, as shown in Figure 5.11. When an object is between the photo transmitter and the photo receiver, the light is blocked and the corresponding output is asserted to 1. By monitoring the events of two sensors, we can determine whether a car is entering or exiting or whether a pedestrian is passing through. For example, the following sequence indicates that a car enters the lot:

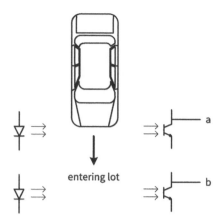

Figure 5.11   Conceptual diagram of gate sensors.

- Initially, both sensors are unblocked (i.e., the a and b signals are 00).
- Sensor a is blocked (i.e., the a and b signals are 10).
- Both sensors are blocked (i.e., the a and b signals are 11).
- Sensor a is unblocked (i.e., the a and b signals are 01).
- Both sensors becomes unblocked (i.e., the a and b signals are 00).

Design a parking lot occupancy counter as follows:

1. Design an FSM with two input signals, a and b, and two output signals, car_enter and car_exit. The car_enter and car_exit signals assert one clock cycle when a car enters and one clock cycle when a car exits the lot, respectively.
2. Derive the HDL code for the FSM.
3. Design a counter with two control signals, inc and dec, which increment and decrement the counter when asserted. Derive the HDL code.
4. Combine the counter and the FSM and seven-segment LED decoding circuits. Use two debounced pushbuttons to mimic operation of the two sensor outputs. Verify operation of the occupancy counter.

CHAPTER 6

FSMD

An FSMD (finite state machine with data path) combines an FSM and regular sequential circuits. The FSMD can be used to implement systems described by *RT (register transfer) operation*, which is a methodology to realize a software algorithm in hardware. In this chapter, we provide an overview of the RT operation and extended ASM chart, discuss the derivation of HDL codes, and use several examples to illustrate the development.

6.1   INTRODUCTION

An FSMD (finite state machine with data path) combines an FSM and regular sequential circuits. The FSM, which is sometimes known as a *control path*, examines the external commands and status and generates control signals to specify operation of the regular sequential circuits, which are known collectively as a *data path*. Algorithms described in *RT (register transfer) operation*, in which the operations are specified as data manipulation and transfer among a collection of registers, can be converted to FSMD and realized in hardware.

6.1.1   Single RT operation

An RT operation specifies data manipulation and transfer for a single destination register. It is represented by the notation

$$r_{dest} \leftarrow f(r_{src1}, r_{src2}, \ldots, r_{srcn})$$

where r_{dest} is the destination register, r_{src1}, r_{src2}, and r_{srcn} are the source registers, and $f(\cdot)$ specifies the operation to be performed. The notation indicates that the contents of the source registers are fed to the $f(\cdot)$ function, which is realized by a combinational circuit, and the result is passed to the input of the destination register and stored in the destination register at the next rising edge of the clock. Following are several representative RT operations:

- r1 ← 0. A constant 0 is stored in the r1 register.
- r1 ← r1. The content of the r1 register is written back to itself.
- r2 ← r2 >> 3. The r2 register is shifted right three positions and then written back to itself.
- r2 ← r1. The content of the r1 register is transferred to the r2 register.
- i ← i + 1. The content of the i register is incremented by 1 and the result is written back to itself.
- d ← s1 + s2 + s3. The summation of the s1, s2, and s3 registers is written to the d register.
- y ← a*a. The a squared is written to the y register.

A single RT operation can be implemented by constructing a combinational circuit for the $f(\cdot)$ function and connecting the input and output of the registers. For example, consider the a ← a-b+1 operation. The $f(\cdot)$ function involves a subtractor and an incrementor. The block diagram is shown in Figure 6.1(a). For clarity, we use the _reg and _next suffixes to represent the input and output of a register. Note that an RT operation is synchronized by an embedded clock. The result from the $f(\cdot)$ function is not stored to the destination register until the next rising edge of the clock. The timing diagram of the previous RT operation is shown in Figure 6.1(b).

6.1.2   ASMD chart

A circuit based on the RT methodology specifies which RT operations should be executed in each step. Since an RT operation is done on a clock-by-clock basis, its timing is similar to a state transition of an FSM. Thus, an FSM is a natural choice to specify the sequencing of an RT algorithm. We extend the ASM chart to incorporate RT operations and call it an *ASMD* (ASM with data path) chart. The RT operations are treated as another type of activity and can be placed where the output signals are used.

A segment of an ASMD chart is shown in Figure 6.2(a). It contains one destination register, r1, which is initialized with 8, added with content of the r2 register, and then shifted left by two positions. Note that the r1 register must be specified in each state. When r1 is not changed, the r1 ← r1 operation should be used to maintain its current content, as in the s3 state. In future discussion, we assume that r ← r is the default RT operation for the r register and do not include it in the ASMD chart. Implementing the RT operations of an ASMD chart involves a multiplexing circuit to route the desired next value to the destination register.

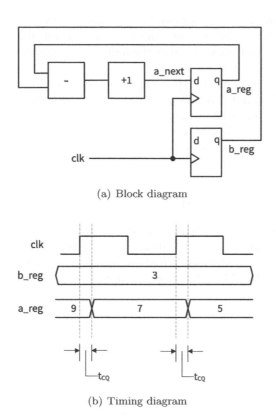

(a) Block diagram

(b) Timing diagram

Figure 6.1   Block and timing diagrams of an RT operation.

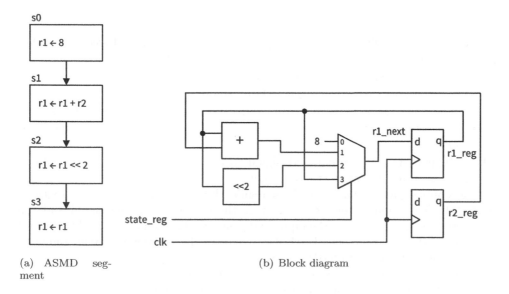

(a) ASMD segment

(b) Block diagram

Figure 6.2   Realization of an ASMD segment.

For example, the previous segment can be implemented by a 4-to-1 multiplexer, as shown in Figure 6.2(b). The current state (i.e., the output of the state register) of the FSM controls the selection signal of the multiplexer and thus chooses the result of the desired RT operation.

An RT operation can also be specified in a conditional output box, as the r2 register shown in Figure 6.3(a). Depending on the a>b condition, the FSMD performs either r2 ← r2+a or r2 ← r2+b. Note that all operations are done in parallel inside an ASMD block. We need to realize the a>b, r2+a, and r2+b operations and use a multiplexer to route the desired value to r2. The block diagram is shown in Figure 6.3(b).

6.1.3 Decision box with a register

The appearance of an ASMD chart is similar to that of a normal flowchart. The main difference is that the RT operation in an ASMD chart is controlled by an embedded clock signal and the destination register is updated *when the FSMD exits the current ASMD block*, but not within the block. The r ← r-1 operation actually means that:

- r_next <= r_reg - 1;
- r_reg <= r_next at the rising edge of the clock (i.e., when the FSMD exits the current block).

This "delayed store" may introduce subtle errors when a register is used in a decision box. Consider the FSMD segment in Figure 6.4(a). The r register is decremented in the state box and used in the decision box. Since the r register is not updated until the FSMD exits the block, the old content of r is used for comparison in the decision box. If the new value of r is desired, we should use the output of the combinational logic (i.e., r_next) in the decision box (i.e., replace the r=0 expression with r_next=0), as shown in Figure 6.4(b). Note that we use the := notation, as in r_next:=r-1, to indicate the immediate assignment of r_next.

Block diagram of an FSMD   The conceptual block diagram of an FSMD is divided into a data path and a control path, as shown in Figure 6.5. The data path performs the required RT operations. It consists of:

- *Data registers*: store the intermediate computation results
- *Functional units*: perform the functions specified by the RT operations
- *Routing network*: routes data between the storage registers and the functional units

The data path follows the `control` signal to perform the desired RT operations and generates the `internal status` signal.

The control path is an FSM. As a regular FSM, it contains a state register, next-state logic, and output logic. It uses the external `command` signal and the data path's `status` signal as the input and generates the `control` signal to control the data path operation. The FSM also generates the `external status` signal to indicate the status of the FSMD operation.

Note that although an FSMD consists of two types of sequential circuits, both circuits are controlled by the same clock, and thus the FSMD is still a synchronous system.

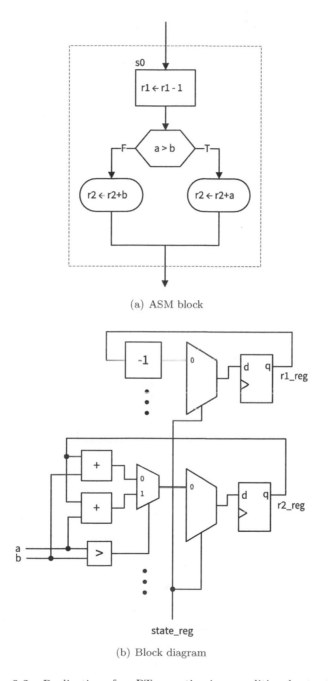

(a) ASM block

(b) Block diagram

Figure 6.3   Realization of an RT operation in a conditional output box.

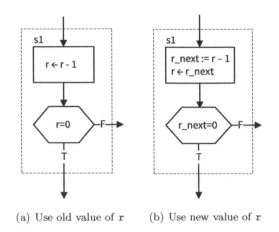

(a) Use old value of r   (b) Use new value of r

Figure 6.4   ASM block affected by a delayed store.

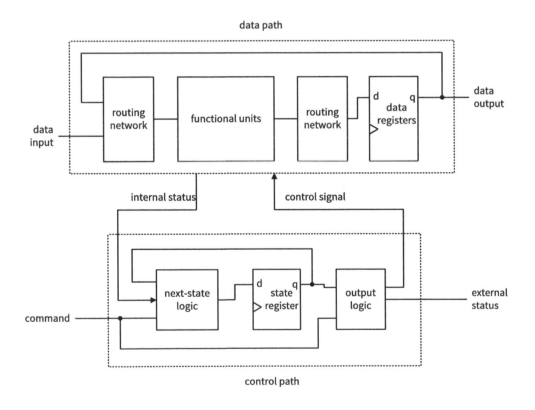

Figure 6.5   Block diagram of an FSMD.

6.2   CODE DEVELOPMENT OF AN FSMD

We use an improved debouncing circuit to demonstrate derivation of the FSMD code. Although the debouncing circuit in Section 5.3.2 uses an FSM and a timer (which is a regular sequential circuit), it is not based on the RT methodology because the two units are running independently and the FSM has no control over the timer. Since the 10-ms enable tick can be asserted at any time, the FSM does not know how much time has elapsed when the first tick is detected in the wait1_1 or wait0_1 state. Thus, the waiting period in this design is between 20 and 30 ms but is not an exact interval. This deficiency can be overcome by applying the RT methodology. In this section, we use this improved debouncing circuit to illustrate FSMD code development.

6.2.1   Debouncing circuit based on RT methodology

With the RT methodology, we can use an FSM to control the initiation of the timer to obtain the exact interval. The ASMD chart is shown in Figure 6.6. The circuit is expanded to include two output signals: db_level, which is the debounced output, and db_tick, which is a one-clock-cycle enable pulse asserted at the zero-to-one transition. The zero and one states mean that the sw input has been stabilized for '0' and '1', respectively. The wait1 and wait0 states are used to filter out short glitches. The sw signal must be stable for a certain amount of time or the transition will be treated as a glitch. The data path contains one register, q, which is 22 bits wide. Assume that the FSMD is originally in the zero state. When the sw input signal becomes '1', the FSMD moves to the wait1 state and initializes q to "$1 \cdots 1$". In the wait1 state, the q decrements in each clock cycle. If sw remains as '1', the FSMD returns to this state repeatedly until q reaches "$0 \cdots 0$" and then moves to the one state.

Recall that the 100M-Hz (i.e., 10-ns period) system clock is used on the prototyping board. Since the FSMD stays in the wait1 state for 2^{22} clock cycles, it is about 40 ms (i.e., $2^{22} * 10$ ns). We can modify the initial value of the q register to obtain the desired wait interval.

There are two ways to derive the HDL code, one with an *explicit description* of the data path components and the other with an *implicit description* of the data path components.

6.2.2   Code with explicit data path components

The first approach to FSMD code development is to separate the control FSM and the key data path components. From an ASMD chart, we first identify the key components in the data path and the associated control signals and then describe these components in individual code segments.

The key data path component of the debouncing circuit ASMD chart is a custom 22-bit decrement counter that can:

- Be initialized with a specific value
- Count downward or pause
- Assert a status signal when the counter reaches 0

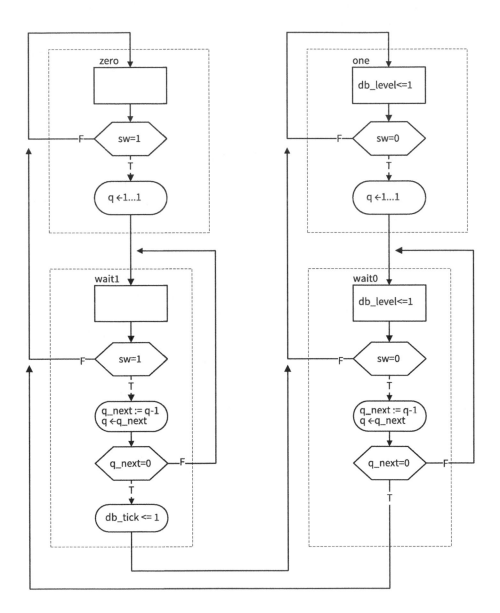

Figure 6.6   ASMD chart of a debouncing circuit.

We can create a binary counter with a `q_load` signal to load the initial value and a q_dec signal to enable the counting. The counter also generates a `q_zero` status signal, which is asserted when the counter reaches zero. The complete data path is composed of the q register and the next-state logic of the custom decrement counter. A comparison circuit is included to generate the `q_zero` status signal. The control path consists of an FSM, which takes the `sw` input and the `q_zero` status and asserts the control signals, `q_load` and q_dec, according to the desired action in the ASMD chart. The HDL code follows the data path specification and the ASMD chart, and is shown in Listing 6.1.

Listing 6.1 Debouncing circuit with an explicit data path component

```vhdl
library ieee;
use ieee.std_logic_1164.all;
use ieee.numeric_std.all;
entity debounce is
   port(
      clk, reset : in  std_logic;
      sw         : in  std_logic;
      db_level   : out std_logic;
      db_tick    : out std_logic
   );
end debounce;

architecture exp_fsmd_arch of debounce is
   constant N: integer : =22;   -- filter of 2^N * 10ns = 40ms
   type state_type is (zero, wait0, one, wait1);
   signal state_reg, state_next : state_type;
   signal q_reg, q_next         : unsigned(N - 1 downto 0);
   signal q_load, q_dec, q_zero : std_logic;
begin
   -- FSMD state & data registers
   process(clk,reset)
   begin
      if reset='1' then
         state_reg <= zero;
         q_reg <= (others=>'0');
      elsif (clk'event and clk='1') then
         state_reg <= state_next;
         q_reg <= q_next;
      end if;
   end process;
   -- FSMD data path (counter) next-state logic
   q_next <= (others=>'1') when q_load='1' else
             q_reg - 1      when q_dec='1' else
             q_reg;
   q_zero <= '1' when q_next=0 else '0';
   -- FSMD control path next-state logic
   process(state_reg,sw,q_zero)
   begin
      q_load  <= '0';
      q_dec   <= '0';
      db_tick <= '0';
      state_next <= state_reg;
      case state_reg is
         when zero =>
            db_level <= '0';
            if (sw='1') then
               state_next <= wait1;
               q_load <= '1';
```

```
                    end if;
                when wait1=>
                    db_level <= '0';
                    if (sw='1') then
                        q_dec <= '1';
                        if (q_zero='1') then
                            state_next <= one;
                            db_tick    <= '1';
                        end if;
                    else --- sw='0'
                        state_next <= zero;
                    end if;
                when one =>
                    db_level <= '1';
                    if (sw='0') then
                        state_next <= wait0;
                        q_load      <= '1';
                    end if;
                when wait0=>
                    db_level <= '1';
                    if (sw='0') then
                        q_dec <= '1';
                        if (q_zero='1') then
                            state_next <= zero;
                        end if;
                    else --- sw='1'
                        state_next <= one;
                    end if;
            end case;
        end process;
end exp_fsmd_arch;
```

6.2.3   Code with implicit data path components

An alternative coding style is to embed the RT operations within the FSM control path. Instead of explicitly defining the data path components, we just list RT operations with the corresponding FSM state. The code of the debouncing circuit is shown in Listing 6.2.

Listing 6.2   Debouncing circuit with an implicit data path component

```
architecture imp_fsmd_arch of debounce is
    constant N: integer:=22;   --- filter of 2^N * 10ns = 40ms
    type state_type is (zero, wait0, one, wait1);
    signal state_reg, state_next : state_type;
    signal q_reg, q_next         : unsigned(N-1 downto 0);
begin
    --- FSMD state & data registers
    process(clk,reset)
    begin
        if reset='1' then
            state_reg <= zero;
            q_reg      <= (others=>'0');
        elsif (clk'event and clk='1') then
            state_reg <= state_next;
            q_reg      <= q_next;
        end if;
    end process;
    --- next-state logic & data path functional units/routing
```

```vhdl
   process(state_reg,q_reg,sw,q_next)
   begin
      state_next <= state_reg;
      q_next    <= q_reg;
      db_tick <= '0';
      case state_reg is
         when zero =>
             db_level <= '0';
             if (sw='1') then
                state_next <= wait1;
                q_next       <= (others=>'1');
             end if;
         when wait1=>
             db_level <= '0';
             if (sw='1') then
                q_next <= q_reg - 1;
                if (q_next=0) then
                   state_next <= one;
                   db_tick      <= '1';
                end if;
             else  -- sw='0'
                state_next <= zero;
             end if;
         when one =>
             db_level <= '1';
             if (sw='0') then
                state_next <= wait0;
                q_next       <= (others=>'1');
             end if;
         when wait0=>
             db_level <= '1';
             if (sw='0') then
                q_next <= q_reg - 1;
                if (q_next=0) then
                   state_next <= zero;
                end if;
             else  -- sw='1'
                state_next <= one;
             end if;
      end case;
   end process;
end imp_fsmd_arch;
```

The code consists of a memory segment and a combinational logic segment. The former contains the state register of the FSM and the data register of the data path. The latter basically specifies the next-state logic of the control path FSM. Instead of generating control signals, the next data register values are specified in individual states. The next-state logic of the data path, which consists of functional units and routing networks, is created accordingly.

6.2.4   Comparison

Code with implicit data path components essentially follows the ASMD chart. We just convert the chart to an HDL description. Although this approach is simpler and more descriptive, we rely on synthesis software for data path construction and have less control. This can best be explained by an example. Consider the ASMD segment in Figure 6.7. The implicit description becomes

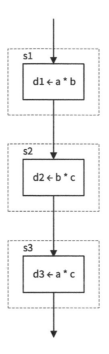

Figure 6.7 ASMD segment with sharing opportunity.

```
case
   when s1
      d1_next <= a * b;
         . . .
   when s2
      d2_next <= b * c;
         . . .
   when s3
      d3_next <= a * c;
         . . .
end case;
```

The synthesis software may infer three multipliers. Since a combinational multiplier is a complex circuit, it is more efficient to share the circuit. We can use an explicit description to isolate the multiplier:

```
case
   when s1
      in1 <= a;
      in2 <= b;
      d1_next <= m_out;
         . . .
   when s2
      in1 <= b;
      in2 <= c;
      d2_next <= m_out;
         . . .
   when s3
```

```
        in1 <= a;
        in2 <= c;
        d3_next <= m_out;
        . . .
end case;
-- explicit  description  of  a  single  multiplier
m_out <= in1 * in2;
```

The code ensures that only one multiplier is inferred during synthesis. The implicit and explicit descriptions can be mixed for a complex FSMD design. We frequently isolate and extract complex data path components for code clarity and efficiency.

6.3 DESIGN EXAMPLES

6.3.1 Fibonacci number circuit

The Fibonacci numbers constitute a sequence defined as

$$fib(i) = \begin{cases} 0 & \text{if } i = 0 \\ 1 & \text{if } i = 1 \\ fib(i-1) + fib(i-2) & \text{if } i > 1 \end{cases}$$

One way to calculate $fib(i)$ is to construct the function iteratively, from 0 to the desired i. This approach requires two temporary registers to store the two most recently calculated values [i.e., $fib(i-1)$ and $fib(i-2)$)] and one index register to keep track of the number of iterations. The ASMD chart is shown in Figure 6.8, in which t1 and t0 are temporary storage registers and n is the index register. In addition to the regular data input and output signals, i and f, we include a command signal, start, which signals the beginning of operation, and two status signals: ready, which indicates that the circuit is idle and ready to take new input, and done_tick, which is asserted for one clock cycle when the operation is completed. Since this circuit, like many other FSMD designs, is probably a part of a larger system, these signals are needed to interface with other subsystems.

The ASMD chart has three states. The idle state indicates that the circuit is currently idle. When start is asserted, the FSMD moves to the op state and loads initial values to three registers. The t0 and t1 registers are loaded with 0 and 1, which represent $fib(0)$ and $fib(1)$, respectively. The n register is loaded with i, the desired number of iterations.

The main computation is iterated through the op state by three RT operations:

- t1 ← t1 + t0
- t0 ← t1
- n ← n - 1

The first two RT operations obtain a new value and store the two most recently calculated values in t1 and t0. The third RT operation decrements the iteration index. The iteration ended when n reaches 1 or its initial value is 0 [i.e., $fib(0)$]. Unlike a regular flowchart, the operations in an ASMD block can be performed concurrently in the same clock cycle. We put all comparison and RT operations in the op state to reduce the computation time. Note that the new values of the t1 and t0 registers are loaded at the same time when the FSMD exits the op state

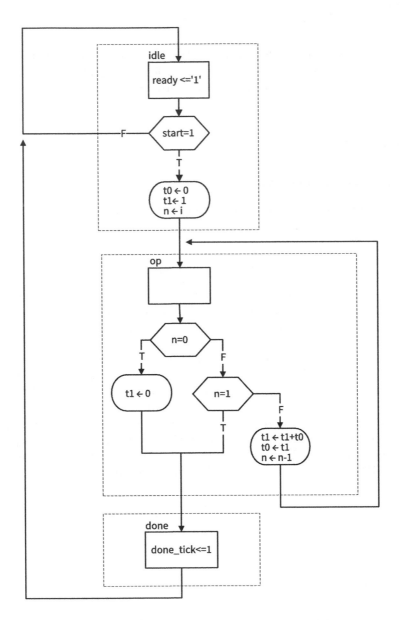

Figure 6.8   ASMD chart of a Fibonacci circuit.

(i.e., at the next rising edge of the clock). Thus, the original value of `t1`, not `t1+t0`, is stored to `t0`. The purpose of the `done` state is to generate the one-clock-cycle `done_tick` signal to indicate completion of the computation. This state can be omitted if this status signal is not needed.

The code follows the ASMD chart and is shown in Listing 6.3. Note that the Fibonacci function grows rapidly and the output signal should be wide enough to accommodate the desired result.

Listing 6.3 Fibonacci number circuit

```vhdl
library ieee;
use ieee.std_logic_1164.all;
use ieee.numeric_std.all;
entity fib is
   port(
      clk        : in   std_logic;
      reset      : in   std_logic;
      start      : in   std_logic;
      i          : in   std_logic_vector(4 downto 0);
      ready      : out std_logic;
      done_tick  : out std_logic;
      f          : out std_logic_vector(19 downto 0)
   );
end fib;

architecture arch of fib is
   type state_type is (idle, op, done);
   signal state_reg      : state_type;
   signal state_next     : state_type;
   signal t0_reg, t0_next : unsigned(19 downto 0);
   signal t1_reg, t1_next : unsigned(19 downto 0);
   signal n_reg, n_next   : unsigned(4 downto 0);
begin
   -- fsmd state and data registers
   process(clk, reset)
   begin
      if reset = '1' then
         state_reg <= idle;
         t0_reg   <= (others => '0');
         t1_reg   <= (others => '0');
         n_reg    <= (others => '0');
      elsif (clk'event and clk = '1') then
         state_reg <= state_next;
         t0_reg   <= t0_next;
         t1_reg   <= t1_next;
         n_reg    <= n_next;
      end if;
   end process;
   -- fsmd next-state logic
   process(state_reg, n_reg, t0_reg, t1_reg, start, i, n_next)
   begin
      ready       <= '0';
      done_tick   <= '0';
      state_next <= state_reg;
      t0_next     <= t0_reg;
      t1_next     <= t1_reg;
      n_next      <= n_reg;
      case state_reg is
         when idle =>
            ready <= '1';
```

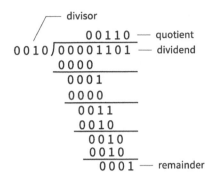

Figure 6.9   Long division of two 4-bit unsigned integers.

```
      if start = '1' then
          t0_next    <= (others => '0');
          t1_next    <= (0 => '1', others => '0');
          n_next     <= unsigned(i);
          state_next <= op;
      end if;
  when op =>
      if n_reg = 0 then
          t1_next    <= (others => '0');
          state_next <= done;
      elsif n_reg = 1 then
          state_next <= done;
      else
          t1_next <= t1_reg + t0_reg;
          t0_next <= t1_reg;
          n_next  <= n_reg - 1;
      end if;
  when done =>
      done_tick  <= '1';
      state_next <= idle;
  end case;
end process;
-- output
f <= std_logic_vector(t1_reg);
end arch;
```

6.3.2   Division circuit

Because of complexity, the division operator cannot be synthesized automatically. We use an FSMD to implement the long-division algorithm in this subsection. The algorithm is illustrated by the division of two 4-bit unsigned integers in Figure 6.9. The algorithm can be summarized as follows:

1. Double the dividend width by appending 0's in front and align the divisor to the leftmost bit of the extended dividend.
2. If the corresponding dividend bits are greater than or equal to the divisor, subtract the divisor from the dividend bits and make the corresponding quotient bit 1. Otherwise, keep the original dividend bits and make the quotient bit 0.

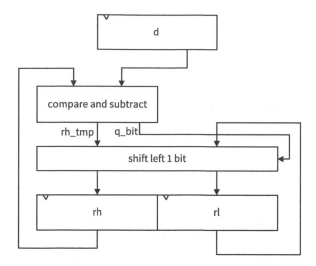

Figure 6.10   Sketch of division circuit's data path.

3. Append one additional dividend bit to the previous result and shift the divisor to the right one position.
4. Repeat steps 2 and 3 until all dividend bits are used.

The sketch of the data path is shown in Figure 6.10. Initially, the divisor is stored in the d register and the extended dividend is stored in the rh and rl registers. In each iteration, the rh and rl registers are shifted to the left by one position. This corresponds to shifting the divisor to the right of the previous algorithm. We can then compare rh and d and perform subtraction if rh is greater than or equal to d. When rh and rl are shifted to the left, the rightmost bit of rl becomes available. It can be used to store the current quotient bit. After we iterate through all dividend bits, the result of the last subtraction is stored in rh and becomes the remainder of the division, and all quotients are shifted into rl.

The ASMD chart of the division circuit is somewhat similar to that of the previous Fibonacci circuit. The FSMD consists of four states, idle, op, last, and done. To make the code clear, we extract the *compare and subtract* circuit to separate code segments. The main computation is performed in the op state, in which the dividend bits and divisor are compared and subtracted and then shifted left by 1 bit. Note that the remainder should not be shifted in the last iteration. We create a separate state, last, to accommodate this special requirement. As in the preceding example, the purpose of the done state is to generate a one-clock-cycle done_tick signal to indicate completion of the computation. The code is shown in Listing 6.4.

Listing 6.4   Division circuit

```
library ieee;
use ieee.std_logic_1164.all;
use ieee.numeric_std.all;
entity div is
   generic(
      W    : integer := 8;
      CBIT : integer := 4    --  CBIT=log2(W)+1
```

```vhdl
   );
   port(
      clk, reset : in  std_logic;
      start      : in  std_logic;
      dvsr, dvnd : in  std_logic_vector(W - 1 downto 0);
      ready      : out std_logic;
      done_tick  : out std_logic;
      quo, rmd   : out std_logic_vector(W - 1 downto 0)
   );
end div;

architecture arch of div is
   type state_type is (idle, op, last, done);
   signal state_reg       : state_type;
   signal state_next      : state_type;
   signal rh_reg, rh_next : unsigned(W - 1 downto 0);
   signal rl_reg, rl_next : std_logic_vector(W - 1 downto 0);
   signal rh_tmp          : unsigned(W - 1 downto 0);
   signal d_reg, d_next   : unsigned(W - 1 downto 0);
   signal n_reg, n_next   : unsigned(CBIT - 1 downto 0);
   signal q_bit           : std_logic;
begin
   -- fsmd state and data registers
   process(clk, reset)
   begin
      if reset = '1' then
         state_reg <= idle;
         rh_reg    <= (others => '0');
         rl_reg    <= (others => '0');
         d_reg     <= (others => '0');
         n_reg     <= (others => '0');
      elsif (clk'event and clk = '1') then
         state_reg <= state_next;
         rh_reg    <= rh_next;
         rl_reg    <= rl_next;
         d_reg     <= d_next;
         n_reg     <= n_next;
      end if;
   end process;
   -- fsmd next-state logic and data path logic
   process(state_reg, n_reg, rh_reg, rl_reg, d_reg,
           start, dvsr, dvnd, q_bit, rh_tmp, n_next)
   begin
      ready      <= '0';
      done_tick  <= '0';
      state_next <= state_reg;
      rh_next    <= rh_reg;
      rl_next    <= rl_reg;
      d_next     <= d_reg;
      n_next     <= n_reg;
      case state_reg is
         when idle =>
            ready <= '1';
            if start = '1' then
               rh_next    <= (others => '0');
               rl_next    <= dvnd;              -- dividend
               d_next     <= unsigned(dvsr);   -- divisor
               n_next     <= to_unsigned(W + 1, CBIT); -- index
               state_next <= op;
            end if;
         when op =>
```

```
                    — shift rh and rl left
               rl_next <= rl_reg(W - 2 downto 0) & q_bit;
               rh_next <= rh_tmp(W - 2 downto 0) & rl_reg(W - 1);
               —decrease index
               n_next  <= n_reg - 1;
               if (n_next = 1) then
                   state_next <= last;
               end if;
           when last =>       — last iteration
               rl_next    <= rl_reg(W - 2 downto 0) & q_bit;
               rh_next    <= rh_tmp;
               state_next <= done;
           when done =>
               state_next <= idle;
               done_tick  <= '1';
       end case;
   end process;
   — compare and subtract
   process(rh_reg, d_reg)
   begin
       if rh_reg >= d_reg then
           rh_tmp <= rh_reg - d_reg;
           q_bit  <= '1';
       else
           rh_tmp <= rh_reg;
           q_bit  <= '0';
       end if;
   end process;
   — output
   quo <= rl_reg;
   rmd <= std_logic_vector(rh_reg);
end arch;
```

6.3.3 Binary-to-BCD conversion circuit

We discussed the BCD format in Section 4.5.2. In this format, a decimal number is represented as a sequence of 4-bit BCD digits. A binary-to-BCD conversion circuit converts a binary number to the BCD format. For example, the binary number "0010 0000 0000" becomes "0101 0001 0010" (i.e., 512_{10}) after conversion.

The binary-to-BCD conversion can be processed by a special BCD shift register, which is divided into 4-bit groups internally, each representing a BCD digit. Shifting a BCD sequence to the left requires adjustment if a BCD digit is greater than 9_{10} after shifting. For example, if a BCD sequence is "0001 0111" (i.e., 17_{10}), it should become "0011 0100" (i.e., 34_{10}) rather than "0010 1110". The adjustment requires subtracting 10_{10} (i.e., "1010") from the right BCD digit and adding 1 (which can be considered as a carry-out) to the next BCD digit. Note that subtracting 10_{10} is equivalent to adding 6_{10} for a 4-bit binary number. Thus, the foregoing adjustment can also be achieved by adding 6_{10} to the right BCD digit. The carry-out bit is generated automatically in this process.

In the actual implementation, it is more efficient to first perform the necessary adjustment on a BCD digit and then shift. We can check whether a BCD digit is greater than 4_{10} and, if this is the case, add 3_{10} to the digit. After all the BCD digits are corrected, we can then shift the entire register to the left by one position. A binary-to-BCD conversion circuit can be constructed by shifting the

Table 6.1 Binary-to-BCD conversion example

Operation		Special BCD shift register			Binary input
		BCD digit 2	BCD digit 1	BCD digit 0	
Initial					111 1111
Bit 6	no adjustment shift left 1 bit			1 (1_{10})	11 1111
Bit 5	no adjustment shift left 1 bit			11 (3_{10})	1 1111
Bit 4	no adjustment shift left 1 bit			111 (7_{10})	1111
Bit 3	BCD digit 0 adjustment shift left 1 bit		1 (1_{10})	1010 0101 (5_{10})	111
Bit 2	BCD digit 0 adjustment shift left 1 bit		1 11 (3_{10})	1000 0001 (1_{10})	11
Bit 1	no adjustment shift left 1 bit		110 (6_{10})	0011 (3_{10})	1
Bit 0	BCD digit 1 adjustment shift left 1 bit	1 (1_{10})	1001 0010 (2_{10})	0011 0111 (7_{10})	

binary input to a BCD shift register bit by bit, from MSB to LSB. Its operation can be summarized as follows:

1. For each 4-bit BCD digit in a BCD shift register, check whether the digit is greater than 4. If this is the case, add 3_{10} to the digit.
2. Shift the entire BCD register left one position and shift in the MSB of the input binary sequence to the LSB of the BCD register.
3. Repeat steps 1 and 2 until all input bits are used.

The conversion process of a 7-bit binary input, "111 1111" (i.e., 127_{10}), is demonstrated in Table 6.1.

The code of a 13-bit conversion circuit is shown in Listing 6.5. It uses a simple FSMD to control the overall operation. When the **start** signal is asserted, the binary input is stored into the **p2s** register. The FSM then iterates through the 13 bits, similar to the process described in previous examples. Four adjustment circuits are used to correct the four BCD digits. For clarity, they are isolated from the next-state logic and described in a separate code segment.

Listing 6.5   Binary-to-BCD conversion circuit

```vhdl
library ieee;
use ieee.std_logic_1164.all;
use ieee.numeric_std.all;
entity bin2bcd is
   port(
      clk         : in  std_logic;
      reset       : in  std_logic;
      start       : in  std_logic;
      bin         : in  std_logic_vector(12 downto 0);
      ready       : out std_logic;
      done_tick   : out std_logic;
      bcd3, bcd2  : out std_logic_vector(3 downto 0);
      bcd1, bcd0  : out std_logic_vector(3 downto 0)
   );
end bin2bcd;

architecture arch of bin2bcd is
   type state_type is (idle, op, done);
   signal state_reg, state_next : state_type;
   signal p2s_reg, p2s_next    : std_logic_vector(12 downto 0);
   signal n_reg, n_next        : unsigned(3 downto 0);
   signal bcd3_reg, bcd3_next  : unsigned(3 downto 0);
   signal bcd2_reg, bcd2_next  : unsigned(3 downto 0);
   signal bcd1_reg, bcd1_next  : unsigned(3 downto 0);
   signal bcd0_reg, bcd0_next  : unsigned(3 downto 0);
   signal bcd3_tmp, bcd2_tmp   : unsigned(3 downto 0);
   signal bcd1_tmp, bcd0_tmp   : unsigned(3 downto 0);
begin
   -- state and data registers
   process(clk, reset)
   begin
      if reset = '1' then
         state_reg <= idle;
         p2s_reg   <= (others => '0');
         n_reg     <= (others => '0');
         bcd3_reg  <= (others => '0');
         bcd2_reg  <= (others => '0');
         bcd1_reg  <= (others => '0');
         bcd0_reg  <= (others => '0');
      elsif (clk'event and clk = '1') then
         state_reg <= state_next;
         p2s_reg   <= p2s_next;
         n_reg     <= n_next;
         bcd3_reg  <= bcd3_next;
         bcd2_reg  <= bcd2_next;
         bcd1_reg  <= bcd1_next;
         bcd0_reg  <= bcd0_next;
      end if;
   end process;
   -- fsmd next-state logic / data path operations
   process(state_reg, start, p2s_reg, n_reg, n_next, bin,
           bcd0_reg, bcd1_reg, bcd2_reg, bcd3_reg, bcd0_tmp,
           bcd1_tmp, bcd2_tmp, bcd3_tmp)
   begin
      state_next <= state_reg;
      ready      <= '0';
      done_tick  <= '0';
      p2s_next   <= p2s_reg;
      bcd0_next  <= bcd0_reg;
      bcd1_next  <= bcd1_reg;
```

```
        bcd2_next   <= bcd2_reg;
        bcd3_next   <= bcd3_reg;
        n_next      <= n_reg;
        case state_reg is
           when idle =>
              ready <= '1';
              if start = '1' then
                 state_next <= op;
                 bcd3_next   <= (others => '0');
                 bcd2_next   <= (others => '0');
                 bcd1_next   <= (others => '0');
                 bcd0_next   <= (others => '0');
                 n_next      <= "1101";    -- index
                 p2s_next    <= bin;       -- input shift register
              end if;
           when op =>
              -- shift in binary bit
              p2s_next   <= p2s_reg(11 downto 0) & '0';
              -- shift 4 BCD digits
              bcd0_next <= bcd0_tmp(2 downto 0) & p2s_reg(12);
              bcd1_next <= bcd1_tmp(2 downto 0) & bcd0_tmp(3);
              bcd2_next <= bcd2_tmp(2 downto 0) & bcd1_tmp(3);
              bcd3_next <= bcd3_tmp(2 downto 0) & bcd2_tmp(3);
              n_next    <= n_reg - 1;
              if (n_next = 0) then
                 state_next <= done;
              end if;
           when done =>
              state_next <= idle;
              done_tick  <= '1';
        end case;
     end process;
     -- data path function units
     bcd0_tmp <= bcd0_reg + 3 when bcd0_reg > 4 else bcd0_reg;
     bcd1_tmp <= bcd1_reg + 3 when bcd1_reg > 4 else bcd1_reg;
     bcd2_tmp <= bcd2_reg + 3 when bcd2_reg > 4 else bcd2_reg;
     bcd3_tmp <= bcd3_reg + 3 when bcd3_reg > 4 else bcd3_reg;
     -- output
     bcd0     <= std_logic_vector(bcd0_reg);
     bcd1     <= std_logic_vector(bcd1_reg);
     bcd2     <= std_logic_vector(bcd2_reg);
     bcd3     <= std_logic_vector(bcd3_reg);
end arch;
```

6.3.4   Period counter

A period counter measures the period of a periodic input waveform. One way to construct the circuit is to count the number of clock cycles between two rising edges of the input signal. Since the frequency of the system clock is known, the period of the input signal can be derived accordingly. For example, if the frequency of the system clock is f and the number of clock cycles between two rising edges is N, the period of the input signal is $N * \frac{1}{f}$.

The design in this subsection measures the period in milliseconds. Its ASMD chart is shown in Figure 6.11. The period counter takes a measurement when the start signal is asserted. We use a rising-edge detection circuit to generate a one-clock-cycle tick, edge, to indicate the rising edge of the input waveform. After start is asserted, the FSMD moves to the waite state to wait for the first rising

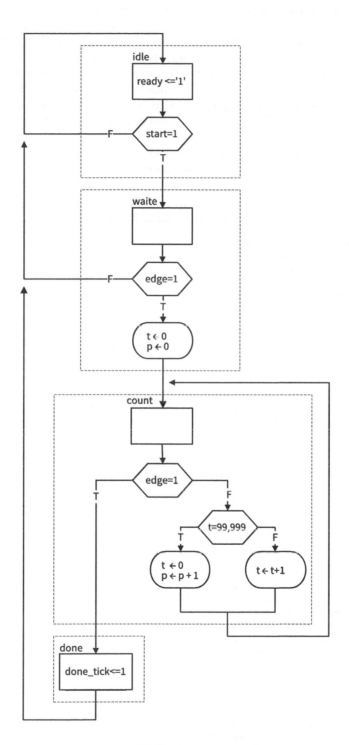

Figure 6.11   ASMD chart of a period counter.

edge of the input. It then moves to the count state when the next rising edge of the input is detected. In the count state, we use two registers to keep track of the time. The t register counts for 100,000 clock cycles, from 0 to 99,999, and then wraps around. Since the period of the system clock is 10 ns, the t register takes 1 ms to circulate through 100,000 cycles. The p register counts in terms of milliseconds. It is incremented once when the t register reaches 99,999. When the FSMD exits the count state, the period of the input waveform is stored in the p register and its unit is milliseconds. The FSMD asserts the done_tick signal in the done state, as in previous examples.

The code follows the ASMD chart and is shown in Listing 6.6. We use a constant, CLK_MS_COUNT, for the boundary of the millisecond counter. It can be replaced if a different measurement unit is desired.

Listing 6.6 Period counter

```vhdl
library ieee;
use ieee.std_logic_1164.all;
use ieee.numeric_std.all;
entity period_counter is
   port(
      clk, reset : in   std_logic;
      start, si  : in   std_logic;
      ready      : out std_logic;
      done_tick  : out std_logic;
      prd        : out std_logic_vector(9 downto 0)
   );
end period_counter;

architecture arch of period_counter is
   constant CLK_MS_COUNT : integer := 100000; -- 1 ms tick
   type state_type is (idle, waite, count, done);
   signal state_reg     : state_type;
   signal state_next    : state_type;
   signal t_reg, t_next : unsigned(16 downto 0); -- up to 100000
   signal p_reg, p_next : unsigned(9 downto 0);  -- up to 1 sec
   signal delay_reg     : std_logic;
   signal edge          : std_logic;
begin
   -- state and data register
   process(clk, reset)
   begin
      if reset = '1' then
         state_reg <= idle;
         t_reg     <= (others => '0');
         p_reg     <= (others => '0');
         delay_reg <= '0';
      elsif (clk'event and clk = '1') then
         state_reg <= state_next;
         t_reg     <= t_next;
         p_reg     <= p_next;
         delay_reg <= si;
      end if;
   end process;
   -- edge detecion
   edge <= (not delay_reg) and si;
   -- data path and next-state logic
   process(start, edge, state_reg, t_reg, t_next, p_reg)
   begin
      ready        <= '0';
```

```
      done_tick  <= '0';
      state_next <= state_reg;
      p_next     <= p_reg;
      t_next     <= t_reg;
      case state_reg is
         when idle =>
            ready <= '1';
            if (start = '1') then
               state_next <= waite;
            end if;
         when waite =>        -- wait for the first edge
            if (edge = '1') then
               state_next <= count;
               t_next     <= (others => '0');
               p_next     <= (others => '0');
            end if;
         when count =>
            if (edge = '1') then        -- 2nd edge arrived
               state_next <= done;
            else                        -- otherwise count
               if t_reg = CLK_MS_COUNT - 1 then -- 1ms tick
                  t_next <= (others => '0');
                  p_next <= p_reg + 1;
               else
                  t_next <= t_reg + 1;
               end if;
            end if;
         when done =>
            done_tick  <= '1';
            state_next <= idle;
      end case;
   end process;
   prd <= std_logic_vector(p_reg);
end arch;
```

6.3.5 Accurate low-frequency counter

A frequency counter measures the frequency of a periodic input waveform. The common way to construct a frequency counter is to count the number of input pulses in a fixed amount of time, say, 1 second. Although this approach is fine for high-frequency input, it cannot measure a low-frequency signal accurately. For example, if the input is around 2 Hz, the measurement cannot tell whether it is 2.123 Hz or 2.567 Hz. Recall that the frequency is the reciprocal of the period (i.e., $frequency = \frac{1}{period}$). An alternative approach is to measure the period of the signal and then take the reciprocal to find the frequency. We use this approach to implement a low-frequency counter in this subsection.

This design example demonstrates how to use the previously designed parts to construct a large system. For simplicity, we assume that the frequency of the input is between 1 and 10 Hz (i.e., the period is between 100 and 1000 ms). The operation of this circuit includes three tasks:

1. Measure the period.
2. Find the frequency by performing a division operation.
3. Convert the binary number to BCD format.

We can use the period counter, division circuit, and binary-to-BCD converter to perform the three tasks and create another FSM as the master control to sequence and coordinate the operation of the three circuits. The block diagram is shown in Figure 6.12(a), and the ASM chart of the master control is shown in Figure 6.12(b). The FSM uses the start and done_tick signals of these circuits to initialize each task and to detect completion of the task. The code is shown in Listing 6.7.

Listing 6.7 Low-frequency counter

```vhdl
library ieee;
use ieee.std_logic_1164.all;
use ieee.numeric_std.all;
entity low_freq_counter is
   port(
      clk, reset : in   std_logic;
      start      : in   std_logic;
      si         : in   std_logic;
      bcd3, bcd2 : out std_logic_vector(3 downto 0);
      bcd1, bcd0 : out std_logic_vector(3 downto 0));
end low_freq_counter;

architecture arch of low_freq_counter is
   type state_type is (idle, count, frq, b2b);
   signal state_reg, state_next    : state_type;
   signal prd                      : std_logic_vector(9 downto 0);
   signal dvsr, dvnd, quo          : std_logic_vector(19 downto 0);
   signal prd_start, prd_done_tick : std_logic;
   signal div_start, div_done_tick : std_logic;
   signal b2b_start, b2b_done_tick : std_logic;
begin
   --*****************************************************************
   -- component instantiation
   --*****************************************************************
   -- instantiate period counter
   prd_count_unit : entity work.period_counter
      port map(
         clk       => clk,
         reset     => reset,
         start     => prd_start,
         si        => si,
         ready     => open,
         done_tick => prd_done_tick,
         prd       => prd
      );
   -- instantiate division circuit
   div_unit : entity work.div
      generic map(
         W    => 20,
         CBIT => 5
      )
      port map(
         clk       => clk,
         reset     => reset,
         start     => div_start,
         dvsr      => dvsr,
         dvnd      => dvnd,
         quo       => quo,
         rmd       => open,
         ready     => open,
         done_tick => div_done_tick
      );
   -- instantiate binary-to-BCD convertor
   bin2bcd_unit : entity work.bin2bcd
```

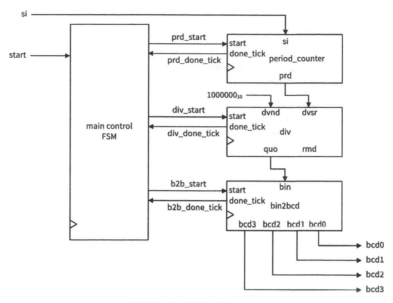

(a) Top-level block diagram

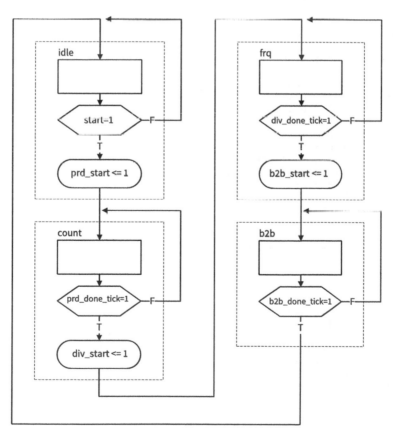

(b) ASM chart of main control

Figure 6.12   Accurate low-frequency counter.

```vhdl
    port map(
        clk        => clk,
        reset      => reset,
        start      => b2b_start,
        bin        => quo(12 downto 0),
        ready      => open,
        done_tick  => b2b_done_tick,
        bcd3       => bcd3,
        bcd2       => bcd2,
        bcd1       => bcd1,
        bcd0       => bcd0
    );
 -- signal width extension
 dvnd <= std_logic_vector(to_unsigned(1000000, 20));
 dvsr <= "0000000000" & prd;

 --*****************************************************************
 -- Master FSM
 --*****************************************************************
 -- register
 process(clk, reset)
 begin
     if reset = '1' then
         state_reg <= idle;
     elsif (clk'event and clk = '1') then
         state_reg <= state_next;
     end if;
 end process;
 -- next-state logic
 process(state_reg, start, prd_done_tick, div_done_tick, b2b_done_tick)
 begin
     state_next <= state_reg;
     prd_start  <= '0';
     div_start  <= '0';
     b2b_start  <= '0';
     case state_reg is
         when idle =>
             if start = '1' then
                 state_next <= count;
                 prd_start  <= '1';
             end if;
         when count =>
             if (prd_done_tick = '1') then
                 div_start  <= '1';
                 state_next <= frq;
             end if;
         when frq =>
             if (div_done_tick = '1') then
                 b2b_start  <= '1';
                 state_next <= b2b;
             end if;
         when b2b =>
             if (b2b_done_tick = '1') then
                 state_next <= idle;
             end if;
     end case;
 end process;
end arch;
```

6.4 BIBLIOGRAPHIC NOTES

The bibliographic information for this chapter is similar to that for Chapter 3.

6.5 SUGGESTED EXPERIMENTS

6.5.1 Early detection debouncing circuit

Consider the alternative debouncing circuit in Experiment 5.5.2. Redesign the circuit using the RT methodology:

1. Derive the ASMD chart for the circuit.
2. Derive the HDL code based on the ASMD chart.
3. Derive a testing circuit similar to that in Section 5.3.3 with the alternative debouncing circuit and verify its operation.

6.5.2 BCD-to-binary conversion circuit

A BCD-to-binary conversion converts a BCD number to the equivalent binary representation. Assume that the input is an 8-bit signal in BCD format (i.e., two BCD digits) and the output is a 7-bit signal in binary representation. Follow the procedure in Section 6.3.3 to design a BCD-to-binary conversion circuit:

1. Derive the conversion algorithm and ASMD chart.
2. Derive the HDL code based on the ASMD chart.
3. Derive a testbench and use simulation to verify operation of the code.
4. Synthesize the circuit, program the FPGA, and verify its operation.

6.5.3 Fibonacci circuit with BCD I/O: design approach 1

To make the Fibonacci circuit more user friendly, we can modify the circuit to use the BCD format for the input and output. Assume that the input is an 8-bit signal in BCD format (i.e., two BCD digits) and the output is displayed as four BCD digits on the seven-segment LED display. Furthermore, the LED will display "9999" if the resulting Fibonacci number is larger than 9999 (i.e., overflow). The operation can be done in three steps: convert input to the binary format, compute the Fibonacci number, and convert the result back to the BCD format.

The first design approach is to follow the procedure in Section 6.3.5. Begin by constructing three smaller subsystems, which are the BCD-to-binary conversion circuit, Fibonacci circuit, and binary-to-BCD conversion circuit, and then use a master FSM to control the overall operation. Design the circuit as follows:

1. Implement the BCD-to-binary conversion circuit in Experiment 6.5.2.
2. Modify the Fibonacci number circuit in Section 6.3.1 to include an output signal to indicate the overflow condition.
3. Derive the top-level block diagram and the master control FSM state diagram.
4. Derive the HDL code.
5. Derive a testbench and use simulation to verify operation of the code.
6. Synthesize the circuit, program the FPGA, and verify its operation.

6.5.4   Fibonacci circuit with BCD I/O: design approach 2

An alternative to the previous "subsystem approach" in Experiment 6.5.3 is to integrate the three subsystems into a single system and derive a customized FSMD for this particular application. The approach eliminates the overhead of the control FSM and provides opportunities to share registers among the three tasks. Design the circuit as follows:

1. Redesign the circuit of Experiment 6.5.3 using one FSMD. The design should eliminate all unnecessary circuits and states, such as the various done_tick signals and the done states, and exploit the opportunity to share and reuse the registers in different steps.
2. Derive the ASMD chart.
3. Derive the HDL code based on the ASMD chart.
4. Derive a testbench and use simulation to verify operation of the code.
5. Synthesize the circuit, program the FPGA, and verify its operation.
6. Check the synthesis report and compare the number of LEs used in the two approaches.
7. Calculate the number of clock cycles required to complete the operation in the two approaches.

6.5.5   Auto-scaled low-frequency counter

The operation of the low-frequency counter in Section 6.3.5 is very restricted. The frequency range of the input signal is limited between 1 and 10 Hz. It loses accuracy when the frequency is beyond this range. Recall that the accuracy of this frequency counter depends on the accuracy of the period counter of Section 6.3.5, which counts in terms of millisecond ticks. We can modify the t counter to generate a microsecond tick (i.e., counting from 0 to 49) and increase the accuracy 1000-fold. This allows the range of the frequency counter to increase to 9999 Hz and still maintain at least four-digit accuracy.

Using a microsecond tick introduces more than four accuracy digits for low-frequency input, and the number must be shifted and truncated to be displayed on the seven-segment LED. An auto-scaled low-frequency counter performs the adjustment automatically, displays the four most significant digits, and places a decimal point in the proper place. For example, according to their range, the frequency measurements will be shown as "1.234", "12.34", "123.4", or "1234.".

The auto-scaled low-frequency counter needs an additional BCD adjustment circuit. It first checks whether the most significant BCD digit (i.e., the four MSBs) of a BCD sequence is zero. If this is the case, the circuit shifts the BCD sequence to the left by four positions and increments the decimal point counter. The operation is repeated until the most significant BCD digit is not "0000".

The complete auto-scaled low-frequency counter can be implemented as follows:

1. Modify the period counter to use the microsecond tick.
2. Extend the size of the binary-to-BCD conversion circuit.
3. Derive the ASMD chart for the BCD adjustment circuit and the HDL code.
4. Modify the control FSM to include the BCD adjustment in the last step.
5. Design a simple decoding circuit that uses the decimal point counter's output to activate the desired decimal point of the seven-segment LED display.
6. Derive a testbench and use simulation to verify operation of the code.

7. Synthesize the circuit, program the FPGA, and verify its operation.

6.5.6 Reaction timer

Eye–hand coordination is the ability of the eyes and hands to work together to perform a task. A reaction timer circuit measures how fast a human hand can respond after a person sees a visual stimulus. This circuit operates as follows:

1. The circuit has three input pushbuttons, corresponding to the `clear`, `start`, and `stop` signals. It uses a single discrete LED as the visual stimulus and displays relevant information on the seven-segment LED display.
2. A user pushes the `clear` button to force the circuit returning to the initial state, in which the seven-segment LED shows a welcome message, "`HI`", and the stimulus LED is off.
3. When ready, the user pushes the `start` button to initiate the test. The seven-segment LED goes off.
4. After a random interval between 2 and 15 seconds, the stimulus LED goes on and the timer starts to count upward. The timer increases every millisecond and its value is displayed in the format of "0000" millisecond on the seven-segment LED.
5. After the stimulus LED goes on, the user should try to push the `stop` button as soon as possible. The timer pauses counting once the `stop` button is asserted. The seven-segment LED shows the reaction time. It should be around 150 to 300 milliseconds for most people.
6. If the `stop` button is not pushed, the timer stops after 1 second and displays "1000".
7. If the `stop` button is pushed before the stimulus LED goes on, the circuit displays "9999" on the seven-segment LED and stops.

Design the circuit as follows:

1. Derive the ASMD chart.
2. Derive the HDL code based on the ASMD chart.
3. Synthesize the circuit, program the FPGA, and verify its operation.

6.5.7 Babbage difference engine emulation circuit

The Babbage difference engine is a mechanical digital computation device designed to tabulate a polynomial function. It was proposed by Charles Babbage, an English mathematician, in the nineteenth century. The engine is based on Newton's method of differences and avoids the need of multiplication. For example, consider a second-order polynomial $f(n) = 2n^2 + 3n + 5$. We can find the difference between $f(n)$ and $f(n-1)$:

$$f(n) - f(n-1) = 4n + 1$$

Assume that n is an integer and $n \geq 0$. The $f(n)$ can be defined recursively as

$$f(n) = \begin{cases} 5 & \text{if } n = 0 \\ f(n-1) + 4n + 1 & \text{if } n > 0 \end{cases}$$

This process can be repeated for the $4n + 1$ expression. Let $g(n) = 4n + 1$. We can find the difference between $g(n)$ and $g(n-1)$:

$$g(n) - g(n-1) = 4$$

The $g(n)$ can be defined recursively as

$$g(n) = \begin{cases} 5 & \text{if } n = 1 \\ g(n-1) + 4 & \text{if } n > 1 \end{cases}$$

and $f(n)$ can be rewritten as

$$f(n) = \begin{cases} 5 & \text{if } n = 0 \\ f(n-1) + g(n) & \text{if } n > 0 \end{cases}$$

Note that only additions are involved in the recursive definitions of $f(n)$ and $g(n)$.

Based on the definition of the last two recursive equations, we can derive an algorithm to compute $f(n)$. Two temporary registers are needed to keep track of the most recently calculated $f(n)$ and $g(n)$, and two additions are needed to update $f(n)$ and $g(n)$. Assume that n is a 6-bit input and interpreted as an unsigned integer. Design this circuit using RT methodology:

1. Derive the ASMD chart.
2. Derive the HDL code based on the ASMD chart.
3. Derive a testbench and use simulation to verify operation of the code.
4. Synthesize the circuit, program the FPGA, and verify its operation.
5. Let $h(n) = n^3 + 2n^2 + 2n + 1$. Use the method above to find the recursive representation of $h(n)$ (note that three levels of recursive equations are needed for a three-order polynomial). Repeat steps 1 to 4.

CHAPTER 7

RAM AND BUFFER OF FPGA

A digital system frequently requires memory for storage. To facilitate this need, FPGA devices contain dedicated embedded memory macro cells. While these modules cannot replace the massive external memory devices, they are useful for applications that require small or intermediate-sized memory. In this chapter, we demonstrate the coding techniques to describe the memory structure and provide several templates to infer the FPGA's internal memory modules.

7.1 EMBEDDED MEMORY OF FPGA DEVICE

A memory component stores data. A sequential circuit discussed in Chapters 4, 5, and 6 contains a register to maintain its internal state. The register can be considered a customized memory component dedicated to the functionality of the specific sequential circuit. A digital system frequently needs "general-purpose" storage as well. The required storage capacity can vary significantly. While a buffer for a serial input data line just needs a few bytes, the main memory of a high-performance computer system requires several hundred million bytes.

An FPGA's logic elements contain a register and thus can provide one-bit memory. However, it takes more than two thousand transistors to implement a logic cell while it just takes five or six transistors to construct a one-bit SRAM (static RAM) cell. Thus, it is a tremendous waste of resources to dedicate the functionalities of a logic cell to implement one-bit memory. To overcome the problem, modern FPGA devices are constructed with prefabricated memory modules for storage.

FPGA Prototyping by VHDL Examples 2nd ed., Pong P. Chu. **145**
Copyright © 2017, John Wiley & Sons, Inc.

These modules are intended for applications that require small or intermediate-sized memory but not for the replacement of the massive external memory devices.

7.1.1 Memory of an Artix device

There are two types of embedded memory in Xilinx FPGA devices: distributed RAM and block RAM (BRAM) A *distributed RAM* is constructed from the logic cell's lookup table (LUT). For example, a 6-input LUT can be configured as a 2^6-by-1 RAM module, and multiple LUTs can be cascaded to form a wider and deeper memory module. The Artix-7 XCA100T device of the Nexys 4 DDR board can provide up to 1188K bits of distributed memory, which is small compared to block RAMs or external memory. Since the distributed RAM is constructed with the logic cells, it competes with the normal logic for resource. In addition, the read data of the distributed memory is not registered (known as *asynchronous*). This type of memory configuration is not used by other FPGA vendors and thus is less portable.

A *block RAM* (*BRAM*) is a prefabricated memory module embedded in an FPGA device and is separated from the regular logic cells. It can be thought of as a fast SRAM wrapped by a synchronous, configurable interface. In an Artix device, each BRAM consists of 32K (2^{15}) data bits plus optional 4K parity bits, totaling 36K bits. It can be organized in different widths, from 32K by 1 (i.e., 2^{15} by 2^0) to 512 by 64 (i.e., 2^9 by 2^6). The 36K-bit BRAM can also be partitioned as two independent 18K-bit BRAMs. The Artix-7 XCA100T device has 135 36K-bit BRAMs, totaling 4,320K usable data bits. This type of memory modules can be found in all modern FPGA devices.

The BRAM is "wrapped" with a synchronous interface, and thus no additional memory controller circuit is needed. It is very flexible and can be configured to perform single- and dual-port access and to support various types of buffering and clocking schemes. We examine four commonly used configurations, including a synchronous dual-port RAM, a "simple" dual-port RAM, a synchronous single-port RAM, and a synchronous ROM, in Section 7.4.

7.1.2 Memory available in a Nexys 4 DDR board

The Artix-7 XCA100T device and external RAM device of the Nexys 4 DDR board provides several storage options. It is a good idea to keep in mind the relative capacities of these options:

- *XCA100T's FFs* (for registers): about 13 KB (13K*8 bits), embedded in logic cells and I/O buffers
- *XCA100T's distributed RAM*: about 150 KB (150K*8 bits), constructed from the logic cells
- *XCA100T's BRAM*: 540 KB , configured as 270 32K-bit modules
- *External SDRAM*: 128 MB, configured as a single rank 8M-by-16 DDR SDRAM (synchronous dynamic RAM) chip

This helps us to decide which option is most suitable for an application at hand.

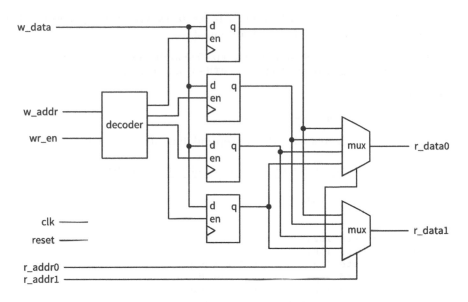

Figure 7.1   Block diagram of a four-word register file.

7.2   GENERAL DESCRIPTION FOR A RAM-LIKE COMPONENT

A register file is a collection of registers with one input port and one or more output ports. It is frequently used inside a processor as a small fast storage unit. We use it in this section to illustrate coding techniques for RAM-like memory components.

7.2.1   Register file

A register file is a collection of registers. The registers share common input ports and output ports and each register is identified by a unique *address*. The write address signal, w_addr, specifies where to store data, and the read address signal, r_addr, specifies where to retrieve data. The register file is generally used as fast, temporary storage. The conceptual diagram of a 4-by-8 (i.e., four words and 8 bits per word) register file is shown Figure 7.1. The design consists of four registers with enable signals, a write decoding circuit, and a read multiplexing circuit.

The write decoding circuit examines the wr_en signal and decodes the write port address. If the wr_en signal is asserted, the decoding circuit functions as a regular 2-to-2^2 binary decoder that asserts one of the four en signals of the corresponding register. The w_data signal will be sampled and stored into the corresponding register at the rising edge of the clock. The read multiplexing circuit consists of a 4-to-1 multiplexer. It utilizes r_addr as the selection signal to route the desired register output to the read port.

The registers are structured as a two-dimensional 4-by-8 array of D FFs and would best be represented by a two-dimensional data type. Since there is no predefined two-dimensional data type in the IEEE std_logic_1164 package, we must create a user-defined data type. Assume that there are ADDR_WIDTH bits in the address (i.e., 2^{ADDR_WIDTH} words) and there are DATA_WIDTH bits per word. The new data type can be defined in a type statement

```
type mem_2d_type is array (0 to 2**ADDR_WIDTH-1) of
       std_logic_vector(DATA_WIDTH-1 downto 0);
```

and then used, as in

```
signal array_reg: mem_2d_type;
```

We can derive the code following the conceptual diagram, as shown in Listing 7.1.

Listing 7.1   Register file with explicit decoding and multiplexing logic

```
library ieee;
use ieee.std_logic_1164.all;
entity reg_file_4x8 is
   port(
      clk    : in  std_logic;
      wr_en  : in  std_logic;
      w_addr : in  std_logic_vector(1 downto 0);
      r_addr : in  std_logic_vector(1 downto 0);
      w_data : in  std_logic_vector(7 downto 0);
      r_data : out std_logic_vector(7 downto 0)
   );
end reg_file_4x8;

architecture explicit_arch of reg_file_4x8 is
   constant ADDR_WIDTH : natural := 2;  -- bits in address
   constant DATA_WIDTH : natural := 8;  -- bits in data
   type mem_2d_type is array (0 to 2**ADDR_WIDTH-1) of
        std_logic_vector(DATA_WIDTH-1 downto 0);
   signal array_reg : mem_2d_type;
   signal en        : std_logic_vector(2**ADDR_WIDTH-1 downto 0);
begin
   -- 4 registers
   process(clk)
   begin
      if (clk'event and clk = '1') then
         if en(3) = '1' then
            array_reg(3) <= w_data;
         end if;
         if en(2) = '1' then
            array_reg(2) <= w_data;
         end if;
         if en(1) = '1' then
            array_reg(1) <= w_data;
         end if;
         if en(0) = '1' then
            array_reg(0) <= w_data;
         end if;
      end if;
   end process;
   -- decoding logic for write address
   process(wr_en, w_addr)
   begin
      if (wr_en = '0') then
         en <= (others => '0');
      else
         case w_addr is
            when "00"   => en <= "0001";
            when "01"   => en <= "0010";
```

```vhdl
            when "10"   => en <= "0100";
            when others => en <= "1000";
         end case;
      end if;
   end process;
   -- read multiplexing
   with r_addr select r_data <=
      array_reg(0) when "00",
      array_reg(1) when "01",
      array_reg(2) when "10",
      array_reg(3) when others;
end explicit_arch;
```

The code consists of a collection of four registers, a decoding logic to generate the enable signals, and a multiplexer to route the desired data to the read port. We can duplicate the decoding logic and multiplexing logic if additional write ports or read ports are needed.

7.2.2 Dynamic array indexing operation

Although the previous code is straightforward, the decoding and multiplexing statements become cumbersome as the size of the register file increases. An alternative method is to use *dynamic indexing*, in which a signal is used as an index to access an element in the array. The code for a parameterized register file is shown in Listing 7.2. The two generics are defined in this design. The DATA_WIDTH generic specifies the number of bits in a word and the ADDR_WIDTH generic specifies the number of address bits, which implies that there are 2^{ADDR_WIDTH} words in the register file.

Listing 7.2   Register file with dynamic indexing

```vhdl
library ieee;
use ieee.std_logic_1164.all;
use ieee.numeric_std.all;
entity reg_file is
   generic(
      ADDR_WIDTH : integer := 2;
      DATA_WIDTH : integer := 8
   );
   port(
      clk    : in  std_logic;
      wr_en  : in  std_logic;
      w_addr : in  std_logic_vector(ADDR_WIDTH-1 downto 0);
      r_addr : in  std_logic_vector(ADDR_WIDTH-1 downto 0);
      w_data : in  std_logic_vector(DATA_WIDTH-1 downto 0);
      r_data : out std_logic_vector(DATA_WIDTH-1 downto 0)
   );
end reg_file;

architecture arch of reg_file is
   type mem_2d_type is array (0 to 2**ADDR_WIDTH-1) of
         std_logic_vector(DATA_WIDTH-1 downto 0);
   signal array_reg : mem_2d_type;
begin
   process(clk)
   begin
      if (clk'event and clk = '1') then
         if wr_en = '1' then
```

```
               array_reg(to_integer(unsigned(w_addr))) <= w_data;
          end if;
       end if;
    end process;
    -- read port
    r_data <= array_reg(to_integer(unsigned(r_addr)));
end arch;
```

Note that the **array_reg(...w_addr...)** **<=** ... and ... **<=** **array_reg(...r_addr...)** statements infer decoding and multiplexing logic, respectively. Although the description is more abstract, Vivado synthesis software recognizes this language construct and can derive the correct implementation accordingly.

The code can be easily revised to accommodate additional read or write ports. For example, a processor may need two read ports to access two operands and one write port to store the result back to a register. The revised code is shown Listing 7.3.

Listing 7.3   Register file two read ports

```
library ieee;
use ieee.std_logic_1164.all;
use ieee.numeric_std.all;
entity reg_file_2_read_port is
   generic(
       ADDR_WIDTH : integer := 2;
       DATA_WIDTH : integer := 8
   );
   port(
       clk     : in std_logic;
       wr_en   : in std_logic;
       w_addr  : in std_logic_vector (ADDR_WIDTH-1 downto 0);
       r_addr1 : in std_logic_vector (ADDR_WIDTH-1 downto 0);
       r_addr2 : in std_logic_vector (ADDR_WIDTH-1 downto 0);
       w_data  : in std_logic_vector (DATA_WIDTH-1 downto 0);
       r_data1 : out std_logic_vector (DATA_WIDTH-1 downto 0);
       r_data2 : out std_logic_vector (DATA_WIDTH-1 downto 0)
   );
end reg_file_2_read_port ;

architecture arch of reg_file_2_read_port is
   type mem_2d_type is array (0 to 2**ADDR_WIDTH-1) of
          std_logic_vector(DATA_WIDTH-1 downto 0);
   signal array_reg: mem_2d_type;
begin
   process(clk)
   begin
      if (clk'event and clk='1') then
         if wr_en='1' then
             array_reg(to_integer(unsigned(w_addr))) <= w_data;
         end if;
      end if;
   end process;
   -- read port 1
   r_data1 <= array_reg(to_integer(unsigned(r_addr1)));
   -- read port 2
   r_data2 <= array_reg(to_integer(unsigned(r_addr2)));
end arch;
```

7.2.3   Key aspects of a RAM module

The basic characteristics of a memory module are specified by several aspects:

- Depth
- Width
- Number of ports
- Direction of a port
- Synchronicity of port access
- Simultaneous address access

The *depth* is the number of the words in the module, which is usually a power of two. In our design, it is derived from the number of bits in the address signal. The ADDR_WIDTH generic in Listing 7.2 specifies the number of address bits and it implies that the depth is 2^{ADDR_WIDTH}. The *width* is the number of bits in a word. It is defined as the DATA_WIDTH generic in Listing 7.2.

The *number of ports* specifies the number of access ports and *direction of a port* indicates whether the port is for read, write, or both. A write enable signal is associated with a port that supports the write operation. The common configuration supports one or two ports. The register file in Listing 7.2 has two ports, one for writing and one for reading.

The *synchronicity* indicates whether the read or write operation is controlled by a clock signal. For example, the write operation of the register file is controlled by a clock signal and thus the write operation is referred to as *synchronous*. On the other hand, its read operation is done by a multiplexer, which is a combinational circuit, and thus the read operation is referred to as *asynchronous*. To make the read operation synchronous, the read data needs to be registered. The revised code of this segment of Listing 7.2 becomes

```
process(clk)
begin
   if (clk'event and clk='1') then
      if wr_en='1' then
         array_reg(to_integer(unsigned(w_addr))) <= w_data;
      end if;
   r_data_reg <= array_reg(to_integer(unsigned(r_addr)));
   end if;
end process;
-- read port w/ registered read data
r_data <= r_data_reg;
```

The *simultaneous address access* defines what happens when a read port and a write port have the same address. The main concern is whether the old or new write data is retrieved in the read operation. In the operation above, the code segment specifies that the old data is retrieved. Note that writing data to the same address of two write ports is usually not allowed.

7.2.4   Genuine ROM

Despite its name, a ROM (read-only memory) is a combinational circuit and has no internal state. Its output depends only on its input (i.e., address). There is no real embedded ROM in an FPGA device, but it can be emulated by a combinational circuit or a RAM with the write operation disabled. The content of the ROM can be expressed as a two-dimensional constant in the HDL code and the values are loaded to the RAM when the device is programmed. The details of this approach are discussed in Section 7.4.5.

A real ROM does not have a buffer or a clock signal. It cannot be realized by FPGA's internal memory module. We include it in this section because its code structure is similar to that of a storage component. The template of a ROM is shown by an example in Listing 7.4. The code is to implement the hex-to-seven segment LED decoder, similar to that in Listing 3.14. The address of the ROM functions as the 4-bit hexadecimal input and its content is the corresponding LED patterns. The content of the ROM is defined by the HEX2LED_LOOK_UP_TABLE constant and is essentially the truth table of this circuit.

Listing 7.4   Template for a true ROM

```vhdl
library ieee;
use ieee.std_logic_1164.all;
use ieee.numeric_std.all;
entity true_rom_template is
   port(
      addr : in std_logic_vector(3 downto 0);
      data : out std_logic_vector(6 downto 0)
   );
end true_rom_template;

architecture arch of true_rom_template is
   constant ADDR_WIDTH : integer := 4;
   constant DATA_WIDTH : integer := 7;
   type rom_type is array (0 to 2**ADDR_WIDTH-1)
        of std_logic_vector(DATA_WIDTH-1 downto 0);
   -- ROM definition
   constant HEX2LED_LOOK_UP_TABLE: rom_type:=(  -- 2^4-by-7
      "1000000",  -- addr 00
      "1111001",  -- addr 01
      "0100100",  -- addr 02
      "0110000",  -- addr 03
      "0011001",  -- addr 04
      "0010010",  -- addr 05
      "0000010",  -- addr 06
      "1111000",  -- addr 07
      "0000000",  -- addr 08
      "0010000",  -- addr 09
      "0001000",  -- addr 10
      "0000011",  -- addr 11
      "1000110",  -- addr 12
      "0100001",  -- addr 13
      "0000110",  -- addr 14
      "0001110"   -- addr 15
   );
begin
   data <= HEX2LED_LOOK_UP_TABLE(to_integer(unsigned(addr)));
end arch;
```

Note that the memory row is defined in ascending order:

```
... array (0 to 2**ADDR_WIDTH-1) of ...
```

and the first row of the HEX2LED_LOOK_UP_TABLE constant corresponds to the address 0000 of the ROM. The rows must be reversed if the rom_type data type is defined in descending order:

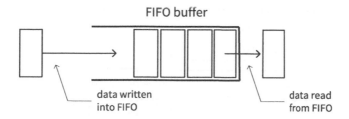

Figure 7.2 Conceptual diagram of a FIFO buffer.

```
... array (2**ADDR_WIDTH-1 downto 0) of ...
```

A ROM is synthesized as a combinational circuit with the logic cells. The code can be considered as another form of a selected signal assignment or case statement. This type of ROM is feasible only for a small table. For larger amount of data, the synchronous ROM template discussed in Section 7.4.5 should be used to take advantage of BRAMs.

7.3 FIFO BUFFER

A FIFO (first-in-first-out) buffer is an "elastic" storage between two subsystems, as shown in the conceptual diagram of Figure 7.2. It can be constructed by "wrapping" a regular memory component with a special controller. We use the register file as the storage and develop a FIFO buffer in this section. More sophisticated BRAM-based implementation is discussed in Section 7.4.6.

7.3.1 FIFO read configuration

A FIFO buffer has two control signals, wr and rd, for write and read operations. When wr is asserted, the input data is written into the *tail* (i.e., end) of the buffer. When rd is asserted, the data is retrieved or removed from the *head* (i.e., front) of the buffer. The date retrieval is based on the order the data written to the buffer and thus is done in a first-in-first-out basis.

One subtle aspect of a FIFO buffer is its "read configuration," which specifies how the data is retrieved and removed from the buffer. In the *FWFT* (*first word fall through*) *configuration*, the current data (i.e., the head of the buffer) is available automatically in the read data port without the assertion of any control signal. When a data word is written to an empty FIFO buffer, it "falls through" to the read data port immediately. The read signal, rd, actually functions as a "removal" signal. When it is asserted, the current head data is deleted from the buffer and the following data item in buffer becomes available in the next clock cycle.

In the *"standard"configuration*, the read signal is used to retrieve the head data. When a data word is written to an empty FIFO buffer, the FIFO's read port remains unchanged. The rd signal functions as a "request" signal. When it is asserted, the current head data is retrieved and becomes available in the next clock cycle. An FWFT FIFO buffer can be converted to a standard FIFO buffer by inserting an extra register, as shown in Figure 7.3.

We use the FWFT FIFO buffer in this book.

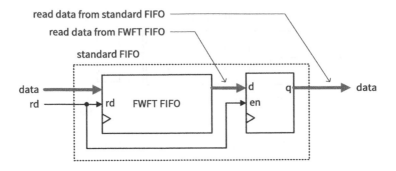

Figure 7.3 FWFT FIFO buffer conversion.

7.3.2 Circular queue implementation

One method to implement a FIFO buffer is to arrange the linear memory space as a *circular queue* with two pointers. The *write pointer* points to the head of the queue and the *read pointer* points to the tail of the queue. The pointer advances one position for each write or read operation. The operation of an eight-word circular queue is shown in Figure 7.4. A *FIFO controller* can be constructed to implement the circular queue algorithm.

The complete FIFO buffer is composed of a FIFO controller and a register file and its top-level block diagram is shown in Figure 7.5. The FWFT configuration is used in this FIFO controller since the configuration is better suited for the asynchronous read operation of the register file.

The FIFO controller generates two status signals, `full` and `empty`, to indicate that the FIFO buffer is full (i.e., cannot be written) and empty (i.e., cannot be read), respectively. One of the two conditions occurs when the read pointer is equal to the write pointer, as shown in Figure 7.4(a), (f), and (i). The most difficult design task of the FIFO controller is to derive a mechanism to distinguish the two conditions. One scheme is to use two FFs to keep track of the empty and full statuses. The FFs are set to '1' and '0' during system initialization and then modified in each clock cycle according to the activities of the `wr` and `rd` signals.

The code of the FIFO controller is shown in Listing 7.5.

Listing 7.5 FIFO controller

```
library ieee;
use ieee.std_logic_1164.all;
use ieee.numeric_std.all;
entity fifo_ctrl is
   generic(ADDR_WIDTH : natural := 4);
   port(
      clk, reset   : in   std_logic;
      rd, wr       : in   std_logic;
      empty, full  : out  std_logic;
      w_addr       : out  std_logic_vector(ADDR_WIDTH-1 downto 0);
      r_addr       : out  std_logic_vector(ADDR_WIDTH-1 downto 0)
   );
end fifo_ctrl;

architecture arch of fifo_ctrl is
   signal w_ptr_reg   : std_logic_vector(ADDR_WIDTH-1 downto 0);
```

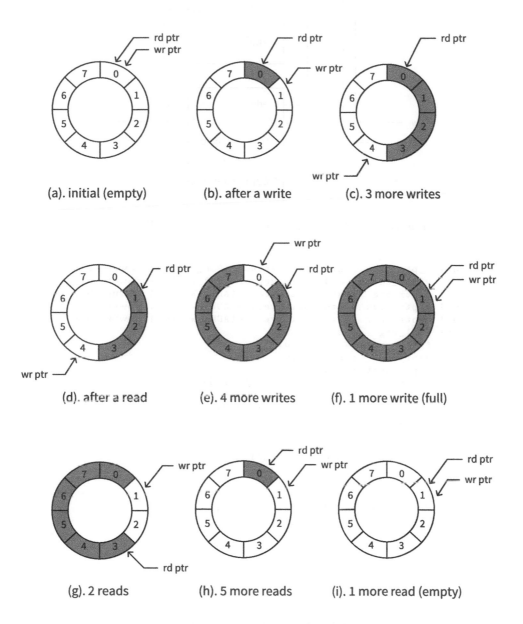

Figure 7.4   FIFO buffer based on a circular queue.

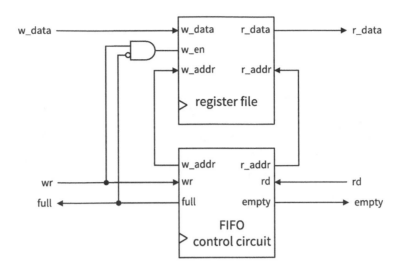

Figure 7.5 Block diagram of a register file based FIFO buffer.

```
signal w_ptr_next : std_logic_vector(ADDR_WIDTH-1 downto 0);
signal w_ptr_succ : std_logic_vector(ADDR_WIDTH-1 downto 0);
signal r_ptr_reg  : std_logic_vector(ADDR_WIDTH-1 downto 0);
signal r_ptr_next : std_logic_vector(ADDR_WIDTH-1 downto 0);
signal r_ptr_succ : std_logic_vector(ADDR_WIDTH-1 downto 0);
signal full_reg   : std_logic;
signal full_next  : std_logic;
signal empty_reg  : std_logic;
signal empty_next : std_logic;
signal wr_op      : std_logic_vector(1 downto 0);
begin
   -- register for read and write pointers
   process(clk, reset)
   begin
      if (reset = '1') then
         w_ptr_reg <= (others => '0');
         r_ptr_reg <= (others => '0');
         full_reg  <= '0';
         empty_reg <= '1';
      elsif (clk'event and clk = '1') then
         w_ptr_reg <= w_ptr_next;
         r_ptr_reg <= r_ptr_next;
         full_reg  <= full_next;
         empty_reg <= empty_next;
      end if;
   end process;

   -- successive pointer values
   w_ptr_succ <= std_logic_vector(unsigned(w_ptr_reg) + 1);
   r_ptr_succ <= std_logic_vector(unsigned(r_ptr_reg) + 1);

   -- next-state logic for read and write pointers
   wr_op <= wr & rd;
   process(w_ptr_reg, w_ptr_succ, r_ptr_reg, r_ptr_succ,
           wr_op, empty_reg, full_reg)
   begin
      w_ptr_next <= w_ptr_reg;
```

```
      r_ptr_next <= r_ptr_reg;
      full_next  <= full_reg;
      empty_next <= empty_reg;
      case wr_op is
         when "00" =>                      -- no op
         when "01" =>                      -- read
            if (empty_reg /= '1') then     -- not empty
               r_ptr_next <= r_ptr_succ;
               full_next  <= '0';
               if (r_ptr_succ = w_ptr_reg) then
                  empty_next <= '1';
               end if;
            end if;
         when "10" =>                      -- write
            if (full_reg /= '1') then      -- not full
               w_ptr_next <= w_ptr_succ;
               empty_next <= '0';
               if (w_ptr_succ = r_ptr_reg) then
                  full_next <= '1';
               end if;
            end if;
         when others =>                    -- write/read;
            w_ptr_next <= w_ptr_succ;
            r_ptr_next <= r_ptr_succ;
      end case;
   end process;
   -- output
   w_addr <= w_ptr_reg;
   r_addr <= r_ptr_reg;
   full   <= full_reg;
   empty  <= empty_reg;
end arch;
```

The controller consists of two pointers and two status FFs. Its next-state logic examines the **wr** and **rd** signals and takes actions accordingly. For example, let us consider the "10" case, which implies that only a write operation occurs. The status FF is checked first to ensure that the buffer is not full. If this condition is met, we advance the write pointer by one position and clear the empty status FF. Storing one extra word to the buffer may make it full. This happens if the new write pointer "catches" the read pointer, which is expressed by the **w_ptr_succ=r_ptr_reg** expression.

Following the diagram in Figure 7.5, we can combine the controller and the register file to construct the complete FIFO buffer. The code is shown in Listing 7.6.

Listing 7.6   FIFO buffer

```
library ieee;
use ieee.std_logic_1164.all;
entity fifo is
   generic(
      ADDR_WIDTH : integer := 2;
      DATA_WIDTH : integer := 8
   );
   port(
      clk, reset : in  std_logic;
      rd, wr     : in  std_logic;
      w_data     : in  std_logic_vector(DATA_WIDTH - 1 downto 0);
      empty      : out std_logic;
      full       : out std_logic;
      r_data     : out std_logic_vector(DATA_WIDTH - 1 downto 0)
```

```
    );
end fifo;

architecture reg_file_arch of fifo is
    signal full_tmp : std_logic;
    signal wr_en    : std_logic;
    signal w_addr   : std_logic_vector(ADDR_WIDTH - 1 downto 0);
    signal r_addr   : std_logic_vector(ADDR_WIDTH - 1 downto 0);
begin
    -- write enabled only when FIFO is not full
    wr_en <= wr and (not full_tmp);
    full  <= full_tmp;
    -- instantiate fifo control unit
    ctrl_unit : entity work.fifo_ctrl(arch)
        generic map(ADDR_WIDTH => ADDR_WIDTH)
        port map(
            clk     => clk,
            reset   => reset,
            rd      => rd,
            wr      => wr,
            empty   => empty,
            full    => full_tmp,
            w_addr  => w_addr,
            r_addr  => r_addr
        );
    -- instantiate register file
    reg_file_unit : entity work.reg_file(arch)
        generic map(
            DATA_WIDTH => DATA_WIDTH,
            ADDR_WIDTH => ADDR_WIDTH)
        port map(
            clk     => clk,
            w_addr  => w_addr,
            r_addr  => r_addr,
            w_data  => w_data,
            r_data  => r_data,
            wr_en   => wr_en
        );
end reg_file_arch;
```

7.4   HDL TEMPLATES FOR MEMORY INFERENCE

An FPGA's embedded memory module is a macro cell and separated from the normal logic cells. There are several methods to incorporate the modules into a design. Our focus is on the behavioral HDL inference and several templates are provided in the following subsections.

7.4.1   Methods to incorporate memory modules

In Vivado Design Suite, embedded memory modules can be incorporated in three ways:

- By HDL instantiation
- By the Block Memory Generator and Distributed Memory Generator utility programs
- By a behavioral HDL inference template

The first two methods are specific for Xilinx devices. In the first method, we copy the instantiation code segment and manually modify the parameters through generics and attributes to obtain the desired configuration. It is tedious and error prone. In the second method, the utility program guides us through the configuration process and generates the instantiation code. It basically automates the process of the first method. However, the generated code is in Xilinx's own format and is difficult to understand.

The third approach is to use *behaviorial HDL templates* to infer the internal memory module. The HDL code is very general and can actually describe the desired memory configuration and needed functionalities. However, the synthesis software may or may not recognize the description and designer's intention. If not recognizing the pattern, the software will synthesize the module from generic logic cells rather than inferring a memory macro cell. To overcome the problem, we can consult the vendor's manual and follow the suggested HDL templates.

The templates are done by behavioral descriptions and contain no device-specific component instantiation. They are easy to understand and can be simulated as regular HDL codes. On the downside, the template approach is based on the ability of the software to recognize the template and infer the proper memory module accordingly. This may not be achieved by other third-party synthesis software. Thus, these templates can best be described as "semi-portable" and "semi-device-independent" behavioral descriptions.

To use behavioral HDL description to infer the memory module, the vendor's suggested templates should be followed closely. In the following subsections, we discuss the behavioral HDL templates for four BRAM-based configurations, including a synchronous dual-port RAM, a "simple" synchronous dual-port RAM, a synchronous single-port RAM, and a synchronous ROM. The discussion does not include templates for the *distributed RAM* since it is less configurable and not portable.

To help the software recognize these templates, the codes should not be mixed with other logic. A template should be confined in an individual file and then instantiated as a component. It is a good idea to check the synthesis report to ensure that the desired memory module is inferred correctly.

7.4.2 Synchronous dual-port RAM

A synchronous dual-port RAM includes two ports for memory access. Each port can conduct read or write operation independently and has its own set of address, data input and output, and control signals. Both read and write operations are synchronous. The template for a synchronous dual-port RAM is shown in Listing 7.7.

Listing 7.7 Template for a synchronous dual-port RAM

```
library ieee;
use ieee.std_logic_1164.all;
use ieee.numeric_std.all;
entity sync_dual_port_ram is
   generic(
      ADDR_WIDTH : integer := 12;
      DATA_WIDTH : integer := 8
   );
   port(
```

```vhdl
      clk     : in    std_logic;
      -- port a
      we_a    : in    std_logic;
      addr_a  : in    std_logic_vector(ADDR_WIDTH-1 downto 0);
      din_a   : in    std_logic_vector(DATA_WIDTH-1 downto 0);
      dout_a  : out   std_logic_vector(DATA_WIDTH-1 downto 0);
      -- port b
      we_b    : in    std_logic;
      addr_b  : in    std_logic_vector(ADDR_WIDTH-1 downto 0);
      din_b   : in    std_logic_vector(DATA_WIDTH-1 downto 0);
      dout_b  : out   std_logic_vector(DATA_WIDTH-1 downto 0)
   );
end sync_dual_port_ram;

architecture beh_arch of sync_dual_port_ram is
   type ram_type is array (0 to 2**ADDR_WIDTH-1) of
        std_logic_vector(DATA_WIDTH-1 downto 0);
   signal ram : ram_type;
begin
   -- port a
   process(clk)
   begin
      if (clk'event and clk = '1') then
         if (we_a = '1') then
            ram(to_integer(unsigned(addr_a))) <= din_a;
         end if;
         dout_a <= ram(to_integer(unsigned(addr_a)));
      end if;
   end process;
   -- port b
   process(clk)
   begin
      if (clk'event and clk = '1') then
         if (we_b = '1') then
            ram(to_integer(unsigned(addr_b))) <= din_b;
         end if;
         dout_b <= ram(to_integer(unsigned(addr_b)));
      end if;
   end process;
end beh_arch;
```

The synchronous dual-port RAM is the most general configuration. When instantiated, it can be converted to the other configurations by connecting its unused inputs to 0's and connecting its unused outputs to the **open** keyword. The "simple" dual-port RAM, single-port RAM, and synchronous ROM can be considered as trimmed versions of the dual-port RAM.

7.4.3 "Simple" synchronous dual-port RAM

A "simple" dual-port RAM contains two ports in which one port is dedicated for the write operation and the other port is dedicated for the read operation. The HDL template is shown in Listing 7.8.

Listing 7.8   Template for a "simple" synchronous dual-port RAM

```vhdl
library ieee;
use ieee.std_logic_1164.all;
use ieee.numeric_std.all;
entity sync_rw_port_ram is
```

```vhdl
   generic(
      ADDR_WIDTH : integer := 10;
      DATA_WIDTH : integer := 12
   );
   port(
      clk    : in std_logic;
      we     : in std_logic;
      addr_w : in std_logic_vector(ADDR_WIDTH-1 downto 0);
      addr_r : in std_logic_vector(ADDR_WIDTH-1 downto 0);
      din    : in std_logic_vector(DATA_WIDTH-1 downto 0);
      dout   : out std_logic_vector(DATA_WIDTH-1 downto 0)
   );
end sync_rw_port_ram;

architecture beh_arch of sync_rw_port_ram is
   type ram_type is array (0 to 2**ADDR_WIDTH-1)
        of std_logic_vector (DATA_WIDTH-1 downto 0);
   signal ram: ram_type;
begin
   process(clk)
   begin
     if (clk'event and clk = '1') then
        if (we = '1') then
           ram(to_integer(unsigned(addr_w))) <= din;
        end if;
        dout <= ram(to_integer(unsigned(addr_r)));
     end if;
   end process;
end beh_arch;
```

Although basic code structure appears to be similar to that of the register file in Listing 7.2, there is a major difference. The read data statement of the dual-port RAM is within the clocked process but the statement of the register file is outside the process. It implies that the read operation of a simple dual-port RAM is controlled by the clock signal and thus is synchronous. Its data becomes available at the rising edge of the next clock.

7.4.4 Synchronous single-port RAM

The code of a synchronous single-port RAM is similar to that of the dual-port RAM except that one port is omitted. The template is shown in Listing 7.9.

Listing 7.9   Template for a synchronous single-port RAM

```vhdl
library ieee;
use ieee.std_logic_1164.all;
use ieee.numeric_std.all;
entity sync_one_port_ram is
   generic(
      ADDR_WIDTH : integer := 12;
      DATA_WIDTH : integer := 8
   );
   port(
      clk    : in  std_logic;
      we_a   : in  std_logic;
      addr_a : in  std_logic_vector(ADDR_WIDTH-1 downto 0);
      din_a  : in  std_logic_vector(DATA_WIDTH-1 downto 0);
      dout_a : out std_logic_vector(DATA_WIDTH-1 downto 0)
   );
```

```
end sync_one_port_ram;

architecture beh_arch of sync_one_port_ram is
    type ram_type is array (0 to 2**ADDR_WIDTH-1) of
        std_logic_vector(DATA_WIDTH-1 downto 0);
    signal ram : ram_type;
begin
    process(clk)
    begin
        if (clk'event and clk = '1') then
            if (we_a = '1') then
                ram(to_integer(unsigned(addr_a))) <= din_a;
            end if;
            dout_a <= ram(to_integer(unsigned(addr_a)));
        end if;
    end process;
end beh_arch;
```

7.4.5 Synchronous ROM

An external RAM devices cannot keep its content when the power is turned off. In other words, the content of a RAM device is undefined when the power is turned on. However, it is possible to define the initial values of FPGA's internal memory modules. When an SRAM-based FPGA device is "programmed," the configuration file is loaded to the device's configuration memory. The initial values of the internal memory modules can be embedded within the configuration file and written into the modules in the programming process. When the configuration is completed, the memory modules are initialized as well. If the content of a memory module is not updated during the operation, it maintains its original values and behaves like a ROM discussed in Section 7.2.4. This is an efficient way to implement large lookup tables or store read-only data.

The read operation of a BRAM is controlled and synchronized by a clock signal and thus the ROM must include a clock signal as well. After an address change, it takes one clock cycle to output the new data. Because of this behavior, it is called a *synchronous ROM*.

In Vivado Design Suite, there are several mechanisms to specify the initial values. The most portable and descriptive way is to define them as a two-dimensional constant in the behavioral HDL code and assign it to the ram signal as its initial value. The value is loaded into the BRAM when the FPGA device is programmed. The previous ROM code in Listing 7.4 is modified for a synchronous ROM and shown in Listing 7.10. It can be used as a template for the synchronous ROM.

Listing 7.10 Template for a synchronous ROM

```
library ieee;
use ieee.std_logic_1164.all;
use ieee.numeric_std.all;
entity sync_rom_template is
    port(
        clk    : in std_logic;
        addr_r : in std_logic_vector(3 downto 0);
        data   : out std_logic_vector(6 downto 0)
    );
end sync_rom_template;
```

```
architecture arch of sync_rom_template is
   constant ADDR_WIDTH : integer:=4;
   constant DATA_WIDTH : integer:=7;
   type rom_type is array (0 to 2**ADDR_WIDTH-1)
        of std_logic_vector(DATA_WIDTH-1 downto 0);
   -- ROM definition
   constant HEX2LED_LOOK_UP_TABLE: rom_type:=(   -- 2^4-by-7
      "1000000",  -- addr 00
      "1111001",  -- addr 01
      "0100100",  -- addr 02
      "0110000",  -- addr 03
      "0011001",  -- addr 04
      "0010010",  -- addr 05
      "0000010",  -- addr 06
      "1111000",  -- addr 07
      "0000000",  -- addr 08
      "0010000",  -- addr 09
      "0001000",  -- addr 10
      "0000011",  -- addr 11
      "1000110",  -- addr 12
      "0100001",  -- addr 13
      "0000110",  -- addr 14
      "0001110"   -- addr 15
   );
   signal rom: rom_type := HEX2LED_LOOK_UP_TABLE;
begin
   process(clk)
   begin
      if (clk'event and clk = '1') then
         data <= rom(to_integer(unsigned(addr_r)));
      end if;
   end process;
end arch;
```

Note that assigning initial values is not just limited to this synchronous ROM template. This can be done for any inferred BRAM modules.

7.4.6 BRAM-based FIFO buffer

The FIFO buffer discussed in Section 7.3 is adequate for simple applications. If a large buffering space is required, it is more efficient to use FPGA's internal memory modules as the storage. An important design issue is the handling of *simultaneous address access*. Recall that the read pointer and write pointer point to the same address when a FIFO buffer is empty. At the time of the first write operation, the write and read operations access the same address at the same time. The memory module's simultaneous address access behavior and timing play a critical role in the FIFO construction. It must be examined carefully to ensure the correct operation of the FIFO buffer.

A FIFO buffer is a commonly used component and synthesis software, including Vivado, provides a variety of pre-designed FIFO buffer cores. In fact, the BRAM module of the Artix device can be configured as a FIFO buffer without any external logic. These cores can be used if a large and robust FIFO buffer is needed. A BRAM-based dual-clock FIFO buffer is used later in Section 21.4.

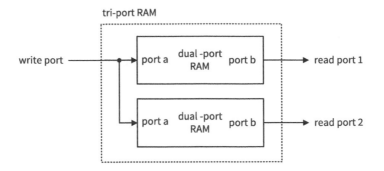

Figure 7.6   Tri-port RAM constructed with two dual-port RAM.

7.4.7   Design considerations

The FPGA's memory modules are very versatile resources and can be used to implement a variety of circuits, such as a faster buffer and a lookup table. However, these modules can only be configured to a limited degree. We may need to manually derive the code for a specific feature. For example, assume that we need to have a synchronous tri-port RAM, which has two read ports and one write port, similar to the register file of Listing 7.3. The HDL description of Listing 7.8 can be expanded to include another port. The internal memory module will not be inferred since it cannot support three access ports. However, closer observation shows that we can duplicate the data in two dual-port memory modules, as shown in Figure 7.6. It can be manually done by instantiating two normal simple dual-port RAMs.

Sometimes the internal RAM cannot support the desired features. For example, BRAM's read operation is synchronous and thus cannot realize the asynchronous operation of the register file in Listing 7.2. To take advantage of BRAMs, we may need to alter the behavior of the register file to match the characteristics of the BRAM. Another example is the synchronous ROM, in which a clock signal is included to match BRAM's synchronous behavior. Of course, the upper-level system that instantiates the register file or ROM needs to be revised to accommodate the change.

Also, since the minimum capacity of a BRAM module is 16K bits, the software may decide to use logic cells to construct a small RAM. After synthesis and implementation, we can check the utilization report to check whether BRAMs are inferred and used.

7.5   OVERVIEW OF THE MEMORY CONTROLLER

The FPGA's internal memory modules are intended for small buffers and lookup tables and their capacity is limited. External memory devices, particularly SDRAM, should be used for massive storage.

A *memory controller* is an interface circuit between the user logic and the physical memory devices. The user logic, such as a processor, issues the desired read or write transactions. The memory controller manages the transaction requests, translates it to proper SDRAM commands, and accesses data from the SDRAM devices. Designing a high-performance memory controller is a complex task. FPGA ven-

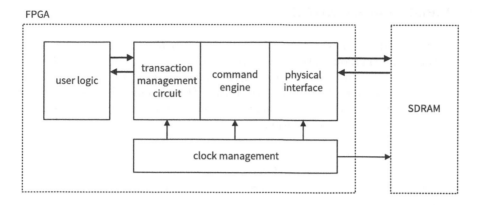

Figure 7.7   Conceptual diagram of an SDRAM controller.

dors usually supply predesigned IP cores for this purpose. Xilinx provides a utility program, Memory Interface Generator (MIG), to generate and configure a memory controller instance that matches the specification of a specific SDRAM module. For the user logic, the memory controller appears as FIFO buffers and it can read data from or write data to the SDRAM device via the FIFO interface.

The conceptual diagram of memory controller is shown in Figure 7.7. It contains a transaction management circuit, a command engine, a physical interface, and a clock management circuit.

An SDRAM's access time is not fixed and its data traffic tends to be bursty. Furthermore, the memory may be shared by several subsystems. The *transaction management circuit* provides buffering, arbitrates the transaction requests, and presents a simpler interface for the user logic.

The *command engine* accepts a transaction request and then generates a sequence of SDRAM commands and control signals to control the address and data buffers and to access the SDRAM module. It is an FSM with various buffering registers and timers.

The *physical interface* translates the "electrical characteristics" between the normal FPGA logic cells and the SDRAM device. In addition, because of the high data rate and off-chip access, the *analog aspect* of the signal, such as noise, cross-talk, and transmission-line reflection, can no longer be ignored. The physical interface helps to maintain the *signal integrity* as well. It is composed of I/O macro cells with special buffering, driving, and termination circuitry.

The *clock management circuit* generates and fine-tunes the frequencies and the phases of the clock signals required by various memory controller circuits.

The implementation of a memory controller depends on the external memory devices, FPGA devices, and even the wire routing on the prototyping boards. Each controller must be individually customized and configured and the resulting instance is tailored to a specific prototyping board. Since the primary focus of this book is on developing portable codes, the derivation of SDRAM controller for the Nexys 4 DDR board is not covered. Its information can be found in bibliographic section.

7.6 BIBLIOGRAPHIC NOTES

Xilinx's user guide UG473, *7 Series FPGAs Memory Resources User Guide*, provides detailed information on the block RAM. Its two product guides, *PG059 Block Memory Generator* and *PG 063 Distributed Memory Generator*, describe the two utility programs. Chapter 3 of the *UG901 Vivado Design Suite User Guide: Synthesis*, titled *HDL Coding Techniques*, includes about two dozen HDL code templates to infer various memory configurations.

A text, titled *Memory Systems: Cache, DRAM, Disk* by B. Jacob et al., has a detailed discussion of SDRAM operation and organization. Xilinx's user guide UG586, *7 Series FPGAs Memory Interface Solutions*, describes the use the MIG utility and product guide PG150, *UltraScale Architecture FPGAs Memory IP*, provides detailed information of the memory controller core.

7.7 SUGGESTED EXPERIMENTS

7.7.1 ROM-based sign-magnitude adder

We can implement any n-input, m-output function with a 2^n-by-m ROM. Consider the sign-magnitude adder discussed in Section 3.7.2 and assume that a and b are 4-bit input signals. Design this circuit as follows:
1. Write a program in a conventional programming language, such as C or Java, to generate a 2^8-by-4 truth table for this circuit.
2. Follow the ROM template in Listing 7.10 to derive the HDL code. Cut and paste the table to the code.
3. Synthesize the circuit and verify its operation.
4. Check the synthesis report and compare the sizes (in terms of the number of logic cells) of the original implementation and the ROM-based implementation.
5. Expand a and b to 8-bit input signals and repeat steps 1 to 4.

7.7.2 ROM-based temperature conversion

Temperature can be measured in Celsius or Fahrenheit scale. Let c and f be a temperature reading in Celsius and Fahrenheit scales. They are related by

$$f = \frac{9}{5} * c + 32$$

The conversion involves multiplication and division operations and direct implementation requires a significant amount of hardware resource. For a simple application, such as a digital thermometer, we can create a lookup table for conversion and store it in a ROM.

Consider a conversion circuit with following specification:
- The range is between 0°C and 100°C (32°F and 212°F).
- The input and output are in 8-bit unsigned format.
- A separate `format` signal indicates whether the input is in Celsius or Fahrenheit scale. The output is to be converted to the other scale.

We can create two lookup tables for the two conversions. Note that because of the small size of these tables, it is possible to store the two tables in a single BRAM module. Design the circuit and verify its operation.

7.7.3   FIFO with data width conversion

In some applications, the widths of the write port and read port of a FIFO buffer may not be the same. For example, a subsystem may write 16-bit data into the FIFO buffer and another subsystem only reads and removes 8-bit data at a time. Assume that the width of the write port is twice the width of the read port. Redesign the FIFO with a modified controller and register file and verify its operation. The DATA_WIDTH generic should be the width of the read port.

7.7.4   Standard FIFO to FWFT FIFO conversion circuit

Some pre-designed FIFO cores only support the standard read configuration. We can transform it to an FWFT FIFO by "wrapping" it with a conversion circuit. Design the conversion circuit and verify its operation. The standard FIFO buffer can be obtained from a pre-designed core or constructed following the block diagram in Figure 7.3.

7.7.5   FIFO buffer with extended status

Two status signals, full and empty, are included in the FIFO buffer of Section 7.3. We want to expand the FIFO buffer with additional signals:
- word_count: indicates the current occupancy (number of data words) of the FIFO buffer
- almost_empty: is asserted when the current FIFO occupancy is below one quarter (25%) of its capacity
- almost_full: is asserted when the current FIFO occupancy is above three quarters (75%) of its capacity

Redesign the FIFO with a modified controller and register file and verify its operation.

7.7.6   Stack

A stack is a last-in-first-out buffer in which the last stored data is retrieved first. Storing a data word to a stack is known as a *push* operation, and retrieving a data word from a stack is known as a *pop* operation. The I/O signals of a stack are similar to those of a FIFO buffer except that we generally use the push and pop signals in place of the wr and rd signals. Design a stack using a register file and verify its operation.

PART II

EMBEDDED SOC I: VANILLA FPRO SYSTEM

CHAPTER 8

OVERVIEW OF EMBEDDED SOC SYSTEMS

The remainder of the book applies the basic hardware design techniques learned in Part I to develop a simple and functional *embedded SoC (System on a Chip)* that contains a video subsystem and a memory-mapped I/O subsystem with general-purpose peripherals, customized hardware accelerators, and a music synthesizer. Our study is still focusing on the hardware design, but within the context of SoC, and it introduces many important design concepts, such as hardware acceleration, bus interface, and software drivers, along the way. In this chapter, we introduce the concept of an embedded SoC, discuss the development flow, explain the simple SoC framework used in this book, and provide an overview of Parts II, III, and IV.

8.1 EMBEDDED SOC

8.1.1 Overview of embedded systems

An *embedded system* (or *embedded computer system*) can be loosely defined as a computer system designed to perform one or a few specific tasks. The computer system is not the end product but a dedicated "embedded" part of a larger system that often includes additional electronic and mechanical parts. By contrast, a *general-purpose computer system*, such as a PC (personal computer), is a general computing platform and itself is the end product. We refer to it as a *desktop-like computer system* in the book. A desktop-like system is designed to be flexible and to support a variety of end-user needs. Application programs are developed based on the available resources of the general-purpose computer system.

Embedded systems are used in a wide range of applications and each application has its own specific requirements. On one hand, a "low-end" system, such as a microwave oven, involves only a simple control function and can be implemented by an 8-bit single-chip microcontroller. On the other hand, a "high-end" system, such as a digital camera, is more complex. It performs two major tasks. The first task involves the general "housekeeping" I/O operations, including processing the button and knob activities, generating a menu on an LCD display, and writing image files to the storage device. These operations are more involved than those of a microwave oven and the system requires a more capable processor. The second task is to process the image and perform data compression to reduce the file size. Because of the large number of pixels and the complexity of the compression algorithm, the task requires a significant amount of computation. An embedded processor is usually not powerful enough to handle the computation-intensive operation. A custom digital circuit, sometimes known as a *hardware accelerator*, can be designed to perform this particular task and take the load off the processor.

8.1.2   FPGA-based SoC

A "high-end" embedded system usually has a processor and simple I/O peripherals to perform general user interface and housekeeping tasks and special hardware accelerators to handle computation-intensive operations. These components can be integrated into a single integrated circuit, commonly referred to as an *SoC* (*system on a chip*).

As the capacity of FPGA devices continues to grow, the same design methodology can be realized in an FPGA chip. Instead of just realizing the system functionalities by *customized software*, we can incorporate *customized hardware* into the embedded system as well. The FPGA technology allows us to tailor the processor, select only the needed I/O peripherals, create a custom I/O interface, and develop specialized hardware accelerators for computation-intensive tasks. The FPGA embedded system provides a new dimension of flexibility because both the hardware and software can be customized to match specific needs. The methodology of exploiting the trade-offs between hardware and software and developing and integrating them concurrently is referred to as *hardware-software co-design*.

8.1.3   IP cores

In SoC development, systems frequently have certain common functionalities and the same building blocks can be reused in different designs. These components are known as *IP (intellectual property) cores*, or simply as *IPs*. They are somewhat like functions in a software library, which can be used in different application programs. The IP cores can be developed by the device manufacturers, third-party vendors, or the users themselves. Unlike software functions, FPGA vendor's IP cores are usually tailored for their own proprietary platforms. They are not portable and frequently delivered as "black boxes" (i.e., without HDL source codes). For example, all companies provide FFT (fast Fourier transform) IP cores. While the cores perform similar functions, their interfaces, timing characteristics, and configuration procedure are different. Therefore, a system must be redesigned or modified if it is re-targeted to a device from a different vendor.

8.2   DEVELOPMENT FLOW OF THE EMBEDDED SOC

The embedded SoC design consists of the following tasks:

- Partition the tasks to software routines and hardware accelerators.
- Design user custom IP cores if needed.
- Develop the hardware.
- Develop the software.
- Implement the hardware and software and perform testing.

These tasks are discussed in the following subsections.

Because of the complexity of modern digital systems, pre-designed IP cores are used extensively in SoC development. Each vendor has its own IP framework, which provides a comprehensive collection of IP cores and supporting software device drivers. The development flow is frequently centered on the IP cores and is shown in Figure 8.1.

8.2.1   Hardware–software partition

Step 1 (labeled 1 in the diagram) is to determine the software–hardware partition. An embedded application usually performs a collection of tasks. In an SoC-based design, a task can be implemented by hardware, software, or both. Based on the performance requirement, complexity, and hardware core availability, we can decide the type of implementation accordingly.

In an ideal scenario, the vendor IP library contains all the needed IP cores for the SoC design. However, in reality, most designs require a certain number of custom IP cores for hardware accelerators and special I/O peripherals. Step 2 is to develop the hardware codes and the corresponding software drivers of these custom IP cores. The details are discussed in Subsection 8.2.5.

8.2.2   Hardware development flow

The left branch represents the hardware design flow. Step 3 is to utilize and integrate the IP cores to construct the system. In Vivado Design Suite, it is done by the IP integrator utility. A user can select IP cores, configure them with the desired characteristics, and connect the cores with a proper interface. IP Integrator will invoke the cores from the library and generate the HDL codes. The top-level HDL file usually resembles the top-level block diagram of an SoC design. During the generation, IP Integrator also produces an auxiliary *hardware platform specification file*, which contains the "definition" of the SoC design, including the processor configuration, memory size and structure, I/O peripheral cores used, memory address mapping, etc.

The top-level HDL file can be treated as a normal HDL file and processed accordingly. Steps 4 and 5 perform synthesis and placement and routing and eventually generate the FPGA configuration file (i.e., the `.bit` file).

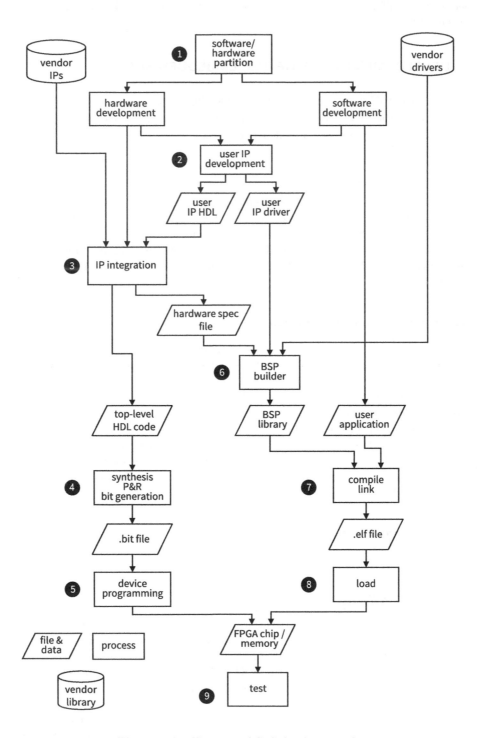

Figure 8.1   IP-centered SoC development flow

8.2.3   Software development flow

The right branch represents the software design flow. A top-level software program usually contains two types of codes. One type is the "system codes," which are pre-designed and provided with the system. They can be functions from libraries, service routines from the operating system, etc. The other type is the "application codes," which are developed by the user to perform the custom tasks. These system functions and service routines are called by the application codes.

In an embedded system, a *BSP (board support package)* is a mechanism to encapsulate the system codes. Since an embedded system is designed to perform a specific task, each system has a different memory structure and contains a unique set of I/O peripherals. BSP is a customized collection of device drivers and initialization routines that support a particular system. The term "board" is used since earlier embedded systems were implemented in a printed circuit board rather than a single silicon device.

An important ingredient of a BSP is the *device drivers*. A device driver is a set of routines that operate or control a particular peripheral device. A driver acts as a "translator" between the hardware peripheral and application programs and enables the application programs to access peripheral functions without needing to know precise details. If an IP core is expected to interface with a processor, a device driver should be developed concurrently.

The embedded SoC flow adopts the BSP mechanism. Step 6 is to build the BSP according to the SoC hardware configuration. Recall that, in Step 3, IP integrator generates a hardware platform specification file. The BSP builder utility examines the IP core information, extracts the pre-build device drivers and initialization routines from the vendor software library, and creates the BSP for the specific SoC design.

The application program can invoke the driver routines in the BSP to access the peripheral I/O cores. Step 7 compiles and links the software routines and BSP library and builds the final software image file (i.e., the .elf file).

8.2.4   Physical implementation and test

Physically implementing the system involves two steps. We first download the FPGA configuration file to the FPGA device (i.e., "program" the device), as in Step 5, and then load the software image into processor's main memory, as in Step 8. The physical system can be tested afterward, as in Step 9.

8.2.5   Custom IP core development

Although FPGA vendors provide a comprehensive collection of pre-designed IP cores, they seldom can cover all the project needs. We usually have to design custom IP cores for special I/O peripherals or less common computation algorithms. The development consists of three tasks:

- Design a custom digital circuit to implement the computation algorithm or special functionality.
- Derive an interface to connect the circuit to the bus or interconnect structure of the vendor's IP framework.

- Develop a device driver to control the new hardware core and integrate it into vendor's software library.

Note that the latter two activities depend on the FPGA vender's IP platform. We need to carefully study the platform's interface protocols and driver structure so that the IP core can be integrated into vendor's framework and used in the IP integration utility. Since each vendor has its own proprietary IP platform, the interface and driver are not portable and must be re-designed for each vendor.

8.3 FPRO SOC PLATFORM

8.3.1 Motivations

While the embedded SoC is powerful methodology, it is not the emphasis of this book. First and foremost, this book focuses on the register-transfer level hardware design rather than the system-level analysis and integration. In addition, a commercial IP platform is not ideal for learning introductory hardware design for several reasons:

- A commercial IP platform is quite complex and thus a significant amount of time will be spent on learning to use the tool rather than doing design.
- Most commercial IP cores are provided as black boxes.
- The interface protocol and driver structure are quite complex.
- The IP framework is proprietary. Thus, learning is tied to a particular platform and the developed IP cores are not portable.

In this book, we define a simple SoC platform and call it *FPro SoC*, (which is abbreviated from the book title "FPGA Prototyping" or can be interpreted as "Fun and Professional"). It contains a video subsystem and a memory-mapped I/O subsystem with general-purpose peripherals, customized hardware accelerators, and a music synthesizer. Our study is still focusing on the hardware design but within the SoC context. The main characteristics of the FPro SoC platform are as follows:

- *Simple.* The FPro SoC platform defines a simple synchronous bus protocol and a straightforward device driver structure. Once a hardware circuit is developed, it can be converted to an IP core by adding a simple interface circuit and a device driver. The core then can be incorporated into the existing embedded system.
- *Functional.* FPro SoC platform provides a variety of I/O peripherals and commonly used serial interfaces (UART, SPI, and I^2C) and includes working device drivers. It resembles a bare-metal 32-bit microprocessor board and can implement real-world projects targeted for this type of boards.
- *Portable.* Except for the processor, FPro SoC's IP cores are developed from scratch in HDL and do not use any vendor's proprietary components. The bus protocol and device drivers are not tied to any specific commercial platform, either. Thus, the IP cores and software codes are portable and can be reused for different FPGA devices and prototyping boards.
- *"Upward compatible."* While the FPro SoC platform is simple, the development follows rigorous and proven design principles and practices. These knowledge and skills can be applied in the future for more complicated commercial platforms and larger projects. In fact, the IP cores and drivers devel-

oped can be easily modified to be incorporated into existing commercial IP frameworks.

- *Fun.* Because the developed system is like a real microprocessor board, it can incorporate existing I/O modules and quickly develop a functional prototyping project. In addition, this platform can provide hardware acceleration capability and thus is more capable and more flexible than any microprocessor board. This give us an opportunity to develop interesting and challenging projects and make studying hardware more "fun" rather than "learning hardware for the sake of hardware."

8.3.2   Platform hardware organization

The top-level diagram of an FPro system is shown in Figure 8.2. It is composed of four major parts:

- Processor module
- FPro bridge and FPro bus
- MMIO (memory mapped I/O) subsystem
- Video subsystem

We only use vendor's IP cores for the processor, memory controller, line buffer, and clock management circuit, which are shown as dotted gray boxes in the figure, and construct all other cores from scratch.

Processor module   The processor module consists of a processor, a memory controller core, and RAM. It is the part that is constructed from the vendor's IP cores. To be used in the FPro SoC platform, the processor core must exhibit the following characteristics:

- 32-bit-wide data path
- 32-bit memory address space
- *Memory-mapped-I/O* scheme for I/O access

Almost all FPGA-based processors support these features. There is no restriction on types of RAM. It can be FPGA's internal memory modules or external memory devices. However, since an FPro system resembles an entry-level 32-bit embedded system, we assume that the size of RAM is limited and software is developed in this context.

FPro bridge and FPro bus   The processor needs to communicate with other cores. This is done by a *bus* or *interconnect structure* specified in the vendor's IP platform. The modern interconnect is designed to accommodate a wide variety of communication and data transfer needs and involves complex protocols. For our learning purposes, we define a simple synchronous bus protocol for the two subsystems and call it *FPro bus*. The *FPro bridge* converts vendor's native bus signals into FPro bus signals. The FPro bus protocol and bridge are discussed in Chapter 10.

MMIO subsystem   In the memory-mapped-I/O scheme, the memory and registers of the I/O peripherals are mapped to the same address space. This means that the processor makes no distinction between the memory and I/O peripherals and uses the same read and write instructions to access the I/O peripherals.

The MMIO subsystem provides a framework to accommodate memory-mapped general-purpose and special I/O peripherals as well as hardware accelerators. For

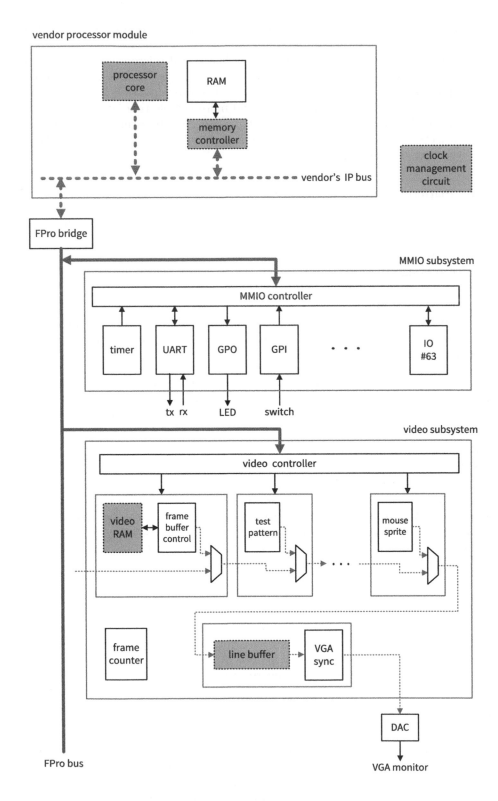

Figure 8.2   Top-level diagram of an FPro system

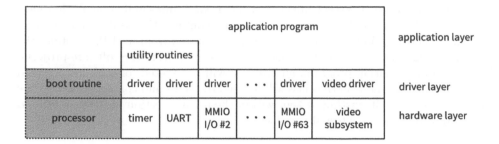

Figure 8.3   Software hierarchy of an FPro SoC system

simplicity, we define a standard *slot interface* that conforms to the FPro bus protocol. The MMIO subsystem consists of a controller to select a specific slot and can accommodate up to 64 instantiated cores. After being "wrapped" with an interface circuit, custom digital logic can be plugged into the FPro platform. About a dozen IP cores are developed and integrated into the MMIO subsystem in the subsequent chapters.

Video subsystem   The video subsystem establishes a framework to coordinate the operation of video cores. A video core generates or processes the video data stream. The cores are arranged as a cascading chain. The data stream is pipelined and "blended" through each stage and eventually displayed on a VGA monitor. The video subsystem demonstrates the principles of handling *stream data*, in which data are generated continuously and passed through a chain of components for processing.

8.3.3   Platform software organization

Since the book focuses on hardware design, we use a simple *bare metal* software scheme for the system. A bare metal system contains no operating system. In its simplest form, the processor boots directly into an infinite main loop, which contains functions to check input, perform computation, and write outputs.

The software hierarchy of an FPro system is shown in Figure 8.3. It contains a *hardware layer*, a *driver layer*, and an *application layer*. A *boot routine* is associated with the processor. It first performs the basic initialization process, such as clearing the caches, configuring the stack and heap segments, and initializing the interrupt, and then transfers control to the main program. The codes are obtained from the vendor, as shown in a dotted gray box in the figure. All other device drivers are constructed from scratch.

To facilitate the software development, we develop several simple utility routines that maintain a system time and assist displaying a debug message on the console. The timer core and UART (universal asynchronous receiver and transmitter) core in slots 0 and 1 are used for this purpose, as shown in Figure 8.2. Thus, the two cores should always be instantiated in the first two slots and not be replaced.

Every I/O core in the FPro system is accompanied by a driver. We select C++ for driver development because of its support of data encapsulation. A C++ class will be created for each core.

Except for accessing system time (via a timer core) and sending debugging messages (via a UART core), a class is largely "self-contained" and does not interact with other classes. When a core is attached or removed from an FPro system, the corresponding driver files should be included or deleted from the software projects. In the main application program, an instance will be created for each instantiated IP core and the methods in the class will be used to access and control the core. The "state" of the core, if existing, is kept within the private section of the instantiated object and involves no external variables.

8.3.4   Modified development flow

The original development flow shown in Figure 8.1 needs to be revised to accommodate the FPro SoC platform. While the basic procedure remains unchanged, we need to manually construct the top-level HDL code and manually include the device driver files in our software application. The modified flow is shown in Figure 8.4 and the new paths are highlighted as thick gray dotted lines in the top half. The main changes are as follows:

- In Step 3, only the processor module, which contains a processor core and RAM, is generated via the IP integration utility. We must manually construct the HDL code for the top-level system, which is composed of the instantiation of the previously generated processor module and the MMIO and video subsystems from Step 2.
- In Step 6, since only the processor module configuration is listed in the hardware specification file, only processor-related codes, such as the boot routine, will be included in the BSP library. We must manually examine the IP cores in the top-level HDL file and include the corresponding driver files in the application software project.
- Since the processor module is the same most of the time, Steps 1 and 6 only need to be executed once. The generated HDL files and BSP library can be use in subsequent designs.

8.4   ADAPTATION ON THE DIGILENT NEXYS 4 DDR BOARD

The book uses the Digilent Nexys 4 DDR prototyping board, which is designed around the Xilinx Artix 7 XC7A100T device, for the experiments and projects. Xilinx provides a soft-core processor, known as *MicroBlaze*, as well as a completely "pre-configured" system, known as *MicroBlaze MCS* (for *MicroBlaze Micro Controller System*). We select MCS as the processor module in Figure 8.2.

MicroBlaze is a 32-bit FPGA-based processor with RISC (reduced instruction set computer) architecture. It is highly configurable and can incorporate an optional floating-point unit, instruction and data caches, a memory management unit, etc. MicroBlaze mainly uses the *AXI (Advanced eXtensible Interface)* protocols from ARM to interface with other IP cores. Hundreds of IPs from Xilinx and third-party vendors, including memory controllers, I/O peripherals, and various types of hardware accelerators, can be integrated with a MicroBlaze to form an SoC design. The flow in Figure 8.1 is targeted for this type of setting.

MicroBlaze MCS is a *complete computer system* that is composed of a pre-configured MicroBlaze processor, a RAM constructed with FPGA's internal mem-

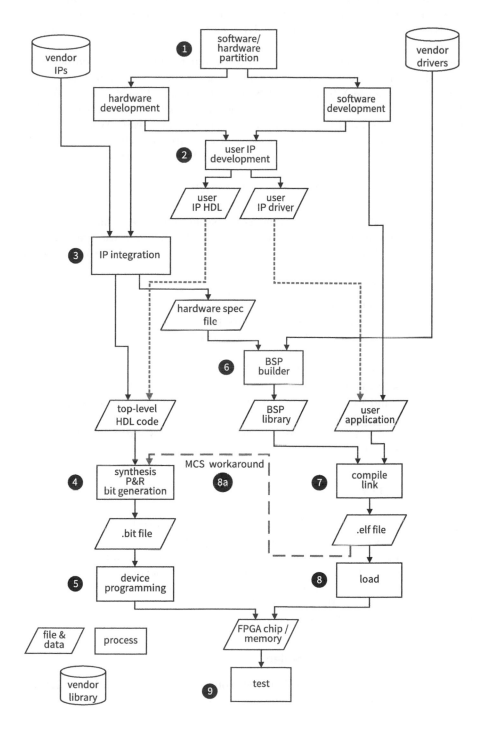

Figure 8.4   FPro SoC development flow

ory, and an I/O module with a standard set of microcontroller peripherals. MicroBlaze MCS provides only a limited degree of configurability. A user can set the size of RAM (between 8 KB and 128 KB) and select a small set of simple I/O peripherals.

Since the focus of the book is hardware design rather than system-level integration, MicroBlaze MCS serves the purpose very well. In addition, many simpler prototyping boards use Xilinx's earlier Spartan devices and must use ISE WebPack for development. At the time of writing, MicroBlaze MCS is free across all Xilinx platforms, including both the ISE WebPack edition and Vivado WebPack edition, but the full-featured MicroBlaze processor is only free for the Vivado WebPack edition. Thus, MicroBlaze MCS can be adopted by more entry-level prototyping boards.

On the down side, support for MicroBlaze MCS is not as comprehensive. The Vivado 2016 edition is used at the time of writing. Step 8 in Figure 8.4 does not function properly. The workaround is to associate the `.elf` file as the "initial values" of FPGA's internal memory and regenerate the configuration file (i.e., `.bit` file). The approach is shown as a thick dashed line in the bottom of Figure 8.4. The revised flow becomes:

- Develop and implement hardware (Steps 1 to 4).
- Develop and implement software (Steps 2 to 7).
- Associate the `.elf` file in the hardware project (Step 8a.).
- Regenerate the configuration `.bit` file with the embedded `.elf` file (i.e., repeating Step 4).
- Program the FPGA device and perform testing (Steps 5 and 9).

This flow is less ideal since regenerating the `.bit` file for each software revision is time-consuming and Vivado software must be invoked during software development.

XC7A100T is a fairly large device and its internal memory modules can accommodate 128 KB RAM for MicroBlaze MCS and 350 KB video RAM for a 9-bit VGA frame buffer. Thus, no external memory device is involved.

8.5   PORTABILITY

A main goal of this book is to develop a portable system to learn hardware design and to introduce SoC practice. Because of the proprietary development software, the IP platform, and IP cores, it is difficult to construct a complete device- and board-independent FPGA-based SoC system. The experiments and projects in this book are constructed and tested on a specific board (Digilent Nexys 4 DDR) that contains a specific FPGA device (Xilinx Artix 7 XC7A100T). The following subsections discuss portability issues.

8.5.1   Processor module and bridge

Since the processor module is constructed from the vendor's proprietary IP cores, it potentially introduces several portability issues:

- Processor
- Memory controller and RAM
- Interface and bridge
- Loading and booting of software

The FPro platform basically requires a 32-bit processor core that supports memory-mapped I/O scheme. Almost all FPGA-based processors satisfy this requirement. The internal and external memory sizes and configurations can vary significantly among different FPGA devices and prototyping boards. However, since the RAM inside the processor module only interacts with the processor core, it does not affect the subsystems directly. In summary, although the proprietary and different memory configurations are used in the processor module, they will not cause serious compatibility issues. The simplest way to create the processor modules is to utilize FPGA's internal memory, as in MicroBlaze MCS. However, older and simpler FPGA devices provide less internal memory. MicroBlaze MCS can be configured with smaller RAM. The size of the RAM, of course, sets the limit on the size of application program.

The FPro bus protocol is designed for simple non-burst synchronous read and write transactions. It can be considered as a very small subset of existing full-featured bus interfaces. Designing a bridge is not very difficult.

While compiling and linking the software code follows a similar tool chain, there is no standard procedure to load an .elf file (Step 8 in the development flow). The process depends on the device, memory configuration, prototyping board, and software development platform. We need to consult the specific manual or user guide to complete this task.

8.5.2   MMIO subsystem

Since the MMIO subsystem's controller and the attached IP cores are constructed from scratch and use no vendor's proprietary IPs or components, the HDL codes are completely portable. They can be implemented as long as a prototyping board has adequate external peripherals. The only exception is the Artix's built-in ADC (analog to digital converter), known as *XADC*, which is only available for newer Xilinx devices.

Since the Nexys 4 DDR board contains all the needed peripherals, all MMIO IP cores can be implemented and tested without any external component. Some peripherals may not be available on other prototyping boards. However, the external circuitries are quite simple and can be easily implemented on a breadboard. The schematics for these peripherals can be found in the Nexys 4 DDR on-line manual and reconstructed accordingly.

8.5.3   Video subsystem

While the majority of the video subsystem is designed from scratch, three components – clock management circuit, line buffer, and frame buffer – utilize vendor's proprietary IP cores. The clock management circuit and line buffer accommodate the VGA synchronization, whose clock rate is different from system clock rate. The former requires a PLL (phase-locked loop) like macro cell and the latter is based on a dual-clock FIFO buffer macro cell. Although these macro cells are proprietary, they are common and can be found in all FPGA devices. The proper macro cells can be instantiated in HDL code directly. Thus, the clock management circuit and line buffer do not lead to serious portability issues.

The frame buffer tends to be the most troublesome and least portable IP core in the FPro framework. The key part of the frame buffer is a dual-port memory that is accessed by the processor and frame control. The latter retrieves data from memory and converts the data into a video stream. The buffer requires a substantial amount of RAM and thus should be implemented by external memory devices. This raises several issues:

- FPGA prototyping boards have different types of memory devices and configurations and some simpler boards may have none.
- Except for simple SRAM devices, a sophisticated proprietary memory controller IP core is needed.
- The frame buffer control must interface with the proprietary memory controller and implement the dual-port access control circuit.
- The same external memory device may be used as processor's RAM and frame buffer at the same time. The partition further complicates the interface and configuration.

Thus, it is difficult to construct a portable frame buffer.

To demonstrate the design principle, the book uses FPGA's internal memory for the video memory. 350 KB of internal RAM is allocated for a frame buffer with a 9-bit VGA resolution. This is doable because the Nexys 4 DDR board contain a large XC7A100T device. It cannot be duplicated in boards with smaller devices. One possible alternative is to reduce the color depth from 9 bits to 1 bit.

Some advanced prototyping boards use HDMI port for the video output. Instead of using a DAC to generate the analog signal, the HDMI interface encodes the output from the line buffer, "serializes" the data, and transmits the video signal digitally through three serial lines. Thus, additional circuits must be added to accommodate the new interface.

8.6 ORGANIZATION

The remaining book consists of three parts. The rest of Part II provides an overview of the hardware architecture and the bare metal embedded software development via the construction of the *vanilla FPro system*, which contains a timer core, a UART core, a GPI (general-purpose input) core, and a GPO (general-purpose output) core. The conceptual diagram is shown in Figure 8.5.

Part III shows how to design an array of MMIO cores for the peripherals on the Nexys 4 DDR prototyping board, including a PWM (pulse width modulation) core, a debouncing core, a seven-segment LED core, a Xilinx XADC controller core, an SPI core, an I^{2}C core, a PS2 core, and a music synthesis module with a DDFS (direct digital frequency synthesis) core and an ADSR (attack-decay-sustain-release) envelope core. Part IV discusses the video subsystem framework and covers construction of relevant IP cores.

8.7 BIBLIOGRAPHIC NOTES

Embedded systems encompass a spectrum of design issues. The two books, *Embedded System Design: A Unified Hardware/Software Introduction* by F. Vahid and T. D. Givargis and *Computers as Components: Principles of Embedded Computing System Design, 2nd edition,* by W. Wolf, provide a comprehensive discussion. Software-hardware co-design is an emerging research area. *A Practical Introduction*

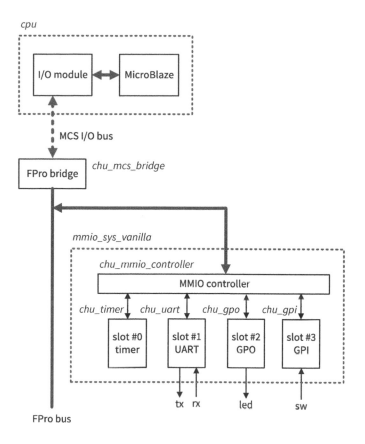

Figure 8.5 Vanilla FPro system.

to Hardware/Software Codesign by P. R. Schaumont addresses the basic concepts and issues of combining hardware and software into a single system.

CHAPTER 9

BARE METAL SYSTEM SOFTWARE DEVELOPMENT

This book uses C/C++ for the software development. We assume that readers have a working knowledge of writing C/C++ code in a desktop-like system. Although the same language is used in both desktop-like and embedded systems, the coding practices are somewhat different. In this chapter, we provide an overview of software development for *bare metal* embedded systems and demonstrate how to write robust and disciplined code to control and access low-level I/O peripherals.

9.1 BARE METAL SYSTEM DEVELOPMENT OVERVIEW

9.1.1 Desktop-like system versus bare metal system

The software development for a *desktop-like* system and a *bare metal* system are very different. A desktop-like system, such as a PC, can be loosely defined as a "fully equipped" computer system. A full-fledged OS (operating system), such as Linux or Windows, runs continuously. The OS serves as a middle layer and shields the hardware details from the application program. The simplified hierarchy is shown in Figure 9.1(a). A bare metal system provides limited resources and requires the application program to interact with hardware (i.e., the "metal") directly. The simplified hierarchy is shown in Figure 9.1(b) and (c).

In C programming context, a desktop-like system constitutes a *hosted environment*. Before starting the `main()` function of a program, the OS (i.e., host) prepares the run-time environment by allocating necessary resources and initializing system

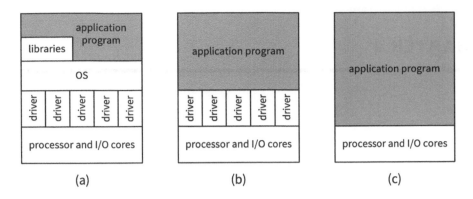

Figure 9.1   Software hierarchy.

services. This means that the application program can assume that all libraries and I/O services are ready to use and no additional work is needed. For example, consider executing a statement like `printf("Hello World!")`. The OS creates a stream channel in advance so that the character stream can be transmitted and displayed on a console. When the execution is completed, the `main()` function "exits" and returns the control to the OS.

In contrast, a *bare metal* system constitutes a *freestanding* environment. Only a minimum set of predefined C libraries, mainly involving data type and constant definitions, is supported. If a service is required, the application program itself needs to set up and manipulate the device. For example, to display the "Hello World!" message on a console via a UART port, the application program should include codes to initialize the device, check the status of the UART data buffer, and write the characters to the buffer sequentially.

In the "barest scenario," as shown in Figure 9.1(c), the application program must create needed I/O services from scratch by directly manipulating the I/O device's registers. A more robust and disciplined approach is to develop a simple driver for each I/O device and access a device via its driver, as shown in Figure 9.1(b). The details of device driver are discussed in Section 9.6.

9.1.2 Basic embedded program architecture

An embedded application consists of a collection of tasks, implemented by hardware accelerators, software routines, or both. Unlike a desktop application, an embedded program may run continuously and does not terminate. The top-level program (i.e., the `main()` function) schedules, coordinates, and manages these tasks. The simplest control architecture is a *super loop*, in which the tasks are executed sequentially in an infinite loop. The pseudo code for a super-loop architecture is

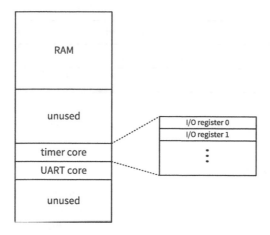

Figure 9.2   Address map of a simple system.

```
main(){
  sys_init();
  while(1){
    task_1();
    task_2();
    ...
    task_n();
  }
}
```

The system runs the sys_init() function once to perform initialization and then enters the super loop and invokes the task functions in turn. This scheme works properly if the overall loop execution time is small and each task can be invoked in a timely manner.

9.2   MEMORY-MAPPED I/O

9.2.1   Overview

An I/O peripheral usually contains a collection of registers for command, status, and data. In the *memory-mapped I/O scheme*, a processor uses the same address space to access memory and the registers of I/O devices. Thus, the load and store instructions used to access memory can also be used to access I/O devices. When an SoC is constructed, chunks of memory space are allocated to the RAM module and I/O cores. The memory address space of a simple computer system containing a timer core and a UART core is shown in Figure 9.2.

The starting address of an assigned chunk is known as the *base address*. The address of a specific memory word or an I/O register can be obtained by adding an *offset* to the base address.

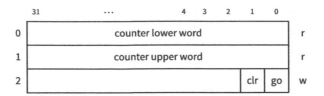

Figure 9.3   I/O register map of a timer core.

9.2.2   Memory alignment

A "32-bit processor" means that the data is transferred and processed in a 32-bit unit. In this book, we define a *word* as a unit of 32 bits (i.e., four bytes). A 32-bit processor normally has a 32-bit address bus and thus can access 2^{32} memory locations.

Although words are used for processing, the address space is usually represented in terms of bytes (i.e., "byte addressable") and the 32-bit address bus implies an addressable space of 2^{32} bytes. To accommodate 32-bit data, four bytes of memory are grouped together to form a word. For easy access, the four bytes are *aligned* at a memory address that is a multiple of four. This implies that the address's two LSBs of the starting byte in a word is always 00 and a word can be accessed by using the 30 MSBs of the address. Thus, memory can be treated as a 32-bit byte addressable space or a 30-bit word addressable space.

9.2.3   I/O register map

An I/O core presents itself to the system as a collection of registers. These registers constitute core's own "addressable memory space." The address value is referred to as a *register offset* or just an *offset*. Unlike homogeneous words in a RAM, the data format and functionality of each register are different. An *I/O register map* depicts properties and fields of these registers .

An I/O register map of the timer core is shown in Figure 9.3. There are three registers. The left number shows the value of the offset (which is 0, 1, and 2) and the right characters indicate the type of access, which can be r, w, or r/w (for read, write, or both read and write). The mnemonics describe the functionality of a register or a field.

After an I/O core is assigned a base address, its registers become part of the memory space. The processor can access a specific register by adding the offset to the base address.

9.2.4   I/O address space of the FPro system

The FPro system consists of two subsystems and many IP cores. To achieve portability, we merge them into a single address space and connect them to the processor's interconnect via a bridge. In other words, from the processor's perspective, the two subsystems, along with all its cores, are treated as a single I/O module. The FPro system's bridge and internal controller do the decoding and multiplexing to access an individual core and its I/O registers.

The combined subsystems require a 24-bit byte address space and appear as an I/O module with 2^{22} 32-bit registers. The large size is mainly due to the frame buffer of the video subsystem. Despite its size, the space only counts for $\frac{1}{2^8}$ (i.e., $\frac{2^{24}}{2^{32}}$) of the processor's total address space, which should not impose a problem. We use the MSB (i.e., bit 23 of the byte address) to distinguish the two subsystems, with 0 for the MMIO subsystem and with 1 for the video subsystem.

The address assignment for the MMIO subsystem is defined as follows:

- The subsystem provides 64 *slots* to connect up to 64 (i.e., 2^6) I/O cores.
- Each I/O core is allocated with 32 (i.e., 2^5) registers.
- Each register is 32 bits wide (i.e., a word).

Thus, the subsystem requires a memory space of 2^{11} (i.e., $2^6 * 2^5$) words or 2^{13} bytes. The 11-bit word address for the MMIO subsystem appears as

$$sss\_sssr\_rrrr$$

in which *ssssss* is the slot number and *rrrrr* is the register offset. When combined with the video subsystem, its 22-bit word address becomes

$$00\_0000\_0000\_0sss\_sssr\_rrrr$$

However, most I/O cores will not use all 32 registers and will not always need 32 data bits. We define it this way to simplify the control circuit and facilitate future expendability. The details are discussed in Section 10.5.

The address assignment for the video subsystems is more involved and its details are discussed in Chapter 21.

9.3   DIRECT I/O REGISTER ACCESS

Accessing an I/O register corresponds to directly read or write a word in a specific memory address. This can be done via C's pointer data type.

9.3.1   Review of C pointer

A C pointer stores a *reference* to an object. The concept of a pointer can be explained by a simple code segment:

```
int x=1, y=5, z=8, *ptr;

ptr = &x;     // ptr gets symbolic address of x
y = *ptr;     // content of y gets content pointed by ptr
*ptr = z;     // content pointed by ptr gets content of z
```

In C, a non-pointer variable can be thought as an abstract memory location identified by the name of the variable and a value is stored to the location in an assignment. A pointer variable is designated with *, as in `int *ptr`, which indicates that `ptr` is a pointer, and it points to a location with the `int` data type. Thus, a reference and a pointer are just an implicit and abstract way to represent a *memory address*.

The operation of the code segment is illustrated in Figure 9.4. A snapshot after the initial declaration and assignment is shown in Figure 9.4(a). We use an arrow

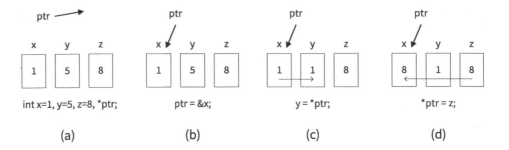

Figure 9.4   Snapshots of pointer operations.

to indicate that `ptr` is a pointer variable. It is pointed to nowhere (i.e., null) since it is unassigned initially. Two unary operators, `&` and `*`, are associated with pointer operations. The `&` operator returns the address of a variable and is known as the *address-of* operator. For example, in statement `ptr = &x`, `&x` returns the address of x, which is then assigned to `ptr`. The result is shown in Figure 9.4(b). The `*` operator returns the content pointed by the pointer and is known as the *dereference* operator. For example, in statement `y = *ptr`, the content pointed by `ptr` is assigned to y and in statement `*ptr = z`, the value of z is stored to the location pointed by `ptr`. The graphical representations are shown in Figure 9.4(c) and (d).

The value of a pointer variable is usually manipulated implicitly, as illustrated by the previous segment. The actual value of `ptr` is system dependent. In a desktop-like programming environment, we usually do not and need not know the explicit value.

9.3.2   C pointer for I/O register

In a bare metal system, an I/O register is assigned with a memory address, which can be thought as a value of a pointer. Unlike the pointer discussed in Section 9.3.1, we know *the explicit value of the address* and must use this value to access the register.

For example, in the vanilla FPro system, a GPO core is instantiated for discrete LEDs of Nexys 4 DDR board and connected to slot 2 of the MMIO subsystem. A GPI core is instantiated for the slide switches and connected to slot 3. Their base addresses are 0xc0000100 and 0xc0000180, respectively, and the offset of the data registers is 0. We can read the value from the switches and write a pattern to LEDs as follows:

```
int sw;
int pattern=0x0055;

sw = *(0xc0000180);
*(0xc0000100) = pattern;
```

The statements are primitive and difficult to comprehend. A more robust method is discussed in the next section.

9.4   ROBUST I/O REGISTER ACCESS

Directly accessing I/O registers is extremely tedious and error prone. It is particularly troublesome for the FPGA-based SoC since each system has its own I/O configuration.

We introduce a robust and disciplined method to pass address assignment information between the hardware and the software driver and to access the I/O registers. It involves three C header files and a VHDL package declaration file:

- `chu_io_map.h`: maintains address mapping information for C/C++ code.
- `chu_io_map.vhd`: maintains address mapping information for VHDL code.
- `inttypes.h`: provides explicitly defined low-level data types.
- `chu_io_rw.h`: provides I/O register read and write macros.

9.4.1   `chu_io_map.h` and `chu_io_map.vhd`

The address assignment of the FPro system is discussed in Section 9.2.4. The assignment is fixed for the slots in the MMIO subsystem and various modules in the video subsystem. Only two parts may change:

- Bridge base address
- Slot assignment in the MMIO subsystem

Instead of using hard literals, we express the information using symbolic constants and record them in `chu_io_map.h` and `chu_io_map.vhd`. The former is a C/C++ header file used for software development and the latter contains a VHDL package declaration to be used in conjunction with hardware development.

Recall that the processor treats the MMIO and video subsystems as a single I/O module and communicates with the module via the bridge. The *bridge base address* is the starting address of the module and is assigned when a system is created. For the MicroBlaze MCS configuration, it is a fixed value of 0xc0000000.

For an IP core attached to the MMIO subsystem, its base address can be calculated using the assigned slot number. Recall that each slot contains 32 words (128 bytes). The base address of slot n is

$$bridge\_base\_address + n * 32 * 4$$

Slot assignment indicates what type of IP core is attached to a particular slot. We use a set of symbolic constants that convey the MMIO configuration. For example, the slot assignment of the vanilla FPro system is

```
#define S0_SYS_TIMER   0    // slot 0
#define S1_UART1       1    // slot 1
#define S2_LED         2    // slot 2
#define S3_SW          3    // slot 3
```

Instead of using a "hard literal," such as 2, the symbolic constant `S2_LED` is more meaningful.

In addition to the address mapping information, `chu_io_map.h` includes a symbolic constant for the system clock frequency, which is needed in certain timing calculations in a driver. The code segment for the vanilla FPro system is shown in Listing 9.1. The actual file also contains additional slot definitions for the cores in Parts III and IV of the book.

Listing 9.1 Slot and constant definitions (in `chu_io_map.h`)

```
#ifndef _CHU_IO_MAP_INCLUDED
#define _CHU_IO_MAP_INCLUDED

#ifdef __cplusplus
extern "C" {
#endif

#define SYS_CLK_FREQ 100

//io base address for microBlaze MCS
#define BRIDGE_BASE 0xc0000000

// slot module definition
#define S0_SYS_TIMER   0
#define S1_UART1       1
#define S2_LED         2
#define S3_SW          3
// ... additional slot definitions for cores in Parts III and IV

#ifdef __cplusplus
} // extern "C"
#endif

#endif   // _CHU_IO_MAP_INCLUDED
```

A VHDL package contains constant definitions, data type definitions, and subprograms to be shared in multiple designs. A custom package is created to facilitate the system development of Parts II, III, and IV. The package is defined in the `chu_io_map.vhd` file and its details are discussed in Section 10.5.1. One portion of the package contains the similar slot definition information and the corresponding code segment is

```
constant S0_SYS_TIMER : integer := 0;
constant S1_UART1     : integer := 1;
constant S2_LED       : integer := 2;
constant S3_SW        : integer := 3;
```

Mismatching slot number between hardware and software can lead to thorny errors. Using the same symbolic names and highlighting them in C/C++ header file and VHDL package declaration can help us maintain consistency and reduce error.

9.4.2 `inttypes.h`

C has many predefined data types, such as `short`, `int`, and `long`. The width (i.e., number of bits) of each data type is left to the compiler and implementation. While interacting with low-level device activities, it is often important to know the exact width and format of registers and data. To facilitate this, C provides a header file, `inttypes.h`, which explicitly specifies the width and format of each data type. These data types are as follows:

- `int8_t`: signed 8-bit integer
- `uint8_t`: unsigned 8-bit integer
- `int16_t`: signed 16-bit integer
- `uint16_t`: unsigned 16-bit integer

- int32_t: signed 32-bit integer
- uint32_t: unsigned 32-bit integer
- int64_t: signed 64-bit integer
- uint64_t: unsigned 64-bit integer

It is good practice to use these data types for low-level coding.

9.4.3 chu_io_rw.h

The I/O access code segment in Section 9.3.2 is tedious and difficult to comprehend. Several modifications can make the operation more robust and less error prone. Let us consider the read statement. First, we can add a type cast,
(volatile uint32_t *), to describe the nature of this value:

```
sw = * (volatile uint32_t *) (0xc0000180);
```

The uint32_t * portion indicates that the constant value is a pointer that points
to an object with the uint32_t data type. The keyword volatile informs the
compiler that the value of the object may be modified without processor interaction and thus certain optimizations should not be performed. Second, we can use
symbolic constants for the base address of the GPI core and the offset of the data
register:

```
#define sw_base    0xc0000180
#define data_reg  0
. . .
sw = * (volatile uint32_t *) (sw_base + 4*data_reg);
```

Third, to maintain modularity and enhance readability, we can define a macro,
io_read(), to encapsulate the type casting and dereference operations:

```
#define io_read(base_addr, offset) \
        (*(volatile uint32_t *)((base_addr) + 4*(offset)))
```

The previous statement becomes

```
#define sw_base    0xc0000180
#define data_reg  0
. . .
sw = io_read(sw_base, data_reg);
```

We also define an auxiliary macro, get_slot_addr(), to calculate the base address
of a given slot

```
#define get_slot_addr(mmio_base, slot) \
        ((uint32_t)((mmio_base) + (slot)*32*4))
```

This allows us to reference an I/O core with the symbolic slot constant defined in
chu_io_map.h:

```
#include "chu_io_map.h"
. . .
#define sw_base    get_slot_addr(BRIDGE_BASE, S3_SW)
#define data_reg  0
. . .
sw = io_read(sw_base, data_reg);
```

A similar macro, io_write(), is created in a similar fashion:

```
#define io_write(base_addr, offset, data) \
    (*(volatile uint32_t *)((base_addr) + 4*(offset)) = (data))
```

The original statement

```
*(0xc0000100) = pattern;
```

can be updated as

```
#include "chu_io_map.h"
. . .
#define led_base  get_slot_addr(BRIDGE_BASE, S2_LED)
#define data_reg  0
. . .
io_write(led_base, data_reg, pattern)
```

These macros are defined in chu_io_rw.h and the code segment is shown in Listing 9.2. The actual file includes an additional address calculation macro for the video cores in Part IV.

Listing 9.2 I/O macros (in chu_io_rw.h)

```
#ifndef _CHU_IO_RW_H_INCLUDED
#define _CHU_IO_RW_H_INCLUDED

#include <inttypes.h>     // to use unitN_t type
#ifdef __cplusplus
extern "C" {
#endif

#define io_read(base_addr, offset) \
    (*(volatile uint32_t *)((base_addr) + 4*(offset)))

#define io_write(base_addr, offset, data) \
    (*(volatile uint32_t *)((base_addr) + 4*(offset)) = (data))

#define get_slot_addr(mmio_base, slot) \
    ((uint32_t)((mmio_base) + (slot)*32*4))

#ifdef __cplusplus
} // extern "C"
#endif

#endif /* _CHU_IO_RW_H_INCLUDED */
```

The macros do not use any system-dependent function and thus are generic. However, they may suffer a subtle problem when a processor contains a data cache. A data cache is a fast buffer between the processor and slow main memory. It stores the recently used data to reduce the need for the processor to access the slow external memory. Since a processor treats an I/O register as regular memory, the relevant data may be temporarily stored in the data cache and can only be read from or written to the I/O register when the corresponding block is deallocated from the cache. To prevent this, the boot routine usually includes a code segment to indicate the memory regions that are not "cacheable."

This is not a issue for MicroBlaze MCS since it has no data cache.

9.5   TECHNIQUES FOR LOW-LEVEL I/O OPERATIONS

Because an embedded program interacts with low-level I/O devices, it frequently needs to manipulate a bit or a field of a data object. We briefly examine some relevant techniques in the following subsections.

9.5.1   Bit manipulation

C has several bitwise operators, including ~ (not), & (and), | (or), and ^ (xor), which operate on one or two operands at bit levels.

The ~ operator inverts all individual bits. For example, if d is 0xb3 (i.e., 1011_0011), ~d becomes 0x8c (i.e., 0100_1100). The statement max = ~0 inverts all bits from 0's to 1's and max becomes the all-one pattern, which corresponds to the largest number in any unsigned data type.

The &, |, and ^ operators can be used to manipulate a bit or a group of bits in a data object. The operation involves a data operand and a mask operand, which specifies the bits to be modified. The operations are shown in the following C segment:

```
uint8_t mask=0x60;  // 0110_0000; mask; bits 6 and 5 asserted
uint8 t d=0xb3;     // 1011_0011; data
uint8_t a0,a1,a2,a3;

a0 = d & mask;      // 0010_0000; isolate bits 6 and 5 from d
a1 = d & ~mask;     // 1001_0011; clear bits 6 and 5 of d to 0
a2 = d | mask;      // 1111_0011; set bits 6 and 5 of d to 1
a3 = d ^ mask;      // 1101_0011; toggle bits 6 and 5 of d
```

In the example, we assume that d is an 8 bit data and bits 6 and 5 represent a special 2-bit field. The mask variable identifies this field by asserting bits 6 and 5. We can isolate this field from d (i.e., clear all other bits to 0) by applying the and operation with the mask, as in d & mask. Conversely, we can clear this field and keep the remaining bits intact by using the inverted mask, as in d & ~mask. Similarly, we can set this field to 11 and keep the remaining bits intact by applying the or operation with the mask, as in d | mask.

The toggle operation is based on the observation that for any 1-bit Boolean variable x, $x \oplus 0 = x$ and $x \oplus 1 = x'$. We can toggle the desired field by applying the xor operation with the mask, as in d ^ mask.

Many manipulations are operated on a single bit of data. An easy way to create a mask for bit n is by shifting 1 to left for n positions, as in

```
1UL << (n)
```

The UL means that 1 is in unsigned long format and is included to prevent overflow.

A set of macros to perform single-bit operation can be defined accordingly:

```
#define bit_set(data, n) ((data) |= (1UL << (n)))
#define bit_clear(data, n) ((data) &= ~(1UL << (n)))
#define bit_toggle(data, n) ((data) ^= (1UL << (n)))
#define bit_read(data, n) (((data) >> (n)) & 0x01)
#define bit_write(data, n, bitvalue) \\
        (bitvalue ? bit_set(data, n) : bit_clear(data, n))
#define bit(n) (1UL << (n))
```

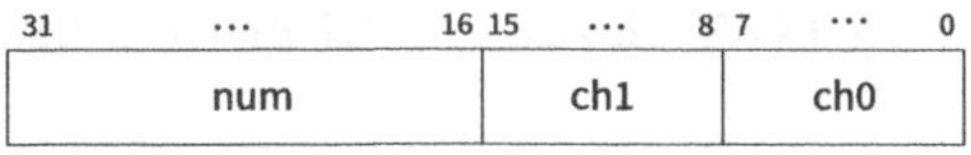

31	...	16	15	...	8	7	...	0
num			ch1			ch0		

Figure 9.5   An I/O register with three fields.

These macros are included in chu_init.h of Section 9.7.2.

9.5.2   Packing and unpacking

To save address space, an I/O register frequently contains multiple fields. These fields are extracted and separated (i.e., *unpacked*) after an application program reads the I/O register. Conversely, these fields need to be *packed* into one object when they are written to the I/O register. The unpacking and packing processes can be done by using the bitwise manipulation and shift operation.

For example, assume that a 32-bit I/O register contains a 16-bit field (for an integer) and two 8-bit fields (for two characters), as shown in Figure 9.5. The code segment to unpack a retrieved I/O word is:

```
uint32_t iodata;
int num;
char ch1, ch0;

iodata = io_read(...);
num = (int)  ((iodata & 0xffff0000) >> 16);
ch1 = (char) ((iodata & 0x0000ff00) >> 8);
ch0 = (char) ((iodata & 0x000000ff));
```

We first apply an **and** mask, such as 0xffff0000, to clear the irrelevant bits, and then shift a proper amount to remove trailing 0's. In this process, the interpretation of a field changes from "a collection of bits" to a specific data type, such as int or char. It is good practice to use type casting to indicate the change of interpretation and data type of the extracted field.

The code segment to pack three fields to an I/O word reverses the previous operation:

```
uint32_t iodata;
int num;
char ch1, ch0;

iodata = (uint32_t)(num);               // num in bits 15:0
iodata = (iodata<<8) | (uint32_t) ch1;  // num in bits 23:8
iodata = (iodata<<8) | (uint32_t) ch0;  // num in bits 31:16
io_write(..., iodata);
```

The first statement puts **num** between bit 15 and bit 0. The second statement first shifts **num** to the left by 8 bits, which makes the 8 LSBs all 0's, and then uses the bitwise **or** operation to fill the 8 LSBs with the value of ch1. The same process is repeated to append the ch0 field. Again, proper type casting should be used in the process.

9.6   DEVICE DRIVERS

9.6.1   Overview

A *device driver* is a layer of routines that operate and control a particular peripheral device. In a bare metal system, a driver acts as an interface between the hardware and an application program and enables the application program to access hardware functions without needing to know precise details. A properly designed driver should be "easy to use correctly and hard to use incorrectly."

We use C++ class for the driver development because of its support for data encapsulation. A class is defined for each IP core and contains the following components:

- *Initialization routine.* This code is associated with the constructor of the class and will be executed when an instance is created.
- *Operations.* The core operations are defined and invoked by the *member functions* (*methods*) of the class.
- *Device state.* The information is maintained in the private section of the class and can only be accessed by the methods defined in the class.

A class can be considered as a "self-contained" stand-alone software unit associated with a specific I/O core.

The application program creates a class instance for each instantiated IP core in the beginning of the code. The process performs initialization on the core and reserves space for its internal state. The subsequent code can just use methods to perform the desired functions. The following subsections demonstrate the construction of drivers for the GPI, GPO, and timer cores and describe the functions of the UART core.

9.6.2   GPO and GPI drivers

It is a good practice to split a C++ class into two files. The class definition goes into the header file (with an extension of .h) and the class implementation goes to the C++ code file (with an extension of .cpp).

We use a class for each core. The class definition of the GPO core is shown in Listing 9.3. We use the naming convention in which the class name starts with a capital letter and has a capital letter for each new word.

Listing 9.3   GpoCore class definition (in gpio_core.h)

```
class GpoCore {
   /* register map */
   enum {
      DATA_REG = 0 //data register
   };
public:
   GpoCore(uint32_t core_base_addr);        //constructor
   ~GpoCore();                              //destructor; not used
   /* methods */
   void write(uint32_t data);               //write a 32-bit word
   void write(int bit_value, int bit_pos);  //write 1 bit
private:
   uint32_t base_addr;
   uint32_t wr_data;                        //same as GPO core data reg
};
```

Despite its simplicity, the definition shows the basic sketch of a device driver used in this book. The I/O register map of the core is listed first with an **enum** declaration. The GPO core only has a data output register, **DATA_REG**, whose offset is 0.

The next part is the public section, which contain the constructor, destructor, and public methods. There are two overloaded write functions. The first method writes a complete data word to the data register and the second method writes a single bit at a specific position to the data register. The destructor is usually not needed for our purposes but is included for completion.

The last part is the private section, which contains two variables. The **base_addr** variable stores the base address of the instantiated core. The **wr_data** variable maintains a copy of the data written to the GPO core and can be considered as the *state* of the core. It becomes useful when only part of the data register is updated.

The class implementation of the GPO core is shown in Listing 9.4.

Listing 9.4   **GpoCore** class implementation (in **gpio_core.cpp**)

```
GpoCore::GpoCore(uint32_t core_base_addr) {
   base_addr = core_base_addr;
   wr_data = 0;
}

GpoCore::~GpoCore() {}

void GpoCore::write(uint32_t data) {
   wr_data = data;
   io_write(base_addr, DATA_REG, wr_data);
}

void GpoCore::write(int bit_value, int bit_pos) {
   bit_write(wr_data, bit_pos, bit_value);
   io_write(base_addr, DATA_REG, wr_data);
}
```

The constructor stores the base address to the private section and clears the write data. The first **write()** method updates the **wr_data** and then writes the word to GPO core's data register. The second **write()** method updates just one bit in core's data register. Note that the GPO core is designed to accept the entire word. Modifying a single bit is achieved with software driver. The **bit_write()** function updates the designated bit in **wr_data** but keeps other bits intact. The entire **wr_data** word is then written to the GPO core's data register.

The class definition and implementation of the GPI core are shown in Listings 9.5 and 9.6. The GPI core only has a data input register, whose offset is 0. They are similar to those of GPO core. However, there is no need to maintain a copy of GPI's data register since it can be retrieved any time.

Listing 9.5   **GpiCore** class definition (in **gpio_core.h**)

```
class GpiCore {
   /* register map */
   enum {
      DATA_REG = 0 //data register
   };
public:
   GpiCore(uint32_t core_base_addr);    //constructor
   ~GpiCore();                          //destructor; not used
   /* methods */
   uint32_t read();                     //read a 32-bit word
```

```
    int read(int bit_pos);              //read 1 bit
private:
    uint32_t base_addr;
};
```

Listing 9.6   GpiCore class implementation (in gpio_core.cpp)

```
GpiCore::GpiCore(uint32_t core_base_addr) {
    base_addr = core_base_addr;
}

GpiCore::~GpiCore() { }

uint32_t GpiCore::read() {
    return (io_read(base_addr, DATA_REG));
}

int GpiCore::read(int bit_pos) {
    uint32_t rd_data = io_read(base_addr, DATA_REG);
    return ((int) bit_read(rd_data, bit_pos));
}
```

In an FPro system, all I/O core's registers are treated as 32-bit words. The actual widths may vary. For example, since the Nexys 4 DDR board contains 16 discrete LEDs and 16 switches, the vanilla FPro system configures the GPO and GPI cores with 16-bit internal registers. In a write operation, the 16 MSBs of the 32-bit word have no effect on the hardware and are ignored. In a read operation, the 16 MSBs of the retrieved 32-bit data contain invalid values. In a more robust implementation, this type of information should be passed from the hardware configuration to the software driver. The methods should check for errors or mask the unused bits. However, this type of integration will introduce a significant amount of overhead and is not implemented in the FPro framework. We need to be aware of the situation occurring and handle it manually in the application program.

9.6.3 Timer driver

The timer IP core interface uses three registers and its I/O register map is shown in Figure 9.3. The core contains a large counter. It runs continuously but can be paused or cleared by the control signals. Registers 0 and 1 correspond to the 32 LSBs and 32 MSBs of the counter. Bits 0 and 1 of register 3 are connected to the enable and clear signals. Writing a 0 to bit 0 will pause the counting and writing a 1 will resume the counting. The bit 1 is not really a memory element. Writing a 1 to bit 1 generates a one-clock tick that clears the counter to 0. The class definition of the timer core is shown in Listing 9.7.

Listing 9.7   TimerCore class definition (in timer_core.h)

```
class TimerCore {
    /* register map */
    enum {
        COUNTER_LOWER_REG = 0,    //lower 32 bits of counter
        COUNTER_UPPER_REG = 1,    //upper 16 bits of counter
        CTRL_REG = 2              //control register
    };
    /* masks */
    enum {
```

```
      GO_FIELD  = 0x00000001,  //bit 0 of ctrl_reg; enable
      CLR_FIELD = 0x00000002   //bit 1 of ctrl_reg; clear
   };
public:
   TimerCore(uint32_t core_base_addr);   //constructor
   ~TimerCore();                         //destructor; not used
   /* methods */
   void pause();                         //pause counter
   void go();                            //resume counter
   void clear();                         //clear the counter to 0
   uint64_t read_tick();                 //retrieve # clocks elapsed
   uint64_t read_time();                 //read time elapsed (in microsecond)
   void sleep(uint64_t us);              //idle for us microseconds
private:
   uint32_t base_addr;
   uint32_t ctrl;                        // current state of ctrl_reg
};
```

The basic structure is similar to that of the GpoCore. The first **enum** definition
uses symbolic names for the three register offsets. Another **enum** definition is added
to specify the masks to extract enable and clear bits. The methods are used to
control the counter, to extract counter's current value, and to implement a "sleep"
function. The private section keeps the base address and contains a variable, **ctrl**,
which maintains a copy of data identical to the content of timer core's control
register.

The class implementation of the timer core is shown in Listing 9.8.

Listing 9.8   TimerCore class implementation (in **timer_core.cpp**)

```
TimerCore::TimerCore(uint32_t core_base_addr) {
   base_addr = core_base_addr;
   ctrl = 0x01;
   io_write(base_addr, CTRL_REG, ctrl);   // enable the timer
}

TimerCore::~TimerCore() { }

void TimerCore::pause() {
   // reset enable bit to 0
   ctrl = ctrl & ~GO_FIELD;
   io_write(base_addr, CTRL_REG, ctrl);
}

void TimerCore::go() {
   // set enable bit to 1
   ctrl = ctrl | GO_FIELD;
   io_write(base_addr, CTRL_REG, ctrl);
}

void TimerCore::clear() {
   uint32_t wdata;

   // write clear_bit to generate a 1-clock pulse
   // clear bit does not affect ctrl
   wdata = ctrl | CLR_FIELD;
   io_write(base_addr, CTRL_REG, wdata);
}

uint64_t TimerCore::read_tick() {
   uint64_t upper, lower;
```

```
    lower = (uint64_t) io_read(base_addr, COUNTER_LOWER_REG);
    upper = (uint64_t) io_read(base_addr, COUNTER_UPPER_REG);
    return ((upper << 32) | lower);
}

uint64_t TimerCore::read_time() {
    // elapsed time in microsecond (SYS_FREQ in MHz)
    return (read_tick() / SYS_CLK_FREQ);
}

void TimerCore::sleep(uint64_t us) {
    uint64_t start_time, now;

    start_time = read_time();
    // busy waiting
    do {
        now = read_time();
    } while ((now - start_time) < us);
}
```

The constructor stores the base address and starts the counter. The `pause()` method resets the enable bit to pause the counting. The `resume()` method sets the enable bit to resume the counting. The `clear()` method writes the clear bit of the control register. Note that the writing is done via creating a temporary `wdata` variable without storing the value to `ctrl`.

The `read_tick()` method retrieves the content of two registers and packs them to obtain the number of clock ticks elapsed since last clear. With the system clock frequency information, the `read_time()` method translates the elapsed clock ticks into elapsed time in microseconds. The `sleep()` method consists of a busy-waiting while loop. It checks the time continuously for us microseconds. The system execution is blocked in this interval and thus it is like forcing the system to sleep.

9.6.4   UART driver

The UART core can establish a serial character communication channel to a host computer via the serial port. Its interface utilizes FIFO buffers instead of I/O registers. Thus, the construction of its driver class, `UartCore`, is somewhat different. The details are discussed in Section 11.4. We just provide an overview of the high-level methods to be used later in this chapter. These methods set the baud rate and display a simple string or number:

- `set_baud_rate(int baud)` sets the baud rate of the UART. When an instance is created, the constructor sets the default baud rate to 9600. The method can be invoked if a different baud rate is desired.
- `disp(const char *str)` transmits a string, which is expected to be displayed on the console.
- `disp(int n, int base, int len)` converts the number n into a string and transmits the string. The `base` parameter can be 2, 8, 10, or 16 and the `len` parameter specifies the number of digits (i.e., length of the string) to be displayed.
- `disp(int n, int base)` is an overloaded version in which the length is determined automatically.

- `disp(int n)` is another overloaded version in which the length is determined automatically and base 10 (i.e., decimal) is used.
- `disp(double f, int digit)` converts a floating-point number f into a string and transmits the string. The `digit` parameter specifies the number of digits in fraction portion.
- `disp(double f)` is an overloaded version in which the `digit` is set to be 3.

The `disp()` method can be used as a primitive version of `printf()`.

9.7 FPRO UTILITY ROUTINES AND DIRECTORY STRUCTURE

A bare metal system runs in a freestanding C environment and cannot access normal C libraries. To facilitate the software development, we develop a set of simple utility routines to access system time and to communicate with a console. The definition and implementation of these routines are in `chu_init.h` and `chu_init.cpp`, respectively.

9.7.1 Minimal hardware requirements

The utility routines imposes the following requirements:
- A timer IP core in slot 0 of MMIO subsystem
- A UART IP core in slot 1 of MMIO subsystem

Thus, when an FPro system is created, the first two slots are reserved and must be connected to timer and UART cores. It is also recommended to use a GPO core for slot 2 and GPI core for slot 3. Recall that this is what constituted a vanilla FPro system in Figure 8.5. Additional peripheral and hardware accelerator IP cores can be added to the remaining slots of the system to meet a specific need.

9.7.2 Utility routines

There are four types of utility routines:
- Console display functions
- Timing functions
- Debugging messaging
- Bit manipulation macros

The definition and implementation files are shown in Listings 9.9 and 9.10

Listing 9.9 FPro utility routine declarations and macros (in `chu_init.h`)

```
// library
#include "chu_io_rw.h"
#include "chu_io_map.h"
#include "timer_core.h"
#include "uart_core.h"

// make uart visible by other code
extern UartCore uart;

// define timer and uart slots
#define TIMER_SLOT  S0_SYS_TIMER   // slot 0
#define UART_SLOT   S1_UART1       // slot 1
```

```
// timing functions
unsigned long now_us();
unsigned long now_ms();
void sleep_us(unsigned long int t);
void sleep_ms(unsigned long int t);

// define debug function
void debug_off();
void debug_on(const char *str, int n1, int n2);

#ifndef _DEBUG
#define debug(str, n1, n2) debug_off()
#endif

#ifdef _DEBUG
#define debug(str, n1, n2) debug_on((str), (n1), (n2))
#endif

// low-level bit-manipulation macros
#define bit_set(data, n) ((data) |= (1UL << (n)))
#define bit_clear(data, n) ((data) &= ~(1UL << (n)))
#define bit_toggle(data, n) ((data) ^= (1UL << (n)))
#define bit_read(data, n) (((data) >> (n)) & 0x01)
#define bit_write(data, n, bitvalue) \\
        (bitvalue ? bit_set(data, n) : bit_clear(data, n))
#define bit(n) (1UL << (n))

#endif
```

Listing 9.10   FPro utility routine implementation (in **chu_init.cpp**)

```
TimerCore _sys_timer(get_slot_addr(BRIDGE_BASE, TIMER_SLOT));
UartCore  uart(get_slot_addr(BRIDGE_BASE, UART_SLOT));

unsigned long now_us() {
   return ((unsigned long) _sys_timer.read_time());
}

unsigned long now_ms() {
   return ((unsigned long) _sys_timer.read_time() / 1000);
}

void sleep_us(unsigned long int t) {
   _sys_timer.sleep(uint64_t(t));
}

void sleep_ms(unsigned long int t) {
   _sys_timer.sleep(uint64_t(1000 * t));
}

void debug_on(const char *str, int n1, int n2) {
   uart.disp("debug: ");
   uart.disp(str);
   uart.disp(n1);
   uart.disp("(0x");
   uart.disp(n1, 16);
   uart.disp(") / ");
   uart.disp(n2);
   uart.disp("(0x");
   uart.disp(n2, 16);
   uart.disp(") \n\r");
```

```
}

void debug_off() {
}
```

Console display functions The console display routines provide primitive function-alities similar to those of `printf()`. An instance, named `uart`, with the `UartCore` class is created in `chu_init.cpp` and is made visible to all external codes. Thus, the instance and the methods of `UartCore` class can be used in external codes to display a string or a number on the console:

- `uart.disp(const char *str)`
- `uart.disp(int n, int base, int len)`
- `uart.disp(int n, int base)`
- `uart.disp(int n)`
- `uart.disp(double f, int digit)`
- `uart.disp(double f)`

Timing functions The timing functions are based on the timer core instantiated in slot 0. An instance with the `TimerCore` class, `_sys_timer`, is created in `chu_init.cpp`. The instance is configured to run continuously from initiation and essentially maintains the *system up time* (simply referred to as *system time*). It is not visible to external files and thus the timer cannot be paused or cleared. Several utility functions are derived from the timer driver:

- `unsigned long now_us()`: returns system time in microseconds.
- `unsigned long now_ms()`: returns system time in milliseconds.
- `void sleep_us(unsigned long int t)`: forces system to idle ("busy waiting") for `t` microseconds.
- `void sleep_ms(unsigned long int t)`: forces system to idle for `t` milliseconds.

Debugging messaging A debugging function, `debug()`, combined with directive `_DEBUG`, provides a simple way to facilitate the software development process. Two versions of `debug()`, which are `debug_off()` and `debug_on()`, are defined in the header file. Based on the status of the `_DEBUG` directive, one is used during preprocessing. The `debug_off()` function contains no code and the `debug_on()` function displays a line in a fixed format, which consists of a string, and two numbers in both decimal and hexadecimal formats. For example, a debugging line is inserted in a loop of the `timer_check()` function in Listing 9.11:

```
for (i = 0; i < 5; i++) {
    . . .
    debug("timer check - (loop #)/now: ", i, now_ms());
}
```

We can turn on the debugging messaging during the initial code development by adding a directive statement in the main program

```
#define _DEBUG
```

When the program execution reaches this line, a debugging message will be displayed on the console:

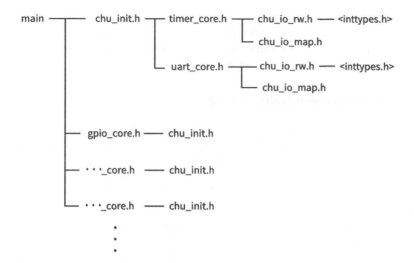

Figure 9.6   Include file hierarchy.

```
debug: timer check - (loop #)/now: 2(0x02) / 386735(0x5e6af)
```

We can turn off the debugging messaging later by simply deleting or commenting out the _DEBUG directive.

Bit manipulation macros   A set of macros are constructed to perform single-bit operations, as discussed in Section 9.5.1. They are repeated here:

- bit_set(data, n): sets bit n of data to 1.
- bit_clear(data, n): clears bit n of data to 0.
- bit_toggle(data, n): toggles bit n of data.
- bit_read(data, n): returns the value of bit n of data.
- bit_write(data, n, bitvalue): updates the bit n of data with a value of bitvalue.
- bit(n): returns a mask with bit n asserted.

9.7.3   Directory structure

The header file hierarchy is shown in Figure 9.6. The chu_init.h file instantiates a timer core and a uart core and thus needs timer_core.h and uart_core.h files. The TimerCore and UartCore classes utilize chu_io_map.h and chu_io_rw for basic I/O operations. The other IP core classes only need to include chu_init.h, which in turn invokes the basic I/O files. The main program should include the header files of the instantiated IP cores and the chu_init.h file as needed.

9.8   TEST PROGRAM

9.8.1   IP core verification routine

After an IP core hardware and driver are developed, we need to develop a software test function to verify their operation. A good test function should exercise all the hardware and driver features and pay particular attention to special conditions and "corner cases." It is a complex and time-consuming task. Since the book's focus is on hardware development, we only include a simple test function to demonstrate the basic functionalities of the core and driver.

9.8.2   Programming with limited memory

One main difference between a desktop computer system and an embedded system is their memory capacities. The former usually has a large physical memory and a sophisticated memory management scheme and a normal application program is provided with a plenty memory space. In contrast, a bare-metal embedded system only has very limited memory and the application program must be carefully crafted to be fitted into the given memory. Since MicroBlaze MCS provides rather limited memory (up to 128 MB), we must be aware of the program size.

After an application program is compiled and linked, its image contains four main segment: *text*, *data/bss*, *stack*, and *heap*. The text segment is the program code and the data/bss contains global (external) and static data. The sizes of these two segments are fixed and known. The stack segment stores the parameters and local variables for function calls and the heap segment provides space for dynamically allocated data. The stack and heap segments grow and shrink dynamically during execution. The program crashes when it runs out of memory.

While the C++ class provides data abstraction for the device driver, creating an instance may require a large block of memory and can lead to consuming too much resource. To alleviate the problem, we adopt the following approaches for the FPro system:

- Create an external (global) instance for each physical core. This allows the object to be allocated in the data/bss segment and be accessed by all routines.
- Avoid excess use of C++ classes and instances and try to use C constructs for the non-driver functionalities.
- Reduce the size and complexity of the device driver.

9.8.3   Test function integration

Incorporating the testing function in a main program involves the following steps:

- Include the core driver header file.
- Develop the test function.
- Create an external instance.
- Call the function in `main()`.

For example, an GPO core is included in the vanilla FPro system for the discrete LEDs. The code for the GPO test function is:

```
#include "gpio_cores.h"
   . . .

// define the test function
void led_check(GpoCore *led_p, int n) {
   int i;

   for (i = 0; i < n; i++) {
      led_p->write(1, i);
```

```
            sleep_ms(200);
            led_p->write(0, i);
            sleep_ms(200);
        }
    }

    // instantiate the core instance as an external variable
    GpoCore led(get_slot_addr(BRIDGE_BASE, S2_LED));
    . . .

    // call the function in main()
    int main() {
        . . .
        led_check(&led, 16);   // invoke the test function
        . . .
    }
```

9.8.4   Test program for the vanilla FPro system

Following the previous guideline, we can develop a test program to verify the operation of the four IP cores in the vanilla FPro system, as shown in Listing 9.11. Note that the UART core and the timer core are used implicitly in the chu_init.h file and thus are not shown in the main program.

Listing 9.11   Vanilla FPro test program (in **main_vanilla_test.cpp**)

```
#define _DEBUG

#include "chu_init.h"
#include "gpio_cores.h"

void timer_check(GpoCore *led_p) {
    int i;

    for (i = 0; i < 5; i++) {
        led_p->write(0xffff);
        sleep_ms(500);
        led_p->write(0x0000);
        sleep_ms(500);
        debug("timer check - (loop #)/now: ", i, now_ms());
    }
}

void led_check(GpoCore *led_p, int n) {
    int i;

    for (i = 0; i < n; i++) {
        led_p->write(1, i);
        sleep_ms(200);
        led_p->write(0, i);
        sleep_ms(200);
    }
}

void sw_check(GpoCore *led_p, GpiCore *sw_p) {
    int i, s;
```

```
    s = sw_p->read();
    for (i = 0; i < 50; i++) {
       led_p->write(s);
       sleep_ms(100);
       led_p->write(0);
       sleep_ms(100);
    }
}

void uart_check() {
   static int loop = 0;

   uart.disp_str("uart test #");
   uart.disp_num(loop);
   uart.disp_str("\n\r");
   loop++;
}

// instantiate switch, led
GpoCore led(get_slot_addr(BRIDGE_BASE, S2_LED));
GpiCore sw(get_slot_addr(BRIDGE_BASE, S3_SW));

int main() {

   while (1) {
      timer_check(&led);
      led_check(&led, 16);
      sw_check(&led, &sw);
      uart_check();
      debug("main - switch value / up time : ", sw.read(), now_ms());
   }
}
```

An instance of `GpiCore` class, `sw`, and an instance of `GpoCore` class, `led`, are created first. The main program follows the super-loop structure discussed in Section 9.1.2. It contains four testing functions, one for each core. The `timer_check()` function blinks all LEDs once per second for five times, `led_check()` turns on and off an individual LED sequentially, `sw_check()` blinks an LED if the corresponding switch input is 1, and `uart_check()` sends a simple message to the serial console.

A `debug()` statement is included in `timer_check()` and the main program and a status message will be transmitted if the _DEBUG directive is not commented out. Note that the instantiated core instances are initialized to their default state via the constructors defined in the classes and thus there is no explicit initialization routine.

9.8.5   Implementation

In Appendix A.5.3, a tutorial is provided to construct the vanilla FPro system in Vivado Design Suite, to develop the software in Xilinx SDK (Software Development Kit) platform, and to download the `.bit` and `.elf` files to the FPGA device on Nexys 4 DDR board for testing. The tutorial is based on the software version at the time of writing. Since Xilinx updates its software in a quarterly basis, the exact procedure may vary in the most current version.

As discussed in Section 8.4, the most troublesome part of using MicroBlaze MCS is downloading the `.elf` file. The MCS's RAM is implemented by the FPGA's internal BRAM modules. From the technical point of view, there should be two ways to download the .elf file:

- The "normal" flow first downloads the .bit file to an FPGA device (i.e., constructing "hardware") and then writes the FPGA's BRAM modules with the .elf file (i.e., loading the software), as shown in Steps 5 and 8 in Figure 8.4.
- The "merged" flow treats the .elf file as the "initial values" of FPGA's BRAM modules, merges them into the .bit file, and then downloads the merged .bit file to an FPGA device, as shown in Steps 8a and 5 in Figure 8.4.

At the time of writing, only the merged flow is functional and Vivado must be invoked to regenerate the bitstream for each software revision.

9.9 BIBLIOGRAPHIC NOTES

Developing a robust and versatile driver is quite involved. *Programming Embedded Systems in C and C ++* by M. Barr introduces the basic concept and illustrates the construction of simple device drivers. *Effective C++* by S. Meyer provides guidelines to produce clear and efficient C++ code. The "easy to use correctly and hard to use incorrectly" principle is one of its guidelines. *Arduino* is a popular micro controller platform and is a representative bare metal system. The bit manipulation macros mirror those in the Arduino library.

9.10 SUGGESTED EXPERIMENTS

9.10.1 Chasing LEDs

The led_check() function of the test program turns on and off an individual LED sequentially. The lit LED appears to move (i.e., chase) along the strip. The chasing LEDs program enhances the function as follows:

1. The 16 discrete LEDs are used as output, one lit at a time.
2. The lit LED moves sequentially in either direction. It changes direction when reaching the rightmost or leftmost position.
3. The slide switch 0 (labeled sw0 on the Nexys 4 DDR board) is used to "initialize" the process. When it is 1, the lit LED is moved to the rightmost position.
4. The next five slide switches (sw1 to sw5) are used to control the chasing speed of the LED. The highest speed should be slow enough for visual inspection.
5. When the chasing speed changes, a one-line message is transmitted to the console via the UART core. The format of the message is "current speed: ddd", where ddd is the value of five speed setting switches.

Develop software and verify its operation.

9.10.2 Collision LEDs

A collision LED fucntion is similar to the chasing LED function in Section 9.10.1 but turns on two LEDs at a time. The two LEDs move independently and change direction when reaching the rightmost or leftmost position or "colliding" in the middle. The detailed specification is as follows:

1. The 16 discrete LEDs are used as output, two lit at a time.
2. The lit LEDs move sequentially in either direction. They change direction when reaching the rightmost or leftmost position or "colliding" in the middle.
3. The slide switch 0 (labeled sw0) is used to "initialize" the first lit LED. When it is 1, the first lit LED is moved to the rightmost position.
4. The slide switch 1 (labeled sw1) is used to "initialize" the second lit LED. When it is 1, the second lit LED is moved to the leftmost position.
5. The two chasing speeds are independent. Slide switches 2 to 5 (sw2 to sw5) are used to control the chasing speed of the first LED and switches 6 to 9 (sw6 to sw9) are used to control the chasing speed of the second LED.

Develop software and verify its operation.

9.10.3   Pulse width modulation

The PWM (pulse width modulation) scheme is described in Section 13.1. Instead of using custom hardware, we can use a software function to generate a periodic pulse with the designated duty cycle. Derive the function for a PWM with eight bits of resolution and use eight slide switches and an LED to verify its operation.

9.10.4   System time display

The system time can be obtained from the timer core. We want to display the time on the console. The format is "up time - mm:ss", where mm and ss are minutes and seconds, respectively. Develop the software and verify its operation.

CHAPTER 10

FPRO BUS PROTOCOL AND MMIO SLOT SPECIFICATION

The FPro system uses a simple synchronous internal bus and defines a fixed slot interface for MMIO I/O cores. The bus and interface provide a flexible mechanism for the processor to communicate with I/O cores and a simple and disciplined way to incorporate custom logic into the system. In this chapter, we explain the bus protocol and slot specification, discuss the design of the bus interface circuit, demonstrate the construction of MMIO I/O cores, and show the derivation of the complete vanilla FPro system.

10.1 FPRO BUS

10.1.1 Overview of the bus

A *bus* is a communication path that connects multiple components. A *system bus* inside a computer is a bus for a processor to exchange data with memory modules and I/O devices. In a simple setting, the processor functions as a *master* and the memory modules and I/O devices are *slaves*. A master initiates the desired transaction, such as a read or write operation, and the designated slave responds accordingly.

Conceptually, a system bus can be thought as a collection of wires shared by all components. The simplified diagram is shown in Figure 10.1. The wires can be classified into three groups:

- Data lines

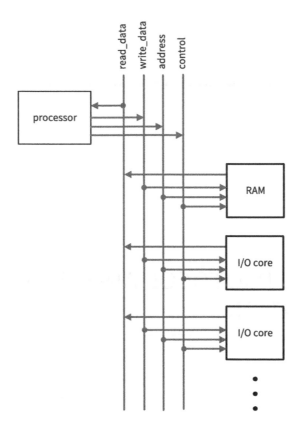

Figure 10.1   Conceptual bus diagram.

- Address lines
- Control lines

The *data lines* carry the data to be transferred. The number of data lines represents the data width. An internal bus usually has separate lines for read data and write data. The *address lines* identify the location of a transaction. With n address lines, up to 2^n distinct locations can be assigned. The upper bits of address lines are used to select a module or device and the lower bits of address lines are used to specify a location within the module or device. The *control lines* consist of command and status signals to control the desired transaction.

10.1.2   SoC interconnect

As the functionalities of a computer system grow, it incorporates more devices and demands faster data transfers. The bus architecture becomes more complex and tends to become the bottleneck. It becomes worse in SoC development since an SoC design contains multiple subsystems interacting and communicating among themselves. Instead of using a single centralized bus, an SoC platform provides a *distributed communication infrastructure*, sometimes referred to as an *interconnect*. The new Xilinx FPGA devices adopt the ARM AMBA AXI protocol for its infrastructure. The AXI protocol specifies a point-to-point interface between a single

AXI master and a single AXI slave via five channels. A master can connect to multiple slaves and vice versa. The AXI protocol does not dictate the implementation and leaves it to the SoC vendors. Xilinx provides a variety of interconnect IP cores in Vivado to implement the desired connection.

While the AXI protocol is capable and flexible, it is too complex for our learning purposes. In addition, although the AXI protocol is an open standard, its implementation relies on vendor's interconnect IP cores and thus can introduce portability problems. We define a much simpler bus protocol for the FPro system.

10.1.3   FPro bus protocol specification

The FPro bus is the communication medium shared by the I/O cores of the MMIO and video subsystems, as shown in Figure 8.2. The FPro bus is a simple synchronous system bus that supports only *read* and *write* operations. The processor is the master and can initiate a read or write transaction via the FPro bridge. The I/O cores in two subsystems are slaves.

In addition to the system clock and reset, the FPro bus contains the following signals:

- `fp_addr` (master to slave). It is a 22-bit address signal used to identify the destination I/O register or a memory location in the MMIO or video subsystems. Note that the memory space of the FPro I/O subsystem is "word addressable," which means that the location specified by `fp_addr` is a 32-bit word.
- `fp_rd_data` (slave to master). It is a 32-bit signal carrying the read data.
- `fp_wr_data` (master to slave). It is a 32-bit signal carrying the write data.
- `fp_rd` (master to slave). It is a 1-bit control signal associated with a read operation.
- `fp_wr` (master to slave). It is a 1-bit control signal to initiate a write operation.
- `fp_mmio_cs` (master to slave). It is a 1-bit enable (i.e., "chip select") signal to activate the MMIO subsystem.
- `fp_video_cs` (master to slave). It is a 1-bit enable signal to activate the video subsystem.

Both FPro bus read and write operations are synchronous and must be completed in one clock cycle. The timing diagram is shown in Figure 10.2. It is assumed that the operation is performed in the MMIO subsystem. The timing of the video system is similar except that `fp_video_cs` is used.

The left portion shows the write cycle. At the rising edge of the clock, marked as *t1*, the bridge completes the translation of the write command from the processor's native bus, places the address and write data on `fp_addr` and `fp_wr_data`, and activates the `fp_mmio_cs` and `fp_wr` signals. The decoding logic decodes the address and activates the enable signal of the designated I/O register. At the rising edge of the next clock, marked as *t4*, the designated register samples and stores the write data.

The right portion shows the read cycle. At the rising edge of the clock, marked as *t5*, the bridge completes the translation of the read command from the processor's native bus, places the address on `fp_addr`, and activates the `fp_mmio_cs` and `fp_rd` signals. The multiplexing logic examines the address and routes the designated I/O register output to `fp_rd_data`. At the rising edge of the next clock, marked

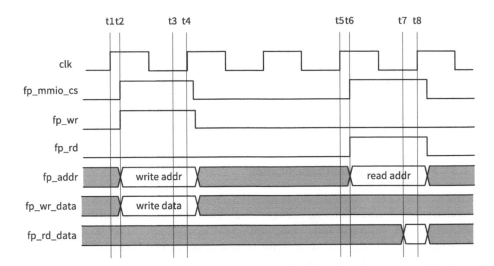

Figure 10.2   Timing diagram of FPro bus.

as *t8*, the bridge samples and stores the data and forwards it to the processor via its native bus. Note that the `fp_rd` signal is not directly involved in retrieving read data. It actually functions as a "removal" signal that gets rid of the old data item in I/O core to make a place for a new one. The use of the removal signal is demonstrated in Section 10.2.4.

The one-cycle operation imposes a strict timing constraint on the cores attached to the bus. However, since these cores are designed from scratch, an adequate buffer can be incorporated in its interface to accommodate the timing requirement. A detailed timing analysis is presented in Section 10.2.5.

10.2   INTERFACE WITH THE BUS

10.2.1   Introduction

A main task in SoC design is to integrate custom logic into the system. A common way is to attach the custom logic to the system's bus and access it as an I/O core. This process involves two parts:

- Add a *wrapping circuit* to the custom logic to form a compatible I/O core that can interface with system's bus.
- Update the *system-level decoding circuit and multiplexing circuit* to identify and access the core.

The *wrapping circuit* makes the core resemble a small memory and thus can be addressed and accessed accordingly. It contains a collection of registers (to mimic memory cells), a *decoding circuit*, and a *multiplexing circuit*. The decoding circuit decodes the address lines to identify and enable the designated destination register. The multiplexing circuit uses the address lines to select the designated source register and routes the data to the bus.

The functionalities of a system-level decoding circuit and a multiplexing circuit are similar to those within the I/O core except that they are used to identify an instantiated I/O core rather than a single register location.

In the memory-mapped I/O scheme, the system assigns a small consecutive memory space to each I/O core. This implies that each core is associated with a specific range of the address space. From this perspective, we can divide the address lines into *module bits* and *offset bits*. The offset bits constitute the lower portion of the address and its width is identical to the width of the I/O core address lines. The module bits constitute the remaining portion of the address. For example, consider a processor with an eight-bit address and an I/O core with four addressable I/O registers (i.e., a 2-bit address). When a system is constructed, we can allocate a four-word memory space, say 100011_00 to 110011_11, to the I/O core (i.e., a four-word block with a base address of 100011_00). The system-level decoding circuit uses the six upper module bits (i.e., 100011) to access this I/O core and the I/O core's internal decoding circuit uses the two lower offset bits to select a specific internal register. Note that in this scheme the offset bits of the base address must be 0's.

We demonstrate the basic design of the wrapping circuit and bus decoding and multiplexing in a simple setting in this section. The discussion assumes that the bus has only eight address lines and an I/O core has four addressable registers. We discuss the actual implementation of the FPro I/O core wrapping circuit and MMIO controller in Sections 10.3 and 10.5.

10.2.2 Write interface and decoding

During a write operation, the master places the data and destination address on bus and activates the write signal. The data is *broadcasted* via the data bus to all memory modules and I/O registers. A sequence of *decoding circuits* takes the address and their outputs assert the enable signals of the designated I/O core and the designated register so that the data will be stored into this particular register.

The block diagram of the write interface circuit is shown in Figure 10.3. Note that the I/O core has a one-bit `cs` (for "chip select") signal to select and enable the core.

Write interface of wrapping circuit   The key parts of the I/O core's wrapping circuit are four registers and a decoding circuit. The function table of the decoding circuit is shown in Table 10.1. It is basically a 2-to-2^2 decoder. The inputs of the decoder are connected to the two LSBs of the address lines of the bus. If the core is selected (i.e., `cs` is 1), the decoded output enables the corresponding register and the write data is sampled and stored into the register at the rising edge of the clock.

System-level decoding circuit   The system-level decoding is tied to the base address of the instantiated core. Assume that a four-word block with a base address of 100011_00 is allocated to this instantiated module. The decoding circuit consists of an "equal-to" comparator that compares the module bits (i.e., 6 MSBs) of the address lines with 100011. When the value matches, it asserts `cs` of the I/O core and activates the core.

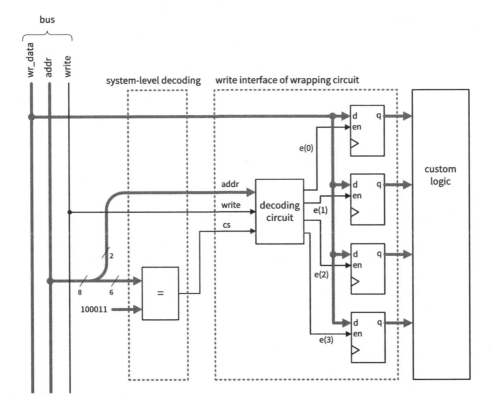

Figure 10.3   Block diagram of a write interface.

Table 10.1   Functional table of a decoding circuit

	input		output
cs	write	addr	e
0	–	–	0000
1	0	–	0000
1	1	00	0001
1	1	01	0010
1	1	10	0100
1	1	11	1000

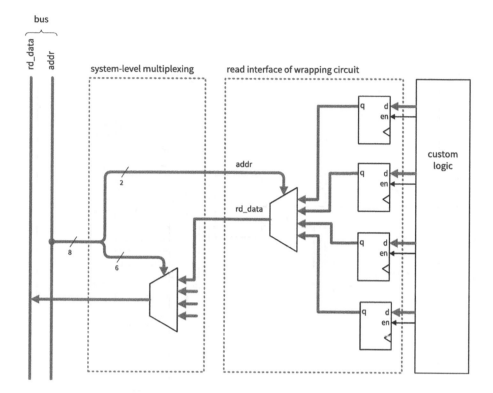

Figure 10.4   Block diagram of a read interface.

10.2.3   Read interface and multiplexing

During a read operation, the master places destination address on the bus and activates the read signal. A sequence of *multiplexing circuits* selects and routes the data from the designated source register and places it on the read data lines of the bus. The master then retrieves the read data. The block diagram of the read interface circuit is shown in Figure 10.4.

Read interface of wrapping circuit   The key parts of the core's wrapping circuit are four registers and a multiplexing circuit. The multiplexing circuit is a standard 2^2-to-1 multiplexer. As in the decoding circuit, the lower two LSBs of the address lines are used to identify the source. They are connected by the selection signal of the multiplexer to route the designated source to the system-level multiplexing circuit. Note that the read signal is not used in this configuration.

System-level multiplexing circuit   The system-level multiplexing circuit constitutes the second level of routing and contains another multiplexer. Its input ports are connected to the readout data from various I/O cores and its output data is connected to the read data lines of bus. The six module bits are used as the selection signal and route the read data from the designated module to the processor.

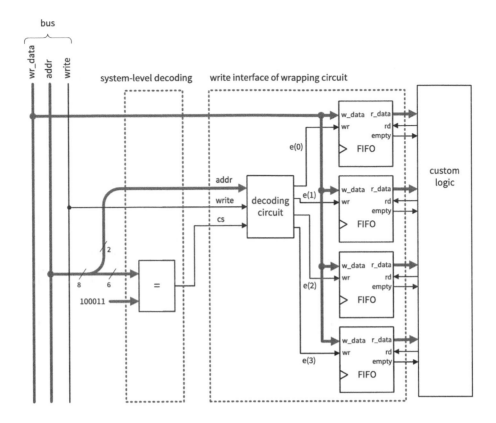

Figure 10.5 Write interface with FIFO buffers.

10.2.4 FIFO buffer as an I/O register

In the memory-mapped I/O scheme, an I/O register of a core is treated as a memory location and the processor reads and writes the register directly. However, there is a difference between a regular memory location and an I/O register. For a memory location, the processor performs both read and write operations and thus always knows the data "validness" of this location. For an I/O register, in contrast, the processor is only responsible for "half" of the access. In a write operation, the processor writes (i.e., produces) the data and an I/O core reads (i.e., consumes) the data. In a read operation, the processor reads (i.e., consumes) the data and an I/O core writes (i.e., produces) the data.

Since a processor usually runs faster than I/O cores, the production rate and consumption rate may not be the same and errors, such as reading the same received data from a slow UART core multiple times, may occur in data access. One way to solve the problem is to replace an I/O register with a more general FIFO buffer. The block diagram of revised write interface is shown in Figure 10.5. The decoded enable signals are connected to the write signals (wr) of the FIFOs. A processor can transmit a "burst" of data into the buffer without worrying about overwriting old data. The I/O core can check the FIFO status with FIFO's empty signal, remove data with FIFO's rd signal, and process the data at its own rate.

The block diagram of revised read interface is shown in Figure 10.6. The design

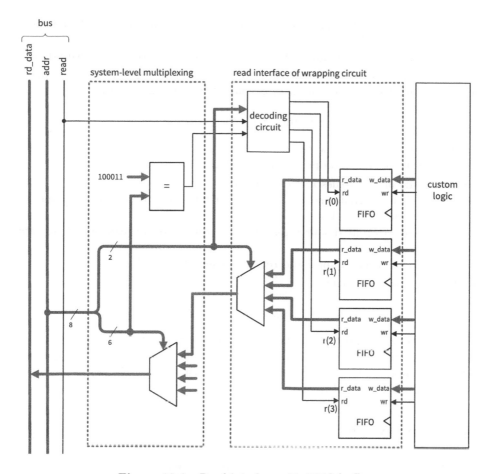

Figure 10.6   Read interface with FIFO buffers.

adds a circuit to remove data from an FIFO after a read operation. The circuit is a decoder identical to that in the write interface except that the `write` signal is replaced by the `read` signal. The decoded output enables the designated FIFO's `rd` signal for one clock and removes the previously retrieved data.

In a more realistic setting, additional circuits can be added to retrieve the FIFO status and to exercise better control. This is demonstrated in the wrapping circuit of the UART core in Section 11.3.2.

10.2.5   Timing consideration

With a better understanding of the interface circuit, the timing of FPro bus's one-clock operation can be examined in more detail. We assume that all the signals of the bus are buffered by output registers. The timing diagram is shown in Figure 10.2. Let us first consider the write operation:

- At $t1$, the master issues a write command.
- At $t2$, all bus signals are stable after t_{CQ} (clock-to-q delay) and the decoding circuit starts.

- At $t3$, the decoded enable signal reaches the designated slave I/O register after the t_{DEC} delay.
- At $t4$, the designated slave I/O register samples and stores the write data.

Let the clock period be t_{CLK} and the setup time of the register be t_{SETUP}. As discussed in Section 4.6, the circuit must satisfy the timing constraint:

$$t_{CQ} + t_{DEC} + t_{SETUP} < t_{CLK}$$

In other words, t_{DEC} must be smaller than $t_{CLK} - (t_{CQ} + t_{SETUP})$.

The analysis for the read operation is similar:

- At $t5$, the master issues a read command.
- At $t6$, all bus signals and I/O register outputs are stable after t_{CQ} and the multiplexing circuit starts.
- At $t7$, the designated source read data is routed to the read data lines after the t_{MUX} delay.
- At $t8$, the master samples and retrieves the data from bus.

The timing constraint becomes

$$t_{CQ} + t_{MUX} + t_{SETUP} < t_{CLK}$$

It implies that t_{MUX} must be smaller than $t_{CLK} - (t_{CQ} + t_{SETUP})$.

Since the decoding and multiplexing operations of the FPro bus are not very complicated, both constraints can be satisfied with a 100M-Hz clock. A slower system clock can be used if necessary.

10.3 MMIO I/O CORE

The MMIO subsystem is composed of an *MMIO controller* and a collection of *MMIO I/O cores*. For simplicity, the MMIO controller defines a *slot interface*. An I/O core complying with the slot interface can be plugged into the controller and accessed by the processor. The block diagram of an MMIO subsystem is shown in Figure 10.7.

10.3.1 MMIO slot interface specification

From the processor's perspective, a *slot* is a 32-word (2^5-word) memory module. The *slot interface* is defined as follows:

- `addr` (bus to core). It is a 5-bit address signal used to identify the 32-bit destination I/O register within the core.
- `rd_data` (core to bus). It is a 32-bit signal carrying the read data.
- `wr_data` (bus to core). It is a 32-bit signal carrying the write data.
- `read` (bus to core). It is a 1-bit control signal activated with the read operation.
- `write` (bus to core). It is a 1-bit control signal to enable the register write.
- `cs` (bus to core). It is a 1-bit enable (i.e., "chip select") signal to select and activate the core.

These characteristics are similar to those of the FPro bus except it contains a smaller 5-bit address.

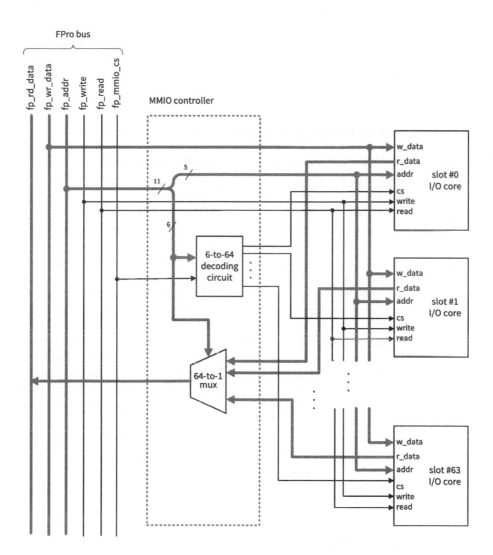

Figure 10.7   Block diagram of an MMIO subsystem.

Note that the widths of the address and data line are fixed. The 32 registers represent the maximum number allowed in a core and some can be left unused. The 32-bit data width is selected to facilitate the processor data access. If a data item is less than 32 bits (e.g., one byte), the unused bits will be ignored. On the other hand, if a data item is more than 32 bits, it must be split into multiple registers and accessed through multiple read or write operations. The software must pack or unpack the data as needed.

An alternative to the fixed-size interface is to let each I/O core define its own address width and data width. This approach will make more efficient use of the memory space. However, because of the variation, a custom system-level decoding circuit and multiplexing circuit are needed for each individual SoC design. For simplicity, we use the fixed interface for the FPro system.

10.3.2   Basic MMIO I/O core construction

Constructing an MMIO I/O cores consists of the following steps:

1. Design the custom digital circuit.
2. Determine the I/O register map for the slot interface.
3. Derive the wrapping circuit.
4. Develop the software driver.

After a custom digital circuit is developed, we need to determine how the processor interacts with the I/O core, such as writing data, reading data, retrieving status, and issuing commands, and to derive a register map accordingly.

A wrapping circuit "wraps" the custom logic to create an interface compatible with the slot specification. The "wrapped logic" then can be inserted into a slot of the MMIO controller.

The basic wrapping circuit design is discussed in Section 10.2. However, the interface of a real I/O core is usually not homogeneous. It contains input and output signals of different widths, access characteristics, and timing constraints. Based on the register map specification, we can add the necessary registers, FIFO buffers, decoding circuits, and multiplexing circuits in the wrapping circuit to match the slot specification.

The entity declaration of an MMIO I/O core should include the slot interface signals and needed external signals. For example, the entity declaration of the UART core is

```
entity chu_uart is
   port(
      clk      : in   std_logic;
      reset    : in   std_logic;
      -- slot interface
      cs       : in   std_logic;
      write    : in   std_logic;
      read     : in   std_logic;
      addr     : in   std_logic_vector(4 downto 0);
      rd_data  : out  std_logic_vector(31 downto 0);
      wr_data  : in   std_logic_vector(31 downto 0);
      -- external signals
      tx       : out  std_logic;   -- transmitting line
      rx       : in   std_logic    -- receiving line
```

```
    );
  end chu_uart;
```

10.3.3 GPO and GPI cores

This book details the development of about a dozen MMIO I/O cores This section illustrates the design of GPO and GPI cores and the next section discussed the derivation of a more sophisticated timer core. The device drivers for GPI and GPO cores are discussed in Section 9.6.2. The derivations of the remaining I/O cores are discussed in the subsequent chapters.

GPO core The GPO core constitutes a general-purpose output port. It is simply a register that maintains the output data. The HDL code is shown in Listing 10.1.

Listing 10.1 GPO core

```
library ieee;
use ieee.std_logic_1164.all;
entity chu_gpo is
   generic(W : integer := 8);    -- width of output port
   port(
      clk     : in   std_logic;
      reset   : in   std_logic;
      -- slot interface
      cs      : in   std_logic;
      write   : in   std_logic;
      read    : in   std_logic;
      addr    : in   std_logic_vector(4 downto 0);
      rd_data : out  std_logic_vector(31 downto 0);
      wr_data : in   std_logic_vector(31 downto 0);
      -- external signal
      dout    : out  std_logic_vector(W-1 downto 0)
   );
end chu_gpo;

architecture arch of chu_gpo is
   signal buf_reg : std_logic_vector(W-1 downto 0);
   signal wr_en : std_logic;
begin
   -- output buffer register
   process(clk, reset)
   begin
      if (reset = '1') then
         buf_reg <= (others => '0');
      elsif (clk'event and clk = '1') then
         if wr_en = '1' then
            buf_reg <= wr_data(W-1 downto 0);
         end if;
      end if;
   end process;
   -- decoding logic
   wr_en <= '1' when write='1' and cs='1' else '0';
   -- slot read interface
   rd_data <= (others => '0');    -- not used
   -- external output
   dout <= buf_reg;
end arch;
```

The core contains an output data register, buf_reg. The decoding is done with the cs='1' and write='1' expression. The write data is stored into the register when the condition is asserted. Since there is only one output register in the core, the addr signal is not used for the decoding. It is possible to assign an address offset, say 0, for the data register and decode this address:

```
wr_en <= '1' when write='1' and cs='1' and addr="00000" else '0';
```

It is not necessary. Although the output signal, rd_data, is not used, it is connected to 0's. The unused signal will be optimized away during synthesis.

GPI core   The GPI core constitutes a general-purpose input port. It contains an input register that samples and stores input data every clock. The HDL code is shown in Listing 10.2.

Listing 10.2   GPI core

```vhdl
llibrary ieee;
use ieee.std_logic_1164.all;
use ieee.numeric_std.all;
entity chu_gpi is
   generic(W : integer := 8);   -- width of input port
   port(
      clk      : in   std_logic;
      reset    : in   std_logic;
      -- slot interface
      cs       : in   std_logic;
      write    : in   std_logic;
      read     : in   std_logic;
      addr     : in   std_logic_vector(4 downto 0);
      rd_data  : out  std_logic_vector(31 downto 0);
      wr_data  : in   std_logic_vector(31 downto 0);
      -- external signal
      din      : in   std_logic_vector(W-1 downto 0)
   );
end chu_gpi;

architecture arch of chu_gpi is
   signal rd_data_reg : std_logic_vector(W-1 downto 0);
begin
   -- input register
   process(clk, reset)
   begin
      if reset = '1' then
         rd_data_reg <= (others => '0');
      elsif (clk'event and clk = '1') then
         rd_data_reg <= din;
      end if;
   end process;
   -- slot read interface
   rd_data(W-1 downto 0) <= rd_data_reg;
   rd_data(31 downto W)  <= (others => '0');
end arch;
```

Since there is only one input source, there is no need for a multiplexing circuit and the addr signal is not used.

10.4   TIMER CORE DEVELOPMENT

A timer core is basically a counter. We can determined the elapsed time based on the counting value. A simple timer core can be constructed following the steps outlined in Section 10.3.2.

10.4.1   Custom logic

The custom logic of the timer core is a counter, similar to those discussed in Section 4.3.2. The following code segment is a 48-bit counter that can be cleared and paused with the clear and go signals:

```
-- register
process(clk, reset)
begin
    if reset = '1' then
        count_reg <= (others => '0');
    elsif (clk'event and clk = '1') then
        count_reg <= count_next;
    end if;
end process;
-- next-state logic
count_next <= (others => '0') when clear = '1' else
              count_reg + 1    when go = '1' else
              count_reg;
```

The 48-bit counter can count up to 2^{48} clocks. With a system clock frequency of 100M Hz, it can count up to about 32 days.

10.4.2   Register map

The processor interacts with the counter as follows:

- retrieve (i.e., read) the 48-bit counter value.
- set or reset (i.e., write) the go signal to resume or pause the counting.
- generate (i.e., write) a pulse to clear the counter to 0.

Based on these interactions, we can define the timer core's register map. There are two read registers. Their address offsets and fields are:

- offset 0 (lower word of the counter)
    - bits 31 to 0: 32 LSBs of the counter
- offset 1 (upper word of the counter)
    - bits 15 to 0: 16 MSBs of the counter

There is one write register. Its address offset and fields are:

- offset 2 (control register)
    - bit 0: the go signal of the counter
    - bit 1: the clear signal of the counter

10.4.3 Wrapping circuit for the slot interface

Based on the register map and the counter's I/O signals, we can derive a wrapping circuit that complies with the slot specification and create the timer core. The HDL code is shown in Listing 10.3.

Listing 10.3 Timer core

```vhdl
library ieee;
use ieee.std_logic_1164.all;
use ieee.numeric_std.all;
entity chu_timer is
   port(
       clk      : in  std_logic;
       reset    : in  std_logic;
       -- slot interface
       cs       : in  std_logic;
       write    : in  std_logic;
       read     : in  std_logic;
       addr     : in  std_logic_vector(4 downto 0);
       rd_data  : out std_logic_vector(31 downto 0);
       wr_data  : in  std_logic_vector(31 downto 0)
   );
end chu_timer;

architecture arch of chu_timer is
   signal count_reg  : unsigned(47 downto 0);
   signal count_next : unsigned(47 downto 0);
   signal ctrl_reg   : std_logic;
   signal wr_en      : std_logic;
   signal clear, go  : std_logic;
begin
   --***************************************************************
   -- counter
   --***************************************************************
   -- register
   process(clk, reset)
   begin
      if reset = '1' then
         count_reg <= (others => '0');
      elsif (clk'event and clk = '1') then
         count_reg <= count_next;
      end if;
   end process;
   -- next-state logic
   count_next <= (others => '0') when clear = '1' else
                 count_reg + 1   when go = '1' else
                 count_reg;

   --***************************************************************
   -- wrapping circuit
   --***************************************************************
   -- ctrl register
   process(clk, reset)
   begin
      if reset = '1' then
         ctrl_reg <= '0';
      elsif (clk'event and clk = '1') then
         if wr_en = '1' then
            ctrl_reg <= wr_data(0);
         end if;
```

```
      end if;
   end process;
   -- decoding logic
   wr_en <=
      '1' when write='1' and cs='1' and addr(1 downto 0)="10" else '0';
   clear <= '1' when wr_en='1' and wr_data(1)='1' else '0';
   go    <= ctrl_reg;
   -- slot read multiplexing
   rd_data <=
      std_logic_vector(count_reg(31 downto 0)) when addr(0)='0' else
      x"0000" & std_logic_vector(count_reg(47 downto 32));
end arch;
```

The wrapping circuit consists of a one-bit control register, a decoding circuit, and a read multiplexing circuit. The control register maintains the value of the go signal. The decoding circuit decodes the two LSBs of addr to generate a write enable signal, wr_en. When it is asserted, bit 0 of wr_data is stored into the control register. Note that bit 1 of wr_data is not stored. It is used in conjunction with wr_en to generate a one-clock clear tick to reset the counter to 0. The multiplexing circuit uses bit 0 of addr (for offsets of 0 or 1) as the selection signal and routes the lower or upper word to rd_data. Since the counter is only 48 bits wide, 16 0's are padded in front of the upper word. The device driver for the timer core is discussed in Section 9.6.3.

10.5   MMIO CONTROLLER

The *MMIO controller* can incorporate up to 64 *slots*. It performs subsystem-level decoding and multiplexing and serves as an interfacing circuit between the FPro bus and the I/O cores. An MMIO I/O core can be plugged into a slot and accessed by the processor.

The MMIO subsystem contains up to 64 (i.e., 2^6) I/O cores, each with up to 32 (i.e., 2^5) I/O registers. From processor's perspective, the entire subsystem can be considered as a stand-alone I/O module with a 2^{11}-bit address space, in which the six MSBs are the module bits used to identify a core and the five LSBs are offset bits used to identify an I/O register within the core.

MMIO controller utilizes the six module bits to select and enable the designated I/O core and to perform subsystem-level decoding and multiplexing. The decoding circuit generates 64 enable signals, each connecting to the chip-select signal (cs) of an I/O core. The multiplexing circuit is a 64-to-1 multiplexer and routes the read data from the designated I/O core to the FPro bus. The block diagram of the controller is shown in Figure 10.7.

10.5.1   chu_io_map.vhd file

A VHDL *package* contains constant definitions, data type definitions, and subprograms to be shared by multiple designs. A package comprises a mandatory *package declaration*, in which the constants, data types, and subprograms are declared, and an optional *package body*, in which the subprogram implementation is defined. The syntax of a *package declaration* is

```
package package_name is
   constant_declaration;
   ...
   data_type_declaration;
   ...
end package_name;
```

After a package is constructed, it can be invoked by other designs with a **use** statement, as in

```
use work.package_name.all;
```

We use a VHDL package, chu_io_map, to facilitate the construction of the MMIO subsystem and the video subsystem of Parts II, III, and IV. The package serves two purposes. First, it maintains a list of symbolic slot names, which help us prevent slot mismatches between hardware and software developments, as discussed in Section 9.4.1. Second, it defines two-dimensional data types to accommodate the output port declarations of the MMIO controller and the video controller. Since no subprogram is defined, no package body is needed. The code of package declaration is shown in Listing 10.4. Only the portion relevant to the vanilla FPro system is included.

Listing 10.4   Data types and constant declarations in chu_io_map package

```
library ieee;
use ieee.std_logic_1164.all;

package chu_io_map is
   -- *************************************************************
   -- 2D data types
   -- *************************************************************
   type slot_2d_data_type is array (63 downto 0) of
        std_logic_vector(31 downto 0);
   type slot_2d_reg_type is array (63 downto 0) of
        std_logic_vector(4 downto 0);
   -- . . . (type definition for the video controller of Part IV);

   -- *************************************************************
   -- Base address for the io_bridge
   -- *************************************************************
   -- for xilinx MCS
   constant BRIDGE_BASE : std_logic_vector(31 downto 0) := X"c7000000";

   -- *************************************************************
   -- slot definition for the "vanilla" and "sampler" MMIO subsystems
   -- avoid changing the first four slots
   -- *************************************************************
   constant S0_SYS_TIMER : integer := 0;
   constant S1_UART1     : integer := 1;
   constant S2_LED       : integer := 2;
   constant S3_SW        : integer := 3;
   constant S4_USER      : integer := 4;
   -- . . .   (symbolic constants for Part III and Part IV)
end chu_io_map;
```

10.5.2 HDL code

The HDL code of MMIO controller is shown in Listing 10.5. Note that the `chu_io_map` package is invoked so that the two-dimensional data types can be used in port declaration .

Listing 10.5   MMIO controller

```
library ieee;
use ieee.std_logic_1164.all;
use ieee.numeric_std.all;
use work.chu_io_map.all;
entity chu_mmio_controller is
    port(
        -- FPro bus
        mmio_cs              : in   std_logic;
        mmio_wr              : in   std_logic;
        mmio_rd              : in   std_logic;
        mmio_addr            : in   std_logic_vector(20 downto 0);
        mmio_wr_data         : in   std_logic_vector(31 downto 0);
        mmio_rd_data         : out  std_logic_vector(31 downto 0);
        -- slot interface
        slot_cs_array        : out  std_logic_vector(63 downto 0);
        slot_mem_rd_array    : out  std_logic_vector(63 downto 0);
        slot_mem_wr_array    : out  std_logic_vector(63 downto 0);
        slot_reg_addr_array  : out  slot_2d_reg_type;
        slot_rd_data_array   : in   slot_2d_data_type;
        slot_wr_data_array   : out  slot_2d_data_type
    );
end chu_mmio_controller;

architecture arch of chu_mmio_controller is
    -- only lower 11 LSB used; 2^6 slots, each  with 2^5 registers
    alias slot_addr : std_logic_vector(5 downto 0) is
        mmio_addr(10 downto 5);
    alias reg_addr  : std_logic_vector(4 downto 0) is
        mmio_addr(4 downto 0);
begin
    -- address decoding
        process(slot_addr, mmio_cs)
        begin
        slot_cs_array <= (others => '0');
        if mmio_cs = '1' then
            slot_cs_array(to_integer(unsigned(slot_addr))) <= '1';
        end if;
    end process;
    -- broadcast to all slots
    slot_mem_rd_array   <= (others => mmio_rd);
    slot_mem_wr_array   <= (others => mmio_wr);
    slot_wr_data_array  <= (others => mmio_wr_data);
    slot_reg_addr_array <= (others => reg_addr);
    -- mux for read data
    mmio_rd_data <= slot_rd_data_array(to_integer(unsigned(slot_addr)));
end arch;
```

The inputs are the signals from the FPro bus and the outputs are 64 slot interfaces connecting to 64 I/O cores. Note that the FPro bus has a 21-bit address line (connected to `mmio_addr`) but only 11 bits are used by the MMIO controller. The 11 bits are divided into slot_addr (i.e., the six module bits) and 5-bit reg_addr (i.e., the five offset bits). The VHDL **alias** construct is used to make the code clear. The statement simply extracts a portion of a signal and gives it a new name.

The key parts of controller are a 6-to-64 decoder and a 64-to-1 multiplexer. They are coded using dynamic indexing, as discussed in Section 7.2.2. The code segment of the multiplexing circuit is

```
mmio_rd_data <= slot_rd_data_array(...slot_addr...);
```

The code segment of the decoding circuit is

```
slot_cs_array <= (others => '0');
if mmio_cs = '1' then
    slot_cs_array(...slot_addr...) <= '1';
end if;
```

All elements of `slot_cs_array` are assigned to '0'. If the MMIO subsystem is enabled (i.e., `mmio_cs = '1'`), the `cs` signal of the designated slot will be overwritten and asserted. Other FPro bus signals, including `mmio_rd`, `mmio_wr`, `mmio_wr_data`, and the five LSBs of `mmio_addr` (i.e., `reg_addr`), are rebroadcasted to all slots.

10.5.3   Vanilla MMIO subsystem

The vanilla MMIO subsystem is composed of MMIO controller and four I/O cores, which are timer core, UART core, GPO core, and GPI core. It is then used to create the vanilla FPro system. The block diagram of the vanilla FPro system is shown in Figure 10.8 and the vanilla MMIO subsystem is in the bottom. The module names in italic font correspond to those in HDL codes.

The HDL code is shown in Listing 10.6. The detailed design and coding of UART core are discussed in Chapter 11.

Listing 10.6   Vanilla MMIO subsystem

```
library ieee;
use ieee.std_logic_1164.all;
use work.chu_io_map.all;
entity mmio_sys_vanilla is
    port(
        -- FPro bus
        clk           : in    std_logic;
        reset         : in    std_logic;
        mmio_cs       : in    std_logic;
        mmio_wr       : in    std_logic;
        mmio_rd       : in    std_logic;
        mmio_addr     : in    std_logic_vector(20 downto 0);
        mmio_wr_data  : in    std_logic_vector(31 downto 0);
        mmio_rd_data  : out   std_logic_vector(31 downto 0);
        -- switches and LEDs
        sw            : in    std_logic_vector(15 downto 0);
        led           : out   std_logic_vector(15 downto 0);
        -- uart
        rx            : in    std_logic;
        tx            : out   std_logic
    );
end mmio_sys_vanilla;

architecture arch of mmio_sys_vanilla is
    signal cs_array        : std_logic_vector(63 downto 0);
    signal reg_addr_array  : slot_2d_reg_type;
```

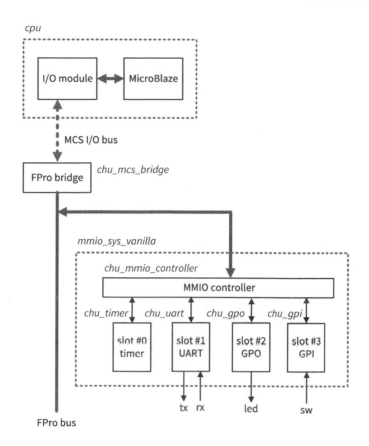

Figure 10.8 Vanilla FPro system.

```
    signal mem_rd_array   : std_logic_vector(63 downto 0);
    signal mem_wr_array   : std_logic_vector(63 downto 0);
    signal rd_data_array  : slot_2d_data_type;
    signal wr_data_array  : slot_2d_data_type;
begin
    --**********************************************************************
    --  MMIO controller instantiation
    --**********************************************************************
    ctrl_unit : entity work.chu_mmio_controller
      port map(
        -- FPro bus interface
        mmio_cs              => mmio_cs,
        mmio_wr              => mmio_wr,
        mmio_rd              => mmio_rd,
        mmio_addr            => mmio_addr,
        mmio_wr_data         => mmio_wr_data,
        mmio_rd_data         => mmio_rd_data,
        -- 64 slot interface
        slot_cs_array        => cs_array,
        slot_reg_addr_array  => reg_addr_array,
        slot_mem_rd_array    => mem_rd_array,
        slot_mem_wr_array    => mem_wr_array,
        slot_rd_data_array   => rd_data_array,
        slot_wr_data_array   => wr_data_array
      );
```

```
--**********************************************************************
-- IO slots instantiations
--**********************************************************************
-- slot 0: system timer
timer_slot0 : entity work.chu_timer
   port map(
      clk            => clk,
      reset          => reset,
      cs             => cs_array(S0_SYS_TIMER),
      read           => mem_rd_array(S0_SYS_TIMER),
      write          => mem_wr_array(S0_SYS_TIMER),
      addr           => reg_addr_array(S0_SYS_TIMER),
      rd_data        => rd_data_array(S0_SYS_TIMER),
      wr_data        => wr_data_array(S0_SYS_TIMER)
   );
-- slot 1: uart1
uart1_slot1 : entity work.chu_uart
   generic map(FIFO_DEPTH_BIT => 6)
   port map(
      clk      => clk,
      reset    => reset,
      cs       => cs_array(S1_UART1),
      read     => mem_rd_array(S1_UART1),
      write    => mem_wr_array(S1_UART1),
      addr     => reg_addr_array(S1_UART1),
      rd_data  => rd_data_array(S1_UART1),
      wr_data  => wr_data_array(S1_UART1),
      -- external signals
      tx       => tx,
      rx       => rx
   );
-- slot 2: GPO for 16 LEDs
gpo_slot2 : entity work.chu_gpo
   generic map(W => 16)
   port map(
      clk      => clk,
      reset    => reset,
      cs       => cs_array(S2_LED),
      read     => mem_rd_array(S2_LED),
      write    => mem_wr_array(S2_LED),
      addr     => reg_addr_array(S2_LED),
      rd_data  => rd_data_array(S2_LED),
      wr_data  => wr_data_array(S2_LED),
      -- external signal
      dout     => led
   );
-- slot 3: input port for 16 slide switches
gpi_slot3 : entity work.chu_gpi
   generic map(W => 16)
   port map(
      clk      => clk,
      reset    => reset,
      cs       => cs_array(S3_SW),
      read     => mem_rd_array(S3_SW),
      write    => mem_wr_array(S3_SW),
      addr     => reg_addr_array(S3_SW),
      rd_data  => rd_data_array(S3_SW),
      wr_data  => wr_data_array(S3_SW),
      -- external signal
      din      => sw
```

```
      );
   -- assign 0's to all unused slot rd_data signals
   gen_unused_slot : for i in 4 to 63 generate
      rd_data_array(i) <= (others => '0');
   end generate gen_unused_slot;
end arch;
```

The code first instantiates the MMIO controller and then instantiates four cores. A core is "inserted into" a controller slot by mapping its interface to a specific element of the controller's 64-slot array. A symbolic constant defined in the `chu_io_map.vhd` file, such `S1_UART1`, is used for the slot number. For the 60 unused slots, a generate statement ties their read ports to 0's. These will be optimized away in the synthesis.

10.6   MCS I/O BUS AND BRIDGE

The Xilinx MicroBlaze MCS processor is used in this book. We provide an overview of the processor and its I/O bus and show the design of the bridge.

10.6.1   Overview of Xilinx MicroBlaze MCS

MicroBlaze is a 32-bit *soft-core processor* provided by Xilinx. The term "soft" here means that the processor is constructed from the logic cells of the FPGA device. MicroBlaze is highly configurable and a variety of optional features, such as instruction cache, data cache, memory management unit, and floating-point unit, can be included. It utilizes the AXI interface to communicate with other IP cores. A large collection of pre-designed IP cores, including memory controllers, general I/O peripherals, and specialized hardware accelerators, can be integrated to form a SoC design.

MicroBlaze MCS (*micro controller system*) is a complete MicroBlaze system that contains a pre-configured MicroBlaze processor, local memory, and a tightly coupled I/O module, which contains a set of standard I/O peripherals and a "bus port." As its name shows, MCS is intended to be used as a 32-bit micro controller.

MicroBlaze MCS can be instantiated with Vivado's the IP catalog utility. For our purposes, it should be configured as follows:

- Set the memory size to 128 KB. This helps to run larger programs.
- Enable the I/O bus port of the I/O module. The bus port will be used to bridge the FPro bus.
- De-select all other I/O peripherals. The I/O modules will be constructed from scratch in the MMIO subsystem. It is a good learning exercise and can maintain the portability.

The detailed steps are shown in the tutorial in Appendix A.4.

10.6.2   MicroBlaze MCS I/O bus

Unlike the full-featured MicroBlaze, MCS does not support the AXI interface. It uses a simple *I/O bus port*, which is port of the I/O module, to communicate with external components. A 30-bit byte address space, from 0xc0000000 to 0xffffffff, is allocated for this purpose.

The MCS I/O bus is a synchronous bus. In addition to the system clock and reset, it contains the following signals:

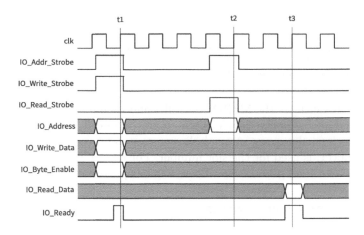

Figure 10.9   Representative timing diagram of the MCS I/O bus.

- IO_Address (master to slave). It is a 32-bit address signal. Note that the memory space of the MCS I/O module is "byte addressable," which means that the location specified by IO_Address is a byte.
- IO_Read_Data (slave to master). It is a 32-bit signal carrying the read data.
- IO_Write_Data (master to slave). It is a 32-bit signal carrying the write data.
- IO_Addr_Strobe (master to slave). It is a 1-bit control signal to indicate whether IO_Address is valid.
- IO_Read_Strobe (master to slave). It is a 1-bit control signal associated with a read operation.
- IO_Write_Strobe (master to slave). It is a 1-bit control signal to initiate a write operation.
- IO_Byte_Enable (master to slave). It is a 4-bit control signal to indicate which bytes of IO_Write_Data are used in the write operation.
- IO_Ready (slave to master). It is a 1-bit status signal from the slave to indicate whether the designated transaction is completed.

The timing diagram of a typical write operation is shown in the left portion of Figure 10.9. The processor places the address and write data on IO_Address and IO_Write_Data lines and asserts the IO_Addr_Strobe and IO_Write_Strobe signals to initiate the operation. The designated I/O peripheral processes the data. After completion, it informs the processor by raising the IO_Ready signal. The IO_Byte_Enable can be used if only part of the word is written.

The timing diagram of a typical read operation is shown in the right portion of Figure 10.9. The processor places the address on the IO_Address lines and asserts the IO_Addr_Strobe and IO_Read_Strobe signals to initiate the operation. The designated I/O peripheral retrieves its internal data. After completion, it places the read data on the IO_Read_Data lines and raises the IO_Ready signal.

The MCS I/O bus utilizes the IO_Ready signal to implement a *handshake protocol*. A slave can hold IO_Ready low when the data is not ready and thus the bus can incorporate slow I/O devices into the system. For example, the I/O device imposes two extra clock cycles in the read operation of Figure 10.9. However, there is a potential issue using handshaking in a prototyping environment. If one I/O core misbehaves and does not assert a ready signal properly, the processor will keep on waiting. This means that one ill-designed I/O core will freeze the entire system. Preventing the deadlock requires additional time-out mechanism and further complicates the bus design.

10.6.3 MCS-to-FPro bridge

The *MCS-to-FPro bridge* "translates" the MCS I/O bus's write and read transactions to the corresponding operations on the FPro bus. Since the FPro bus protocol is simpler than the MCS I/O bus protocol, the bridge design is rather straightforward. It mainly performs *address translation* and *control line conversion*. The HDL code is shown in Listing 10.7.

Listing 10.7 MCS-to-FPro bridge

```vhdl
library ieee;
use ieee.std_logic_1164.all;
use ieee.numeric_std.all;
use work.chu_io_map.all;
entity chu_mcs_bridge is
   generic(BRG_BASE : std_logic_vector(31 downto 0) := x"C0000000");
   port(
      -- uBlaze MCS I/O bus
      io_addr_strobe   : in   std_logic;   -- not used
      io_read_strobe   : in   std_logic;
      io_write_strobe  : in   std_logic;
      io_byte_enable   : in   std_logic_vector(3 downto 0);
      io_address       : in   std_logic_vector(31 downto 0);
      io_write_data    : in   std_logic_vector(31 downto 0);
      io_read_data     : out  std_logic_vector(31 downto 0);
      io_ready         : out  std_logic;
      -- FPro bus
      fp_video_cs      : out  std_logic;
      fp_mmio_cs       : out  std_logic;
      fp_wr            : out  std_logic;
      fp_rd            : out  std_logic;
      fp_addr          : out  std_logic_vector(20 downto 0);
      fp_wr_data       : out  std_logic_vector(31 downto 0);
      fp_rd_data       : in   std_logic_vector(31 downto 0)
   );
end chu_mcs_bridge;

architecture arch of chu_mcs_bridge is
   signal mcs_bridge_en : std_logic;
   signal word_addr     : std_logic_vector(29 downto 0);
begin
   -- address translation and decoding
   -- 2 LSBs are "00" due to word alignment
   word_addr      <= io_address(31 downto 2);
   mcs_bridge_en <=
      '1' when io_address(31 downto 24)=BRG_BASE(31 downto 24) else '0';
   fp_video_cs    <=
      '1' when mcs_bridge_en='1' and io_address(23)='1' else '0';
   fp_mmio_cs     <=
      '1' when mcs_bridge_en='1' and io_address(23)='0' else '0';
   fp_addr        <= word_addr(20 downto 0);
   -- control line conversion
   fp_wr          <= io_write_strobe;
   fp_rd          <= io_read_strobe;
```

```
    io_ready        <= '1';  -- not used; transaction done in 1 clock
    -- data line conversion
    fp_wr_data      <= io_write_data;
    io_read_data    <= fp_rd_data;
end arch;
```

The address translation circuit decodes the MCS I/O bus's 32-bit byte address and converts it to the FPro bus's 21-bit word address and chip-select signals. From MCS processor's perspective, the FPro I/O system is a single I/O module with a 24-bit byte addressable space (i.e., 22-bit word addressable space) at a base address of 0xc0000000. The address translation circuit decodes the 32-bit address as follows:

- bits 31 to 24: used to decode the FPro I/O module base address.
- bit 23: used to distinguish the two subsystems and generate the chip-select signals.
- bits 22 to 2: used to identify an I/O register or a memory location in the MMIO and video subsystems. They form the 21-bit FPro address.
- bits 1 to 0: not used because FPro system uses "word address."

The base address is passed by a generic constant, BRG_BASE, which is 0xc0000000 and is defined in the chu_io_map package. The eight MSBs of the address, which can be considered as the module bits, are used for the system-level decoding. The mcs_bridge_en is asserted if the module bits of I/O bus match those of the base address. It is then used to enable the chip-select signal of the MMIO subsystem or video subsystem (fp_mmio_cs or fp_video_cs).

The control line conversion circuit passes the read and write signals directly from the MCS I/O bus to the FPro bus. It asserts IO_Ready continuously (i.e., connects it to 1). From the master's perspective, after it initiates a read or write operation, IO_Ready is asserted immediately in the next rising edge of the clock. It implies that operation is completed in one clock cycle and involves no handshake mechanism. The IO_Addr_Strobe signal is not checked since we assume that MCS I/O module always asserts it in a read or write operation. The IO_Byte_Enable signal is also not used since we assume that the I/O transaction is always performed on a word basis, as defined in the chu_io_rw.h in Section 9.4.3.

To meet the timing requirement discussed in Section 10.2.5, the output signals of the bridge should be buffered with output registers. However, the documentation of MCS I/O module indicates that the I/O bus signals are already buffered and thus the buffer in this particular bridge is omitted.

10.7 VANILLA FPRO SYSTEM CONSTRUCTION

The vanilla FPro system is a basic but functional configuration. It does not include the video subsystem and its MMIO subsystem contains only the four most essential I/O cores. It is the base for more sophisticated systems. The block diagram is shown in Figure 10.8.

The HDL code can be derived by following the hierarchy of the block diagram. The vanilla MMIO subsystem contains the MMIO controller and four I/O cores and is discussed in Section 10.5.3. The top-level FPro system consists of an MCS processor, the MCS-to-FPro bridge, and the vanilla MMIO subsystem. The HDL code is shown in Listing 10.8.

Listing 10.8   Vanilla FPro system

```vhdl
library ieee;
use ieee.std_logic_1164.all;
use work.chu_io_map.all;
entity mcs_top_vanilla is
   generic(BRIDGE_BASE : std_logic_vector(31 downto 0) := x"C0000000");
   port(
      clk     : in  std_logic;
      reset_n : in  std_logic;
      -- switches and LEDs
      sw      : in  std_logic_vector(15 downto 0);
      led     : out std_logic_vector(15 downto 0);
      -- uart
      rx      : in  std_logic;
      tx      : out std_logic

   );
end mcs_top_vanilla;

architecture arch of mcs_top_vanilla is
   component cpu
      port(
         clock_rtl       : in  std_logic;
         reset_rtl       : in  std_logic;
         io_addr_strobe  : out std_logic;
         io_read_strobe  : out std_logic;
         io_write_strobe : out std_logic;
         io_address      : out std_logic_vector(31 downto 0);
         io_byte_enable  : out std_logic_vector(3 downto 0);
         io_write_data   : out std_logic_vector(31 downto 0);
         io_read_data    : in  std_logic_vector(31 downto 0);
         io_ready        : in  std_logic
      );
   end component;
   signal clk_100M         : std_logic;
   signal reset_sys        : std_logic;
   -- MCS IO bus
   signal io_addr_strobe   : std_logic;
   signal io_read_strobe   : std_logic;
   signal io_write_strobe  : std_logic;
   signal io_byte_enable   : std_logic_vector(3 downto 0);
   signal io_address       : std_logic_vector(31 downto 0);
   signal io_write_data    : std_logic_vector(31 downto 0);
   signal io_read_data     : std_logic_vector(31 downto 0);
   signal io_ready         : std_logic;
   -- fpro bus
   signal fp_mmio_cs       : std_logic;
   signal fp_wr            : std_logic;
   signal fp_rd            : std_logic;
   signal fp_addr          : std_logic_vector(20 downto 0);
   signal fp_wr_data       : std_logic_vector(31 downto 0);
   signal fp_rd_data       : std_logic_vector(31 downto 0);
begin
   -- clock and reset
   clk_100M  <= clk;     -- 100 MHz external clock
   reset_sys <= not reset_n;
   -- instantiate microBlaze MCS
   mcs_0 : cpu
      port map(
         clock_rtl      => clk_100M,
         reset_rtl      => reset_sys,
```

```
        io_addr_strobe   => io_addr_strobe ,
        io_read_strobe   => io_read_strobe ,
        io_write_strobe  => io_write_strobe ,
        io_byte_enable   => io_byte_enable ,
        io_address       => io_address ,
        io_write_data    => io_write_data ,
        io_read_data     => io_read_data ,
        io_ready         => io_ready
     );
  -- instantiate MCS IO bus to FPro bus bridge
  bridge_unit : entity work.chu_mcs_bridge
     generic map(BRG_BASE => BRIDGE_BASE)
     port map(
        io_addr_strobe   => io_addr_strobe ,
        io_read_strobe   => io_read_strobe ,
        io_write_strobe  => io_write_strobe ,
        io_byte_enable   => io_byte_enable ,
        io_address       => io_address ,
        io_write_data    => io_write_data ,
        io_read_data     => io_read_data ,
        io_ready         => io_ready ,
        fp_video_cs      => open,
        fp_mmio_cs       => fp_mmio_cs ,
        fp_wr            => fp_wr ,
        fp_rd            => fp_rd ,
        fp_addr          => fp_addr ,
        fp_wr_data       => fp_wr_data ,
        fp_rd_data       => fp_rd_data
     );
  -- instantiate vanilla MMIO subsystem
  mmio_sys_unit : entity work.mmio_sys_vanilla
     generic map(
        N_LED=>16 ,
        N_SW=>16
     )
     port map(
        clk          => clk_100M ,
        reset        => reset_sys ,
        mmio_cs      => fp_mmio_cs ,
        mmio_wr      => fp_wr ,
        mmio_rd      => fp_rd ,
        mmio_addr    => fp_addr ,
        mmio_wr_data => fp_wr_data ,
        mmio_rd_data => fp_rd_data ,
        sw           => sw ,
        led          => led ,
        rx           => rx ,
        tx           => tx
     );
end arch;
```

The code follows the top-level diagram and instantiates the three components. The tutorial in Appendix A.5 shows the detailed steps to incorporate these HDL files in a project and to synthesize and implement the vanilla FPro system.

10.8 BIBLIOGRAPHIC NOTES

The interconnect structure becomes a key element in a computer system and in an SoC design. *Computer Organization and Architecture* by W. Stallings provides an overview of the bus interface in a traditional computer system. *On-Chip Communication Architectures* by N. Dutt and S. Pasricha gives a comprehensive discussion of the on-chip SoC interconnect structure.

Xilinx's product guides, *MicroBlaze Micro Controller System (PG116)* and *I/O Module (PG111)*, provide detailed information on the MicroBlaze MCS processor and its internal I/O module.

10.9 SUGGESTED EXPERIMENTS

10.9.1 FPro bus with a byte-lane enable signal

The FPro bus write operation always writes a 32-bit word. A more elaborate scheme is to include an additional four-bit *byte-lane enable* signal, `bytelane`, to indicate which byte (or bytes) is to be modified. For example, if `bytelane` is 0001, only the least significant byte of the word will be updated, and if `bytelane` is 1100, the two most significant bytes of the word will be updated. Show the needed modification for the write interface in Figure 10.3.

10.9.2 Seven-segment control with a GPO core

It is possible to use a GPO core to control a four-digit seven-segment LED display. The basic design and verification procedure is as follows:
- Expand the vanilla FPro system to include a 12-bit GPO core in slot 4.
- Use eight bits of the GPO port for the decimal point and seven LED segments.
- Use four bits of the GPO port for four time-multiplexed enable signals.
- Derive and synthesize the new FPro system.
- Develop a software routine to generate the four periodic time-multiplexing signal.
- Derive a testing program and and verify its operation.

Note that a custom seven-segment control core is discussed in Chapter 14.

10.9.3 GPIO core

A GPIO (general-purpose input-output) core supports bidirectional port and can be configured as either an input port or an output port. The core should have tristate buffers and three registers, which are an input data register, an output data register, and a *direction register*. The output of the direction register controls the enable signal of the tristate buffers. The processor can write the direction register to specify the direction of the signal flow (i.e., input or output) of individual bits. The basic design and verification procedure is as follows:
- Determine the register map and derive the wrapping circuit.
- Derive the HDL code.
- Derive the device driver.
- Expand the vanilla MMIO subsystem to include a 32-bit GPIO core in slot 4.
- Modify the vanilla FPro system to connect both `led` and `sw` signals to the GPIO core and synthesize the new system.
- Derive a testing program and and verify its operation.

10.9.4   Blinking-LED core

A blinking-LED core can turn on and off LEDs at specific rates. The core has a four-bit output signal connected to four discrete LEDs. It has four 16-bit registers that specify the values of the individual blinking intervals in millisecond. With the blinking-LED core, the processor only needs to write the registers. The basic design and verification procedure is as follows:

- Design the blinking circuit for one LED and duplicate it four times.
- Determine the register map and derive the wrapping circuit.
- Derive the HDL code.
- Derive the device driver.
- Expand the vanilla MMIO subsystem to include a blinking-LED core in slot 4.
- Modify the vanilla FPro system to connect the `led` signal to the blinking-LED core and synthesize the new system.
- Derive a testing program and verify its operation.

10.9.5   Timer core with a programmable period

One useful feature of a timer is to generate a periodic tick as an output signal. The tick can drive other circuits or function as a periodic interrupt source or a watchdog timer. This can be achieved by revising the counter of the timer core in Section 10.4. The original counter runs continuously to the maximum value (i.e., $2^{48} - 1$) and wraps around, which implies that the period of this counter is $2^{48} - 1$ clock cycles. An additional circuit can be added to compare the designated maximum value and thus the counter will wrap around after a "programmable" interval. A one-clock tick signal will be generated each time the counter reaches the maximum value. The basic design and verification procedure is as follows:

- Design the new counter.
- Determine the register map and derive the wrapping circuit.
- Derive the HDL code.
- Derive the device driver.
- Expand the vanilla MMIO subsystem to include a new timer core in slot 4.
- Connect the tick signal to an FPGA pin in the vanilla FPro system and synthesize the new system.
- Derive a testing program.
- Use an oscilloscope or a logic analyzer to check the tick signal and verify its operation.

10.9.6   Timer core with a *run-once* mode

Another useful feature of a timer is to function as a *one-shot circuit*, which is used to generate a single output pulse of a specified width after a trigger. This can be achieved by incorporating a *run-once* mode into the timer core in Experiment 10.9.5. The previous counter operates in a *continuous* mode, in which it runs to maximum value and then wraps around and repeats. The modified counter can run once to the maximum after a start signal is asserted and then pause. An additional control signal specifies the operation mode. The basic design and verification procedure is as follows:

- Design the new counter with continuous and run-once modes.

- Determine the register map and derive the wrapping circuit.
- Derive the HDL code.
- Derive the device driver.
- Expand the vanilla MMIO subsystem to include a new timer core in slot 4.
- Connect the pulse signal to an FPGA pin in the vanilla FPro system and synthesize the new system.
- Derive a testing program.
- Use an oscilloscope or a logic analyzer to check the output pulse and verify its operation.

CHAPTER 11

UART CORE

Serial communication uses a single data line to exchange information between two systems. The transmitting system converts the parallel data to a serial stream and the receiving system reassembles the serial data back to its original parallel format. A *UART* (*universal asynchronous receiver and transmitter*) is the most commonly used scheme. In this chapter, we describe the construction of a UART controller, the development of its slot wrapping circuit, the derivation of the software driver, and its potential application.

11.1 INTRODUCTION

11.1.1 Overview of serial communication

A system frequently needs to communicate with another system that does not reside in the same device. To reduce the number of I/O pins and external wiring and cabling, the two systems can transfer data over a single *serial* line, one bit at a time. The transmitting system performs the parallel-to-serial conversion and then sends the serial data via a single line. The receiving system performs the serial-to-parallel conversion and restores the original parallel data.

Serial communication can be used for both high-speed and low-speed data transfer. In a high-speed interface, such as USB and gigabit Ethernet, the data rate can reach several hundred million bits per second or more. The transmitting and receiving circuits are commonly referred to as *SerDes* (for serializer-deserializer) blocks.

Figure 11.1   Transmission of a byte.

Its protocol and specification are very complex and its design is beyond the scope of this book.

In a low-speed interface, the data rate ranges from several thousand to several hundred thousand bits per second. It is adequate for most general I/O peripherals and for data acquisition and control tasks. Since the data rate is much slower than the clock rate of the FPGA board, these schemes can be realized by an FPGA's generic logic elements. We discuss the *UART* scheme in this chapter and cover the I^2C, *SPI*, and *PS2* schemes in Chapters 16, 15, and 17.

11.1.2   Overview of UART

A basic UART controller includes a transmitter and a receiver. The transmitter is a special shift register that loads data in parallel and then shifts it out bit by bit at a specific rate. The receiver, on the other hand, shifts in data bit by bit and then reassembles the data. The serial line is 1 when it is idle. The transmission starts with a *start bit*, which is 0, followed by *data bits* and an optional *parity bit*, and ends with *stop bits*, which are 1. The number of data bits can be 6, 7, or 8. The optional parity bit is used for error detection. For odd parity, it is set to 0 when the data bits have an odd number of 1's. For even parity, it is set to 0 when the data bits have an even number of 1's. The number of stop bits can be 1, 1.5, or 2. The transmission with 8 data bits, no parity, and 1 stop bit is shown in Figure 11.1. Note that the LSB of the data word is transmitted first.

No clock information is conveyed through the serial line. Before the transmission starts, the transmitter and receiver must agree on a set of parameters in advance, which include the *baud rate* (i.e., number of bits per second), the number of data bits and stop bits, and use of the parity bit. With the predetermined parameters, the receiver uses an *oversampling* scheme to retrieve the data bits. The commonly used baud rates are 9,600 and 19,200 baud.

11.1.3   Oversampling procedure

The most commonly used sampling rate is 16 times the baud rate, which means that each serial bit is sampled 16 times. For a communication with N data bits and M stop bits, the oversampling scheme works as follows:

1. Wait until the incoming signal becomes 0, the beginning of the start bit, and then start the sampling tick counter.
2. When the counter reaches 7, the incoming signal reaches the middle point of the start bit. Clear the counter to 0 and restart.
3. When the counter reaches 15, the incoming signal progresses for one bit and reaches the middle of the first data bit. Retrieve its value, shift it into a register, and restart the counter.

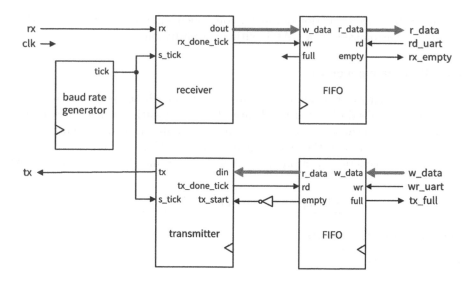

Figure 11.2   Block diagram of a complete UART.

4. Repeat step 3 N–1 more times to retrieve the remaining data bits.
5. If the optional parity bit is used, repeat step 3 one time to obtain the parity bit.
6. Repeat step 3 M more times to obtain the stop bits.

The oversampling scheme basically performs the function of a clock signal. Instead of using the rising edge to indicate when the input signal is valid, it utilizes sampling ticks to estimate the middle point of each bit. While the receiver has no information about the exact onset time of the start bit, the estimation can be off by at most $\frac{1}{16}$. The subsequent data bit retrievals are off by at most $\frac{1}{16}$ from the middle point as well. Because of the oversampling, the baud rate can only be a small fraction of the system clock rate, and thus this scheme is not appropriate for a high data rate.

11.2 UART CONSTRUCTION

11.2.1 Conceptual design

The top-level diagram of a UART system is shown in Figure 11.2. It consists of five major components. The baud rate generator generates an oversampling tick. The receiver performs the serial-to-parallel conversion and the transmitter performs the parallel-to-serial conversion. Two FIFO buffers are used between the UART receiver and transmitter and the processor as cushions.

The FIFO buffers are needed because the data processing rate of a UART is much slower than the rate of FPGA's soft-core processor. A buffer allows the processor to process a *burst* of data. For example, instead of waiting for completion for each individual byte, a processor can write several bytes in a burst to a transmitting buffer and continue other tasks.

We illustrate the construction of the key components in the following subsections. The design is customized for a UART without a parity bit.

11.2.2   Baud rate generator

The baud rate generator generates a sampling signal whose frequency is exactly 16 times the UART's designated baud rate. To avoid creating a new clock domain and violating the synchronous design principle, the sampling signal should function as enable ticks rather than the clock signal to the UART receiver.

The baud rate generator is a programmable counter and the HDL code is shown in Listing 11.1.

Listing 11.1   Baud rate generator

```vhdl
library ieee;
use ieee.std_logic_1164.all;
use ieee.numeric_std.all;
entity baud_gen is
   port(
      clk   : in std_logic;
      reset : in std_logic;
      dvsr  : in std_logic_vector(10 downto 0);
      tick  : out std_logic
   );
end baud_gen;

architecture arch of baud_gen is
   constant N    : integer := 11;
   signal r_reg  : unsigned(N - 1 downto 0);
   signal r_next : unsigned(N - 1 downto 0);
begin
   -- register
   process(clk, reset)
   begin
      if (reset = '1') then
         r_reg <= (others => '0');
      elsif (clk'event and clk = '1') then
         r_reg <= r_next;
      end if;
   end process;
   -- next-state logic
   r_next <= (others=>'0') when r_reg=unsigned(dvsr) else r_reg + 1;
   -- output logic
   tick <= '1' when r_reg=1 else '0'; -- not use 0 because of reset
end arch;
```

The code is similar to the parameterized mod-m counter discussed in Section 4.3.2 except that the original comparison expression, r_reg=(M-1), is replaced with r_reg=unsigned(dvsr). The dvsr (for divisor) term is an external signal and thus its value can be set dynamically during operation. In addition, dvsr (instead of dvsr-1) is used to reduce hardware complexity. If the value of dvsr is v, the counter counts from 0 to v and wraps around. Thus, it is a mod-$(v + 1)$ counter.

The value v depends on the desired baud rate and system clock rate. Let the baud rate be b and the system clock rate be f. The desired sampling rate becomes $16 * b$ and the counter should count $\frac{f}{16*b}$ and wrap around. It means that

$$v + 1 = \frac{f}{16 * b}$$

and v becomes $\frac{f}{16*b} - 1$.

11.2.3   UART receiver

The *UART receiver* obtains the data byte from the serial line via oversampling. It uses the tick from the baud rate generator to estimate the middle points of transmitted bits and then retrieves them at these points accordingly. The overall receiving operation can be described by an ASMD chart, as shown in Figure 11.3. To accommodate future modification, two generics are used in the description. DBIT indicates the number of data bits and SB_TICK indicates the number of ticks needed for the stop bits, which is 16, 24, and 32 for 1, 1.5, and 2 stop bits, respectively.

The ASMD chart follows the steps discussed in Section 11.1.3 and includes three major states, start, data, and stop, which represent the processing of the start bit, data bits, and stop bit. The s_tick signal is the enable tick from the baud rate generator and there are 16 ticks in a bit interval. Note that the FSMD stays in the same state unless the s_tick signal is asserted. There are two counters, represented by the s and n registers. The s register keeps track of the number of sampling ticks and counts to 7 in the start state, to 15 in the data state, and to SB_TICK in the stop state. The n register keeps track of the number of data bits received in the data state. The retrieved bits are shifted into and reassembled in the b register. A status signal, rx_done_tick, is included. It is asserted for one clock cycle after the receiving process is completed.

The corresponding code is shown in Listing 11.2. Since the incoming rx signal is not driven by the system clock, an additional two-FF synchronizer is added.

Listing 11.2   UART receiver

```
library ieee;
use ieee.std_logic_1164.all;
use ieee.numeric_std.all;
entity uart_rx is
   generic(
      DBIT    : integer := 8;    -- # data bits
      SB_TICK : integer := 16    -- # ticks for stop bits
   );
   port(
      clk, reset   : in  std_logic;
      rx           : in  std_logic;
      s_tick       : in  std_logic;
      rx_done_tick : out std_logic;
      dout         : out std_logic_vector(7 downto 0)
   );
end uart_rx;

architecture arch of uart_rx is
   type state_type is (idle, start, data, stop);
   signal state_reg     : state_type;
   signal state_next    : state_type;
   signal s_reg, s_next : unsigned(4 downto 0);
   signal n_reg, n_next : unsigned(2 downto 0);
   signal b_reg, b_next : std_logic_vector(7 downto 0);
   signal sync1_reg     : std_logic;
   signal sync2_reg     : std_logic;
   signal sync_rx       : std_logic;

begin
```

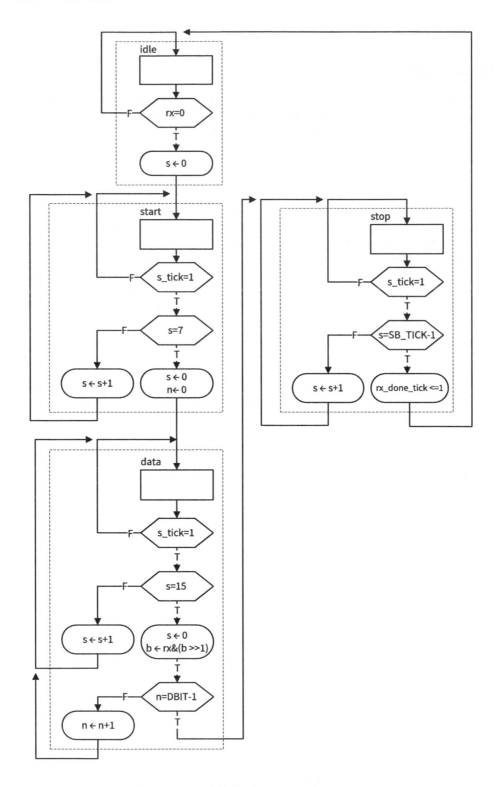

Figure 11.3 ASMD chart of a UART receiver.

```vhdl
-- synchronization for rx
process(clk, reset)
begin
   if reset = '1' then
      sync1_reg <= '0';
      sync2_reg <= '0';
   elsif (clk'event and clk = '1') then
      sync1_reg <= rx;
      sync2_reg <= sync1_reg;
   end if;
end process;
sync_rx <= sync2_reg;
-- FSMD state & data registers
process(clk, reset)
begin
   if reset = '1' then
      state reg  <= idle;
      s_reg      <= (others => '0');
      n_reg      <= (others => '0');
      b_reg      <= (others => '0');
   elsif (clk'event and clk = '1') then
      state_reg  <= state_next;
      s_reg      <= s_next;
      n_reg      <= n_next;
      b_reg      <= b_next;
   end if;
end process;
-- next-state logic & data path
process(state_reg, s_reg, n_reg, b_reg, s_tick, sync_rx)
begin
   state_next  <= state_reg;
   s_next      <= s_reg;
   n_next      <= n_reg;
   b_next      <= b_reg;
   rx_done_tick <= '0';
   case state_reg is
      when idle =>
         if sync_rx = '0' then
            state_next <= start;
            s_next     <= (others => '0');
         end if;
      when start =>
         if (s_tick = '1') then
            if s_reg = 7 then
               state_next <= data;
               s_next     <= (others => '0');
               n_next     <= (others => '0');
            else
               s_next <= s_reg + 1;
            end if;
         end if;
      when data =>
         if (s_tick = '1') then
            if s_reg = 15 then
               s_next <= (others => '0');
               b_next <= sync_rx & b_reg(7 downto 1);
               if n_reg = (DBIT - 1) then
                  state_next <= stop;
               else
                  n_next <= n_reg + 1;
               end if;
```

```
                else
                    s_next <= s_reg + 1;
                end if;
            end if;
        when stop =>
            if (s_tick = '1') then
                if s_reg = (SB_TICK - 1) then
                    state_next  <= idle;
                    rx_done_tick <= '1';
                else
                    s_next <= s_reg + 1;
                end if;
            end if;
        end case;
    end process;
    dout <= b_reg;
end arch;
```

11.2.4 UART transmitter

The UART transmitter is essentially a shift register that shifts out data bits at a specific rate. The rate is controlled by the same enable tick signal generated by the baud rate generator. Each bit lasts for 16 ticks. The FSMD of the UART transmitter is similar to that of the UART receiver. After assertion of the tx_start signal, the FSMD loads the data word and then gradually progresses through the start, data, and stop states to shift out the corresponding bits. It signals completion by asserting the tx_done_tick signal for one clock cycle. A one-bit buffer, tx_reg, is used to filter out any potential glitch. The corresponding code is shown in Listing 11.3.

Listing 11.3   UART transmitter

```
library ieee;
use ieee.std_logic_1164.all;
use ieee.numeric_std.all;
entity uart_tx is
    generic(
        DBIT    : integer := 8;     -- # data bits
        SB_TICK : integer := 16     -- # ticks for stop bits
    );
    port(
        clk, reset    : in  std_logic;
        tx_start      : in  std_logic;
        s_tick        : in  std_logic;
        din           : in  std_logic_vector(7 downto 0);
        tx_done_tick  : out std_logic;
        tx            : out std_logic
    );
end uart_tx;

architecture arch of uart_tx is
    type state_type is (idle, start, data, stop);
    signal state_reg          : state_type;
    signal state_next         : state_type;
    signal s_reg, s_next      : unsigned(4 downto 0);
    signal n_reg, n_next      : unsigned(2 downto 0);
    signal b_reg, b_next      : std_logic_vector(7 downto 0);
    signal tx_reg, tx_next    : std_logic;
```

```
begin
   -- FSMD state & data registers
   process(clk, reset)
   begin
      if reset = '1' then
         state_reg <= idle;
         s_reg      <= (others => '0');
         n_reg      <= (others => '0');
         b_reg      <= (others => '0');
         tx_reg     <= '1';
      elsif (clk'event and clk = '1') then
         state_reg <= state_next;
         s_reg      <= s_next;
         n_reg      <= n_next;
         b_reg      <= b_next;
         tx_reg     <= tx_next;
      end if;
   end process;
   -- next-state logic & data path
   process(state_reg,s_reg,n_reg,b_reg,s_tick,tx_reg,tx_start,din)
   begin
      state_next    <= state_reg;
      s_next        <= s_reg;
      n_next        <= n_reg;
      b_next        <= b_reg;
      tx_next       <= tx_reg;
      tx_done_tick <= '0';
      case state_reg is
         when idle =>
            tx_next <= '1';
            if tx_start = '1' then
               state_next <= start;
               s_next      <= (others => '0');
               b_next      <= din;
            end if;
         when start =>
            tx_next <= '0';
            if (s_tick = '1') then
               if s_reg = 15 then
                  state_next <= data;
                  s_next      <= (others => '0');
                  n_next      <= (others => '0');
               else
                  s_next <= s_reg + 1;
               end if;
            end if;
         when data =>
            tx_next <= b_reg(0);
            if (s_tick = '1') then
               if s_reg = 15 then
                  s_next <= (others => '0');
                  b_next <= '0' & b_reg(7 downto 1);
                  if n_reg = (DBIT - 1) then
                     state_next <= stop;
                  else
                     n_next <= n_reg + 1;
                  end if;
               else
                  s_next <= s_reg + 1;
               end if;
            end if;
```

```
        when stop =>
            tx_next <= '1';
            if (s_tick = '1') then
                if s_reg = (SB_TICK - 1) then
                    state_next    <= idle;
                    tx_done_tick <= '1';
                else
                    s_next <= s_reg + 1;
                end if;
            end if;
    end case;
  end process;
  tx <= tx_reg;
end arch;
```

11.2.5 Top-level HDL codes

The top-level HDL code follows the diagram in Figure 11.2 and instantiates the five major components. The FIFO buffer component is the one designed in Section 7.3. The code is shown in Listing 11.4.

Listing 11.4   UART top-level description

```
library ieee;
use ieee.std_logic_1164.all;
use ieee.numeric_std.all;
entity uart is
   generic(
       DBIT    : integer := 8;    -- # data bits
       SB_TICK : integer := 16;   -- # ticks for stop bits, 16 per bit
       FIFO_W  : integer := 4     -- # FIFO addr bits (depth: 2^FIFO_W)
   );
   port(
       clk, reset : in  std_logic;
       rd_uart    : in  std_logic;
       wr_uart    : in  std_logic;
       dvsr       : in  std_logic_vector(10 downto 0);
       rx         : in  std_logic;
       w_data     : in  std_logic_vector(7 downto 0);
       tx_full    : out std_logic;
       rx_empty   : out std_logic;
       r_data     : out std_logic_vector(7 downto 0);
       tx         : out std_logic
   );
end uart;

architecture str_arch of uart is
   signal tick               : std_logic;
   signal rx_done_tick       : std_logic;
   signal tx_fifo_out        : std_logic_vector(7 downto 0);
   signal rx_data_out        : std_logic_vector(7 downto 0);
   signal tx_empty           : std_logic;
   signal tx_fifo_not_empty  : std_logic;
   signal tx_done_tick       : std_logic;
begin
   baud_gen_unit : entity work.baud_gen(arch)
       port map(
           clk  => clk, reset => reset, dvsr => dvsr,
           tick => tick
       );
```

```vhdl
    uart_rx_unit : entity work.uart_rx(arch)
        generic map(DBIT => DBIT, SB_TICK => SB_TICK)
        port map(
            clk           => clk,
            reset         => reset,
            rx            => rx,
            s_tick        => tick,
            rx_done_tick  => rx_done_tick,
            dout          => rx_data_out
        );
    uart_tx_unit : entity work.uart_tx(arch)
        generic map(DBIT => DBIT, SB_TICK => SB_TICK)
        port map(
            clk           => clk,
            reset         => reset,
            tx_start      => tx_fifo_not_empty,
            s_tick        => tick,
            din           => tx_fifo_out,
            tx_done_tick  => tx_done_tick,
            tx            => tx
        );
    fifo_rx_unit : entity work.fifo(reg_file_arch)
        generic map(DATA_WIDTH => DBIT, ADDR_WIDTH => FIFO_W)
        port map(
            clk    => clk,
            reset  => reset,
            rd     => rd_uart,
            wr     => rx_done_tick,
            w_data => rx_data_out,
            empty  => rx_empty,
            full   => open,
            r_data => r_data
        );
    fifo_tx_unit : entity work.fifo(reg_file_arch)
        generic map(DATA_WIDTH => DBIT, ADDR_WIDTH => FIFO_W)
        port map(
            clk    => clk,
            reset  => reset,
            rd     => tx_done_tick,
            wr     => wr_uart,
            w_data => w_data,
            empty  => tx_empty,
            full   => tx_full,
            r_data => tx_fifo_out
        );
    tx_fifo_not_empty <= not tx_empty;
end str_arch;
```

The generics specify the number of data bits, the number of stop bits, and the size of the FIFO buffers.

11.3 UART CORE DEVELOPMENT

We can follow the procedure in Section 10.3.2 to add a wrapping circuit for the UART controller and create an MMIO IP core.

11.3.1   Register map

The processor interacts with the UART controller as follows:

- set (i.e., write) the divisor value of the baud rate generator.
- receive (i.e., read) a data byte from the receiving FIFO buffer.
- generate (i.e., write) a pulse to remove a data byte from the receiving FIFO buffer.
- check (i.e., read) the `rx_empty` signal to determine whether a data byte is in the receiving FIFO buffer.
- transmit (i.e., write) a data byte to the transmitting FIFO buffer.
- check (i.e., read) the `tx_full` signal to determine whether the transmitting FIFO buffer is available.

Based on these interactions, we can define the UART core's register map. For clarity, we separate the read and write operations into different registers. There is one read register. Its address offset and fields are:

- offset 0 (read data and status register)
  - bits 7 to 0: 8-bit received data
  - bit 8: empty status of the receiving FIFO buffer
  - bit 9: full status of the transmitting FIFO buffer

There are three write registers. Their address offsets and fields are:

- offset 1 (baud rate divisor register)
  - bits 10 to 0: 11-bit divisor value
- offset 2 (write data register)
  - bits 7 to 0: 8-bit transmitted data
- offset 3 (read data removal register)
  - dummy data write to generate a pulse to remove a data byte from the receiving FIFO buffer

Note that the term "register" here is just used to express a conceptual location within the I/O core. It may or may not correspond to a physical register.

11.3.2   Wrapping circuit for the slot interface

Based on the register map and the I/O signals of the UART controller, we can derive a wrapping circuit that complies with the slot specification and create the UART core. The HDL code of the core is shown in Listing 11.5.

Listing 11.5   UART core

```
library ieee;
use ieee.std_logic_1164.all;
use ieee.numeric_std.all;
entity chu_uart is
   generic(
      FIFO_DEPTH_BIT : integer := 8 — # FIFO addr bits
   );
   port(
      clk      : in   std_logic;
      reset    : in   std_logic;
      — slot interface
      cs       : in   std_logic;
      write    : in   std_logic;
```

```vhdl
      read     : in  std_logic;
      addr     : in  std_logic_vector(4 downto 0);
      rd_data  : out std_logic_vector(31 downto 0);
      wr_data  : in  std_logic_vector(31 downto 0);
      -- external signals
      tx       : out std_logic;
      rx       : in  std_logic
   );
end chu_uart;

architecture arch of chu_uart is
   signal wr_en    : std_logic;
   signal wr_uart  : std_logic;
   signal rd_uart  : std_logic;
   signal wr_dvsr  : std_logic;
   signal tx_full  : std_logic;
   signal rx_empty : std_logic;
   signal r_data   : std_logic_vector(7 downto 0);
   signal dvsr_reg : std_logic_vector(10 downto 0);
begin
   -- instantiate uart controller
   uart_unit : entity work.uart(str_arch)
      generic map(
         DBIT    => 8,
         SB_TICK => 16,
         FIFO_W  => FIFO_DEPTH_BIT
      )
      port map(
         clk      => clk,
         reset    => reset,
         rd_uart  => rd_uart,
         wr_uart  => wr_uart,
         dvsr     => dvsr_reg,
         rx       => rx,
         tx       => tx,
         w_data   => wr_data(7 downto 0),
         r_data   => r_data,
         tx_full  => tx_full,
         rx_empty => rx_empty
      );
   -- baud rate register
   process(clk, reset)
   begin
      if (reset = '1') then
         dvsr_reg <= (others => '0');
      elsif (clk'event and clk = '1') then
         if wr_dvsr = '1' then
            dvsr_reg <= wr_data(10 downto 0);
         end if;
      end if;
   end process;
   -- write decoding
   wr_en   <= '1' when write = '1' and cs = '1' else '0';
   wr_dvsr <= '1' when addr(1 downto 0)="01" and wr_en = '1' else '0';
   wr_uart <= '1' when addr(1 downto 0)="10" and wr_en = '1' else '0';
   rd_uart <= '1' when addr(1 downto 0)="11" and wr_en = '1' else '0';
   -- read multiplexing
   rd_data <= x"00000" & "00" & tx_full & rx_empty & r_data;
end arch;
```

The wrapping circuit consists of a UART controller instance, a register to store the value of the baud rate divisor, and a write decoding circuit. Since the UART controller already has built-in receiving and transmitting FIFO buffers, no additional I/O register is created for the receiving and transmitting data. Note that DBIT and SB_TICK generics are assigned to 8 and 16 in this design, which leads to an UART with eight data bits and one stop bit, the most widely used configuration.

The decoding circuit uses the two LSBs of addr and cs to generate three enable signals. The wr_dvsr and wr_uart signals write data to the divisor register and the transmitting FIFO buffer, respectively. The rd_uart signal acts as a one-clock data removal tick.

The read data is constructed by the data from the receiving FIFO buffer and two FIFO status signals. Since there is only one read register in the register map, there is no need for multiplexing.

Note that the control of the FIFO buffer is somewhat different from that of Section 10.2.4. When reading a data item from the FIFO buffer, we can either remove the item or keep it intact. The latter approach provides more flexibility for software development but requires a separate instruction to perform the removal operation. We use this approach in the design. When the processor writes a dummy data to I/O register 3, rd_uart is asserted for one clock cycle and and the data byte in the head of the receiving buffer is removed.

Alternatively, rd_uart can be tied to a read operation:

```
rd_uart <= '1' when read = '1' and cs = '1' else '0';
```

The data item will be automatically removed during a read operation and no separate write instruction is needed. However, since the FIFO buffer can be empty, it is necessary to check the rx_empty status to verify that the read data is valid.

11.4 UART DRIVER

UART driver consists of two sets of routines. The first set accesses the I/O registers and performs basic operations such as transmitting a byte, receiving a byte, and checking status. The second set transmits and displays a string or a number on a console. The routines can be considered as a primitive version of printf(). However, they are very simple and do not require much memory space. If more formatting and printing functionalities are desired, the second set can be separated into a new class with more sophisticated methods.

11.4.1 Class definition

The class definition of the UART core is shown in Listing 11.6.

Listing 11.6   UartCore class definition (in uart_core.h)

```
#ifndef _UART_CORE_H_INCLUDED
#define _UART_CORE_H_INCLUDED

#include "chu_io_rw.h"
#include "chu_io_map.h"
class UartCore {
   /* register map */
   enum {
```

```
        RD_DATA_REG = 0,
        DVSR_REG = 1,
        WR_DATA_REG = 2,
        RM_RD_DATA_REG = 3   // remove read data
    };
    /* masks */
    enum {
        TX_FULL_FIELD = 0x00000200,
        RX_EMPT_FIELD = 0x00000100,
        RX_DATA_FIELD = 0x000000ff
    };
public:
    /* methods */
    UartCore(uint32_t core_base_addr);
    ~UartCore();
    // basic I/O access
    void set_baud_rate(int baud);
    int rx_fifo_empty();
    int tx_fifo_full();
    void tx_byte(uint8_t byte);
    int rx_byte();
    // display methods
    void disp(char ch);
    void disp(const char *str);
    void disp(int n, int base, int len);
    void disp(int n, int base);
    void disp(int n);
    void disp(double f, int digit);
    void disp(double f);
private:
    uint32_t base_addr;
    int baud_rate;
    void disp_str(const char *str);
};

#endif   // _UART_CORE_H_INCLUDED
```

The first **enum** definition uses symbolic names for the four register offsets. The second **enum** definition specifies the masks to extract the data byte and status bits from the RD_DATA_REG register.

11.4.2 Basic methods

The class implementation of the constructor and the first set of methods is shown in Listing 11.7.

Listing 11.7 UartCore basic methods (in **uart_core.cpp**)

```
UartCore::UartCore(uint32_t core_base_addr) {
    base_addr = core_base_addr;
    set_baud_rate(9600);       //default baud rate
}

void UartCore::set_baud_rate(int baud) {
    uint32_t dvsr;

    dvsr = SYS_CLK_FREQ*1000000 / 16 / baud - 1;
    io_write(base_addr, DVSR_REG, dvsr);
}
```

```
int UartCore::rx_fifo_empty() {
   uint32_t rd_word;
   int empty;

   rd_word = io_read(base_addr, RD_DATA_REG);
   empty = (int) (rd_word & RX_EMPT_FIELD) >> 8;
   return (empty);
}

int UartCore::tx_fifo_full() {
   uint32_t rd_word;
   int full;

   rd_word = io_read(base_addr, RD_DATA_REG);
   full = (int) (rd_word & TX_FULL_FIELD) >> 9;
   return (full);
}

void UartCore::tx_byte(uint8_t byte) {
   while (tx_fifo_full()) { };   // busy waiting
   io_write(base_addr, WR_DATA_REG, (uint32_t )byte);
}

int UartCore::rx_byte() {
   uint32_t data;

   if (rx_fifo_empty())
      return (-1);
   else {
      data = io_read(base_addr, RD_DATA_REG) & RX_DATA_FIELD;
      io_write(base_addr, RM_RD_DATA_REG, 0); //dummy write
      return ((int) data);
   }
}
```

The `UartCore()` constructor saves the base address and sets the baud rate to a default value of 9600. The `set_baud_rate()` method calculates the divisor value to the desired baud rate and writes it to the divisor register. This method can be invoked if the desired baud rate is different from the default value. The `rx_fifo_empty()` and `tx_fifo_full()` methods extract and return the corresponding status bit.

The `tx_byte()` method transmits a data byte. It "busy-waits" until a space in the transmitting FIFO buffer is available (i.e., not full) and then writes the data to buffer. The `rx_byte()` method tries to retrieve a data byte. It first checks whether the receiving FIFO buffer is empty and returns -1 as if the condition is true. Otherwise, it retrieves the data byte and deletes it from the FIFO buffer. Since the external source may not send a data byte in a timely manner, this method intentionally avoids the busy waiting scheme.

11.4.3 ASCII code

A main application of a UART is to communicate with an external host and display the relevant information in a command console. The information is transmitted as characters in *ASCII code*, which is 7 bits wide and consists of 128 code words, including regular alphabets, digits, punctuation symbols, and nonprintable control characters. The characters and their code words (in hexadecimal format) are shown in Table 11.1. The nonprintable characters are shown enclosed in parentheses, such as (del). Several nonprintable characters may introduce special action when received:

- (nul): null byte, which is the all-zero pattern
- (bel): generate a bell sound, if supported
- (bs): backspace
- (ht): horizontal tab
- (nl): new line
- (vt): vertical tab
- (np): new page
- (cr): carriage return
- (esc): escape
- (sp): space
- (del): delete, which is also the all-one pattern

Many low-level driver functionalities involve serial communication. The following observations help us to manipulate and process the ASCII code:

- When the first hex digit in a code word is 0x0 or 0x1, the corresponding character is a control character.
- When the first hex digit in a code word is 0x2 or 0x3, the corresponding character is a digit or punctuation.
- When the first hex digit in a code word is 0x4 or 0x5, the corresponding character is generally an uppercase letter.
- When the first hex digit in a code word is 0x6 or 0x7, the corresponding character is generally a lowercase letter.
- If the first hex digit in a code word is 0x3, the lower hex digit represents the corresponding decimal digit.
- The upper- and lowercase letters differ in a single bit and can be converted to each other by adding or subtracting 0x20 or inverting the sixth bit.

Note that the ASCII code uses only 7 bits, but a data word is normally composed of 8 bits (i.e., a byte). The PC uses an extended set in which the MSB is 1 and the characters are special graphics symbols. This code, however, is not part of the ASCII standard.

11.4.4 Display methods

The class implementation of the second set of routines is shown in Listing 11.8. They are overloaded methods to transmit and display a string and a number. A private method, `disp_str()`, is used to facilitate the processing.

Listing 11.8 UartCore display methods (in `uart_core.cpp`)

```
void UartCore::disp_str(const char *str) {
   while ((uint8_t) *str) {
      tx_byte(*str);
      str++;
   }
}

void UartCore::disp(char ch) {
    tx_byte(ch);
}

void UartCore::disp(const char *str) {
```

Table 11.1 ASCII codes

Code	Char	Code	Char	Code	Char	Code	Char
00	(nul)	20	(sp)	40	@	60	'
01	(soh)	21	!	41	A	61	a
02	(stx)	22	"	42	B	62	b
03	(etx)	23	#	43	C	63	c
04	(eot)	24	$	44	D	64	d
05	(enq)	25	%	45	E	65	e
06	(ack)	26	&	46	F	66	f
07	(bel)	27	'	47	G	67	g
08	(bs)	28	(	48	H	68	h
09	(ht)	29	)	49	I	69	i
0a	(nl)	2a	*	4a	J	6a	j
0b	(vt)	2b	+	4b	K	6b	k
0c	(np)	2c	,	4c	L	6c	l
0d	(cr)	2d	-	4d	M	6d	m
0e	(so)	2e	.	4e	N	6e	n
0f	(si)	2f	/	4f	O	6f	o
10	(dle)	30	0	50	P	70	p
11	(dc1)	31	1	51	Q	71	q
12	(dc2)	32	2	52	R	72	r
13	(dc3)	33	3	53	S	73	s
14	(dc4)	34	4	54	T	74	t
15	(nak)	35	5	55	U	75	u
16	(syn)	36	6	56	V	76	v
17	(etb)	37	7	57	W	77	w
18	(can)	38	8	58	X	78	x
19	(em)	39	9	59	Y	79	y
1a	(sub)	3a	:	5a	Z	7a	z
1b	(esc)	3b	;	5b	[	7b	{
1c	(fs)	3c	¡	5c	\	7c	—
1d	(gs)	3d	=	5d	]	7d	}
1e	(rs)	3e	¿	5e	^	7e	~
1f	(us)	3f	?	5f	_	7f	(del)

```
      disp_str(str);
}

void UartCore::disp(int n, int base, int len) {
   char buf[33];            // 32 bit #
   char *str, ch, sign;
   int rem, i;
   unsigned int un;

   /* error check */
   if (base != 2 && base != 8 && base != 16)
      base = 10;
   if (len > 32)   // error check
      len = 32;
   /* handle neg decimal # */
   if (base == 10 && n < 0) {
      un = (unsigned) -n;
      sign = '-';
   } else {
      un = (unsigned) n; // interpreted as unsigned for hex/bin conversion
      sign = ' ';
   }
   /* convert # to string */
   str = &buf[33];
   *str = '\0';
   i = 0;
   do {
      str--;
      rem = un % base;
      un = un / base;
      if (rem < 10)
         ch = (char) rem + '0';
      else
         ch = (char) rem - 10 + 'a';
      *str = ch;
      i++;
   } while (un);
   /* attach - sign for neg decimal # */
   if (sign == '-') {
      str--;
      *str = sign;
      i++;
   }
   /* pad with blank */
   while (i < len) {
      str--;
      *str = ' ';
      i++;
   };
   disp_str(str);
}

void UartCore::disp(int n, int base) {
   disp(n, base, 0);

}void UartCore::disp(int n) {
   disp(n, 10, 0);
}

void UartCore::disp(double f, int digit) {
   double fa, frac;
```

```
    int n, i, i_part;

    // obtain absolute value of f
    fa = f;
    if (f < 0.0) {
        fa = -f;
        disp_str("-");
    }
    // display integer portion
    i_part = (int) fa;
    disp(i_part);
    disp_str(".");
    // display fraction part
    frac = fa - (double) i_part;
    for (n = 0; n < digit; n++) {
        frac = frac * 10.0;
        i = (int) frac;
        disp(i);
        frac = frac - i;
    }
}

void UartCore::disp(double f) {
    disp(f, 3);
}
```

The `disp_str()` method transmits a string one character at a time, which is expected to be displayed on a console. The overloaded `disp()` methods display a character, a string or a number. The `disp(char ch)` simply transmits `ch` as a raw byte without processing. The `disp(const char *str)` method simply calls `disp_str()`. The `disp(int n, int base, int len)` method converts a number, `n`, into a displayable string. The `base` parameter specifies the base, which can be 2 (for binary), 8 (for octal), 10 (for decimal), or 16 (for hexadecimal). The `len` parameter specifies the number of digits in the string (i.e., length). Extra 0's will be padded in front as needed. No padding 0's will be added if `len` is 0. The other two relevant overloaded `disp()` methods use no padding 0's and base-10 representation. The `disp(double f, int digit)` method converts a floating-point number, `f`, into a displayable string. The string contains an integer portion and a fraction portion. The `digit` parameter specifies the number of digits in the fraction portion. The other relevant method sets `digit` to be 3.

11.4.5   Test

The testing program shown in Listing 9.11 includes a simple function, `uart_check()`, to verify the operation of the UART core and driver.

11.5   ADDITIONAL PROJECT IDEAS

11.5.1   Original serial port

A UART is sometimes referred to as a *serial port*. The original serial port was used in conjunction with the *RS-232 standard*, which formally defines the electrical characteristics and physical dimension of the connectors. The valid signals are either in the range of +3 to +15 volts or −3 to −15 volts and the connection requires

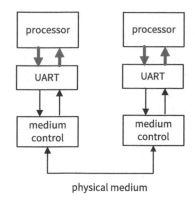

Figure 11.4   Layered model of an emulated serial port.

a bulky nine-pin D-sub connector. These voltage ranges are not compatible with today's FPGA devices, whose I/O pins are usually configured using LVTTL or LVCOMS standard with an output voltage range between 0 and 3.3 volts.

The oversampling scheme limits the performance of a UART. While it is not adequate for high-speed communication, its simple construction is still a cost-effective way to transfer low-rate data. An embedded processor usually contains one or more serial ports

11.5.2   Emulated serial port

Because of the bulky connector and thick cable, very few systems support the original RS-232 port now. Various types of "adaptors" are used in conjunction with a UART and function like an *emulated serial port*. We can treat the emulated port as a layered framework, as shown in Figure 11.4. The layers are:

- UART layer: convert a byte stream to a bitstream
- Medium control layer: control the bit transfer over a *physical medium*
- Physical medium

In the original serial port, the medium control layer is a voltage level shifting device, such as MAX232, which converts the signals to and from different voltage levels, and the physical medium is an RS-232 cable.

The emulated serial port can be implemented by several schemes:

- Direct connection
- USB-to-UART adaptor
- Bluetooth adaptor
- ZigBee adaptor
- WiFi adaptor

The following subsections discuss their potential applications.

11.5.3   Direct connection

If the voltage range is compatible, two UARTs can be connected directly. The wires constitute the media control layer and the physical medium. The most common application is to establish a wired link between two FPGA boards or between an FPGA board and an external module.

One interesting application is to use direct connection to control *iRobot Create*, a robot based on vacuum cleaning iRobot Roomba. The iRobot Create constitutes a versatile "ready-to-use" mobile platform with built-in motors and wheels, a chargeable battery, and several dozen sensors. It has a built-in processor and a serial port. The FPGA board can control the platform and retrieve sensor data via a UART. For example, setting Create's speed can be done by sending a five-byte command, `0x7d`, `0xmm`, `0xmm`, `0xnn`, and `0xnn`, where `0x7d` is the code for "drive," and `mmmm` and `nnnn` are two 16-bit numbers specifying the speeds of the left and right wheels, respectively. The FPGA board, along with other sensors and components, can be mounted on the cargo bay.

11.5.4   USB-to-UART adaptor

Most today's desktop and lap computers use a single type of serial interface - the USB port - for general I/O peripherals. The USB implementation needs a complicated software protocol stack and involves licensing. A device, FT232, manufactured by FTDI (Future Technology Devices International), can be used as an adaptor between a UART and a USB port. The device converts the UART signals to USB data streams and functions as a USB client. The corresponding software driver in the desktop computer translates the USB data back to the UART byte streams and creates an emulated serial port. In this scheme, the adaptor constitutes the medium control layer and the USB cable is the physical medium.

The Nexys 4 DDR board contains an FT2232 (similar to FT232) device and connects the UART signals through a USB port. For an FPGA board without such device, a separate USB-to-UART adaptor module can be used. The UART's `rx` and `tx` signals can be routed to two normal I/O pins and then connected to the adaptor module.

From the desktop computer's point of view, the behavior of an emulated serial port is identical to that of a normal serial port. Thus, any application or software library designed for a serial port can be used for the emulated port. The terminal emulator, such as PuTTY, is the most common application. It establishes a serial link between the FPGA board and the desktop computer to exchange information. The terminal emulator is too primitive for complex interaction. A more useful software tool is the *Processing* language and IDE (integrated development environment). It is based on a simple graphics programming model and can be used to capture and display serial data and create a GUI to interact with the FPGA board.

11.5.5   Wireless adaptor

A wireless serial port adaptor contains a processor, radio, and firmware that implements a wireless protocol stack. *Bluetooth*, *Zigbee*, and *Wifi* are three commonly used wireless technology. In these schemes, the wireless adaptor constitutes the medium control layer and the radio wave is the physical medium.

A Bluetooth adaptor implements the *Bluetooth SPP* (serial port profile) protocol. Its main application is to establish a wireless link to connect (i.e., pair) two devices. Most tablet computers and smart phones support Bluetooth and thus the Bluetooth adaptor provides an easy way for an FPGA board to communicate with these devices. The simplest way to exchange information is to run a terminal emulator app in a tablet or a smart phone. Creating a custom app to interact with the FPGA board requires a substantial amount of programming skill. In the Android platform, an alternative is to use the *App Inventor* framework, which provides a simple graphic environment to build an Android app. The "programming" is done by dragging and dropping visual objects that look like jigsaw pieces and the finished program resembles a completed jigsaw puzzle. The framework can be used to access sensors of an Android device and create a custom GUI. It can be used to develop apps to interact with an FPGA board.

The *ZigBee* standard specifies a collection of protocols based on IEEE 802.15.4 and is intended to create a low data rate mesh wireless network. An *XBee* is an adaptor based on the ZigBee standard and manufactured by Digi International. It is wrapped with a serial port and thus can be connected to a UART of an FPGA board. An XBee module can work in either transparent mode or API (application programming interface) mode. The former is used for point-to-point communication and the link establishes an emulated serial port. The API mode uses a frame, which is composed of address, command, and various data payload fields, to communicate between nodes.

The *WiFi* technology is specified by IEEE 802.11 standard and is the method to connect a device wirelessly to the internet. In an embedded system, the connection is usually done with a WiFi adaptor, which contains an embedded processor, radio, and firmware that implements the TCP/IP protocol stack and supports commonly used internet protocols. The adaptor establishes a link to an access point (usually a wireless router) and sets up a TCP/IP socket. It is wrapped with a serial port and thus can be accessed by an FPGA board. The World Wide Web follows a client-server model, in which the client initiates the request and the server sends the responses back to the client. With a WiFi adaptor, the FPGA board can transmit and receive http messages via a UART. A common setting is to configure the FPGA board and the WiFi module as a server. A client (i.e., a web browser) can access the FPGA board by requesting information or posting commands through the internet.

11.6   BIBLIOGRAPHIC NOTES

Although the serial port is very old, it still provides a simple and reliable low-speed communication link. The *Wikipedia* website has a good overview article and several useful links on the subject (search with the RS232 and `serial port` keywords). *Serial Port Complete* by Jan Axelson provides comprehensive information on programming, interfacing, and using serial ports. The `disp()` method is based on the `print()` function of the Arduino platform and its codes can be found in Arduino library.

The *iRobot Create Open Interface* standard defines command and response structures of iRobot Create. *Processing: A Programming Handbook for Visual Designers and Artists* by C. Reas and B. Fry provides a detailed discussion of the Processing language and *App Inventor for Android: Build Your Own Apps* by J. Tyler describes the App Inventor platform.

11.7   SUGGESTED EXPERIMENTS

11.7.1   UART-controlled chasing LEDs

Consider the chasing LEDs circuit in Experiment 9.10.2. With the UART, we can use a PC's terminal emulator, such as PuTTY, to control its operation:

- When a q or Q ASCII code is received, "initialize" the first lit LED by moving it to the rightmost position.
- When a w or W ASCII code is received, "initialize" the second lit LED by moving it to the left position.
- When a "Rddd" sequence, where "ddd" is three decimal digits representing a number, is received, set the speed of the first LED to ddd milliseconds (i.e., move after ddd milliseconds).
- When a "Lddd" sequence is received, set the speed of the second LED to ddd milliseconds.
- All other ASCII codes will be ignored.

Derive the program and verify its operation.

11.7.2   Alternative read configuration

An alternative way to retrieve a byte from the receiving FIFO buffer is to use the read signal to remove the byte from the buffer, as discussed in Section 11.3.2. We can modify the HDL code of the wrapping circuit to implement this scheme and revise the rx_byte() method accordingly. Design the new UART core, synthesize the circuit, derive the method, and verify its operation.

11.7.3   UART controller with a parity bit

The UART designed in Section 11.3 is tailored for eight data bits and no parity. Revise the design for seven data bits and one parity bit. The type of parity (i.e., even or odd) is specified by an additional control signal. Develop the new UART core, synthesize the circuit, and verify its operation.

11.7.4   UART core with an error status

Several kinds of errors may occur in UART operation and detected by the UART receiver:

- *Parity error.* If the parity bit is included, the receiver can check the correctness of the received parity bit.
- *Frame error.* A UART *frame* is composed of a start bit, data bits, a parity bit, and a stop bit. The receiver samples the stop bit in the stop state and the value should always be 1. The frame error occurs if it is not the case. The error is usually due to the mismatched UART parameters.
- *Buffer overrun error.* This happens when the processor does not retrieve the received words in a timely manner. The UART receiver can check FIFO's full signal when the newly arrived data is ready to be stored (i.e., when the rx_done_tick signal is generated). A data overrun occurs if the full signal is still asserted.

We wish to expand the UART core in Experiment 11.7.3 to detect these errors. The new UART controller should have an additional three-bit output signal in which the bits indicate the existence of the parity error, frame error, and data overrun error. The status signal can constitute an additional field in the read data register of the wrapping circuit and be extracted by a new method. Design the new UART core, synthesize the circuit, derive the methods, and verify the operation.

11.7.5   Configurable UART core

For the UART designed in Experiment 11.7.3 in Section 11.2, only the baud rate can be dynamically configured. Other parameters are defined with generics and thus are fixed after synthesis. To make it fully configurable, additional control signals can be incorporated to specify the number of data bits, the number of stop bits, and the type of parity bit:

- d_num: 1-bit input signal specifying the number of data bits, which can be 7 or 8
- s_num: 2-bit input signal specifying the number of stop bits, which can be 1, 1.5, or 2
- par: 2-bit input signal specifying the desired parity scheme, which can be no parity, even parity, or odd parity

The dvsr_reg register in wrapping circuit will be expanded with additional fields to accommodate the new control signals.

Expand the UART core in Experiment 11.7.4 to include these features, synthesize the circuit, derive the methods, and verify the operation.

11.7.6   UART core with automatic baud rate detection

The most commonly used number of data bits of a serial connection is eight, which corresponds to a byte. When a regular ASCII code is used in communication (as we type in a terminal emulator), only seven LSBs are used and the MSB is 0. If the UART is configured as 8 data bits, 1 stop bit, and no parity, the received frame is in the form of 0_dddd_ddd0_1, in which d is a data bit and it can be 0 or 1. Assume that there is sufficient time between the first word and subsequent transmissions. We can determine the baud rate by measuring the time interval between the first 0 and last 0 of the frame. Based on this observation, we can derive a UART core with an automatic baud rate detection. The new UART core can be designed as follows:

- Design a automatic baud detection circuit that returns the time interval between the first 0 and last 0 of a frame.
- Include this circuit in the top-level diagram of the UART controller as the sixth component.
- Expand the wrapping circuit of the UART core to include a register for this value.
- Add a method to calculate the baud rate and set the divisor value accordingly.
- During the operation, the external system should first send a "synchronization" byte (any ASCII character). The method then can be called to set the baud rate automatically.

Design the new UART core, synthesize the circuit, derive the methods, and verify the operation.

11.7.7 UART core with enhanced automatic baud rate detection

The automatic baud rate detection scheme discussed in Experiment 11.7.6 requires the external system to send an explicit "synchronization" byte in the beginning. The byte is just for the detection purposes and its actual content (i.e., data bits) cannot be retrieved. We can design an enhanced UART core that can recover the data bits of the first frame. With this, the automatic rate detection scheme becomes "transparent." The external system just needs to send normal ASCII scheme and the baud rate will be detected and set automatically. Expand the previous UART core and driver and repeat Experiment 11.7.6.

11.7.8 UART core with an automatic baud rate and a parity detection circuit

The automatic baud rate detection scheme discussed in Experiment 11.7.7 can be extended to detect parity scheme as well. We assume that the UART is configured with an 8 data bit and its parity scheme can be no parity, even parity, or odd parity. Expand the previous UART core and driver to detect the parity configuration and repeat Experiment 11.7.7.

PART III

EMBEDDED SOC II: BASIC I/O CORES

CHAPTER 12

XILINX XADC CORE

An ADC (analog-to-digital convertor) is a circuit that digitizes a continuous analog signal by converting its voltage level to to a discrete digital quantity. Xilinx 7 series FPGA devices contain a macro cell, *XADC*, which provides basic analog-to-digital conversion functionality. This chapter provides an overview of XADC and illustrates the procedure to integrate it into an FPro system.

12.1 OVERVIEW OF XADC

12.1.1 Block diagram

The conceptual block diagram of an XADC macro cell is shown in Figure 12.1. It consists of five major parts:
- Dual ADCs
- On-chip sensors and alarm
- Analog multiplexers
- Control register and status register
- DRP (dynamic reconfiguration port) interface

The key components are two *ADCs*. The ADC has a resolution of 12 bits and a sampling rate of 100M SPS (samples per second). It utilizes a *differential input* structure, in which the voltage difference between two pins (the plus and minus inputs) is used for measurement, and the maximum swing of the input signal is

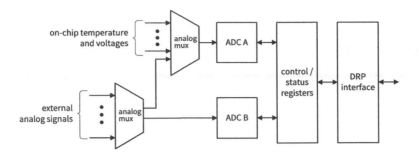

Figure 12.1 Conceptual block diagram of XADC

1.0 V. The input can be configured as *unipolar mode*, in which the voltage range is between 0.0 V and +1.0 V, or as *bipolar mode*, in which the voltage range is between −0.5 V and +0.5 V.

Modern FPGA devices impose strict constraints on power supplies. The *on-chip sensors* measure the die temperature and various supplied voltage levels. These analog readings can be passed to the ADCs and the digitized data can be retrieved by the user or used to set up alarms.

An *analog multiplexer* selects a specific input signal and connects it to the output. It allows multiple analog input channels to be routed to a single ADC. One useful application is to share a fast ADC with multiple low-data rate channels in a round-robin fashion, which is known as *sequencer mode* in Xilinx literature. For example, four input channels can be multiplexed though a 100M SPS ADC and the measurement and conversion are performed in turn. Each input channel is sampled at a rate of 25M SPS and the system appears as four 25M SPS ADCs. The XADC can accommodate one dedicated analog channel, 16 auxiliary analog channels, and internal temperature and voltage sensor readings. In sequencer mode, the needed channels and on-chip sensors can be selected. The I/O pins for the unused auxiliary analog channels can be configured as regular digital I/O pins.

The XADC contains 64 16-bit *control registers* and 64 16-bit *status registers*. The control registers store the configuration information, such as channel selection. The status registers contain the conversion results (i.e., digitized data) of input channels and on-chip sensors as well as the maximum and minimum values of various internal sensors. The address range of 0x00 to 0x1f is used for the conversion results.

The *DRP interface* provides an interface to access the internal registers. Its basic specification is similar to the MicroBlaze MCS I/O bus discussed in Section 10.6.2.

In addition to these parts, the XADC provides a collection of status signals (not related to status registers) that indicate the analog channel used in conversion and the completion of conversion.

12.1.2 Configuration

The XADC is versatile and flexible and many aspects can be configured. The configuration information is stored in its control registers. The "power-on" configuration can be specified via the registers' initial values, which are embedded in the bitstream and loaded into the control registers when an FPGA device is programmed. If desired, a write interface circuit can be included and the XADC can be reconfigured in real time by writing new values into the control registers.

12.2   XADC CORE DEVELOPMENT

The Nexys 4 DDR board utilizes four analog channels (channels 2, 3, 10, and 11) and connects them to a PMOD labeled JXADC. We develop an FPro XADC core that appears as four independent ADCs that sample four analog channels continuously. In addition, the core also provides the readings of die temperature and core voltage. The development involves creating a properly configured XADC instance and designing a wrapping circuit to retrieve the readings.

12.2.1   XADC instantiation

To accommodate the FPro XADC core, an XADC instance should be configured as follows:

- Set the channel selection and operation to *sequencer* mode to multiplex multiple channels and on-chip sensors.
- Set *sequencer* operation to *continuous* mode to perform conversion continuously (i.e., not "one-time" operation).
- Set timing to *continuous* to automatically trigger the conversion (i.e., not based on "events").
- Select the on-chip die temperature and core voltage sensors and analog channels 2, 3, 10, and 11 as for conversion.
- Select the *unipolar* mode for the input range of 0.0 V to 1.0 V.

The instance can be created and configured by Xilinx's XADC Wizard, which is illustrated in Appendix A.4.

The resulting HDL file, xadc_fpro.vhd, can also be downloaded from the book's companion website. The entity declaration of the configured XADC instance is

```vhdl
entity xadc_fpro is
   port
   (
   -- clock and reset
   dclk_in    : in   std_logic;
   reset_in   : in   std_logic;
   -- DRP interface signals
   daddr_in   : in   std_logic_vector (6 downto 0);
   den_in     : in   std_logic;
   di_in      : in   std_logic_vector (15 downto 0);
   dwe_in     : in   std_logic;
   do_out     : out  std_logic_vector (15 downto 0);
   drdy_out   : out  std_logic;
   -- dedicated analog input channel
   vp_in      : in   std_logic;
   vn_in      : in   std_logic;
   -- auxiliary analog input channel
   vauxp2     : in   std_logic;
   vauxn2     : in   std_logic;
   vauxp3     : in   std_logic;
```

```
        vauxn3      : in    std_logic;
        vauxp10     : in    std_logic;
        vauxn10     : in    std_logic;
        vauxp11     : in    std_logic;
        vauxn11     : in    std_logic;
        -- conversion status signals
        busy_out    : out   std_logic;
        channel_out : out   std_logic_vector (4 downto 0);
        eoc_out     : out   std_logic;
        eos_out     : out   std_logic;
        -- alarm output
        alarm_out   : out std_logic
    );
    end xadc_fpro;
```

The signals are for the DRP interface, four auxiliary analog input channels, and the XADC conversion status.

The dedicated analog input channel, vp_in and vn_in, and the alarm output, alarm_out, are not used in our design.

12.2.2 Basic wrapping circuit design

The instantiated XADC is configured to run automatically. It measures and converts the two on-chip sensor readings and four analog channels in turn and stores the most recently converted results in the corresponding status registers. The main functionality of the wrapping circuit is to read the designated status register via the DRP interface.

The basic specification of the DRP interface is similar to the MicroBlaze MCS I/O bus discussed in Section 10.6.2. The read operation performs as follows:

- The external control circuit places the register address on daddr_in.
- The external control circuit sets the register enable signal, den_in, to 1 and sets the write enable signal, dwe_in, to 0 (i.e., not write).
- The XADC retrieves data from the designated register.
- When the operation is completed, the XADC places data on do_out and asserts drdy_out to signal that the data is ready.

The DRP read operation may take multiple clock cycles and drdy_out functions as an acknowledgment. This is not compatible with the FPro bus specification, in which read is completed in one clock cycle.

One way to design the wrapping circuit is to access the DRP interface directly. The circuit will include a flag register associated with drdy_out and the software driver polls the register to check the validity of read data. This approach is left as an exercise in Section 12.7.5.

To simplify the interface and driver, we use an alternative wrapping circuit. The design contains a collection of external registers to maintain the digitized readings and uses the XADC status signals to refresh their contents. The block diagram is shown in Figure 12.2. The operation is performed in two phases. In the first phase, the XADC completion status signal activates a read operation to retrieve the newly converted data. In the second phase, the DRP's ready signal enables and stores the data into the designated external register.

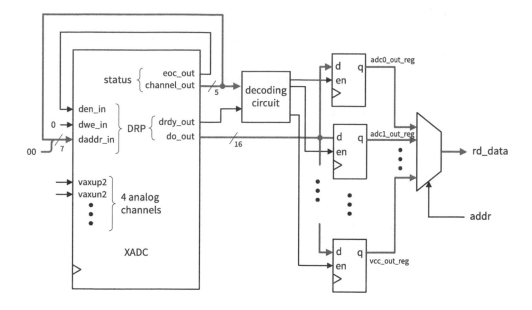

Figure 12.2   Block diagram of the XADC core wrapping circuit.

When a conversion is completed, the XADC core stores the result in the corresponding status register. At the same time, it places the channel number, which is the same as the five LSBs of the status register's address, on `channel_out`, and asserts the `eoc_out` (for "end of conversion") signal for one clock cycle. We use `eoc_out` signal as the DRP enable signal and the extended channel number as the DRP address to initiate a read operation, which retrieves the newly converted data.

When DRP completes the read operation, it places data on `do_out` and asserts the ready signal, `drdy_out`, for one clock cycle. The `drdy_out` signal in conjunction with the channel number enables and stores the newly converted data into the designated external register.

In summary, the wrapping circuit appears as six registers that maintain the current readings of two on-chip sensors and four analog inputs. The data is continuously and automatically updated. A read multiplexing circuit routes the selected reading to output.

12.2.3   Register map

From the processor's point of view, the XADC FPro core is a collection of six registers that supplies ADC readings. We arrange the register address offsets as follows:

- offset 0 to 3: analog inputs 0 to 3, which correspond to XADC's auxiliary analog input channels 3, 10, 2, and 11, respectively
- offset 4: on-chip die temperature reading
- offset 5: on-chip internal core voltage reading

All registers are 16 bits wide and the 12 MSBs are the ADC readings.

12.2.4   HDL code

The HDL code of XADC core follows the diagram in Figure 12.2 and is shown in
Listing 12.1.

Listing 12.1   XADC core

```
library ieee;
use ieee.std_logic_1164.all;
entity chu_xadc_core is
    port(
        clk     : in   std_logic;
        reset   : in   std_logic;
        -- slot interface
        cs      : in   std_logic;
        write   : in   std_logic;
        read    : in   std_logic;
        addr    : in   std_logic_vector(4 downto 0);
        rd_data : out  std_logic_vector(31 downto 0);
        wr_data : in   std_logic_vector(31 downto 0);
        -- external signals
        adc_p   : in   std_logic_vector(3 downto 0);
        adc_n   : in   std_logic_vector(3 downto 0)
    );
end chu_xadc_core;

architecture arch of chu_xadc_core is
    component xadc_fpro
        port(
            di_in       : in   std_logic_vector(15 downto 0);
            daddr_in    : in   std_logic_vector(6 downto 0);
            den_in      : in   std_logic;
            dwe_in      : in   std_logic;
            drdy_out    : out  std_logic;
            do_out      : out  std_logic_vector(15 downto 0);
            dclk_in     : in   std_logic;
            reset_in    : in   std_logic;
            vp_in       : in   std_logic;
            vn_in       : in   std_logic;
            vauxp2      : in   std_logic;
            vauxn2      : in   std_logic;
            vauxp3      : in   std_logic;
            vauxn3      : in   std_logic;
            vauxp10     : in   std_logic;
            vauxn10     : in   std_logic;
            vauxp11     : in   std_logic;
            vauxn11     : in   std_logic;
            channel_out : out  std_logic_vector(4 downto 0);
            eoc_out     : out  std_logic;
            alarm_out   : out  std_logic;
            eos_out     : out  std_logic;
            busy_out    : out  std_logic
        );
    end component;
    signal channel      : std_logic_vector(4 downto 0);
    signal daddr_in     : std_logic_vector(6 downto 0);
    signal eoc          : std_logic;
    signal rdy          : std_logic;
    signal adc_data     : std_logic_vector(15 downto 0);
    signal adc0_out_reg : std_logic_vector(15 downto 0);
    signal adc1_out_reg : std_logic_vector(15 downto 0);
    signal adc2_out_reg : std_logic_vector(15 downto 0);
```

```vhdl
    signal adc3_out_reg : std_logic_vector(15 downto 0);
    signal tmp_out_reg  : std_logic_vector(15 downto 0);
    signal vcc_out_reg  : std_logic_vector(15 downto 0);
begin
   -- instantiate cusomized xadc core
   xadc_unit : xadc_fpro
      port map(
         dclk_in       => clk,
         reset_in      => reset,           --reset,
         di_in         => (others => '0'),
         daddr_in      => daddr_in,
         den_in        => eoc,
         dwe_in        => '0',             -- read only
         drdy_out      => rdy,
         do_out        => adc_data,
         vp_in         => '0',
         vn_in         => '0',
         vauxp2        => adc_p(2),
         vauxn2        => adc_n(2),
         vauxp3        => adc_p(0),
         vauxn3        => adc_n(0),
         vauxp10       => adc_p(1),
         vauxn10       => adc_n(1),
         vauxp11       => adc_p(3),
         vauxn11       => adc_n(3),
         channel_out   => channel,
         eoc_out       => eoc,
         eos_out       => open,
         busy_out      => open,
         alarm_out     => open
      );
   -- form xadc DRP address
   daddr_in <= "00" & channel;
   -- registers and decoding
   process(clk, reset)
   begin
      if reset = '1' then
         adc0_out_reg <= (others => '0');
         adc1_out_reg <= (others => '0');
         adc2_out_reg <= (others => '0');
         adc3_out_reg <= (others => '0');
         tmp_out_reg  <= (others => '0');
         vcc_out_reg  <= (others => '0');
      elsif (clk'event and clk = '1') then
         if rdy = '1' and channel = "10011" then
            adc0_out_reg <= adc_data;
         end if;
         if rdy = '1' and channel = "11010" then
            adc1_out_reg <= adc_data;
         end if;
         if rdy = '1' and channel = "10010" then
            adc2_out_reg <= adc_data;
         end if;
         if rdy = '1' and channel = "11011" then
            adc3_out_reg <= adc_data;
         end if;
         if rdy = '1' and channel = "00000" then
            tmp_out_reg <= adc_data;
         end if;
         if rdy = '1' and channel = "00001" then
            vcc_out_reg <= adc_data;
```

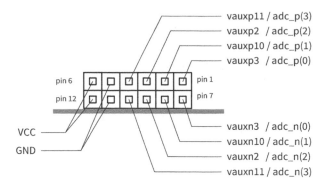

Figure 12.3 Analog input pin arrangement of Nexys 4 DDR JXADC PMOD port.

```
          end if;
       end if;
    end process;
    -- read multiplexing
    with addr(2 downto 0) select
       rd_data <=
          x"0000" & adc0_out_reg when "000",
          x"0000" & adc1_out_reg when "001",
          x"0000" & adc2_out_reg when "010",
          x"0000" & adc3_out_reg when "011",
          x"0000" & tmp_out_reg  when "100",
          x"0000" & vcc_out_reg  when others;
end arch;
```

Note that XADC has 128 registers with an address range between 0x00 and 0x7f and the ADC readings are assigned to registers 0x00 to 0x1f. The 16 auxiliary analog input channel readings are stored in 0x10 to 0x1f, and the die temperature and core voltage readings are stored in 0x00 and 0x01, respectively.

The strange mapping between the XADC's original auxiliary channel numbers (with prefixes of **vauxn** and **vauxp**) and FPro core's "logical channel numbers" (with prefixes of **adc_n** and **adc_p**) is due to the layout of Nexys 4 DDR board. The physical pin arrangement of **JXADC** PMOD port and naming convention are shown in Figure 12.3.

12.3 XADC CORE DEVICE DRIVER

The driver routines retrieve the six ADC measurements, including the die temperature and core voltage of the FPGA device and the voltages of the four external channels.

12.3.1 Class definition

The class definition of the XADC core is shown in Listing 12.2.

Listing 12.2 XadcCore class definition (in xadc_core.h)

```
#ifndef _XADC_CORE_H_INCLUDED
#define _XADC_CORE_H_INCLUDED
```

```
#include "chu_init.h"
class XadcCore {
public:
   enum {
      ADC_0_REG = 0,
      TMP_REG = 4,
      VCC_REG = 5,
   };
   XadcCore(uint32_t core_base_addr);
   ~XadcCore(); // not used
   uint16_t read_raw(int n);
   double read_adc_in(int n);
   double read_fpga_vcc();
   double read_fpga_temp();
private:
   uint32_t base_addr;
};

#endif   // _XADC_CORE_H_INCLUDED
```

The first **enum** definition uses symbolic names for the register offsets of the first analog input channel, the on-chip die temperature, and the core voltage register.

12.3.2 Class implementation

The class implementation of the constructor and the methods is shown in Listing 12.3.

Listing 12.3   XadcCore class implementation (in xadc_core.cpp)

```
XadcCore::XadcCore(uint32_t core_base_addr) {
   base_addr = core_base_addr;
}

XadcCore::~XadcCore() {}

uint16_t XadcCore:: read_raw(int n) {
   uint16_t rd_data;

   rd_data= (uint16_t) io_read(base_addr, ADC_0_REG+n) & 0x0000ffff;
   return (rd_data);
}

double  XadcCore::read_adc_in(int n){
   uint16_t raw;

   raw =  read_raw(n) >> 4;
   return((double)raw/4096.0);
}

double  XadcCore::read_fpga_vcc(){
   return(read_adc_in(VCC_REG)*3.0);
}

double  XadcCore::read_fpga_temp(){
   return(read_adc_in(TMP_REG)*503.975 - 273.15);
}
```

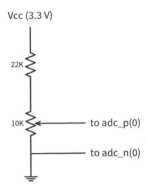

Figure 12.4 Voltage divider for an analog input.

The `XadcCore()` constructor saves the base address and the other methods retrieve and process the ADC readings. The `read_raw()` method retrieves and returns the 16-bit raw data, in which the 12 MSBs are the ADC reading with a range between 0x000 and 0xfff. For an analog channel, the range corresponds to 0.0 V and 1.0 V. The `read_adc_in()` method converts the 12-bit unsigned representation to the voltage. The on-chip voltage reading is scaled to one third of the actual value. The `read_fpga_vcc()` method performs the adjustment and returns the actual voltage. The relationship between the die temperature and the voltage reading is

$$temperature(^{\circ}C) = voltage * 503.975 - 273.15$$

The `read_fpga_temp()` method calculates and returns the temperature.

12.3.3 Testing for the XADC core

To test an analog input channel, a variable analog input voltage source with a range of 0.0 V to 1.0 V is needed. A simple way to achieve this is to use PMOD port's 3.3 V supply (labeled as VCC with pins 6 and 12) and a voltage divider. The diagram is shown in Figure 12.4. The voltage divider is composed of a 22K Ω resistor and a 10K Ω potentiometer. The minimum and maximum voltage levels crossing the potentiometer are 0.00 V ($\frac{0K}{22K+10K} * 3.3V$) and 1.03 V ($\frac{10K}{22K+10K} * 3.3V$), which provide the desired range. The potentiometer can be considered as another form of input device that provides increment adjustment over about 4000 values.

A simple test function is used to verify the operation of the XADC core, as in Listing 12.4. It retrieves the six readings and displays the values on the UART console. The measurements are repeated for five times.

Listing 12.4 XADC test function (in `main_sampler_test.cpp`)

```
void adc_check(XadcCore *adc_p, GpoCore *led_p) {
   double reading;
   int n, i;
   uint16_t raw;

   for (i = 0; i < 5; i++) {
      // display 12-bit channel 0 reading in LED
      raw = adc_p->read_raw(0);
```

```
        raw = raw >> 4;
        led_p->write(raw);
        // display on-chip sensor and 4 channels in console
        uart.disp("FPGA vcc/temp: ");
        reading = adc_p->read_fpga_vcc();
        uart.disp(reading, 3);
        uart.disp(" / ");
        reading = adc_p->read_fpga_temp();
        uart.disp(reading, 3);
        uart.disp("\n\r");
        for (n = 0; n < 4; n++) {
            uart.disp("analog channel/voltage: ");
            uart.disp(n);
            uart.disp(" / ");
            reading = adc_p->read_adc_in(n);
            uart.disp(reading, 3);
            uart.disp("\n\r");
        } // end for
        sleep_ms(200);
    }
}
```

12.4 SAMPLER FPRO SYSTEM

After a core is developed, it needs to be integrated into an FPro system and tested. Four MMIO cores are introduced in Part II and nine more cores are developed in this and subsequent chapters. It is tedious to construct a new FPro system for each core. We create a system that incorporates all these cores and call it a *sampler FPro system*. The general testing procedure and the sampler system are discussed in the following subsections.

12.4.1 Testing procedure of an FPro core

Following is the general procedure to verify the operation of an FPro MMIO core:

- Assign a new slot number for the core instance. Record the number as a symbolic constant in both **chu_io_map.h** and **chu_io_map.vhd** files, as discussed in Section 9.4.1.
- Attach the core to the designated slot of the MMIO controller and create a new MMIO subsystem, similar to the one in Section 10.5.3.
- Construct a new top-level FPro system that incorporates the MMIO subsystems, similar to the one in Section 10.7, and perform synthesis and placement and routing.
- Derive the device driver for the core, as discussed in Section 9.6.
- Develop a software verification program and obtain the .elf file, as discussed in Section 12.4.4.
- Regenerate the bit file, program the FPGA device, and verify the core's physical operation.

12.4.2 System configuration

The sampler FPro system contains 13 cores developed in Parts I and II of the book. The slot assignments and the corresponding symbolic names are shown in

Table 12.1   Slot assignment

Slot	Symbolic name	Description
0	S0_SYS_TIMER	system timer core
1	S1_UART1	UART core #1
2	S2_LED	GPO core for discrete LEDs
3	S3_SW	GPI core for discrete switches
4	S4_USER	user prototyping core
5	S5_XADC	Xilinx ADC core
6	S6_PWM	pulse width modulation core
7	S7_BTN	debouncing core for buttons
8	S8_SSEG	LED multiplexing core for seven-segment LED display
9	S9_SPI	SPI core
10	S10_I2C	I2C core
11	S11_PS2	PS2 core for keyboard or mouse
12	S12_DDFS	direct digital frequency synthesis core
13	S13_ADSR	ADSR envelope generator core

Table 12.1. The names are used in the `chu_io_map.vhd` and `chu_io_map.h` files. The first four slots are for the four basic cores, identical to the vanilla FPro system in Section 10.7. Slot 4 is reserved to facilitate new core prototyping. A new user-defined core can be inserted into this slot without affecting the sampler configuration. Slots 5 to 13 are used for the nine cores developed in Part III.

12.4.3 Hardware derivation

The code of the sampler MMIO subsystem is shown in Listing 12.5. The MMIO cores are instantiated based on the slot assignment in Table 12.1.

Listing 12.5   Sampler MMIO subsystem

```
library ieee;
use ieee.std_logic_1164.all;
use work.chu_io_map.all;
entity mmio_sys_sampler is
   port(
      -- FPro bus
      clk           : in    std_logic;
      reset         : in    std_logic;
      mmio_cs       : in    std_logic;
      mmio_wr       : in    std_logic;
      mmio_rd       : in    std_logic;
      mmio_addr     : in    std_logic_vector(20 downto 0);
      mmio_wr_data  : in    std_logic_vector(31 downto 0);
      mmio_rd_data  : out   std_logic_vector(31 downto 0);
      -- switches and LEDs
      sw            : in    std_logic_vector(15 downto 0);
      led           : out   std_logic_vector(15 downto 0);
      -- uart
      rx            : in    std_logic;
      tx            : out   std_logic;
      -- 4 analog input pair
      adc_p         : in    std_logic_vector(3 downto 0);
```

```vhdl
      adc_n           :  in     std_logic_vector(3 downto 0);
      -- pwm
      pwm             :  out    std_logic_vector(7 downto 0);
      -- btn
      btn             :  in     std_logic_vector(4 downto 0);
      -- 8-digit 7-seg LEDs
      an              :  out    std_logic_vector(7 downto 0);
      sseg            :  out    std_logic_vector(7 downto 0);
      -- spi accelerator
      acl_sclk        :  out    std_logic;
      acl_mosi        :  out    std_logic;
      acl_miso        :  in     std_logic;
      acl_ss          :  out    std_logic;
      -- i2c temperature sensor
      tmp_i2c_scl     :  out    std_logic;
      tmp_i2c_sda     :  inout  std_logic;
      -- ps2
      ps2d            :  inout  std_logic;
      ps2c            :  inout  std_logic;
      --ddfs square wave output
      ddfs_sq_wave    :  out    std_logic;
      -- 1-bit dac
      pdm             :  out    std_logic
   );
end mmio_sys_sampler;

architecture arch of mmio_sys_sampler is
   signal cs_array        :  std_logic_vector(63 downto 0);
   signal reg_addr_array  :  slot_2d_reg_type;
   signal mem_rd_array    :  std_logic_vector(63 downto 0);
   signal mem_wr_array    :  std_logic_vector(63 downto 0);
   signal rd_data_array   :  slot_2d_data_type;
   signal wr_data_array   :  slot_2d_data_type;
   signal adsr_env        :  std_logic_vector(15 downto 0);
begin
   --******************************************************************
   --  MMIO controller instantiation
   --******************************************************************
   ctrl_unit : entity work.chu_mmio_controller
      port map(
         -- FPro bus interface
         mmio_cs               => mmio_cs,
         mmio_wr               => mmio_wr,
         mmio_rd               => mmio_rd,
         mmio_addr             => mmio_addr,
         mmio_wr_data          => mmio_wr_data,
         mmio_rd_data          => mmio_rd_data,
         -- 64 slot interface
         slot_cs_array         => cs_array,
         slot_reg_addr_array   => reg_addr_array,
         slot_mem_rd_array     => mem_rd_array,
         slot_mem_wr_array     => mem_wr_array,
         slot_rd_data_array    => rd_data_array,
         slot_wr_data_array    => wr_data_array
      );
   --******************************************************************
   --  IO slots instantiations
   --******************************************************************
   -- slot 0: system timer
   timer_slot0 : entity work.chu_timer
      port map(
```

```
        clk      => clk,
        reset    => reset,
        cs       => cs_array(S0_SYS_TIMER),
        read     => mem_rd_array(S0_SYS_TIMER),
        write    => mem_wr_array(S0_SYS_TIMER),
        addr     => reg_addr_array(S0_SYS_TIMER),
        rd_data  => rd_data_array(S0_SYS_TIMER),
        wr_data  => wr_data_array(S0_SYS_TIMER)
    );
 -- slot 1: uart1
 uart1_slot1 : entity work.chu_uart
    generic map(FIFO_DEPTH_BIT => 6)
    port map(
        clk      => clk,
        reset    => reset,
        cs       => cs_array(S1_UART1),
        read     => mem_rd_array(S1_UART1),
        write    => mem_wr_array(S1_UART1),
        addr     => reg_addr_array(S1_UART1),
        rd_data  => rd_data_array(S1_UART1),
        wr_data  => wr_data_array(S1_UART1),
        -- external signals
        tx       => tx,
        rx       => rx
    );
 -- slot 2: GPO for 16 LEDs
 gpo_slot2 : entity work.chu_gpo
    generic map(W => 16)
    port map(
        clk      => clk,
        reset    => reset,
        cs       => cs_array(S2_LED),
        read     => mem_rd_array(S2_LED),
        write    => mem_wr_array(S2_LED),
        addr     => reg_addr_array(S2_LED),
        rd_data  => rd_data_array(S2_LED),
        wr_data  => wr_data_array(S2_LED),
        -- external signal
        dout     => led
    );
 -- slot 3: input port for 16 slide switches
 gpi_slot3 : entity work.chu_gpi
    generic map(W => 16)
    port map(
        clk      => clk,
        reset    => reset,
        cs       => cs_array(S3_SW),
        read     => mem_rd_array(S3_SW),
        write    => mem_wr_array(S3_SW),
        addr     => reg_addr_array(S3_SW),
        rd_data  => rd_data_array(S3_SW),
        wr_data  => wr_data_array(S3_SW),
        -- external signal
        din      => sw
    );
 -- slot 4: reserved for user defined
 -- user_slot4 : entity work.
 -- port map(
 --    clk         => clk,
 --    reset       => reset,
 --    cs          => cs_array(S4_USER),
```

```
--      read     => mem_rd_array(S4_USER),
--      write    => mem_wr_array(S4_USER),
--      addr     => reg_addr_array(S4_USER),
--      rd_data  => rd_data_array(S4_USER),
--      wr_data  => wr_data_array(S4_USER)
--   );
rd_data_array(4) <= (others => '0');
-- slot 5: xadc
xadc_slot5 : entity work.chu_xadc_core
   port map(
      clk     => clk,
      reset   => reset,
      cs      => cs_array(S5_XADC),
      read    => mem_rd_array(S5_XADC),
      write   => mem_wr_array(S5_XADC),
      addr    => reg_addr_array(S5_XADC),
      rd_data => rd_data_array(S5_XADC),
      wr_data => wr_data_array(S5_XADC),
      -- external signal
      adc_p   => adc_p,
      adc_n   => adc_n
   );
-- slot 6: pwm
pwm_slot6 : entity work.chu_io_pwm_core
   generic map(
      W => 8,
      R => 10)
   port map(
      clk     => clk,
      reset   => reset,
      cs      => cs_array(S6_PWM),
      read    => mem_rd_array(S6_PWM),
      write   => mem_wr_array(S6_PWM),
      addr    => reg_addr_array(S6_PWM),
      rd_data => rd_data_array(S6_PWM),
      wr_data => wr_data_array(S6_PWM),
      -- external interface
      pwm_out => pwm
   );
-- slot 7: push button
debounce_slot7 : entity work.chu_debounce_core
   generic map(
      W => 5,
      N => 20
   )
   port map(
      clk     => clk,
      reset   => reset,
      cs      => cs_array(S7_BTN),
      read    => mem_rd_array(S7_BTN),
      write   => mem_wr_array(S7_BTN),
      addr    => reg_addr_array(S7_BTN),
      rd_data => rd_data_array(S7_BTN),
      wr_data => wr_data_array(S7_BTN),
      -- external interface
      din     => btn
   );
-- slot 8: 7-seg LED
sseg_led_slot8 : entity work.chu_led_mux_core
   port map(
      clk     => clk,
```

```
        reset   => reset,
        cs      => cs_array(S8_SSEG),
        read    => mem_rd_array(S8_SSEG),
        write   => mem_wr_array(S8_SSEG),
        addr    => reg_addr_array(S8_SSEG),
        rd_data => rd_data_array(S8_SSEG),
        wr_data => wr_data_array(S8_SSEG),
        -- external interface
        an      => an,
        sseg    => sseg
    );
-- slot 9 SPI
spi_slot9 : entity work.chu_spi_core
    generic map(S => 1)
    port map(
        clk         => clk,
        reset       => reset,
        cs          => cs_array(S9_SPI),
        read        => mem_rd_array(S9_SPI),
        write       => mem_wr_array(S9_SPI),
        addr        => reg_addr_array(S9_SPI),
        rd_data     => rd_data_array(S9_SPI),
        wr_data     => wr_data_array(S9_SPI),
        spi_sclk    => acl_sclk,
        spi_mosi    => acl_mosi,
        spi_miso    => acl_miso,
        spi_ss_n(0) => acl_ss
    );
-- slot 10: i2C
i2c_slot10 : entity work.chu_i2c_core
    port map(
        clk     => clk,
        reset   => reset,
        cs      => cs_array(S10_I2C),
        read    => mem_rd_array(S10_I2C),
        write   => mem_wr_array(S10_I2C),
        addr    => reg_addr_array(S10_I2C),
        rd_data => rd_data_array(S10_I2C),
        wr_data => wr_data_array(S10_I2C),
        scl     => tmp_i2c_scl,
        sda     => tmp_i2c_sda
    );
-- slot 11: PS2
ps2_slot11 : entity work.chu_ps2_core
    generic map(W_SIZE => 6)
    port map(
        clk     => clk,
        reset   => reset,
        cs      => cs_array(S11_PS2),
        read    => mem_rd_array(S11_PS2),
        write   => mem_wr_array(S11_PS2),
        addr    => reg_addr_array(S11_PS2),
        rd_data => rd_data_array(S11_PS2),
        wr_data => wr_data_array(S11_PS2),
        -- external interface
        ps2d    => ps2d,
        ps2c    => ps2c
    );
-- slot 12: ddfs
ddfs_slot12 : entity work.chu_ddfs_core
    port map(
```

```
        clk          => clk,
        reset        => reset,
        cs           => cs_array(S12_DDFS),
        read         => mem_rd_array(S12_DDFS),
        write        => mem_wr_array(S12_DDFS),
        addr         => reg_addr_array(S12_DDFS),
        rd_data      => rd_data_array(S12_DDFS),
        wr_data      => wr_data_array(S12_DDFS),
        -- external interface
        focw_ext     => (others => '0'),
        pha_ext      => (others => '0'),
        env_ext      => adsr_env,
        pcm_out      => open,
        digital_out  => ddfs_sq_wave,
        pdm_out      => pdm
    );
    -- slot 13: adsr
    adsr_slot13 : entity work.chu_adsr_core
      port map(
        clk       => clk,
        reset     => reset,
        cs        => cs_array(S13_ADSR),
        read      => mem_rd_array(S13_ADSR),
        write     => mem_wr_array(S13_ADSR),
        addr      => reg_addr_array(S13_ADSR),
        rd_data   => rd_data_array(S13_ADSR),
        wr_data   => wr_data_array(S13_ADSR),
        -- external interface
        adsr_env => adsr_env
      );
    -- assign 0's to all unused slot rd_data signals
    gen_unused_slot : for i in 14 to 63 generate
      rd_data_array(i) <= (others => '0');
    end generate gen_unused_slot;
end arch;
```

The sampler FPro system can be created with the expanded MMIO subsystem. Its construction is similar to that of the vanilla FPro system discussed in Section 10.7. The HDL code is shown in Listing 12.6.

Listing 12.6 Sampler FPro subsystem

```
library ieee;
use ieee.std_logic_1164.all;
use work.chu_io_map.all;
entity mcs_top_sampler is
   generic(BRIDGE_BASE : std_logic_vector(31 downto 0) := x"C0000000");
   port(
      clk          : in    std_logic;
      reset_n      : in    std_logic;
      -- switches and LEDs
      sw           : in    std_logic_vector(15 downto 0);
      led          : out   std_logic_vector(15 downto 0);
      -- uart
      rx           : in    std_logic;
      tx           : out   std_logic;
      -- xadc
      adc_p        : in    std_logic_vector(3 downto 0);
      adc_n        : in    std_logic_vector(3 downto 0);
      -- rgb leds
      rgb_led1     : out   std_logic_vector(2 downto 0);
      rgb_led2     : out   std_logic_vector(2 downto 0);
```

```
      -- buttons
      btn            : in     std_logic_vector(4 downto 0);
      -- 4-digit 7-seg LEDs
      an             : out    std_logic_vector(7 downto 0);
      sseg           : out    std_logic_vector(7 downto 0);
      -- spi accelerator
      acl_sclk       : out    std_logic;
      acl_mosi       : out    std_logic;
      acl_miso       : in     std_logic;
      acl_ss_n       : out    std_logic;
      -- i2c temperature sensor
      tmp_i2c_scl    : out    std_logic;
      tmp_i2c_sda    : inout  std_logic;
      -- ps2
      ps2d           : inout  std_logic;
      ps2c           : inout  std_logic;
      -- nexys 4 audio
      audio_on       : out    std_logic;
      audio_pdm      : out    std_logic;
      -- PMOD JA (divided into top row and bottom row
      ja_top         : out    std_logic_vector(4 downto 1);
      ja_btm         : out    std_logic_vector(10 downto 7)
   );
end mcs_top_sampler;

architecture arch of mcs_top_sampler is
   component cpu
      port(
         clk              : in   std_logic;
         reset            : in   std_logic;
         io_addr_strobe   : out  std_logic;
         io_read_strobe   : out  std_logic;
         io_write_strobe  : out  std_logic;
         io_address       : out  std_logic_vector(31 downto 0);
         io_byte_enable   : out  std_logic_vector(3 downto 0);
         io_write_data    : out  std_logic_vector(31 downto 0);
         io_read_data     : in   std_logic_vector(31 downto 0);
         io_ready         : in   std_logic
      );
   end component;
   signal io_addr_strobe   : std_logic;
   signal io_read_strobe   : std_logic;
   signal io_write_strobe  : std_logic;
   signal io_byte_enable   : std_logic_vector(3 downto 0);
   signal io_address       : std_logic_vector(31 downto 0);
   signal io_write_data    : std_logic_vector(31 downto 0);
   signal io_read_data     : std_logic_vector(31 downto 0);
   signal io_ready         : std_logic;
   signal mmio_cs          : std_logic;
   signal mmio_wr          : std_logic;
   signal mmio_rd          : std_logic;
   signal mmio_addr        : std_logic_vector(20 downto 0);
   signal mmio_wr_data     : std_logic_vector(31 downto 0);
   signal mmio_rd_data     : std_logic_vector(31 downto 0);
   -- clk/reset related
   signal clk_100M         : std_logic;
   signal reset_sys        : std_logic;
   -- pwm
   signal pwm              : std_logic_vector(7 downto 0);
   -- ddfs/audio pdm
   signal pdm              : std_logic;
```

```vhdl
    signal ddfs_sq_wave    : std_logic;
begin
   -- clock and reset
   clk_100M             <= clk;              -- 100 MHz external clock
   reset_sys            <= not reset_n;
   -- audio
   audio_pdm            <= pdm;
   audio_on             <= '1';
   -- rgb leds
   rgb_led2             <= pwm(5 downto 3);
   rgb_led1             <= pwm(2 downto 0);
   -- PMOD JA
   ja_top(1)            <= ddfs_sq_wave;
   ja_top(2)            <= pdm;
   ja_top(4 downto 3)   <= pwm(7 downto 6);
   ja_btm               <= "0000";
   -- instantiate microBlaze MCS
   mcs_0 : cpu
      port map(
         clk              => clk_100M,
         reset            => reset_sys,
         io_addr_strobe   => io_addr_strobe,
         io_read_strobe   => io_read_strobe,
         io_write_strobe  => io_write_strobe,
         io_byte_enable   => io_byte_enable,
         io_address       => io_address,
         io_write_data    => io_write_data,
         io_read_data     => io_read_data,
         io_ready         => io_ready
      );
   -- instantiate MCS IO bus to FPro bus bridge
   bridge_unit : entity work.chu_mcs_bridge
      generic map(BRG_BASE => BRIDGE_BASE)
      port map(
         io_addr_strobe   => io_addr_strobe,
         io_read_strobe   => io_read_strobe,
         io_write_strobe  => io_write_strobe,
         io_byte_enable   => io_byte_enable,
         io_address       => io_address,
         io_write_data    => io_write_data,
         io_read_data     => io_read_data,
         io_ready         => io_ready,
         fp_video_cs      => open,
         fp_mmio_cs       => mmio_cs,
         fp_wr            => mmio_wr,
         fp_rd            => mmio_rd,
         fp_addr          => mmio_addr,
         fp_wr_data       => mmio_wr_data,
         fp_rd_data       => mmio_rd_data
      );
   -- instantiate sampler MMIO subsystem
   mmio_sys_unit : entity work.mmio_sys_sampler
      port map(
         clk              => clk_100M,
         reset            => reset_sys,
         mmio_cs          => mmio_cs,
         mmio_wr          => mmio_wr,
         mmio_rd          => mmio_rd,
         mmio_addr        => mmio_addr,
         mmio_wr_data     => mmio_wr_data,
         mmio_rd_data     => mmio_rd_data,
```

```
        sw            => sw,
        led           => led,
        rx            => rx,
        tx            => tx,
        adc_p         => adc_p,
        adc_n         => adc_n,
        pwm           => pwm,
        btn           => btn,
        an            => an,
        sseg          => sseg,
        tmp_i2c_scl   => tmp_i2c_scl,
        tmp_i2c_sda   => tmp_i2c_sda,
        acl_sclk      => acl_sclk,
        acl_mosi      => acl_mosi,
        acl_miso      => acl_miso,
        acl_ss        => acl_ss_n,
        ps2d          => ps2d,
        ps2c          => ps2c,
        ddfs_sq_wave  => ddfs_sq_wave,
        pdm           => pdm
    );
end arch;
```

12.4.4   Software verification program

The test program for the sampler FPro system is shown in Listing 12.7. It expands the test program of the vanilla FPro system discussed in Section 9.8.4 for the additional cores. In addition to the testing functions, the main program also includes an auxiliary function, show_test_id(), which flashes the binary pattern in discrete LEDs for a few seconds to indicate the number of the current test. The bodies of these test functions can be found in the respective chapters.

Listing 12.7   Sampler FPro test program (in main_sampler_test.cpp)

```cpp
#include "chu_init.h"
#include "gpio_cores.h"
#include "xadc_core.h"
#include "sseg_core.h"
#include "spi_core.h"
#include "i2c_core.h"
#include "ps2_core.h"
#include "ddfs_core.h"
#include "adsr_core.h"

void timer_check(GpoCore *led_p) { ... }
void led_check(GpoCore *led_p, int n) { ... }
void sw_check(GpoCore *led_p, GpiCore *sw_p) { ... }
void uart_check() { ... }
void adc_check(XadcCore *adc_p, GpoCore *led_p) { ... }
void pwm_3color_led_check(PwmCore *pwm_p) { ... }
void debounce_check(DebounceCore *db_p, GpoCore *led_p) { ... }
void sseg_check(SsegCore *sseg_p) { ... }
void gsensor_check(SpiCore *spi_p, GpoCore *led_p) { ... }
void adt7420_check(I2cCore *adt7420_p, GpoCore *led_p) { ... }
void ps2_check(Ps2Core *ps2_p) { ... }
void ddfs_check(DdfsCore *ddfs_p, GpoCore *led_p) { ... }
void adsr_check(AdsrCore *adsr_p, GpoCore *led_p, GpiCore *sw_p) { ... }

void show_test_id(int n, GpoCore *led_p) {
```

```
    int i, ptn;

    ptn = n;
    for (i = 0; i < 20; i++) {
        led_p->write(ptn);
        sleep_ms(30);
        led_p->write(0);
        sleep_ms(30);
    }
}

// instantiate cores
GpoCore led(get_slot_addr(BRIDGE_BASE, S2_LED));
GpiCore sw(get_slot_addr(BRIDGE_BASE, S3_SW));
XadcCore adc(get_slot_addr(BRIDGE_BASE, S5_XADC));
PwmCore pwm(get_slot_addr(BRIDGE_BASE, S6_PWM));
DebounceCore btn(get_slot_addr(BRIDGE_BASE, S7_BTN));
SsegCore sseg(get_slot_addr(BRIDGE_BASE, S8_SSEG));
SpiCore spi(get_slot_addr(BRIDGE_BASE, S9_SPI));
I2cCore adt7420(get_slot_addr(BRIDGE_BASE, S10_I2C));
Ps2Core ps2(get_slot_addr(BRIDGE_BASE, S11_PS2));
DdfsCore ddfs(get_slot_addr(BRIDGE_BASE, S12_DDFS));
AdsrCore adsr(get_slot_addr(BRIDGE_BASE, S13_ADSR), &ddfs);

int main() {
    timer_check(&led);
    while (1) {
        show_test_id(1, &led);
        led_check(&led, 16);
        sw_check(&led, &sw);
        show_test_id(3, &led);
        uart_check();
        show_test_id(5, &led);
        adc_check(&adc, &led);
        show_test_id(6, &led);
        pwm_3color_led_check(&pwm);
        show_test_id(7, &led);
        debounce_check(&btn, &led);
        show_test_id(8, &led);
        sseg_check(&sseg);
        show_test_id(9, &led);
        gsensor_check(&spi, &led);
        show_test_id(10, &led);
        adt7420_check(&adt7420, &led);
        show_test_id(11, &led);
        ps2_check(&ps2);
        show_test_id(12, &led);
        ddfs_check(&ddfs, &led);
        show_test_id(13, &led);
        adsr_check(&adsr, &led, &sw);
    } //while
} //main
```

12.5   ADDITIONAL PROJECT IDEAS

An ADC component allows the FPGA device to read analog input and to access a
variety of sensors. One type of sensors outputs continuous voltage levels. Following
are a few examples:

- Analog temperature sensor (such as TMP36): It outputs an analog voltage proportional to the ambient temperature
- IR (infrared) analog distance sensor (such as Sharp GY2Y0A series): It bounces an IR signal to an object and produces a voltage based on the distance of an object.
- Amplified microphone: It converts the ambient sound into electrical signal whose level is based on the amplitude (loudness) of the sound.

Another type of device has variable resistance and its value can be measured by a simple voltage divider circuit similar to that in Figure 12.4. The simplest device is a potentiometer discussed in Section 12.3.3. An interesting variation is a two-axis joystick. It is constructed with two independent potentiometers (one per axis), whose resistances vary according to the position of the "stick."

There are a larger number of "resistive sensors." They function as a variable resistor and their resistance changes according to certain physical stimuli. Following are some example devices and their stimuli:

- Photo resistor: light
- Force-sensitive resistor: physical pressure or weight
- Membrane potentiometer ("SoftPot"): touch position
- Flex sensor: bending in one direction
- Epoxy thermistor: temperature
- eTape liquid level sensor: liquid level

12.6   BIBLIOGRAPHIC NOTES

Xilinx includes the XADC macro cell in many FPGA device families. Its *7 Series FPGAs and Zynq-7000 All Programmable SoC XADC Dual 12-Bit 1 MSPS Analog-to-Digital Converter User Guide (UG480)* contains detailed information.

There is a wide variety of sensors available. A processor usually accesses sensors via an ADC or a serial SPI or I^2C interface (to be discussed in Chapters 15 and 16). *Sensors: A Hands-On Primer for Monitoring the Real World with Arduino and Raspberry Pi* by Tero Karvinen et al. provides a broad introduction of sensors. The projects and ideas in the book can be applied to the FPro system as well.

12.7   SUGGESTED EXPERIMENTS

12.7.1   Real-time voltage display

A LED-mux core developed in Chapter 14 controls the eight-digit seven-segment LED display on the Nexys 4 DDR board. We can use it to display the real-time voltage and the temperature reading of the FPGA device. Derive the program and verify its operation.

12.7.2   Potentiometer-controlled chasing LEDs

Consider the chasing LEDs circuit in Experiment 9.10.2. Instead of sliding switches, we can use a potentiometer and XADC in Section 12.3.3 to control the chasing speed. Derive the program and verify its operation.

12.7.3   Potentiometer-controlled LED dimmer

The PWM (pulse width modulation) core in Chapter 13 can vary the power delivered to an LED. Use a potentiometer and XADC to control the brightness of the LED. Derive the program and verify its operation.

12.7.4   Enhanced wrapping circuit I

The wrapping circuit in Section 12.2.2 is based on a fixed configuration. We want to dynamically configure the following:

- The channels in the sequencer.
- The input mode, which can be either *bipolar* or *unipoloar* mode.

Design a write interface circuit and incorporate it into the wrapping circuit and verify its operation.

12.7.5   Enhanced wrapping circuit II

Instead of accessing XADC's registers in an ad hoc way, we can develop a wrapping circuit for the DRP interface. Note that DRP has a seven-bit address space, which is wider than the FPro slot's five-bit address. For simplicity, the wrapping circuit only reads XADC's external analog input channels (with address between 0x10 to 0x1f) and only writes XADC's configuration register (with address between 0x40 to 0x4f). Design the new wrapping circuit, derive the new driver routines, and verify its operation.

CHAPTER 13

PULSE WIDTH MODULATION CORE

PWM (pulse-width modulation) is a scheme to encode analog signal levels with "digital pulses." It is frequently used to control the power supplied to a device, such as a motor or light. In this chapter, we develop an MMIO core to generate multiple PWM channels.

13.1 INTRODUCTION

13.1.1 PWM as analog output

A rectangle wave oscillates between zero and one. The *duty cycle* is the proportion of *"on (logic 1) interval"* within a period. Consider a periodic rectangle wave with a period of t_{PWM} and an on interval of t_{ON}, as shown in Figure 13.1(a). The duty cycle is defined as

$$duty\ cycle = \frac{t_{ON}}{t_{PWM}}$$

The waveforms of different duty cycles are illustrated in Figure 13.1(b).

The PWM is a scheme to adjust the duty cycle of a wave. The PWM signal can be considered as a specific form of analog signal. Let the voltages of logic 0 and logic 1 be 0 and Vcc and the duty cycle be d. The average voltage level is $d * Vcc$ (i.e., $d * Vcc + (1 - d) * 0$). Thus, while the instantaneous output voltage values are "digital" (i.e., either 0 or Vcc), the *average* voltage is "analog" and can

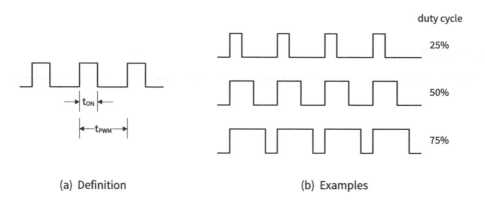

(a) Definition   (b) Examples

Figure 13.1   Definition and examples of duty cycle.

be incremented gradually from 0 to Vcc. The PWM scheme can be thought of as a special digital-to-analog conversion method that embeds the analog value in its duty cycle.

A main application of PWM is to control the power supplied to a device since the power delivery concerns less about the instantaneous value. If a PWM signal switches fast enough, the delivered power perceived by the load appears to be smooth and proportional to the duty cycle.

13.1.2   Main characteristics

The two important characteristics of PWM are the *switching frequency* and *resolution*. The *switching frequency* is the frequency of the rectangle wave, which is $\frac{1}{t_{PWM}}$ in Figure 13.1. The exact switching frequency depends on the nature of the application. For example, we can use PWM to control the brightness of an LED. If the switching frequency is too low (e.g., 10 Hz), the LED may appear to be flickering. On the other hand, if the switching frequency is too high (e.g., 1M Hz), the LED may not have enough time to completely turn on or off. It is about a few KHz to tens of KHz for a motor drive and into the tens or hundreds of KHz in audio amplifiers and power supplies.

The *resolution* describes the "granularity" (i.e., the minimum incremental step) of the duty cycle. It is specified in terms of *number of bits*. For a resolution of R bits, there are 2^R levels in the duty cycle and the minimum incremental step is $\frac{1}{2^R}$. For example, the granularity of an 8-bit PWM is $\frac{1}{256}$ (i.e., $\frac{1}{2^8}$) and its duty cycles are $\frac{0}{256}, \frac{1}{256}, \frac{2}{256}, \frac{3}{256}, \cdots, \frac{255}{256}, \frac{256}{256}$.

13.2   PWM DESIGN

13.2.1   Basic design

The basic PWM circuit consists of a binary counter and a comparator, as shown in Figure 13.2. The desired duty cycle input, `duty`, sets the threshold for the comparator and the comparator's output is asserted when the counter's value is below the threshold. For example, assume that the resolution of the PWM is eight

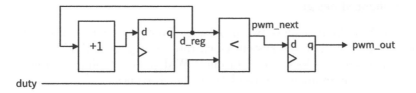

Figure 13.2   Block diagram of basic PWM circuit.

bits. The counter counts from 0 to 255 and wraps around. If the value of the `duty` signal is d, the comparator's output is asserted when the counter's value is 0, 1, 2, $\cdots$, $d - 1$. Thus, d out of 256 clock cycles is asserted, and the duty cycle is $\frac{d}{256}$. A D FF is added to the comparator's output to filter out potential glitches.

The HDL code of the basic PWM circuit follows the diagram in Figure 13.2 and is shown in Listing 13.1.

Listing 13.1   Basic PWM circuit

```vhdl
library ieee;
use ieee.std_logic_1164.all;
use ieee.numeric_std.all;
entity pwm_basic is
   generic(
      R : integer := 8    -- # bits of PWM resolution (i.e., 2^R levels)
   );
   port(
      clk     : in  std_logic;
      reset   : in  std_logic;
      duty    : in  std_logic_vector(R - 1 downto 0);
      pwm_out : out std_logic
   );
end pwm_basic;

architecture arch of pwm_basic is
   signal d_reg, d_next    : unsigned(R - 1 downto 0);
   signal pwm_reg, pwm_next : std_logic;
begin
   process(clk, reset)
   begin
      if reset = '1' then
         d_reg <= (others => '0');
         pwm_reg <= '0';
      elsif (clk'event and clk = '1') then
         d_reg <= d_next;
         pwm_reg <= pwm_next;
      end if;
   end process;
   -- duty cycle counter
   d_next   <= d_reg + 1;
   -- comparison circuit
   pwm_next <= '1' when d_reg < unsigned(duty) else '0';
   pwm_out  <= pwm_reg;
end arch;
```

A generic, R, represents the number of bits in counter and thus defines the resolution of the PWM.

13.2.2   Enhanced design

The basic PWM circuit suffers two problems. First, it cannot control the PWM switching frequency. Let the system clock rate be f_{sys} and PWM resolution be R bits. There are 2^R clock cycles in one PWM period (t_{PWM} in Figure 13.1). Thus, the switching frequency of the PWM wave becomes:

$$f_{pwm} = \frac{f_{sys}}{2^R}$$

For a PWM with a 100M-Hz system clock and 8-bit resolution, f_{pwm} is about 390K Hz, which is too high for many applications. We can use a *prescaler* counter, similar to the tick-generation scheme discussed in Section 4.5.2, to produce a one-clock tick to drive the duty counter. If the desired PWM frequency is f_p, the divisor value should satisfy

$$f_p = \frac{\frac{f_{sys}}{dvsr}}{2^R}$$

Thus, the divisor value should be

$$dvsr = \frac{f_{sys}}{2^R * f_p}$$

The second problem is the 100% duty cycle. Consider an eight-bit PWM in which the value of the `duty` signal is d. The comparator's output is asserted when the counter's value is 0, 1, 2, $\cdots$, $d-1$ and thus its duty cycle is $\frac{d}{256}$. However, since the `duty` signal is eight bits wide and the maximum value is 255, the maximum duty cycle can be achieved is $\frac{255}{256}$, which is not 100% (i.e., $\frac{256}{256}$). Closer examination shows that d contains 257 values (i.e., 0, 1, 2, $\cdots$, 254, 255, 256). The `duty` signal should be extended by one extra bit to accommodate the additional value; i.e., the input range should be extended from [0x00, 0xff] to [0x000, 0x100].

The enhanced PWM design incorporates the prescaler counter and the extended comparison circuit and extends the width of the `duty` port. Its HDL code is shown in Listing 13.2. To achieve flexibility, the prescaler value is specified by an external signal, `dvsr`.

Listing 13.2   Enhanced PWM circuit

```
library ieee;
use ieee.std_logic_1164.all;
use ieee.numeric_std.all;
entity pwm_enhanced is
   generic(
      R : integer := 10
   );
   port(
      clk     : in  std_logic;
      reset   : in  std_logic;
      dvsr    : in  std_logic_vector(31 downto 0);
      duty    : in  std_logic_vector(R downto 0);
      pwm_out : out std_logic
   );
end pwm_enhanced;

architecture arch of pwm_enhanced is
   signal q_reg, q_next    : unsigned(31 downto 0);
```

```
   signal d_reg, d_next      : unsigned(R - 1 downto 0);
   signal d_ext              : unsigned(R downto 0);
   signal pwm_reg, pwm_next  : std_logic;
   signal tick               : std_logic;
begin
   process(clk, reset)
   begin
      if reset = '1' then
         q_reg   <= (others => '0');
         d_reg   <= (others => '0');
         pwm_reg <= '0';
      elsif (clk'event and clk = '1') then
         q_reg   <= q_next;
         d_reg   <= d_next;
         pwm_reg <= pwm_next;
      end if;
   end process;
   — "prescaler" counter
   q_next <= (others => '0') when q_reg=unsigned(dvsr) else q_reg + 1;
   tick   <= '1' when q_reg = 0 else '0';
   — duty cycle counter
   d_next <= d_reg + 1 when tick = '1' else d_reg;
   d_ext  <= '0' & d_reg;
   — comparison circuit
   pwm_next <= '1' when d_ext < unsigned(duty) else '0';
   pwm_out  <= pwm_reg;
end arch;
```

13.3   PWM CORE DEVELOPMENT

The GPO core discussed in Section 10.3.3 generates multiple outputs. Each output assumes a value of logic 0 (i.e., 0 V) or logic 1 (i.e., Vcc) and thus is "digital" in nature. The PWM circuit generates a signal with an adjustable duty cycle, which can be treated as a one-bit "analog" output. A PWM core contains multiple PWM circuits and can be considered as a core that generates multiple "analog" outputs. For simplicity, we assume that the PWM outputs are used to drive similar devices (e.g., LEDs) and thus the same switching frequency is applied to all PWM circuits in the core.

13.3.1   Register map

The processor interacts with the PWM circuits as follows:

- specify (i.e., write) the switching frequency via the dvsr signal.
- set (i.e., write) the value of the duty cycle for each output.

The register map of the PWM core is somewhat different from other cores. Because setting the duty cycle requires multiple bits (i.e., the resolution of the PWM circuit), each PWM output needs a register. Thus, the number of registers in the core is not fixed but depends on the number of PWM outputs.

Assume that a PWM core has $w + 1$ outputs. The address offsets and fields of the registers are:

- offset 0x00 (divisor register)
  - bits 31 to 0: divisor value of the prescaler counter

- offset 0x10 (duty cycle register for PWM bit 0)
  - bits R to 0: duty cycle value (where R is resolution of the PWM circuit)
- offset 0x11 (duty cycle register for PWM bit 1)
  - bits R to 0: duty cycle value
- offset 0x12 (duty cycle register for PWM bit 2)
  - bits R to 0: duty cycle value

  $\vdots$

- offset 0x1w (duty cycle register for PWM bit w)
  - bits R to 0: duty cycle value

Clearly, the maximum number of PWM outputs is 16.

13.3.2 Wrapped PWM circuit

To comply to the slot specification, we can construct a wrapping circuit that contains registers and a decoding circuit to store the divisor and duty cycle values. The HDL code of the PWM core is shown in Listing 13.3.

Listing 13.3   PWM core

```vhdl
library ieee;
use ieee.std_logic_1164.all;
use ieee.numeric_std.all;
entity chu_io_pwm_core is
   generic(
      W : integer := 8;      -- width (# bits) of output port
      R : integer := 8       -- # bits of PWM resolution (2^R levels)
   );
   port(
      clk    : in  std_logic;
      reset  : in  std_logic;
      -- slot interface
      cs     : in  std_logic;
      write  : in  std_logic;
      read   : in  std_logic;
      addr   : in  std_logic_vector(4 downto 0);
      rd_data : out std_logic_vector(31 downto 0);
      wr_data : in  std_logic_vector(31 downto 0);
      -- external signals
      pwm_out : out std_logic_vector(W - 1 downto 0)
   );
end chu_io_pwm_core;

architecture arch of chu_io_pwm_core is
   type reg_file_type is array (W - 1 downto 0) of
         std_logic_vector(R downto 0);
   signal duty_2d_reg    : reg_file_type;
   signal wr_en, dvsr_en : std_logic;
   signal duty_array_en  : std_logic;
   signal q_reg, q_next  : unsigned(31 downto 0);
   signal d_reg, d_next  : unsigned(R - 1 downto 0);
   signal d_ext          : unsigned(R downto 0);
   signal pwm_next       : std_logic_vector(W - 1 downto 0);
   signal pwm_reg        : std_logic_vector(W - 1 downto 0);
   signal tick           : std_logic;
   signal dvsr_reg       : std_logic_vector(31 downto 0);
begin
```

```vhdl
   --*********************************************************************
   -- wrapping circuit
   --*********************************************************************
   -- decoding logic
   wr_en          <= '1' when write = '1' and cs = '1' else '0';
   duty_array_en <= '1' when wr_en = '1' and addr(4) = '1' else '0';
   dvsr_en        <= '1' when wr_en = '1' and addr = "00000" else '0';
   -- register for divisor
   process(clk, reset)
   begin
      if (reset = '1') then
         dvsr_reg <= (others => '0');
      elsif (clk'event and clk = '1') then
         if dvsr_en = '1' then
            dvsr_reg <= wr_data;
         end if;
      end if;
   end process;
   -- register file for duty cycles
   process(clk, reset)
   begin
      if (reset = '1') then
         duty_2d_reg <= (others => (others => '0'));
      elsif (clk'event and clk = '1') then
         if duty_array_en = '1' then
            duty_2d_reg(to_integer(unsigned(addr(3 downto 0))))<=wr_data;
         end if;
      end if;
   end process;
   --*********************************************************************
   -- multi-bit PWM
   --*********************************************************************
   process(clk, reset)
   begin
      if reset = '1' then
         q_reg    <= (others => '0');
         d_reg    <= (others => '0');
         pwm_reg <= (others => '0');
      elsif (clk'event and clk = '1') then
         q_reg    <= q_next;
         d_reg    <= d_next;
         pwm_reg <= pwm_next;
      end if;
   end process;
   -- "prescale" counter
   q_next <= (others=>'0') when q_reg=unsigned(dvsr_reg) else q_reg + 1;
   tick    <= '1' when q_reg = 0 else '0';
   -- duty cycle counter
   d_next <= d_reg + 1 when tick = '1' else d_reg;
   d_ext  <= '0' & d_reg;
   -- comparison circuit
   gen_comp_cell : for i in 0 to W - 1 generate
      pwm_next(i) <= '1' when d_ext<unsigned(duty_2d_reg(i)) else '0';
   end generate;
   pwm_out <= pwm_reg;
   -- read data not used
   rd_data <= (others => '0');
end arch;
```

The code is composed of two major parts. The first part is the wrapping circuit. It contains a decoding circuit, a register for the divisor, and a two-dimensional register file for the duty cycles. The size of the register file is specified by the W generic and thus can be adjusted to accommodate the desired number of duty cycle registers.

The second part is the multiple-bit PWM circuit. The code is similar to that in Listing 13.2 except that multiple comparison circuits are inferred from the generate statement. Note that the prescaler counter and duty counter are shared by all PWM bits.

13.4 PWM DRIVER

The PWM driver consists of a set of routines to set the PWM switching frequency and the duty cycle of individual PWM channels. Since the PWM signal is considered a form of "analog" output, the class definition and implementation are included in the gpio_core.h and gpio_core.cpp files.

13.4.1 Class definition

The class definition of the PWM core is shown in Listing 13.4.

Listing 13.4 PwmCore class definition (in gpio_core.h)

```
class PwmCore {
   enum {
      DVSR_REG = 0x00,
      DUTY_REG_BASE = 0x10
   };
   enum {
      RESOLUTION_BITS = 10,
      MAX = 1 << RESOLUTION_BITS
   };
public:
   PwmCore(uint32_t core_base_addr);
   ~PwmCore();
   void set_freq(int freq);
   void set_duty(int duty, int channel);
   void set_duty(double f, int channel);
private:
   uint32_t base_addr;
};
```

The enum definition uses symbolic names for the divisor register offsets and the base offset of the duty cycle registers. The second enum definition specifies the PWM resolution, which is defined in HDL code when the PWM core is instantiated, and the maximum duty cycle value, which is $2^{RESOLUTION\_BITS}$. The PWM resolution is set to 10 bits in our instantiated core.

13.4.2 Class implementation

The class implementation of the constructor and methods is shown in Listing 13.5.

Listing 13.5 PwmCore class implementation (in gpio_core.cpp)

```
PwmCore::PwmCore(uint32_t core_base_addr) {
   base_addr = core_base_addr;
   set_freq(1000);
}
```

```
PwmCore::~PwmCore() { }

void PwmCore::set_freq(int freq) {
   uint32_t dvsr;
   dvsr = (uint32_t) SYS_CLK_FREQ * 1000000 / MAX / freq;
   io_write(base_addr, DVSR_REG, dvsr);
}

void PwmCore::set_duty(int duty, int channel) {
   uint32_t d;

   if (duty > MAX) {
      d = MAX;
   } else {
      d = duty;
   }
   io_write(base_addr, DUTY_REG_BASE + channel, d);
}

void PwmCore::set_duty(double f, int channel) {
   int duty;

   duty = (int) (f * MAX);
   //debug("set_duty_f: ", f, duty);
   set_duty(duty, channel);
}
```

The `PwmCore()` constructor saves the base address and sets the default PWM switching frequency to 1000 Hz. The `set_freq()` method calculates the divisor value from the desired switching frequency and writes the value to the divisor register. The `set_duty()` methods write the duty cycle value to the designated register. The methods are overloaded. The duty cycle can be expressed as an integer (as with the `duty` parameter) with a range between 0 and `MAX`, and the raw data will be used directly. Alternatively, the duty cycle can be expressed as a real number integer (as with the `f` parameter) with a range between 0.0 and 1.0, representing 0 and 100% duty cycle. The latter method converts the real number to integer and writes the integer to the designated register. This method is preferred since the duty cycle value is independent of the resolution.

13.5 TESTING

The sampler FPro system instantiates an eight-bit PWM core instance. Six output channels are used to drive two tricolor LEDs on the Nexys 4 DDR board and two channels are connected to the two pins of a PMOD port.

The testing routine demonstrates the basic PWM operation by gradually increasing the brightness of tricolor LEDs by incrementing the values of duty cycle. The code is shown in Listing 13.6.

Listing 13.6   PWM test function (in `main_sampler_test.cpp`)

```
void pwm_3color_led_check(PwmCore *pwm_p) {
   int i, n;
   double bright, duty;
   const double P20 = 1.2589;   // P20=100^(1/20); i.e., P20^20=100
```

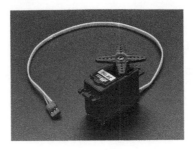

Figure 13.3   Servo motor.

```
pwm_p->set_freq(50);
for (n = 0; n < 3; n++) {
    bright = 1.0;
    for (i = 0; i < 20; i++) {
        bright = bright * P20;
        duty = bright / 100.0;
        pwm_p->set_duty(duty, n);
        pwm_p->set_duty(duty, n + 3);
        sleep_ms(100);
    }
    sleep_ms(300);
    pwm_p->set_duty(0.0, n);
    pwm_p->set_duty(0.0, n + 3);
}
}
```

13.6   PROJECT IDEAS

Servo motors (or simply *servos*) are DC motors equipped with a feedback mechanism to control the angular position. A common range of rotation is 180° (from −90° to +90°). The servo motors are used for precision positioning, such as sensor scanners and robotic arms and legs. A picture of a servo motor is shown in Figure 13.3. Servo motors have three wires, which are power, ground, and "signal." The angle of rotation is controlled by sending pulses of variable "on" duration through the "signal" wire. For a typical servo motor, the period of the pulse is 20 ms (millisecond) and the "on" duration is between 1.0 ms and 2.0 ms, corresponding to the rotation angle from −90° to +90°. The timing diagram of different angles is shown in Figure 13.4.

A PWM pulse can be used to control a servo motor by setting its switching frequency to 50 Hz (i.e., a period of 20 ms). The "on" duration corresponds to a duty cycle between 5% (i.e., $\frac{1\ ms}{20\ ms}$) to 10% (i.e., $\frac{2\ ms}{20\ ms}$). For a 10-bit PWM, there are about 50 steps within the 1-ms duration (i.e., $2^{10} * \frac{1\ ms}{20\ ms}$), which corresponds to a minimum angle increment of 3.6° (i.e., $\frac{180°}{50}$). Finer angle increment can be obtained with a better PWM resolution.

Since the PWM core can be easily configured to obtain the desired PWM channels and resolution, it can support robot projects that involve a large number of servo motors, such as a robotic arm or a multi-leg spider-like robot.

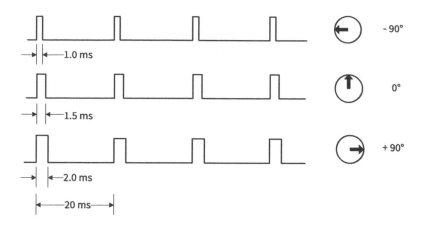

Figure 13.4   Servo motor timing diagram.

While most FPGA prototyping boards provide some voltage sources, they should not be used to drive motors. First, a motor can draw a significant amount of current, which may exceed the board's capability. Second, a motor is an *inductive load*. When it is turned on or changes direction, the sudden change of current may lead to a large transient voltage surge that can affect the power supply. It is recommended that a separate power supply is used for the motors.

13.7   SUGGESTED EXPERIMENTS

13.7.1   Police dash light

A police car dash light flashes various red and blue patterns. Search on line to find a few such patterns and use the two tri-color LEDs to emulate these patterns. Derive the program and verify its operation.

13.7.2   Rainbow night light

A rainbow-like spectrum is shown in Figure 13.5 (if needed, the color diagram can be downloaded from the companion website). The spectrum can be generated by varying the intensity of the red, green, and blue inputs of the tri-color LEDs. A rainbow light uses the potentiometer and XADC core (discussed in Section 12.3.3) as the input to specify the desired color on a tricolor LED. Derive the program and verify its operation.

13.7.3   Enhanced PWM core: part I

The resolution of the PWM core designed in Section 13.3 is determined by the generic R and thus is fixed after synthesis. To have more flexibility, we want to make the number of resolution bits dynamically configurable. Design the new PWM core, revise the driver as needed, and verify its operation.

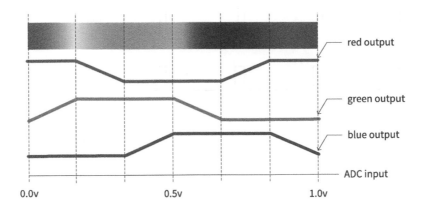

Figure 13.5 "Rainbow" spectrum.

13.7.4 Enhanced PWM core: part II

The PWM core designed in Section 13.3 can accommodate multiple channels. We want to have more control and independently specify the switching frequency for each channel. Design the new PWM core, revise the driver as needed, and verify its operation.

13.7.5 Enhanced GPIO core

A bi-direction GPIO core is defined in Experiment 10.9.3. We want to enhance the core by using PWM for its output; i.e., instead of "digital output" with two discrete levels, the core can generate a PWM output pulse with different power levels. Repeat Experiment 10.9.3 with the enhanced output potion.

13.7.6 Servo motor driver

Derive the driver routines for the servo motor discussed in Section 13.6. Write a testing program and verify its operation.

CHAPTER 14

DEBOUNCING CORE AND LED-MUX CORE

We develop a debouncing circuit in Chapter 5 and an LED multiplexing circuit in Chapter 4. In this chapter, we demonstrate how to convert them into MMIO cores and incorporate them into an FPro system.

14.1 DEBOUNCING CORE

A debouncing circuit filters out the transient contact bounces from a mechanical switch. A debouncing core is like a general-purpose input core but with "cleaned" input transitions.

14.1.1 Multi-bit debouncing circuit

A debouncing circuit is introduced in Section 5.3.2 and revisited in Section 6.2.1. The former uses a separate timer with an FSM and the latter integrates the counting functionality into an FSMD. To facilitate the core development, the circuit must be expanded to support multiple input bits. A simple way to achieve this is to replicate the original one-bit circuit multiple times with a generate statement. However, this approach is not very efficient. When the circuit is synthesized, the timer, which counts for the 10-ms interval, infers a large incrementor. Replicating it multiple times is a waste of hardware resources.

For the FSM design discussed in Section 5.3.2, a timer is separated from the FSM circuit and thus can be shared as needed. A multi-bit debouncing circuit

thus only needs to replicate the FSM. To facilitate the development, the original
code in Listing 5.6 is split into an FSM module and a timer module, as shown in
Listings 14.1 and 14.2.

Listing 14.1   Debouncing FSM

```vhdl
library ieee;
use ieee.std_logic_1164.all;
use ieee.numeric_std.all;
entity debounce_fsm is
   port(
      clk, reset : in   std_logic;
      sw         : in   std_logic;
      m_tick     : in   std_logic;
      db         : out  std_logic
   );
end debounce_fsm;

architecture arch of debounce_fsm is
   type eg_state_type is
      (zero, wait1_1, wait1_2, wait1_3, one, wait0_1, wait0_2, wait0_3);
   signal state_reg, state_next : eg_state_type;
begin
   -- state register
   process(clk, reset)
   begin
      if (reset = '1') then
         state_reg <= zero;
      elsif (clk'event and clk = '1') then
         state_reg <= state_next;
      end if;
   end process;
   -- next-state/output logic
   process(state_reg, sw, m_tick)
   begin
      state_next <= state_reg;
      db         <= '0';
      case state_reg is
         when zero =>
            if sw = '1' then
               state_next <= wait1_1;
            end if;
         when wait1_1 =>
            if sw = '0' then
               state_next <= zero;
            else
               if m_tick = '1' then
                  state_next <= wait1_2;
               end if;
            end if;
         when wait1_2 =>
            if sw = '0' then
               state_next <= zero;
            else
               if m_tick = '1' then
                  state_next <= wait1_3;
               end if;
            end if;
         when wait1_3 =>
            if sw = '0' then
               state_next <= zero;
            else
```

```
                    if m_tick = '1' then
                        state_next <= one;
                    end if;
                end if;
            when one =>
                db <= '1';
                if sw = '0' then
                    state_next <= wait0_1;
                end if;
            when wait0_1 =>
                db <= '1';
                if sw = '1' then
                    state_next <= one;
                else
                    if m_tick = '1' then
                        state_next <= wait0_2;
                    end if;
                end if;
            when wait0_2 =>
                db <= '1';
                if sw = '1' then
                    state_next <= one;
                else
                    if m_tick = '1' then
                        state_next <= wait0_3;
                    end if;
                end if;
            when wait0_3 =>
                db <= '1';
                if sw = '1' then
                    state_next <= one;
                else
                    if m_tick = '1' then
                        state_next <= zero;
                    end if;
                end if;
        end case;
    end process;
end arch;
```

Listing 14.2   Debouncing timer

```vhdl
library ieee;
use ieee.std_logic_1164.all;
use ieee.numeric_std.all;
entity debounce_counter is
    generic(N : integer := 20);    -- 2^N * 10ns = 10ms tick
    port(
        clk, reset : in  std_logic;
        m_tick     : out std_logic
    );
end debounce_counter;

architecture arch of debounce_counter is
    signal q_reg, q_next : unsigned(N - 1 downto 0);
begin
    process(clk, reset)
    begin
        if (reset = '1') then
            q_reg <= (others => '0');
        elsif (clk'event and clk = '1') then
```

```
            q_reg <= q_next;
        end if;
    end process;
    q_next <= q_reg + 1;
    --output tick
    m_tick <= '1' when q_reg = 0 else '0';
end arch;
```

14.1.2   Register map and the slot wrapping circuit

The debouncing core is like a general-purpose input core that filters the unwanted transient bounces. To make it more versatile, the core provides both untreated and debounced input signals. The register map can be defined accordingly. There are two read registers. Their address offsets and fields are:

- offset 0 (raw input data register)
    - bits W-1 to 0: W-bit untreated inputs, where W is the VHDL generic value used in core instantiation
- offset 1 (debounced input data register)
    - bits W-1 to 0: W-bit debounced inputs

Based on the register map, we can derive a wrapping circuit that complies with the slot specification and create the debouncing core. The HDL code of the core is shown in Listing 14.3. It consists of an input data register for untreated inputs, an instance of a timer, and multiple instances of the debouncing FSMs, and a read multiplexing circuit. The multiple FSMs are obtained by a generate statement and they share the same timer.

Listing 14.3   Debouncing core

```
library ieee;
use ieee.std_logic_1164.all;
entity chu_debounce_core is
   generic(
       W : integer := 8;      -- width of input port
       N : integer := 20      -- # bit for 10-ms tick 2^N * clk period
   );
   port(clk       : in  std_logic;
        reset     : in  std_logic;
        -- io bridge interface
        cs        : in  std_logic;
        write     : in  std_logic;
        read      : in  std_logic;
        addr      : in  std_logic_vector(4 downto 0);
        rd_data   : out std_logic_vector(31 downto 0);
        wr_data   : in  std_logic_vector(31 downto 0);
        -- external signal
        din       : in  std_logic_vector(W-1 downto 0));
end chu_debounce_core;

architecture arch of chu_debounce_core is
   signal rd_data_reg : std_logic_vector(W-1 downto 0);
   signal m_tick      : std_logic;
   signal db_out      : std_logic_vector(W-1 downto 0);
begin
   --********************************************************************
   -- input register
```

```
   --*****************************************************************
   process(clk, reset)
   begin
      if reset = '1' then
         rd_data_reg <= (others => '0');
      elsif (clk'event and clk = '1') then
         rd_data_reg <= din;
      end if;
   end process;
   --*****************************************************************
   -- instantiate one counter and W debouncing FSMs
   --*****************************************************************
   db_counter_unit : entity work.debounce_counter
      generic map(N => N)
      port map(
         clk   => clk,
         reset => reset,
         m_tick => m_tick
      );
   gen_fsm_cell : for i in 0 to W-1 generate
      db_fsm_unit : entity work.debounce_fsm
         port map(
            clk   => clk,
            reset => reset,
            sw    => din(i),
            m_tick => m_tick,
            db    -> db_out(i)
         );
   end generate;
   --*****************************************************************
   -- read multiplexing
   --*****************************************************************
   rd_data(W-1 downto 0) <= rd_data_reg when addr(0)='0' else db_out;
   rd_data(31 downto W)  <= (others => '0');
end arch;
```

14.1.3 Driver

The debouncing core driver contains methods to read the untreated input and the debounced input. The class definition and implementation are shown in Listings 14.4 and 14.5. Since a debounced input is a specific form of input, the class definition and implementation are included in the gpio_core.h and gpio_core.cpp files.

Listing 14.4 DebounceCore class definition (in gpio_core.h)

```
class DebounceCore {
   /* register map */
   enum {
      NORMAL_DATA_REG = 0,
      DB_DATA_REG = 1
   };
public:
   DebounceCore(uint32_t core_base_addr);
   ~DebounceCore();
   uint32_t read();
   int read(int bit_pos);
   uint32_t read_db();
   int read_db(int bit_pos);
```

```
private :
   uint32_t base_addr ;
};
```

Listing 14.5 DebounceCore class implementation (in gpio_core.cpp)

```
DebounceCore::DebounceCore(uint32_t core_base_addr) {
   base_addr = core_base_addr ;
}
DebounceCore::~DebounceCore() {}

uint32_t DebounceCore::read() {
   return (io_read(base_addr , NORMAL_DATA_REG));
}

int DebounceCore::read(int bit_pos) {
   uint32_t rd_data = io_read(base_addr , NORMAL_DATA_REG);
   return ((int) bit_read(rd_data , bit_pos));
}

uint32_t DebounceCore::read_db() {
   return (io_read(base_addr , DB_DATA_REG));
}

int DebounceCore::read_db(int bit_pos) {
   uint32_t rd_data = io_read(base_addr , DB_DATA_REG);
   return ((int) bit_read(rd_data , bit_pos));
}
```

14.1.4 Test

The sampler FPro system instantiates a five-bit debouncing core instance and connects the inputs to the five push buttons on the Nexys 4 DDR board.

The testing routine counts the transitions on five buttons using both untreated input and debounced inputs and shows the results on the discrete LEDs. The code is shown in Listing 14.6. The number of bounces depends on the individual switch. Some switches may exhibit very few bounces.

Listing 14.6 Debouncing test function (in main_sampler_test.cpp)

```
void debounce_check(DebounceCore *db_p, GpoCore *led_p) {
   long start_time ;
   int btn_old , db_old , btn_new , db_new ;
   int b = 0;
   int d = 0;
   uint32_t ptn ;

   start_time = now_ms ();
   btn_old = db_p->read ();
   db_old = db_p->read_db ();
   do {
      btn_new = db_p->read ();
      db_new = db_p->read_db ();
      if (btn_old != btn_new) {
         b = b + 1;
         btn_old = btn_new ;
      }
      if (db_old != db_new) {
```

```
        d = d + 1;
        db_old = db_new;
      }
      ptn = d & 0x0000000f;
      ptn = ptn | (b & 0x0000000f) << 4;
      led_p->write(ptn);
   } while ((now_ms() - start_time) < 5000);
}
```

14.2 LED-MUX CORE

An LED multiplexing circuit performs time-multiplexing to reduce the number of
I/O pins. A multiplexing circuit for a four-digit seven-segment LED display is
introduced in Section 4.5.1. We expand the circuit to accommodate the eight-
digit seven-segment LED display of the Nexys 4 DDR board and convert it into
an MMIO core for the FPro system. The core can still control a four-digit LED
display. However, the data written to the four most significant digits are ignored.

14.2.1 Eight-digit seven-segment LED display multiplexing circuit

The code for the expanded eight-digit multiplexing circuit is shown in Listing 14.7.
The design utilizes the three MSBs of the counter to perform eight-to-one multiplexing and to generate the active-low enable signal.

Listing 14.7   Eight-digit seven-segment display multiplexing circuit

```
use ieee.std_logic_1164.all;
use ieee.numeric_std.all;
entity led_mux8 is
   port(
      clk, reset         : in  std_logic;
      in3, in2, in1, in0 : in  std_logic_vector(7 downto 0);
      in7, in6, in5, in4 : in  std_logic_vector(7 downto 0);
      an                 : out std_logic_vector(7 downto 0);
      sseg               : out std_logic_vector(7 downto 0)
   );
end led_mux8;

architecture arch of led_mux8 is
   -- refreshing rate around 1600 Hz (100MHz/2^16)
   constant N          : integer := 18;
   signal q_reg, q_next : unsigned(N - 1 downto 0);
   signal sel          : std_logic_vector(2 downto 0);
begin
   -- register
   process(clk, reset)
   begin
      if reset = '1' then
         q_reg <= (others => '0');
      elsif (clk'event and clk = '1') then
         q_reg <= q_next;
      end if;
   end process;
   -- next-state logic for the counter
   q_next <= q_reg + 1;
   -- 3 MSBs of counter to control 8-to-1 multiplexing
```

```vhdl
   -- and to generate active-low enable signal
   sel <= std_logic_vector(q_reg(N - 1 downto N - 3));
   process(sel, in0, in1, in2, in3, in4, in5, in6, in7)
   begin
      case sel is
         when "000" =>
            an   <= "11111110";
            sseg <= in0;
         when "001" =>
            an   <= "11111101";
            sseg <= in1;
         when "010" =>
            an   <= "11111011";
            sseg <= in2;
         when "011" =>
            an   <= "11110111";
            sseg <= in3;
         when "100" =>
            an   <= "11101111";
            sseg <= in4;
         when "101" =>
            an   <= "11011111";
            sseg <= in5;
         when "110" =>
            an   <= "10111111";
            sseg <= in6;
         when others =>
            an   <= "01111111";
            sseg <= in7;
      end case;
   end process;
end arch;
```

14.2.2   Register map and the slot wrapping circuit

The eight-digit multiplexing circuit contains eight 8-bit input ports, each representing a seven-segment LED pattern plus the decimal point. We pack them into two 32-bit words. The LED-mux core's register map can be defined accordingly. There are two write registers. Their address offsets and fields are:

- offset 0 (output data register for the lower four digits of the seven-segment LED display)
  - bits 7 to 0: LED digit 0 (rightmost digit on Nexys 4 DDR board)
  - bits 15 to 8: LED digit 1
  - bits 23 to 16: LED digit 2
  - bits 31 to 24: LED digit 3
- offset 1 (output data register for the upper four digits of the seven-segment LED display)
  - bits 7 to 0: LED digit 4
  - bits 15 to 8: LED digit 5
  - bits 23 to 16: LED digit 6
  - bits 31 to 24: LED digit 7

Based on the register map, we can derive a wrapping circuit that complies with the slot specification and create the debouncing core. The HDL code of the core is shown in Listing 14.8. It consists of two output data registers to store the eight patterns and a write decoding circuit.

Listing 14.8 LED-mux core

```vhdl
library ieee;
use ieee.std_logic_1164.all;

entity chu_sseg8 is
   port(
      clk      : in  std_logic;
      reset    : in  std_logic;
      -- io bridge interface
      cs       : in  std_logic;
      write    : in  std_logic;
      read     : in  std_logic;  -- not used
      addr     : in  std_logic_vector(4 downto 0);
      rd_data  : out std_logic_vector(31 downto 0);
      wr_data  : in  std_logic_vector(31 downto 0);
      -- external interface
      an       : out std_logic_vector(7 downto 0);
      sseg     : out std_logic_vector(7 downto 0)
   );
end chu_sseg8;

architecture arch of chu_sseg8 is
   signal d0_reg, d1_reg : std_logic_vector(31 downto 0);
   signal wr_en          : std_logic;
   signal wr_d0          : std_logic;
   signal wr_d1          : std_logic;
begin
   -- instantiate LED multiplexing circuit
   led_mux_unit : entity work.led_mux8
      port map(
         clk   => clk,
         reset => reset,
         in7   => d1_reg(31 downto 24),
         in6   => d1_reg(23 downto 16),
         in5   => d1_reg(15 downto 8),
         in4   => d1_reg(7 downto 0),
         in3   => d0_reg(31 downto 24),
         in2   => d0_reg(23 downto 16),
         in1   => d0_reg(15 downto 8),
         in0   => d0_reg(7 downto 0),
         an    => an,
         sseg  => sseg
      );
   -- 2 write registers
   process(clk, reset)
   begin
      if reset = '1' then
         d0_reg <= (others => '0');
         d1_reg <= (others => '1');
      elsif (clk'event and clk = '1') then
         if wr_d0 = '1' then
            d0_reg <= wr_data(31 downto 0);
         end if;
         if wr_d1 = '1' then
            d1_reg <= wr_data(31 downto 0);
         end if;
      end if;
   end process;
```

```
  — decoding
  wr_en <= '1' when write = '1' and cs = '1' else '0';
  wr_d0 <= '1' when addr(0) = '0' and wr_en = '1' else '0';
  wr_d1 <= '1' when addr(0) = '1' and wr_en = '1' else '0';
  — unused
  rd_data <= (others => '0');
end arch;
```

14.2.3 Driver

The LED-mux core driver contains methods to write the desired patterns on the eight-digit seven-segment LED display. The class definition is shown in Listing 14.9.

Listing 14.9   SsegCore class definition (in sseg_core.h)

```
#ifndef _SSEG_CORE_H_INCLUDED
#define _SSEG_CORE_H_INCLUDED

#include "chu_init.h"

class SsegCore {
public:
    enum {
        DATA_LOW_REG = 0,
        DATA_HIGH_REG = 1
    };
    SsegCore(uint32_t core_base_addr);
    ~SsegCore(); // not used
    void write_1ptn(uint8_t pattern, int pos);
    void write_8ptn(uint8_t *ptn_array);
    void set_dp(uint8_t pt);
    uint8_t h2s(int hex);
private:
    uint32_t base_addr;
    uint8_t ptn_buf[8];     // led pattern buffer
    int8_t dp;              // decimal point
    /* methods */
    void write_led();       // write patterns to reg
};

#endif   // _SSEG_CORE_H_INCLUDED
```

A single-digit seven-segment LED display is repeated in Figure 14.1(a). To facilitate the processing, we store the LED patterns in two private variables. The ptn_buf[8] variable is an eight-element array that stores eight seven-segment LED patterns. An element is eight bits wide and is defined as $0gfedcba$, where g, f, e, d, c, b, and a are the segments. The display on Nexys 4 DDR board is configured as active low and thus an LED segment is turned on when the corresponding value is 0. The dp variable stores the eight decimal points. A decimal point is turned on if the corresponding bit in dp is 0. The private write_led() method combines and packs the data from the two variables into two 32-bit words and writes the two words to the LED-mux core.

The write_1ptn() method sets a pattern for one specific digit of the display, the write_8ptn() method sets all eight digits, and the set_dp() method specifies the decimal points. The h2s() method converts a hexadecimal digit into a seven-segment pattern defined in Figure 14.1(b).

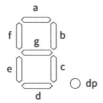

(a) Seven-segment LED display

(b) Hexadecimal patterns

Figure 14.1   Seven-segment LED display and hexadecimal patterns.

The class implementation is shown in Listings 14.10.

Listing 14.10   SsegCore class implementation (in **sseg_core.cpp**)

```
#include "sseg_core.h"

SsegCore::SsegCore(uint32_t core_base_addr) {
   // pattern for "HI"; the order in array is reversed in 7-seg display
   // i.e., HI_PTN[0] is the leftmost led
   const uint8_t HI_PTN[]={0xff,0xf9,0x89,0xff,0xff,0xff,0xff,0xff};

   base_addr = core_base_addr;
   write_8ptn((uint8_t*) HI_PTN);
   set_dp(0x02);
}

SsegCore::~SsegCore() {}

void SsegCore::write_8ptn(uint8_t *ptn_array) {
   int i;

   for (i = 0; i < 8; i++) {
      ptn_buf[i] = *ptn_array;
      ptn_array++;
   }
   write_led();
}

void SsegCore::write_1ptn(uint8_t pattern, int pos) {
   ptn_buf[pos] = pattern;
   write_led();
}

void SsegCore::set_dp(uint8_t pt) {
   dp = ~pt;      // active low
   write_led();
}

// convert a hex digit to
uint8_t SsegCore::h2s(int hex) {
```

```
/* active-low hex digit 7-seg patterns (0-9,a-f); MSB assigned to 1 */
static const uint8_t PTN_TABLE[16] =
   {0xc0, 0xf9, 0xa4, 0xb0, 0x99, 0x92, 0x82, 0xf8, 0x80, 0x90, //0-9
    0x88, 0x83, 0xc6, 0xa1, 0x86, 0x8e };                       //a-f
uint8_t ptn;

if (hex < 16)
    ptn = PTN_TABLE[hex];
else
    ptn = 0xff;
return (ptn);
}

void SsegCore::write_led() {
   int i, p;
   uint32_t word=0;

   // pack left 4 patterns into a 32-bit word
   // ptn_buf[0] is the leftmost led
   for (i = 0; i < 4; i++) {
       word = (word << 8) | ptn_buf[3 - i];
   }
   // incorporate decimal points (bit 7 of pattern)
   for (i = 0; i < 4; i++) {
       p = bit_read(dp, i);
       bit_write(word, 7 + 8 * i, p);
   }
   io_write(base_addr, DATA_LOW_REG, word);
   // pack right 4 patterns into a 32-bit word
   for (i = 0; i < 4; i++) {
       word = (word << 8) | ptn_buf[7 - i];
   }
   // incorporate decimal points
   for (i = 0; i < 4; i++) {
       p = bit_read(dp, 4 + i);
       bit_write(word, 7 + 8 * i, p);
   }
   io_write(base_addr, DATA_HIGH_REG, word);
}
```

The constructor stores the base address and shows a "HI" pattern on the LED display. The write_1ptn(), write_8ptn(), and set_dp() methods first update the corresponding private variables and then call write_led() to download the patterns to the core for display. The h2s() method uses a 16-element lookup table to store the seven-segment patterns to do the conversion. The write_led() method packs four 8-bit patterns into a word and then incorporates the decimal points into the word. It then writes the 32-bit word to a register. The operation is repeated twice for eight digits.

14.2.4 Test

The sampler FPro system instantiates an LED-mux core instance and connects the output to the eight-digit seven-segment LED display on the Nexys 4 DDR board.

The testing routine rotates the 16 hexadecimal digits through the display and also verifies the operation of the decimal points. The code is shown in Listing 14.11.

Figure 14.2   An 8-by-8 LED matrix

Listing 14.11   Seven-segment test function (in `main_sampler_test.cpp`)

```cpp
void sseg_check(SsegCore *sseg_p) {
   int i, n;
   uint8_t dp;

   //turn off led
   for (i = 0; i < 8; i++) {
      sseg_p->write_1ptn(0xff, i);
   }
   //turn off all decimal points
   sseg_p->set_dp(0x00);

   // display 0x0 to 0xf in 4 epochs
   // upper 4  digits mirror the lower 4
   for (n = 0; n < 4; n++) {
      for (i = 0; i < 4; i++) {
         sseg_p->write_1ptn(sseg_p->h2s(i + n * 4), 3 - i);
         sseg_p->write_1ptn(sseg_p->h2s(i + n * 4), 7 - i);
         sleep_ms(300);
      } // for i
   }  // for n
      // shift a decimal point 4 times
   for (i = 0; i < 4; i++) {
      bit_set(dp, 3 - i);
      sseg_p->set_dp(1 << (3 - i));
      sleep_ms(300);
   }
   //turn off led
   for (i = 0; i < 8; i++) {
      sseg_p->write_1ptn(0xff, i);
   }
   //turn off all decimal points
   sseg_p->set_dp(0x00);
}
```

14.3 PROJECT IDEAS

An *LED matrix* is a device that arranges a collection of discrete LEDs in a square or rectangular grid. An 8-by-8 matrix is shown in Figure 14.2. The schematic of the 8-by-8 matrix is shown in Figure 14.3(a). The LED grid is controlled by eight row signals (r7, r6, $\cdots$, r0) and eight column signals (c7, c6, $\cdots$, c0). A specific LED can be turned on by creating a forward bias between the corresponding row and column signals, as demonstrated in Figure 14.3(b).

Although the appearances of the eight-digit seven-segment LED display and the 8-by-8 LED matrix are very different, their schematics are similar. Instead of

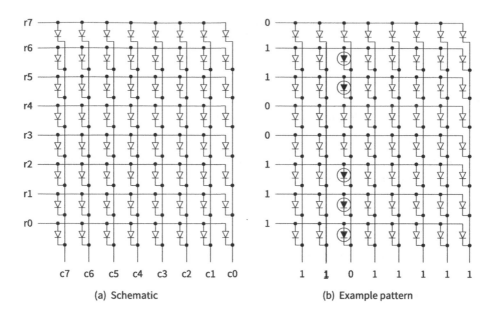

Figure 14.3   LED matrix diagram.

arranging LEDs as segment bars and the decimal point, the LED matrix places eight discrete LEDs in a single column. Thus, the LED-mux core can be used to control the LED matrix as well. The column signals in Figure 14.3(a) function as the eight-bit **an** signal and are active low. The row signals function as **sseg** and specify the vertical patterns (except that they are active high). For example, in Figure 14.3(b), the column c5 is enabled and a vertical pattern of "01100111" is shown. As in the 8-digit seven-segment LED display, the LED-mux core performs time multiplexing to show the eight vertical patterns. The LED-mux core can be easily modified to accommodate different dimensions and polarity requirements.

Since turning on eight vertical LEDs at the same time draws a large current, the column signals cannot be driven by FPGA's I/O pins directly. Transistor amplifiers are needed. Some ICs, such as ULN2803, contain multiple Darlington drivers and can be used for this purpose.

14.4   SUGGESTED EXPERIMENTS

14.4.1   Area comparison of two debouncing circuits

The debouncing core designed in Section 14.1.1 uses a shared timer. An alternative is to replicate the one-bit FSMD based design of Section 6.2.1 multiple times. Derive the core, instantiate an eight-bit instance of the two design approaches, and compare the number of logic cells (which reflect the circuit area) of the two approaches.

14.4.2   Enhanced debouncing core: part I

The debouncing core designed in Section 14.1.1 uses a 10-ms timer tick to obtain the needed 20-ms "settling time." Since the timer runs independently, the actual settling time is between 20 ms and 30 ms. We can improve the resolution by using a 1-ms timer tick, which leads to a settling time between 20 ms and 21 ms. Design the new debouncing core and verify its operation.

14.4.3   Enhanced debouncing core: part II

The debouncing circuit can generate a "delayed response" or an "early response," as shown in Figure 5.8. We want to enhance the debouncing core to accommodate both modes of responses and include a control signal to set the mode. Design the new debouncing core, expand the driver to include a method to select the mode, and verify its operation.

14.4.4   Rotating square pattern revisited

Generate the rotating square pattern in Experiment 4.8.3 with the LED mux core. Derive the program and verify its operation.

14.4.5   Heartbeat pattern revisited

Generate the heartbeat pattern in Experiment 4.8.4 with the LED-mux core. Derive the program and verify its operation.

14.4.6   Stopwatch

We want to design a stopwatch with the following features:
- Show the time in mm.ss.tt format in the seven-segment LED display, where mm is minutes, ss is seconds, and tt is 0.01 seconds.
- A "clear" button clears the count to 00.00.00 when pressed.
- A "go" button pauses and resumes counting alternatively each time it is pressed.

The stopwatch can be implemented by the debouncing and LED-mux cores. Derive the program and verify its operation.

14.4.7   Enhanced LED-mux core

We can incorporate PWM capability into the LED-mux core to control the brightness of the LED display. Design the new core, expand the driver to include a method to control the brightness, and verify its operation.

CHAPTER 15

SPI CORE

The *SPI (serial peripheral interface)* standard is a serial data transfer protocol originally developed by Motorola. The SPI bus is composed of three lines, including two lines for transmitting and receiving serial data and one line for a "clock" signal. One master device and multiple slave devices can be attached to the bus. The master generates the clock signal and initiates the data transfer. The SPI standard is widely used in embedded system to connect peripheral modules. In this chapter, we develop an SPI core and driver and demonstrate its operation by retrieving acceleration reading via the SPI interface of the ADXL362 device.

15.1 OVERVIEW

The SPI standard specifies the protocol to exchange data between two devices via serial lines. Instead of using UART's oversampling scheme, the SPI interface includes a third line to control the shifting and sampling of serial data. The activities are performed at the *transition edge* of this signal. The role is similar to the clock signal of a synchronous system and thus this line is referred to as the *SPI clock*.

Unlike the UART setting, in which two systems are symmetric and both can initiate a transmission, the SPI standard uses a master-slave configuration. The master controls the overall operation and generates the SPI clock signal. Only the master can initiate a data transfer.

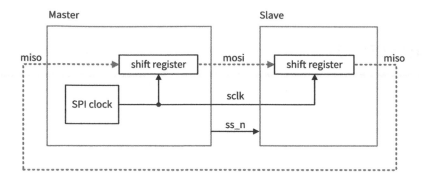

Figure 15.1   Conceptual diagram of an SPI bus.

Despite its name, the SPI clock is not a real system clock and should not be used to drive any register directly. The system clock rate of the SPI controller is much faster than the rate of the SPI clock. From SPI controller's point of view, the SPI clock is just another control signal.

15.1.1   Conceptual architecture

The conceptual diagram of an SPI bus with two devices is shown in Figure 15.1. Both the master and the slave have a shift register inside. The two shift registers are connected as a ring via the mosi (for "master-out-slave-in") and miso (for "master-in-slave-out") lines and their operation is coordinated by the same SPI clock signal, sclk. The mosi and miso signals are somewhat like the UART's transmitting signal (tx) and receiving signal (rx). We assume that both registers are eight bits wide and data transfer is done on a byte-by-byte basis. In the beginning of the operation, both master and slave load data into the registers. During the data transfer, data in both registers is shifted to the right by one bit in each sclk cycle. After eight sclk cycles, eight data bits are shifted and the master and slave have exchanged register values. The master and slave then can process the received data. This operation can be interpreted that the master writes data to and reads data from the slave simultaneously, which is known as *full-duplex* operation.

In addition to the mosi, miso, and sclk lines, a slave device may also have an active-low chip select input, ss_n (for "slave select"). The signal is somewhat like the cs (chip select) signal defined in the slot interface in Section 10.3.1. It can be used for the master to select the desired slave device if there are multiple slave devices on the bus. Many SPI devices also use ss_n for certain control functionality and it cannot be omitted, even in a single-slave configuration.

15.1.2   Multiple device configuration

The SPI standard supports a multiple-slave configuration, in which a master device can control more than one slave device. There are two basic schemes, which are the *parallel configuration* and the *daisy-chain configuration*. A three-slave example is shown in Figure 15.2.

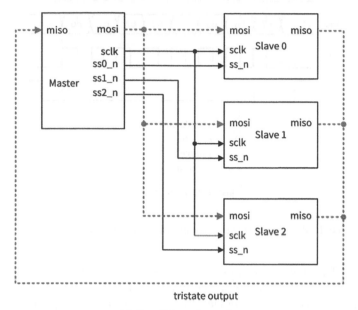

tristate output

(a) Parallel configuration

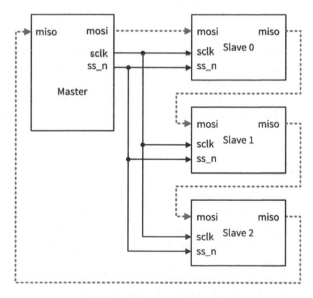

(b) Daisy-chain configuration

Figure 15.2   Multiple-slave configuration.

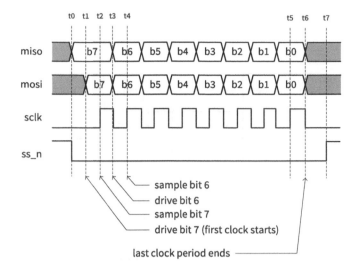

Figure 15.3   Representative timing diagram of an SPI data transfer.

The parallel configuration uses a dedicated ss_n line for each slave device, as shown in Figure 15.2(a). An ss_n line functions as the chip select signal and the master can select the desired device by asserting the corresponding line. This configuration can accommodate a master and "independent" slave devices. Since the miso lines of the slaves are tied together, the miso line must be driven by a tristate buffer and its output should be in a high-impedance state when the slave device is not selected.

The daisy-chain configuration connects the mosi and miso lines into a cascading chain, as shown in Figure 15.2(b). A single ss_n line is used to control all slave devices. Conceptually, the chain forms a large shift register and the data is transferred serially from device to device. The devices in this configuration must be "cooperative" and follow the same protocol to transmit, insert, and extract data bytes.

15.1.3   Basic timing

The SPI bus uses the *edges* of SPI clock (sclk) to control and synchronize the bit data transfer. To facilitate our discussion, we define two activities during a bit transfer: *driving* (i.e., shifting) a new bit to the data line and *sampling* (i.e., latching) a bit from the data line. The driving and sampling are completed in the same SPI clock cycle but take place in opposite clock edges.

A representative timing diagram is shown in Figure 15.3. Initially the bus is idle and the sclk line is 0. At $t0$, the master asserts ss_n and the designated slave places the first data bit (bit 7) on miso line. At $t1$, the master starts the SPI clock and drives the bit b7 on mosi line. Since the first half of the SPI clock period is 0, the value on sclk remains unchanged. At $t2$, the master raises the SPI clock and progresses to the second half of the clock period. At the 0-to-1 transition edge, the master samples the data on miso and the slave samples the data on mosi. At $t3$, the first SPI clock cycle is completed. The master starts the second SPI clock period

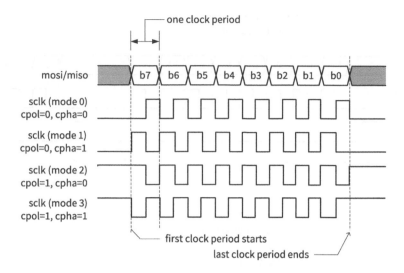

Figure 15.4 SPI modes.

and lowers sclk. Both master and slave drive new bits to the data line. At $t4$, the master raises sclk again and both master and slave sample the new data bits. The driving and sampling activities are repeated until eight data bits are transferred. The last data are sampled at $t5$ and sclk returns to 0 at $t6$. At $t7$, the master de-asserts ss_n. Note that all samplings are performed at the rising edges of sclk and all drivings (with the exception of the initial one) are performed at the falling edges of sclk.

15.1.4 Operation modes

The SPI's *operation mode* defines the relationships between the SPI clock edges and driving and sampling activities on the data lines. There are four modes. The modes depend on two parameters, which are *clock polarity* (abbreviated as *cpol*) and *clock phase* (abbreviated as *cpha*). The clock polarity is defined as the value of sclk when it is idle, which can be either 0 or 1. The clock phase is harder to define. One interpretation is whether a clock edge is used in driving the first data bit. If *cpha* is 1, the master drives the bit at the *first transition edge*. If *cpha* is 0, the master drives the bit at the *zeroth transition edge* (which means no edge or not the first edge).

Based on the two parameters, the SPI mode is defined as follows:

- Mode 0: *cpol*=0 and *cpha*=0
- Mode 1: *cpol*=0 and *cpha*=1
- Mode 2: *cpol*=1 and *cpha*=0
- Mode 3: *cpol*=1 and *cpha*=1

Note that the timing diagram in Figure 15.3 corresponds to mode 0 since sclk is 0 when it is idle and the first bit is not driven by the first transition edge.

The timing diagram of four modes is shown in Figure 15.4. Mode 0 is the most commonly used mode. In this mode, the idle value is 0 and the clock cycle begins with 0. Since the idle value and clock's starting value are the same, the first data

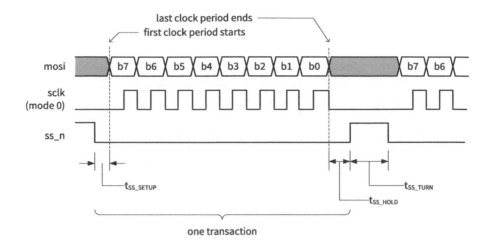

Figure 15.5 Timing with the ss_n signal.

bit is driven before the first transition edge. In mode 1, the idle value is also 0, but the clock cycle begins with 1. The starting value of 1 leads to a 0-to-1 transition edge and thus the first bit is driven at the first edge. Note that in these two modes the clock period and the starting time are the same but their values are out of phase.

The sclk idle value in modes 2 and 3 are 1. The sclk waveform in mode 2 is the exact opposite of that in mode 0 and the waveform in mode 3 is the exact opposite of that in mode 1. Again, note that the clock period and the starting time are the same for all modes.

15.1.5 Undefined aspects

The SPI interface was developed by Motorola and has become a *de facto standard.* There is no governing body or organization overseeing the standard. Several important aspects are not defined in the standard.

The first aspect is the use of the ss_n signal. The ss_n signal mainly acts as an enable or chip-select signal. A slave device is disabled if its ss_n is not asserted. In many devices, the ss_n also functions as a control signal. The data exchange is done on a *transaction-by-transaction* basis:

- The master asserts ss_n.
- The master and the selected slave transfer data bits.
- The master de-asserts ss_n.

A transaction is shown in Figure 15.5. The edges caused by asserting and de-asserting ss_n are used to activate certain actions, such as driving a bit or latching parallel data, in the slave device. This implies that ss_n must be connected to the master even if there is only one slave device; in other words, simply tying it to 0 will not work. The SPI standard does not explicitly define the role of the ss_n signal or protocol on the transaction. In addition, the timing requirement of "setup time" of ss_n, t_{SS_SETUP}, which is the interval between the ss_n assertion and clock initiation, "hold time" of ss_n, t_{SS_HOLD}, which is the interval between the

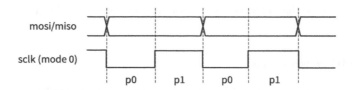

Figure 15.6   Timing of data shift operation.

ss_n de-assertion and clock termination, and the turn-around time between two transactions, t_{SS_TURN}, are not specified.

The second undefined aspect is the number of bits in one data exchange. There are eight bits transferred in Figure 15.3. However, the SPI standard does not specify the number of bits transferred in a transaction.

Finally, the SPI standard does not specify the bit order of transmission; i.e., whether the MSB or LSB of a data byte or data word is transferred first. "MSB first" is commonly used but not warranted.

Because of these undefined aspects, we must consult the device's data sheet and tailor the access for each device. This is commonly done by software driver and application.

15.2   SPI CONTROLLER

15.2.1   Basic design

We develop an SPI master controller in this section. The design assumes that eight data bits are transferred with MSB first. Since the operation of the ss_n is not directly related to the transfer, it is handled in the upper level and not included in the controller design.

The fundamental function of the controller is to shift in and shift out data bit by bit at a specific rate and to generate the SPI clock. The simplified timing diagram of shifting a bit is shown in Figure 15.6. To facilitate the driving and sampling actions, the shifting operation is divided into two equal parts, labeled as p0 and p1 (for part 0 and part 1). The sclk signal oscillates between the two parts. The waveform of mode 0 is shown in Figure 15.6.

At the beginning of p0, the controller shifts out a new bit (i.e., drives) in the mosi line. It lowers the sclk line at the same time, which leads to a 1-to-0 transition edge. The falling edge causes the slave to shift out a new bit in the miso line as well. After reaching the half point (at the end of p0), the controller samples the miso line and shifts in one new data bit. It also raises the sclk line at the same time, which leads to a 0-to-1 transition edge. The rising edge signals the slave to sample the new bit in the mosi line. These activities repeat eight times and the controller transmits and receives eight bits of data.

The FPGA's system clock rate is much faster than the SPI clock rate. From the controller's point of view, the SPI clock (sclk) is just an output signal based on the "part" of a data bit. For example, in mode 0, the controller sets the sclk line to 0 in the p0 portion and sets it to 1 in the p1 portion. Note that the data-shifting operation is the same regardless of the operation mode, as demonstrated in Figure 15.4.

15.2.2 FSMD construction

The SPI controller is constructed by an FSMD and special output logic to generate the SPI clock. The ASMD chart is shown in Figure 15.7. The p0 and p1 states represent the two parts labeled under the timing diagram in Figure 15.6. Asserting the start signal initiates the operation. The FSM loops the p0 and p1 states eight times to transfer eight bits of data.

The time spent in the p0 and p1 states depends on the SPI clock rate. Let the clock rates of SPI bus and the FPGA system be f_{spi} and f_{sys}. There are $\frac{f_{sys}}{f_{spi}}$ system clocks in one SPI clock period. Since an SPI cycle is composed of two parts, there are $\frac{f_{sys}}{2*f_{spi}}$ system clocks in the p0 or p1 state. One possible scheme to obtain the desired interval is the baud rate generator discussed in in Section 11.2. However, since the SPI clock rate is not fixed, a more general approach is needed. We make the desired interval as an input signal, dvsr (for clock divisor), whose value is equal to $\frac{f_{sys}}{2*f_{spi}} - 1$, and use a counter, labeled c in the FSMD chart, to keep track of the number of elapsed clock cycles. The c register is cleared to 0 when the FSM exits the current state and starts incrementing in the new state. The FSM stays in the same state until c reaches dvsr.

There are three other data registers. The n register keeps track of the number of bits processed and the si and so registers are shifting registers for the serial input and output.

A separate output control logic generates the SPI clock based on the FSM states. The exact value of sclk depends on the SPI operation mode. Close observation of Figure 15.4 shows the following:

- Mode 0: sclk is 1 in p1 state and is 0 otherwise.
- Mode 1: sclk is 1 in p0 state and is 0 otherwise.
- Mode 2: sclk is the inverted version of mode 0.
- Mode 3: sclk is the inverted version of mode 1.

15.2.3 HDL implementation

The HDL code is shown in Listing 15.1. The din port is the byte data to be transmitted and the dout port is the received byte. The SPI operation mode is specified by the cpol and cpha ports, as defined in Section 15.1.4. The ready signal indicates that the controller is idle and ready to transfer a new data byte. The done_tick signal is asserted for one clock cycle after the controller completes processing a transfer.

Listing 15.1 SPI controller

```
library ieee;
use ieee.std_logic_1164.all;
use ieee.numeric_std.all;
entity spi is
    port(
        clk, reset   : in   std_logic;
        din          : in   std_logic_vector(7 downto 0);
        dvsr         : in   std_logic_vector(15 downto 0);
```

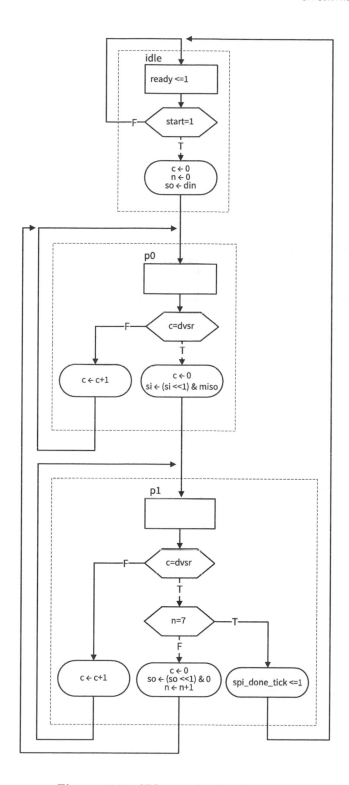

Figure 15.7 SPI controller ASMD chart.

```vhdl
      start           : in  std_logic;
      cpol, cpha      : in  std_logic;
      dout            : out std_logic_vector(7 downto 0);
      spi_done_tick   : out std_logic;
      ready           : out std_logic;
      sclk            : out std_logic;
      miso            : in  std_logic;
      mosi            : out std_logic
   );
end spi;

architecture arch of spi is
   type statetype is (idle, p0, p1);
   signal state_reg        : statetype;
   signal state_next       : statetype;
   signal p_clk            : std_logic;
   signal c_reg, c_next    : unsigned(15 downto 0);
   signal spi_clk_next     : std_logic;
   signal spi_clk_reg      : std_logic;
   signal n_reg, n_next    : unsigned(2 downto 0);
   signal si_reg, si_next  : std_logic_vector(7 downto 0);
   signal so_reg, so_next  : std_logic_vector(7 downto 0);
begin
   -- registers
   process(clk, reset)
   begin
      if reset = '1' then
         state_reg   <= idle;
         si_reg      <= (others => '0');
         so_reg      <= (others => '0');
         n_reg       <= (others => '0');
         c_reg       <= (others => '0');
         spi_clk_reg <= '0';
      elsif (clk'event and clk = '1') then
         state_reg   <= state_next;
         si_reg      <= si_next;
         so_reg      <= so_next;
         n_reg       <= n_next;
         c_reg       <= c_next;
         spi_clk_reg <= spi_clk_next;
      end if;
   end process;
   -- next-state logic and data path
   process(state_reg,si_reg,so_reg,n_reg,c_reg,din,dvsr,start,cpha,miso)
   begin
      state_next      <= state_reg;
      ready           <= '0';
      spi_done_tick   <= '0';
      si_next         <= si_reg;
      so_next         <= so_reg;
      n_next          <= n_reg;
      c_next          <= c_reg;
      case state_reg is
         when idle =>
            ready <= '1';
            if start = '1' then
               so_next     <= din;
               n_next      <= (others => '0');
               c_next      <= (others => '0');
               state_next  <= p0;
            end if;
```

```
      when p0 =>
          if c_reg = unsigned(dvsr) then  ── sclk 0-to-1
              state_next <= p1;
              si_next       <= si_reg(6 downto 0) & miso;
              c_next        <= (others => '0');
          else
              c_next <= c_reg + 1;
          end if;
      when p1 =>
          if c_reg = unsigned(dvsr) then  ── sclk 1-to-0
              if n_reg = 7 then
                  spi_done_tick <= '1';
                  state_next    <= idle;
              else
                  so_next       <= so_reg(6 downto 0) & '0';
                  state_next <= p0;
                  n_next        <= n_reg + 1;
                  c_next        <= (others => '0');
              end if;
          else
              c_next <= c_reg + 1;
          end if;
   end case;
end process;
── lookahead output decoding
p_clk <= '1' when ((state_next = p1 and cpha = '0') or
                   (state_next = p0 and cpha = '1')) else '0';
spi_clk_next <= p_clk when (cpol = '0') else not p_clk;
── output
dout  <= si_reg;
mosi  <= so_reg(7);
sclk  <= spi_clk_reg;
end arch;
```

The HDL code follows the ASMD chart. A special output logic is used to obtain the SPI clock, as discussed in the previous subsection. The first statement

```
p_clk <= '1' when (state_next=p1 and cpha='0') or
                  (state_next=p0 and cpha='1')
            else '0';
```

assumes an SPI clock polarity of 0 and generates an intermediate signal based on the SPI clock phase (i.e., modes 0 and 1). The second statement

```
spi_clk_next <= p_clk when (cpol='0') else not p_clk;
```

inverts the intermediate clock if the polarity is 1. The signal is then fed to a buffer to remove any potential glitch and the output of the buffer is connected to the sclk line.

A special "lookahead output decoding" scheme is applied to obtain the sclk signal. Instead of *state_reg*, the *state_next* signal is used to generate an "intermediate" output signal. The signal is then fed to a register to obtain the final output signal. Since the output signal is obtained directly from a register, it is glitch-free and better suited to drive the SPI clock line.

15.3 SPI CORE DEVELOPMENT

The SPI master may have multiple slave select (ss_n) signals, as shown in Figure 15.2(a). These signals can be implemented as general-purpose output registers and can be asserted and de-asserted according to the specification of a specific SPI device. In fact, it is possible to instantiate a separate GPO core and use its outputs as slave select signals. For better integration, we include an output register for slave select signals in the SPI core.

15.3.1 Register map

The processor interacts with the SPI controller as follows:
- set (i.e., write) the value of dvsr.
- set (i.e., write) the SPI mode via cpol and cpha.
- activate and deactivate (i.e., write) the slave select signals.
- specify (i.e., write) the output data and start the data transfer.
- retrieve (i.e., read) the received data byte.
- check (i.e., read) the ready signal to determine whether the controller is ready.

Based on these interactions, we can define the SPI core's register map. There is one read register. Its address offset and fields are:
- offset 0 (read data and status register)
  - bits 7 to 0: 8-bit received data
  - bit 8: ready status

There are three write registers. Their address offsets and fields are:
- offset 1 (slave select register)
  - bits S-1 to 0: S-bit ss_n signal to control S slave devices, where S is the VHDL generic value used in SPI core instantiation
- offset 2 (write data and command register)
  - bits 7 to 0: 8-bit data
  - Writing register 1 also initiates the SPI controller to transfer a data byte.
- offset 3 (control register)
  - bits 15 to 0: 16-bit divisor value.
  - bit 16: 1-bit cpol value
  - bit 17: 1-bit cpha value

15.3.2 Wrapping circuit for the slot interface

Based on the register map and the I/O signals of the SPI controller, we can derive a wrapping circuit that complies with the slot specification and create the SPI core. The HDL code of the core is shown in Listing 15.2. It consists of an instance of SPI controller, a control register, and a decoding circuit.

Listing 15.2   SPI core

```
library ieee;
use ieee.std_logic_1164.all;
use ieee.numeric_std.all;
entity chu_spi_core is
   generic(S : integer := 2);
```

```vhdl
    port(
        clk        : in   std_logic;
        reset      : in   std_logic;
        -- io bridge interface
        cs         : in   std_logic;
        write      : in   std_logic;
        read       : in   std_logic;
        addr       : in   std_logic_vector(4 downto 0);
        rd_data    : out  std_logic_vector(31 downto 0);
        wr_data    : in   std_logic_vector(31 downto 0);
        -- external signals
        spi_sclk : out std_logic;
        spi_mosi : out std_logic;
        spi_miso : in  std_logic;
        spi_ss_n : out std_logic_vector(S - 1 downto 0)
    );
end;

architecture arch of chu_spi_core is
    signal wr_en, wr_ss : std_logic;
    signal wr_ctrl       : std_logic;
    signal wr_spi        : std_logic;
    signal ctrl_reg      : std_logic_vector(31 downto 0);
    signal ss_n_reg      : std_logic_vector(S - 1 downto 0);
    signal spi_out       : std_logic_vector(7 downto 0);
    signal spi_ready     : std_logic;
    signal dvsr          : std_logic_vector(15 downto 0);
    signal cpol          : std_logic;
    signal cpha          : std_logic;
begin
    -- instantiate SPI unit
    spi_unit : entity work.spi
        port map(
            clk           -> clk,
            reset         => reset,
            din           => wr_data(7 downto 0),
            dvsr          => dvsr,
            start         => wr_spi,
            cpol          => cpol,
            cpha          => cpha,
            dout          => spi_out,
            sclk          => spi_sclk,
            miso          => spi_miso,
            mosi          => spi_mosi,
            spi_done_tick => open,
            ready         => spi_ready
        );
    --registers
    process(clk, reset)
    begin
        if reset = '1' then
            ctrl_reg <= x"00000200";        -- dvsr=1028 (50 KHz sclk)
            ss_n_reg <= (others => '1');    -- de-assert all ss_n
        elsif (clk'event and clk = '1') then
            if (wr_ctrl = '1') then
                ctrl_reg <= wr_data;
            end if;
            if (wr_ss = '1') then
                ss_n_reg <= wr_data(S - 1 downto 0);
            end if;
        end if;
```

```
    end process;
    -- decoding logic
    wr_en    <= '1' when cs='1' and write='1' else '0';
    wr_ss    <= '1' when wr_en='1' and addr(1 downto 0)="01" else '0';
    wr_spi   <= '1' when wr_en='1' and addr(1 downto 0)="10" else '0';
    wr_ctrl  <= '1' when wr_en='1' and addr(1 downto 0)="11" else '0';
    -- control signals
    dvsr     <= ctrl_reg(15 downto 0);
    cpol     <= ctrl_reg(16);
    cpha     <= ctrl_reg(17);
    spi_ss_n <= ss_n_reg;
    -- read output
    rd_data  <= x"00000" & "000" & spi_ready & spi_out;
end arch;
```

Note that the write byte data is connected to the din port directly. When the processor writes register 2, the wr_spi signal is asserted and the write data is sampled at the same time.

15.4 SPI DRIVER

The driver consists of a set of routines to specify the SPI clock frequency, set the mode, control the slave select signals, and perform a data transfer.

15.4.1 Class definition

The class definition of the SPI core is shown in Listing 15.3.

Listing 15.3 SpiCore class definition (in spi_core.h)

```
#ifndef _SPI_CORE_H_INCLUDED
#define _SPI_CORE_H_INCLUDED

#include "chu_init.h"

class SpiCore {
public:
    enum {
        RD_DATA_REG = 0,
        SS_REG = 1,
        WRITE_DATA_REG = 2,
        CTRL_REG = 3
    };
    enum {
        READY_FIELD = 0x00000100,
        RX_DATA_FIELD = 0x000000ff
    };
    SpiCore(uint32_t core_base_addr);
    ~SpiCore(); // not used
    int ready();
    void set_freq(int freq);
    void set_mode(int icpol, int icpha);
    void write_ss_n(uint32_t data);
    void write_ss_n(int bit_value, int bit_pos);
    void assert_ss(int n);
    void deassert_ss(int n);
    uint8_t transfer(uint8_t wr_data);
private:
```

```
    /* variable to keep track of current status */
    uint32_t base_addr;
    uint32_t ss_n_data;
    uint16_t dvsr;
    int cpol;
    int cpha;
};

#endif    // _SPI_CORE_H_INCLUDED
```

The first **enum** definition uses symbolic names for the register offsets and the second **enum** definition specifies the fields of the read data register. The variables in the private section store the base address and keep track of various control parameters and slave select signals.

15.4.2   Class implementation

The implementation of the constructor and the methods is shown in Listing 15.4.

Listing 15.4   SpiCore class implementation (in **spi_core.cpp**)

```
#include "spi_core.h"

SpiCore::SpiCore(uint32_t core_base_addr) {
    base_addr = core_base_addr;
    // set default spi configuration to be 400K Hz, mode 0
    set_freq(100000);
    set_mode(0, 0);
    write_ss_n(0xffffffff);   // de-assert all ss_n signals
}
SpiCore::~SpiCore() {}

int SpiCore::ready() {
    uint32_t rd_word;
    int rdy;

    rd_word = io_read(base_addr, RD_DATA_REG);
    rdy = (int) (rd_word & READY_FIELD) >> 8;
    return (rdy);
}

void SpiCore::set_freq(int freq) {
    uint32_t ctrl_word;

    dvsr = (uint16_t) (SYS_CLK_FREQ * 1000000 / (2 * freq));
    dvsr = dvsr - 1;    // counts 0 to dvsr-1
    ctrl_word = cpha << 17 | cpha << 16 | dvsr;
    io_write(base_addr, CTRL_REG, ctrl_word);

}

void SpiCore::set_mode(int icpol, int icpha) {
    uint32_t ctrl_word;

    cpol = icpol;
    cpha = icpha;
    ctrl_word = cpha << 17 | cpha << 16 | dvsr;
    io_write(base_addr, CTRL_REG, ctrl_word);
}
```

```
void SpiCore::write_ss_n(uint32_t data) {
   ss_n_data = data;
   io_write(base_addr, SS_REG, ss_n_data);
}

void SpiCore::write_ss_n(int bit_value, int bit_pos) {
   bit_write(ss_n_data, bit_pos, bit_value);
   io_write(base_addr, SS_REG, ss_n_data);
}

void SpiCore::assert_ss(int n) {
   write_ss_n(0, n);
}

void SpiCore::deassert_ss(int n) {
   write_ss_n(1, n);
}

uint8_t SpiCore::transfer(uint8_t wr_data) {
   uint32_t rd_data;

   while (!ready()) {};
   io_write(base_addr, WRITE_DATA_REG, (uint32_t) wr_data);
   while (!ready()) {};
   rd_data = io_read(base_addr, RD_DATA_REG) & RX_DATA_FIELD;
   return ((uint8_t) rd_data);
}
```

The SpiCore() constructor saves the base address, sets the SPI clock frequency to 100K Hz and the operation mode to 0, and deactivates all slave devices. The ready() method reads the register and extracts the ready bit. The set_freq() method sets the SPI clock frequency. It calculates the divisor value with the $\frac{f_{sys}}{2*f_{spi}}$ formula and writes the value to the control register. The set_mode() method specifies the operation mode via the cpol and cpha parameters and writes the values to the control register.

The SPI core's ss_n register is the same as the GPO core's output register. The first write_ss_n() method writes a word to this register and and the second one writes a bit to a specific position. The assert_ss() and deassert_ss() methods use write_ss_n() to activate and deactivate a specific slave device.

A data transfer on the SPI bus performs both the write operation (i.e., data transmitted from the master to a slave) and the read operation (i.e., data transmitted from a slave to the master) at the same time. The transfer() method performs a data transfer. The wr_data parameter is the write data to be transferred and the returned result is the received read data. If only a write operation is needed, the returned result will be ignored. If only a read operation is needed, the master will write a "dummy data," which is ignored by the slave.

15.5   TEST

The Nexys 4 DDR board contains an ADXL362 device, which is an accelerometer with an SPI interface. It is used to verify the operation of the SPI core.

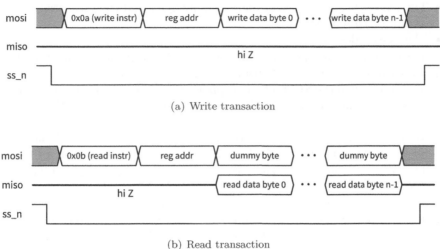

(a) Write transaction

(b) Read transaction

Figure 15.8   ADXL362 read and write transactions.

15.5.1   ADXL362 accelerometer

The ADXL362 device is a low-power high-resolution three-axis MEMS accelerometer made by Analog Devices. It measures both dynamic acceleration, resulting from motion or shock, and static acceleration, due to gravity. The measurement is digitalized internally and stored in internal registers. Some features, such as the measurement range and output data rate, can be configured.

The digitized acceleration measurement, status, and configuration are stored in ADXL362's 36 internal registers, which can be accessed via the SPI interface. The ADXL362 device uses SPI mode 0 with the MSB transferred first. The protocol of reading and writing registers from ADXL362 is shown in Figure 15.8. The sequence of a write transaction is as follows:

- The master asserts ss_n (i.e., lowers the line to 0).
- The master issues a "write instruction byte," 0x0a.
- The master issues the register address byte (i.e., the register number).
- The master transmits one or more write data bytes.
- The master de-asserts ss_n (i.e., raises the line to 1).

Note that ADXL362 performs an *automatic address increment* for multiple-byte operations. After storing the first data byte into the designated register, it automatically increments the register address for each additional byte in the transaction. The process is repeated until ss_n is de-asserted. It can be thought that a write transaction is done in a "burst mode," in which the "starting register address" is specified first and a burst of write data follows afterwards. The ADXL362 device does not send feedback in a write transaction and thus the miso line is in high-impedance state.

The sequence of a read transaction is similar:

- The master asserts ss_n.
- The master issues a "read instruction byte," 0x0b.
- The master issues the register address byte.

- The master transmits one or more dummy bytes and the slave transmits one or more read data bytes.
- The master de-asserts ss_n.

As in a write transaction, the read transaction can be performed in burst mode and the master can receive multiple bytes in a single transaction.

Note that although the SPI protocol supports the duplex operation, most devices only transfer data in one direction, in a fashion similar to the operation shown in Figure 15.8.

To verify the operation of the SPI core, our test reads the device signature and eight-bit three-axis acceleration. Following is the relevant information for the test:

- The ADXL362 signature (i.e., part id) is stored in register 2 and its value is 0xf2.
- The eight-bit x-axis acceleration is stored in register 8 in signed format.
- The eight-bit y-axis acceleration is stored in register 9 in signed format.
- The eight-bit z-axis acceleration is stored in register 10 in signed format.
- The default range is $\pm 2g$ (where g is Earth's gravitational acceleration) and the default output data rate is 100 Hz.

More detailed usage and configuration information can be found in the ADXL362 data sheet.

15.5.2 Test program

A simple test function is derived to access the ADXL362 acceleration measurement via the SPI core. The code is shown in Listing 15.5. It first checks the ADXL362's part id to verify the existence of the device and then retrieves the three-axis acceleration measurements. The reading is converted to values in terms of g and displayed on the UART console. When the board is placed on a flat surface, the reading of the z-axis should be close to -1.0 g and the readings of the two other axes should be close to 0 g.

Listing 15.5   ADXL362 SPI test function (in `main_sampler_test.cpp`)

```
void gsensor_check(SpiCore *spi_p, GpoCore *led_p) {
   const uint8_t RD_CMD = 0x0b;
   const uint8_t PART_ID_REG = 0x02;
   const uint8_t DATA_REG = 0x08;
   const float raw_max = 127.0 / 2.0;   //128 max 8-bit reading for +/-2g
   int8_t xraw, yraw, zraw;
   float x, y, z;
   int id;

   spi_p->set_freq(400000);
   spi_p->set_mode(0, 0);
   // check part id
   spi_p->assert_ss(0);            // activate
   spi_p->transfer(RD_CMD);        // for read operation
   spi_p->transfer(PART_ID_REG);   // part id address
   id = (int) spi_p->transfer(0x00);
   spi_p->deassert_ss(0);
   uart.disp("read ADXL362 id (should be 0xf2): ");
   uart.disp(id, 16);
   uart.disp("\n\r");
   // read 8-bit x/y/z g values once
   spi_p->assert_ss(0);            // activate
```

```
    spi_p->transfer(RD_CMD);          // for read operation
    spi_p->transfer(DATA_REG);
    xraw = spi_p->transfer(0x00);
    yraw = spi_p->transfer(0x00);
    zraw = spi_p->transfer(0x00);
    spi_p->deassert_ss(0);
    x = (float) xraw / raw_max;
    y = (float) yraw / raw_max;
    z = (float) zraw / raw_max;
    uart.disp("x/y/z axis g values: ");
    uart.disp(x, 3);
    uart.disp(" / ");
    uart.disp(y, 3);
    uart.disp(" / ");
    uart.disp(z, 3);
    uart.disp("\n\r");
}
```

15.6 PROJECT IDEAS

The SPI standard is used as an I/O interface in a wide variety of peripherals, including sensors, ADCs (analog-to-digital converters), DACs (digital-to-analog converters), LCDs (liquid crystal displays), and EPROM (electrically erasable programmable ROM) and flash memory. An FPro system can instantiate multiple SPI cores and incorporate any number of desired peripherals into the system. Two useful peripherals are the SD card and TFT (thin-film-transistor) LCD module.

15.6.1 SD card

A *flash memory card* is a device that contains flash memory and a controller. It is frequently used as an external massive storage for embedded applications. The *SD card standard* is a widely used standard developed by the SD Card Association. An SD card can operate in either *SD mode* or *SPI mode*. The latter uses the SPI interface as its "physical layer" to transfer bit-level data. The standard defines the physical and electrical characteristics as well as the upper-level protocols to initialize a card, read a data block, and write a data block. For example, the basic protocol of reading a block in SPI mode is shown in Figure 15.9.

The Nexys 4 DDR board contains a microSD card socket. We can instantiate an SPI core for the socket and access the SD card. The software should be constructed with a three-layer hierarchy. The bottom layer is the SPI driver discussed in Section 15.4. The middle layer is the routines that implement the SD card protocol to perform card initialization and retrieve and store data blocks. The top layer should be a *file system*, which specifies how to map a logical file to a physical massive storage and organize multiple files in the storage. The variants of FAT (file allocation table) system are widely used for simple embedded applications.

15.6.2 TFT LCD module

A complete video subsystem is constructed and discussed in Part IV of the book. For some projects, an interesting and simple alternative is to use a low-resolution *TFT LCD module*. A typical low-resolution module contains a TFT LCD display

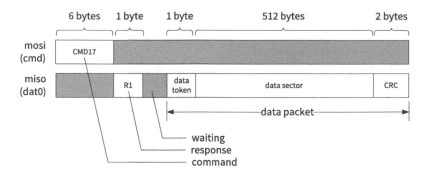

Figure 15.9   Protocol to read a data block in SPI mode.

and a TFT video controller with integrated video memory. Some may also include an SD card socket and a touch screen. These modules are frequently equipped with an SPI interface and a processor can write the video memory through the interface. Although the data transfer rate is very limited, it is still adequate for applications that do not require a high-resolution display or animation.

15.7   BIBLIOGRAPHIC NOTES

The SPI interface is a de facto standard and there is no official documentation. Wikipedia website provides a comprehensive overview. The ADXL362 data sheet from Analog Devices, Inc. contains detailed information for the SPI timing, command, data transfer protocol, and register usage.

The SPI interface is widely used in sensor modules and additional information can be found in the bibliographic section in Chapter 12. The author's other text, *Embedded SoPC Design with Nios II Processor and VHDL Examples*, includes discussions on the SD card initialization and identification process, data transfer protocol, and FAT16 file structure and access.

15.8   SUGGESTED EXPERIMENTS

15.8.1   Inclination sensing

For a still object, the only acceleration stimulus is associated with gravity. The tilting or inclination of an object can be determined by examining the acceleration in three axes. One application of accelerometer is to determine the orientation of a handheld device, as in the portrait mode or landscape mode of a smart phone. Pretend that the prototyping board has a screen on the front as on a smart phone. Derive a program to determine the orientation of the board, which can be $0°$, $90°$, $180°$, or $270°$, and use four LEDs to show the result.

15.8.2 "Tapping" detection

When we quickly tap an object, the force introduces "spikes" in acceleration. With an accelerometer, the tapping can be used as an input method. This can be achieved as follows:

- Derive a program to capture the acceleration value during a "tapping event."
- Record the tapping several times and derive a "signature" of tapping.
- Derive a function to detect the specific signature.

After completion, test the function with an LED by turning it on and off alternatively with tapping.

15.8.3 ADXL362 C++ class

The ADXL362 is versatile device with many features. Consult the data sheet and create an ADXL362 C++ class with the following methods:

- Read the device id, part id, and revision id.
- Perform a software reset.
- Perform a self-test.
- Set the output data rate.
- Set the measurement range.
- Read the 8-bit acceleration data.
- Read the 12-bit acceleration data.
- Read the 12-bit temperature data.
- Read the acceleration data from the internal FIFO buffer.

Write a test program to verify the operation of these methods.

15.8.4 Enhanced SPI controller: part I

Many SPI devices use the ss_n signal for transaction-based data transfer, as discussed in Section 15.1.5. We can incorporate this feature into the SPI controller in Section 15.2 as follows:

- Generate an additional output to control the ss_n signal.
- Expand the FSM with an additional state to accommodate the ss_n setup time, hold time, and turn-around time (t_{SS_SETUP}, t_{SS_HOLD}, and t_{SS_TURN} shown in Figure 15.5).
- Add three extra input ports, ss_s_cycle, ss_h_cycle, and ss_t_cycle, which specify the number of system clock cycles in t_{SS_SETUP}, t_{SS_HOLD}, and t_{SS_TURN} interval.

Design the new SPI controller, revise the SPI core and driver as needed, and verify its operation with the ADXL362 device.

15.8.5 Enhanced SPI controller: part II

The number of data bits and the bit order (i.e., MSB or LSB first) of the SPI controller designed in Section 15.2 are fixed. We can extend the SPI controller to make these two aspects configurable and use two additional input ports to specify the number of bits to be transferred (between 8 and 64) and the bit order. The FSMD can be modified to accommodate the new features. Design the new SPI controller, revise the SPI core and driver as needed, and verify its operation with the ADXL362 device.

15.8.6   Automatic-read ADXL362 wrapper: part I

An automatic reading wrapping circuit retrieves the acceleration data from ADXL362 continuously and returns three-axis acceleration as its outputs. An external system can access the data directly without worrying about the SPI protocol. The entity declaration is

```
entity adxl362_wrapper is
   port (
      clk, reset: in  std_logic;
      -- 3-axis acceleration reading
      x_acc: out std_logic_vector (7 downto 0);
      y_acc: out std_logic_vector (7 downto 0);
      z_acc: out std_logic_vector (7 downto 0);
      -- SPI interface connected to ADXL362
      ss_n: out std_logic;
      sclk: out std_logic;
      miso: in std_logic;
      mosi: out std_logic
   );
end adxl362_wrapper;
```

The wrapping circuit can be constructed by an FSMD on top of the SPI controller. The FSMD should follow the ADXL362 protocol to issue command and retrieve the measurement data. Derive the wrapping circuit and verify its operation.

15.8.7   Automatic-read ADXL362 wrapper: part II

Repeat the automatic wrapping circuit in Experiment 15.8.6 to generate 12-bit acceleration measurement.

15.8.8   Flash memory access

The Nexys 4 DDR board has a 16-MB flash memory device (Spansion S25FL128S). The device communicates to an external host via an SPI interface. About 4 MB is used for Artix-7's configuration file (loaded to the FPGA device during power on). We can instantiate an SPI core to access the remaining space for user data. Study the data sheets of the board and the device, derive software routines to perform the read and write operations, and develop a test program to verify its operation.

15.8.9   SPI slave controller: part I

An SPI slave controller accepts commands from an SPI master and responds accordingly. With an SPI slave interface on an FPGA board, other prototyping boards can access the FPGA board and exchange information. We want to construct a slave controller following the transaction protocol similar to the one shown in Figure 15.8. The controller can be designed and tested as follows:

- Instantiate a 2^4-by-8 register file discussed in Section 7.3.

- Develop an FSMD following the read and write protocol. Its data path should contain an address register to maintain the current address and a data register to shift in or out data.
- Connect the SPI signals to the Pmod port of the Nexys 4 DDR board.
- Synthesize and implement the slave controller.
- Use a separate prototyping board as the SPI master and connect the SPI bus signals. The second board can be another FPGA board or a microprocessor-based boards, such as Arduino or Raspberry Pi.
- Derive a testing program in the SPI master board to access the register file and verify the SPI slave operation.

15.8.10   SPI slave controller: part II

Experiment 15.8.9 essentially creates an "SPI slave wrapping circuit" for the register file. The wrapping circuit allows other prototyping boards to access the register file. The same concept can be applied to other components as well. Follow the steps in Experiment 15.8.9 and derive an SPI slave interface for the following cores:

- XADC core in Chapter 12
- PWM core in Chapter 13
- Seven-segment LED core in Chapter 14
- DDFS core in Chapter 18
- ADSR core in Chapter 19

CHAPTER 16

I²C CORE

The I^2C (for *inter-IC*) standard is a serial data transfer protocol developed by Philips Semiconductors (now NXP Semiconductors). The I²C bus is composed of two lines, one for bidirectional data transfer and one for a "clock" signal. Multiple devices can be attached to the bus and the designated master device generates the clock signal and controls the data transfer. Along with the SPI standard, the I²C standard is also widely used in embedded systems to connect peripheral modules. In this chapter, we develop an I²C core and driver and demonstrate its operation by retrieving temperature readings via the I²C interface of the ADT7420 device.

16.1 OVERVIEW

The I²C protocol is a low-speed serial bus for efficient communication between devices. The I²C bus consists of two bidirectional lines, sda (for "serial data") and scl (for "serial clock"), for data and clock, respectively. During the operation, one device on the bus functions as the *master* and other devices function as *slaves*. The master generates the clock on the scl line and also initiates and terminates the data transfer. A slave listens to the bus and responds when addressed. The master and the designated slave then exchange data via the sda line.

Similar to the SPI protocol, the I²C master generates a clock signal and controls the transaction. However, there are several differences between them. First, the I²C's data line is bidirectional and is used for both read and write operation. Sec-

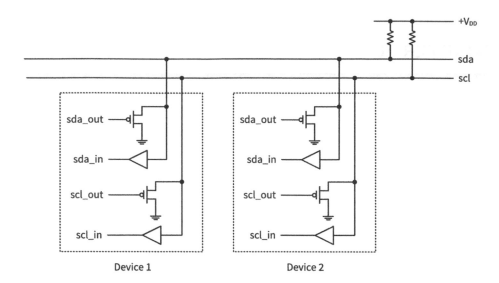

Figure 16.1 Conceptual diagram of I^{2}C bus.

ond, I^{2}C protocol does not use a device selection signal (as did the ss_n signal in SPI). Instead, each device on the I^{2}C bus is assigned a unique address and can be identified and accessed via the address. Third, the I^{2}C protocol supports multiple masters and defines a mechanism to arbitrate contention.

The I^{2}C standard originally specified an scl rate between 0 and 100K Hz (known as the *standard mode*) and later extended its upper limit to 400K Hz (known as the *fast mode*) and to 3.4M Hz (known as the *high-speed mode*).

16.1.1 Electrical characteristics

The physical connection of the I^{2}C bus is shown in Figure 16.1. It uses *open-drain* technology, which means that the output stage of a device must have an open-drain structure. Both sda and scl lines are connected to the voltage source (V_{DD}) via pull-up resistors and are high when the bus is idle. A line becomes low as soon as one device's output turns to low. It thus performs the *wired-and* function.

16.1.2 Basic bus protocol

The I^{2}C bus is shared by the attached devices. The data transfer is performed on a *transaction* basis. A master device can generate a *start condition* and then take over the bus. It then transmits or receives the data for the designated slave byte by byte. Each byte is followed by an *acknowledge bit* in the ninth clock cycle. The number of bytes in a transfer is unrestricted. After completion, the master generates a *stop condition* and releases the bus.

The detailed sequence of writing two bytes of data is shown in the top part of Figure 16.2. The basic steps are summarized below:

1. The master initiates the transfer by creating a start condition and takes over the bus.

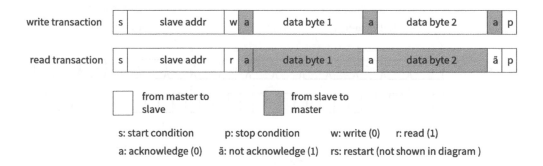

Figure 16.2   Conceptual sequence of reading and writing two bytes of data.

2. The master broadcasts a byte, in which the seven MSBs are the slave's seven-bit address (i.e., device id) plus a *direction bit* of 0. The direction bit indicates the direction of data flow, in which a 1 is for read operation (i.e., from slave to master) and a 0 is for write operation (i.e., from master to slave).
3. The slave with the matched address acknowledges with a 0 in the ninth clock cycle.
4. The master transmits eight bits of data (i.e., the first data byte).
5. The slave acknowledges with a 0 in the next clock cycle.
6. The master transmits eight bits of data (i.e., the second data byte).
7. The slave acknowledges with a 0 in the next clock cycle.
8. The master terminates the transfer by creating a *stop condition* and releases the bus.

The sequence of reading two bytes of data is shown in the bottom part of Figure 16.2. The master initiates the transfer by creating a start condition and broadcasts the designated slave address with a direction bit of 1. The designated slave then takes over the bus and transmits the data bits. After receiving the eight data bits, the master acknowledges with a 0 in the ninth clock cycle. The read-byte operation is repeated until the last byte is received (i.e., the second data byte in this example). The master acknowledges the last byte with a 1 (i.e., a "not acknowledge") in the ninth clock and then terminates the transfer by creating a stop condition.

When a master generates a stop condition, it releases the bus and other master devices may seize control afterwards. If a master wants to perform multiple transactions, it can generate a *restart condition* (also referred to as a *repeated start condition*) after each transaction and only issues a stop condition after the last transaction to free the bus. The restart condition replaces the normal stop condition to terminate a transaction but without giving up the bus.

16.1.3   Basic timing

The timing diagram of writing two bytes is shown in Figure 16.3. Initially, the bus is in idle state, in which both `sda` and `scl` lines are high. The master creates the start condition, in which `sda` transits from high to low while `scl` is high, and takes over the bus. It then generates the clock signal over the `scl` line and transmits

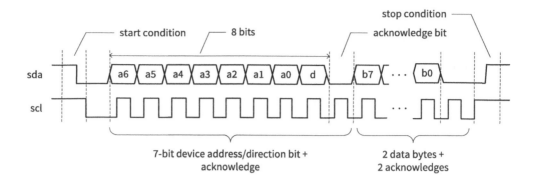

Figure 16.3 Timing diagram of an I²C write transaction.

the first byte, which contains the device address and the direction bit of 0. The data bit over sda must be stable when scl is high. Since switching to a new data bit may cause transition on sda, this implies that new data can only be placed on sda when scl is 0. This requirement ensures that the start and stop conditions will never be confused as data.

After completing transmission of eight bits, the master releases the sda line and reverses to receiving mode in the ninth clock cycle. The addressed slave should generate an active-low acknowledge (i.e., 0) in this period and the master can read this bit to verify that the previous transmitted byte was properly received by the designated slave. The master then repeats the write-byte operation twice to transmit two data bytes.

The master terminates the transaction by creating the stop condition, in which sda changes from low to high while scl is high, and releases the bus.

The timing diagram of reading two bytes is similar except that the roles of receiver and transmitter are reversed for the last two bytes. The master now receives data bits and acknowledges in the ninth cycle.

16.1.4 Additional features

The I²C protocol is comprehensive and flexible and includes mechanisms to accommodate slow slave devices and multiple master devices. The *clock stretching* scheme allows a slow slave device to hold down the scl line and thus enables it to reduce the clock rate or even suspend communication for a while. The *arbitration* mechanism resolves the contention between two masters, which occurs when two master devices initiate the transactions at the same time. These features require the master to continuously monitor the scl and sda lines and to utilize the open-drain output structure.

16.2   I²C CONTROLLER

16.2.1   Basic design

We develop an I²C master controller in this section. The design is intended for a single-master configuration (i.e., only one master on bus) and does not support clock stretching.

The I²C protocol only defines the starting and terminating conditions of transaction and bit-level timing. It does not specify the number of bytes in a transaction and thus designing a general-purpose I²C controller at the transaction level is tedious. For simplicity, we develop a controller that supports basic I²C *bus actions* and use the software device driver to "assemble" the needed actions to form a transaction specified by a specific I²C device. There are five basic actions performed by an I²C master controller:

- Write eight bits and check the acknowledge bit.
- Read eight bits and assert an acknowledge or negative-acknowledge bit.
- Generate a start condition.
- Generate a stop condition.
- Generate a restart condition.

Based on the needed transitions, each action can be divided into several phases. The phases of the start, restart, and stop conditions are shown in Figure 16.4(a). Note that both `scl` and `sda` lines are high when the bus is idle and both are low after start and restart conditions and after completing transmitting or receiving a byte. We assume that the I²C clock rate is f_{i2c} and its period is t_{I2C} (which is $\frac{1}{f_{i2c}}$). Each phase is half of t_{I2C}. The start and stop conditions take one I²C clock cycle. The restart condition is similar to the start condition. The controller simply first raises the `scl` and `sda` lines to high and then generates a normal start condition. It takes 1.5 I²C clock cycles.

The phases of processing a bit in read or write actions are shown in Figure 16.4(b). Each bit takes one I²C clock cycle and the period is divide into four phases, labeled as `data1`, `data2`, `data3`, and `data4`. An `scl` clock pulse is generated in this period. The I²C protocol specifies that the data on `sda` must be stable when `scl` is high. Correct data exchange can be achieved by placing the data bit on `sda` in the beginning of the `data1` phase in a write operation and retrieving the data bit in the transition between the `data2` and `data3` phases in a read operation. After nine bits are processed, the `scl` and `sda` lines are lowered in preparation for the next action, which is labeled as `data_end`.

The I²C controller is constructed by an FSMD and a special output control logic to handle the data bit flow and acknowledgment.

16.2.2   Conceptual FSMD construction

The sketch of the control FSM is shown in Figure 16.5. The `cmd` input specifies the action, whose value can be `RD_CMD` (for a read operation), `WR_CMD` (for a write operation), `START_CMD` (for the start condition), `RESTART_CMD` (for the restart condition), or `STOP_CMD` (for the stop condition). The FSM states represent the phases of various actions labeled under the timing diagram in Figure 16.4(a) and (b). Based on the state, it generates the applicable `scl` and `sda` signals.

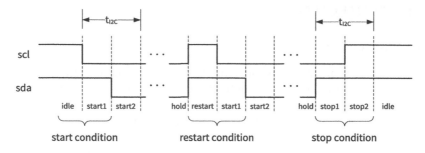

(a) Division of control conditions

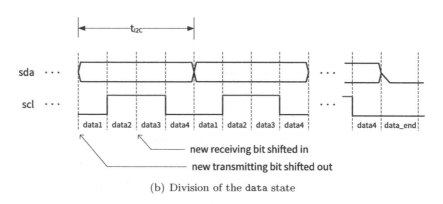

(b) Division of the `data` state

Figure 16.4 State division of I²C transactions.

A transaction begins with `START_CMD`. It initiates the FSM to move through the `start1` and `start2` states to generate a start condition. A `WR_CMD` or `RD_CMD` command should follow. Both read and write actions go through the same data states. The `data1`, `data2`, `data3`, and `data4` states process a data bit or an acknowledge bit. They are circulated through nine times and `bit_reg` is used as a counter to keep track of the number of iterations. A separate output control logic controls the direction of data flow based on the type of command. After nine bits are processed, the FSM moves to the `data_end` state, which lowers the `scl` and `sda` lines for a quarter of `scl` clock. It then moves to the `hold` state.

After completing a command, the FSM stays in the `hold` state waiting for the next command. The state represents an intermediate "holding point" in a transaction and both `scl` and `sda` lines are low in this state. Note that the FSM generates a `ready` status signal to indicate whether the controller is ready to accept a new command. It is asserted in the `idle` and `hold` states.

The transaction is ended with a `STOP_CMD` or `RESTART_CMD` command. The former makes the FSM move through the `stop1` and `stop2` states to generate a stop condition and the latter forces the FSM to regenerate a start condition and begin a new transaction.

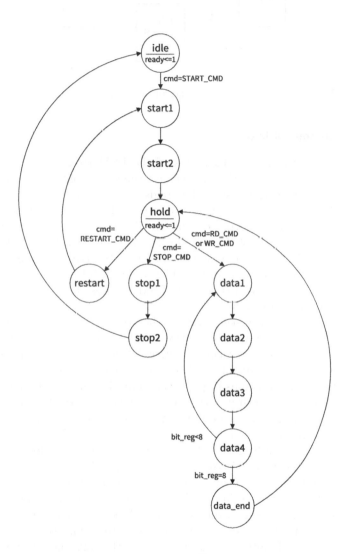

Figure 16.5 Sketch of an I²C controller FSM.

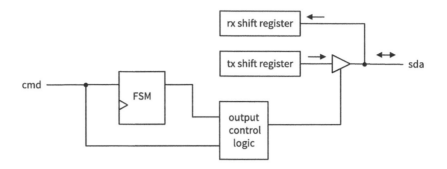

Figure 16.6   Conceptual diagram of output control logic.

16.2.3   Output control logic

The FSMD utilizes the four `data` states of the FSM to generate the `scl` clock and to transfer data between the master and slave. It contains a receiving shift register and a transmitting register. The registers are nine bits wide and used to handle eight data bits and one acknowledge bit. The same shift operations are performed for both read and write operation. The receiving shift register samples the `sda` input at the middle of the `scl` clock, when the FSM transits from `state2` to `state3`. The FSM loads the nine-bit word into the transmitting shift register before the initiation of the `data` states. The bits are shifted out at the beginning of the `data1` state one bit at a time. The FSM iterates the four `data` states nine times to complete the processing.

A separate *output control logic* controls the tristate buffer of the `sda` I/O pin and sets the correct direction of data flow. The conceptual diagram is shown in Figure 16.6. During a write operation, an eight-bit data and a dummy acknowledge bit is loaded into the transmitting register. The control logic enables the tristate buffer in the first eight iterations to place the data bits on the `sda` line. It disables the tristate buffer in the ninth iteration to release the line so that the slave device can issue an acknowledge. In the end of a write operation, nine bits are shifted out from the transmitting register but the last dummy acknowledge bit is irrelevant. Similarly, nine bits are shifted into the receiving register. The first eight bits are the loopback from the transmitted data and are irrelevant. The last bit is the acknowledge from the slave and can be used to verify the status of the transmission.

During a read operation, a dummy byte and a designated acknowledge bit are loaded into the transmitting register. The control logic reverses the previous direction. It disables the tristate buffer in the first eight iterations so that the receiving register can sample and shift eight data bits from the `sda` line. It enables the tristate buffer in the ninth iteration so that the master can place an acknowledge bit on the `sda` line.

16.2.4   I²C bus clock generation

The clock of I²C bus is the pulses on the `scl` line during data transfer. The clock rate is between 0 and 100K Hz in I²C's standard mode and can be up to 3.4M Hz in other modes. In our design, the clock is generated in FSM's four data states, as

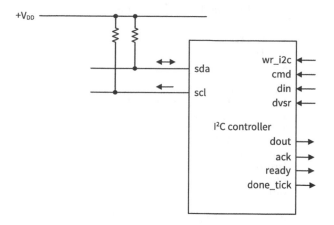

Figure 16.7   Top-level diagram of the I²C controller.

shown in Figure 16.4(b). The desired I²C clock rate can be obtained by adjusting the interval of a data state.

Let the clock rates of I²C bus and the FPGA system be f_{i2c} and f_{sys}. There are $\frac{f_{sys}}{f_{i2c}}$ system clocks in one I²C clock period. Since a cycle is composed of four phases, there are $\frac{f_{sys}}{4*f_{i2c}}$ system clocks in each **data** state. A counter in the FSMD data path can be used to keep track of the number of elapsed clock cycles and to make the transition at the desired time.

The phases of start, restart, and stop condition are half of the I²C clock period, as shown in Figure 16.4(a). There are $\frac{f_{sys}}{2*f_{i2c}}$ system clocks in each phase and the same counter can be used to obtain the desired time interval.

16.2.5   HDL implementation

The top-level sketch of I²C controller is shown in Figure 16.7. The `scl` and `sda` ports are connected to the I²C bus. There are three other input ports. The `dvsr` port is the clock divisor to obtain a quarter of an I²C clock period and should be equal to $\frac{f_{sys}}{4*f_{i2c}}$. The `cmd` port is for the three-bit command that specifies the desired action, which can be start (000), write (001), read (010), stop (011), or restart (100). The `din` port is for the input data. In a write operation, it is the data byte to be transmitted. In a read operation, the LSB (i.e., `din(0)`) is the acknowledge bit to be transmitted by the I²C master. The `wr_i2c` port is a control signal that writes the data and command into the registers and starts the action.

There are four output ports. The `dout` port is the received data byte from a read operation. The `ack` port is the received acknowledge bit from a write operation and it should be 0. The `ready` signal indicates that the controller is idle and ready to take a new command. The `done_tick` signal is asserted for one clock cycle after the controller completes processing nine bits in a read or write action.

The HDL code follows the FSM in Figure 16.5 and generates the desired `scl` and `sda` level in Figure 16.4. It is shown in Listing 16.1.

Listing 16.1 I²C controller

```vhdl
library ieee;
use ieee.std_logic_1164.all;
use ieee.numeric_std.all;
entity i2c_master is
    port(
        clk, reset : in      std_logic;
        din        : in      std_logic_vector(7 downto 0);
        cmd        : in      std_logic_vector(2 downto 0);
        dvsr       : in      std_logic_vector(15 downto 0);
        wr_i2c     : in      std_logic;
        scl        : out     std_logic;
        sda        : inout   std_logic;
        ready      : out     std_logic;
        done_tick  : out     std_logic;
        ack        : out     std_logic;
        dout       : out     std_logic_vector(7 downto 0)
    );
end i2c_master;

architecture arch of i2c_master is
    constant START_CMD   : std_logic_vector(2 downto 0) := "000";
    constant WR_CMD      : std_logic_vector(2 downto 0) := "001";
    constant RD_CMD      : std_logic_vector(2 downto 0) := "010";
    constant STOP_CMD    : std_logic_vector(2 downto 0) := "011";
    constant RESTART_CMD : std_logic_vector(2 downto 0) := "100";
    type statetype is (
        idle, hold, start1, start2, data1, data2, data3, data4, data_end,
        restart, stop1, stop2);
    signal state_reg          : statetype;
    signal state_next         : statetype;
    signal c_reg, c_next      : unsigned(15 downto 0);
    signal qutr, half         : unsigned(15 downto 0);
    signal tx_reg, tx_next    : std_logic_vector(8 downto 0);
    signal rx_reg, rx_next    : std_logic_vector(8 downto 0);
    signal cmd_reg, cmd_next  : std_logic_vector(2 downto 0);
    signal bit_reg, bit_next  : unsigned(3 downto 0);
    signal sda_out, scl_out   : std_logic;
    signal sda_reg, scl_reg   : std_logic;
    signal into               : std_logic;
    signal nack               : std_logic;
    signal data_phase         : std_logic;
begin
    --*****************************************************************
    -- output control logic
    --*****************************************************************
    -- buffer for sda and scl lines
    process(clk, reset)
    begin
        if reset = '1' then
            sda_reg <= '1';
            scl_reg <= '1';
        elsif (clk'event and clk = '1') then
            sda_reg <= sda_out;
            scl_reg <= scl_out;
        end if;
    end process;
    -- only master drives scl line
    scl  <= 'Z' when scl_reg = '1' else '0';
    -- sda are with pull-up resistors
    -- and becomes high when not driven
```

```
into <= '1' when
   (data_phase = '1' and cmd_reg = RD_CMD and bit_reg < 8) or
   (data_phase = '1' and cmd_reg = WR_CMD and bit_reg = 8) else '0';
sda   <= 'Z' when into = '1' or sda_reg = '1' else '0';
-- output
dout <= rx_reg(8 downto 1);
ack   <= rx_reg(0);    -- obtained from slave in write operation
nack <= din(0);        -- used by master in read operation
--*******************************************************************
-- fsmd
--*******************************************************************
-- registers
process(clk, reset)
begin
   if reset = '1' then
      state_reg <= idle;
      c_reg     <= (others => '0');
      bit_reg   <= (others => '0');
      cmd_reg   <= (others => '0');
      tx_reg    <= (others => '0');
      rx_reg    <= (others => '0');
   elsif (clk'event and clk = '1') then
      state_reg <= state_next;
      c_reg     <= c_next;
      bit_reg   <= bit_next;
      cmd_reg   <= cmd_next;
      tx_reg    <= tx_next;
      rx_reg    <= rx_next;
   end if;
end process;
--   intervals
qutr <= unsigned(dvsr);
half <= qutr(14 downto 0) & '0';   -- half = 2*qutr
-- next-state logic
process(state_reg, bit_reg, tx_reg, c_reg, rx_reg, cmd_reg,
        cmd, din, wr_i2c, sda, nack, qutr, half)
begin
   state_next <= state_reg;
   c_next     <= c_reg + 1;   -- timer counts continuously
   bit_next   <= bit_reg;
   tx_next    <= tx_reg;
   rx_next    <= rx_reg;
   cmd_next   <= cmd_reg;
   done_tick  <= '0';
   ready      <= '0';
   scl_out    <= '1';
   sda_out    <= '1';
   data_phase <= '0';
   case state_reg is
      when idle =>
         ready <= '1';
         if wr_i2c = '1' and cmd = START_CMD then  -- start
            state_next <= start1;
            c_next     <= (others => '0');
         end if;
      when start1 =>    -- start condition
         sda_out <= '0';
         if c_reg = half then
            c_next     <= (others => '0');
            state_next <= start2;
         end if;
```

```vhdl
            when start2 =>
               sda_out <= '0';
               scl_out <= '0';
               if c_reg = qutr then
                  c_next      <= (others => '0');
                  state_next <= hold;
               end if;
            when hold =>      -- in progress; prepare for the next op
               ready   <= '1';
               sda_out <= '0';
               scl_out <= '0';
               if wr_i2c = '1' then
                  cmd_next <= cmd;
                  c_next   <= (others => '0');
                  case cmd is
                     when RESTART_CMD | START_CMD =>
                        state_next <= restart;
                     when STOP_CMD =>
                        state_next <= stop1;
                     when others =>      -- read/write a byte
                        bit_next   <= (others => '0');
                        state_next <= data1;
                        tx_next    <= din & nack; --nack used in read
                  end case;
               end if;
            when data1 =>
               sda_out    <= tx_reg(8);
               scl_out    <= '0';
               data_phase <= '1';
               if c_reg = qutr then
                  c_next     <= (others => '0');
                  state_next <= data2;
               end if;
            when data2 =>
               sda_out    <= tx_reg(8);
               data_phase <= '1';
               if c_reg = qutr then
                  c_next     <= (others => '0');
                  state_next <= data3;
                  rx_next    <= rx_reg(7 downto 0) & sda;
               end if;
            when data3 =>
               sda_out    <= tx_reg(8);
               data_phase <= '1';
               if c_reg = qutr then
                  c_next     <= (others => '0');
                  state_next <= data4;
               end if;
            when data4 =>
               sda_out    <= tx_reg(8);
               scl_out    <= '0';
               data_phase <= '1';
               if c_reg = qutr then
                  c_next <= (others => '0');
                  if bit_reg = 8 then      -- done with 8 data bits + 1 ack
                     state_next <= data_end;
                     done_tick  <= '1';
                  else
                     tx_next    <= tx_reg(7 downto 0) & '0';
                     bit_next   <= bit_reg + 1;
                     state_next <= data1;
```

```
                end if;
            end if;
        when data_end =>
            sda_out <= '0';
            scl_out <= '0';
            if c_reg = qutr then
                c_next      <= (others => '0');
                state_next <= hold;
            end if;
        when restart =>    -- generate idle condition
            if c_reg = half then
                c_next      <= (others => '0');
                state_next <= start1;
            end if;
        when stop1 =>     -- stop condition
            sda_out <= '0';
            if c_reg = half then
                c_next      <= (others => '0');
                state_next <= stop2;
            end if;
        when stop2 =>     -- turnaround time
            if c_reg - half then
                state_next <= idle;
            end if;
    end case;
  end process;
end arch;
```

There are several registers in the data path. The `c_reg` register is used as a counter to keep track of the amount of time spent in each state. The `qutr` and `half` are derived from the `dvsr` input signal and are used to represent the number of system clock cycles for a quarter and a half of the I²C clock period. The `c_reg` register counts continuously and is cleared to zero when the FSM exits the previous state. The FSM can only move out from the current state when `c_reg` reaches the designated value. The `bit_reg` register is used to keep track of the number of data bits processed and the `cmd_reg` register stores the current command. The `tx_reg` and `rx_reg` registers are the transmitting and receiving shift registers.

The output control logic determines the direction of data flow on the `sda` line and generates a signal, `into`, to control the tristate buffer. The data flows into the controller under the following conditions:

- The execution of the first eight bits of the RD_CMD
- The execution of the ninth bit of the WR_CMD

These conditions translate into the following logic expression:

```
(data_phase = '1' and cmd_reg = RD_CMD and bit_reg < 8) or
(data_phase = '1' and cmd_reg = WR_CMD and bit_reg = 8)
```

Finally, the design uses FPGA I/O pin's tristate buffer to accommodate the open-drain structure of the I²C bus. The HDL code for the `scl` port, in which the signal flows out of the device, is

```
scl <= 'Z' when scl_reg = '1' else '0';
```

In this scheme, the FPGA device turns off the tristate buffer (i.e., changes the output to a high-impedance state) when a desired bus line level is 1. Since the bus line is connected to V_{DD} via a pull-up resistor, it is driven to 1 implicitly when all devices output 1 (i.e., all are in high-impedance state).

The HDL code for the `sda` port is

```
sda <= 'Z' when into = '1' or sda_reg = '1' else '0';
```

The first condition, `into='1'`, is to turn off the tristate buffer for the slave device to transmit data and the second condition, `sda_reg='1'`, is to generate an output of 1 via the implicit pull-up resistor circuit.

16.3   I²C CORE DEVELOPMENT

We can follow the procedure in Section 10.3.2 to add a wrapping circuit for the I²C controller and create an MMIO IP core.

16.3.1   Register map

The processor interacts with the I²C controller as follows:
- set (i.e., write) the value of `dvsr`.
- specify (i.e., write) the action and data and start the operation.
- receive (i.e., read) a data byte.
- check (i.e., read) the `ready` signal to determine whether the controller is ready to accept a new command.
- verify (i.e., read) the acknowledge bit of write action.

Based on these interactions, we can define the I²C core's register map. There is one read register. Its address offset and fields are:
- offset 0 (read data and status register)
  - bits 7 to 0: 8-bit received data
  - bit 8: ready status
  - bit 9: acknowledge bit

There are two write registers. Their address offsets and fields are:
- offset 0 (dvsr register)
  - bits 15 to 0: 16-bit divisor value
- offset 1 (data and command register)
  - bits 7 to 0: 8-bit data
  - bits 10 to 8: 3-bit command
  - writing register 1 also initiates the I²C controller operation

16.3.2   Wrapping circuit for the slot interface

Based on the register map and the I/O signals of the I²C controller, we can derive a wrapping circuit that complies to the slot specification and create the I²C core. The HDL code of the core is shown in Listing 16.2. It consists of an instance of I²C controller, a register to store the divisor value, and a decoding circuit.

Listing 16.2   I²C core

```
library ieee;
use ieee.std_logic_1164.all;
use ieee.numeric_std.all;
entity chu_i2c_core is
  port(
```

```vhdl
      clk     : in std_logic;
      reset   : in std_logic;
      -- slot interface
      cs      : in std_logic;
      write   : in std_logic;
      read    : in std_logic;
      addr    : in std_logic_vector(4 downto 0);
      rd_data : out std_logic_vector(31 downto 0);
      wr_data : in std_logic_vector(31 downto 0);
      -- external signals
      scl     : out std_logic;
      sda     : inout std_logic
   );
end chu_i2c_core;

architecture arch of chu_i2c_core is
   signal ready, ack : std_logic;
   signal dout       : std_logic_vector(7 downto 0);
   signal dvsr_reg   : std_logic_vector(15 downto 0);
   signal wr_en      : std_logic;
   signal wr_i2c     : std_logic;
   signal wr_dvsr    : std_logic;
begin
   -- instantiate codec controller
   i2c_unit : entity work.i2c_master(arch)
      port map(
         clk       => clk,
         reset     => reset,
         din       => wr_data(7 downto 0),
         cmd       => wr_data(10 downto 8),
         dvsr      => dvsr_reg,
         wr_i2c    => wr_i2c,
         scl       => scl,
         sda       => sda,
         ready     => ready,
         done_tick => open,
         ack       => ack,
         dout      => dout);
   -- registers
   process(clk, reset)
   begin
      if reset = '1' then
         dvsr_reg <= (others => '0');
      elsif (clk'event and clk = '1') then
         if wr_dvsr = '1' then
            dvsr_reg <= wr_data(15 downto 0);
         end if;
      end if;
   end process;
   --decoding logic
   wr_en   <= '1' when write='1' and cs='1' else '0';
   wr_dvsr <= '1' when addr(0)='0' and wr_en='1' else '0';
   wr_i2c  <= '1' when addr(0)='1' and wr_en='1' else '0';
   -- read data
   rd_data <= x"00000" & "00" & ack & ready & dout;
end arch;
```

Note that the command and write data are connected to the `cmd` and `din` ports directly. When the processor writes register 1, the `wr_i2c` signal is asserted and the command and write data are written to I²C controller's registers at the same time.

16.4 I²C DRIVER

The driver consists of a set of routines to set I²C bus clock frequency, to issue a command, and to perform a read or write transaction.

16.4.1 Class definition

The class definition of the I²C core is shown in Listing 16.3.

Listing 16.3 I2cCore class definition (in i2c_core.h)

```
#ifndef _I2C_CORE_H_INCLUDED
#define _I2C_CORE_H_INCLUDED

#include "chu_init.h"

class I2cCore
{
    enum {
        DVSR_REG = 0,
        WR_REG   = 1,
        RD_REG   = 0
    };
    /* command to i2c controller */
    enum {
        I2C_START_CMD   = 0x00<<8,
        I2C_WR_CMD      = 0x01<<8,
        I2C_RD_CMD      = 0x02<<8,
        I2C_STOP_CMD    = 0x03<<8,
        I2C_RESTART_CMD = 0x04<<8
    };
public:
    I2cCore(uint32_t core_base_addr);
    ~I2cCore();
    void set_freq(int freq);
    int ready();
    void start();
    void restart();
    void stop();
    int write_byte(uint8_t data);
    int read_byte(int last);
    int read_transaction(uint8_t dev, uint8_t *bytes, int num, int repeat);
    int write_transaction(uint8_t dev, uint8_t *bytes,int num, int repeat);
private:
    uint32_t base_addr;
};
#endif  //_I2C_CORE_H_INCLUDED
```

The first enum definition uses symbolic names for the register offsets. The second enum definition specifies the commands in the designed field (bits 10 to 8).

16.4.2 Class implementation

The class implementation of the constructor and the methods is shown in Listing 16.4.

Listing 16.4 I2cCore class implementation (in i2c_core.cpp)

```cpp
I2cCore::I2cCore(uint32_t core_base_addr) {
   base_addr = core_base_addr;
   set_freq(100000);   // default 100K Hz
}

I2cCore::~I2cCore() { }

void I2cCore::set_freq(int freq) {
   uint32_t dvsr;

   dvsr = (uint32_t) (SYS_CLK_FREQ * 1000000 / freq / 4);
   io_write(base_addr, DVSR_REG, dvsr);
}

int I2cCore::ready() {
   return ((int) (io_read(base_addr,RD_REG) >> 8) & 0x01);
}

void I2cCore::start() {
   while (!ready()) {}
   io_write(base_addr, WR_REG, I2C_START_CMD);
}

void I2cCore::restart() {
   while (!ready()) {}
   io_write(base_addr, WR_REG, I2C_RESTART_CMD);
}

void I2cCore::stop() {
   while (!ready()) {}
   io_write(base_addr, WR_REG, I2C_STOP_CMD);
}

int I2cCore::write_byte(uint8_t data) {
   int ack, acc_data;

   acc_data = data | I2C_WR_CMD;
   while (!ready()) {}
   io_write(base_addr, WR_REG, acc_data);
   while (!ready()) {}
   ack = (io_read(base_addr, RD_REG) & 0x0200) >> 9;
   if (ack == 0)
      return (0);
   else
      return (-1);
}

int I2cCore::read_byte(int last) {
   int acc_data;

   acc_data = last | I2C_RD_CMD;
   while (!ready()) {}
   io_write(base_addr, WR_REG, acc_data);
   while (!ready()) {}
   return (io_read(base_addr, RD_REG) & 0x00ff);
}

int I2cCore::read_transaction(uint8_t dev, uint8_t *bytes,
                              int num, int rstart){
   uint8_t dev_byte;
```

```
    int ack1;
    int i;

    dev_byte = (dev << 1) | 0x01;   // LSB=1 for I2c read
    start();
    ack1 = write_byte(dev_byte);    // send device id/read
    for (i = 0; i < (num - 1); i++) {
        *bytes = read_byte(0);
        bytes++;
    }
    *bytes = read_byte(1);    // last byte in read cycle
    if (rstart==1){
        restart();
    }else{
        stop();
    }
    return (ack1);
}

int I2cCore::write_transaction(uint8_t dev, uint8_t *bytes,
                               int num, int rstart){
    uint8_t dev_byte;
    int ack1, ack;
    int i;

    dev_byte = (dev << 1);        // LSB=0 for I2c write
    start();
    ack = write_byte(dev_byte);   // send device id/write
    for (i = 0; i < num; i++) {
        ack1 = write_byte(*bytes);
        ack = ack + ack1;
        bytes++;
    }
    if (rstart==1){
        restart();
    }else{
        stop();
    }
    return (ack);
}
```

The I2cCore() constructor saves the base address and sets the default scl frequency to 100K Hz. The set_freq() method sets the I²C bus clock frequency. It calculates the divisor value with the $\frac{f_{sys}}{4*f_{i2c}}$ formula and writes the value to the divisor register.

The ready() method reads the register and extracts the ready bit. The start(), restart(), and stop() methods wait until controller is ready and then issue the corresponding command.

The write_byte() method writes a byte of data and returns the status. It concatenates the data and write command and writes them to core's register when it is ready. The method waits until the write operation is completed and then retrieves the acknowledge bit. If the slave device does not generate a proper acknowledge bit, a −1 is returned to indicate the write failure.

The read_byte() method retrieves a byte of data from the slave device. The parameter, last, whose value should be either 0 or 1, is used to flag whether the read operation is the last one in a transaction.

The `read_transaction()` method performs a read transaction. The `dev` parameter specifies the slave device address (i.e., id) and the `num` parameter specifies the number of byte. The retrieved data bytes are stored in an array pointed to by `bytes`. The `rstart` parameter indicates whether a restart condition or a stop condition is generated at the end of the transaction. The method follows the read transaction sequence similar to that in Figure 16.2 and issues the following commands:

- Start.
- Write with device address.
- Read first `num-1` bytes (with acknowledge of 0).
- Read read last byte (with acknowledge of 0).
- Restart or stop.

The `write_transaction()` method are similar except that multiple write commands are issued.

16.5   TEST

The Nexys 4 DDR board contains an ADT7420 device, which is a temperature sensor with the I^{2}C interface. It is used to verify the operation of the I^{2}C core.

16.5.1   ADT7420 temperature sensor

The ADT7420 device is a high-accuracy digital temperature sensor made by Analog Devices. It contains an internal band gap reference, a temperature sensor, and an ADC. The digitized reading is stored in internal registers. In addition, the device contains additional output pins that become active when the temperature exceeds predefined thresholds. Some features, such as the values of the thresholds and the ADC resolution, can be configured.

The digitized temperature, status, and configuration are stored in ADT7420's 14 internal registers, which can be accessed via the I^{2}C interface. The protocol of reading two consecutive registers from ADT7420 (as in reading the registers 0 and 1) is shown in Figure 16.8. The I^{2}C master first issues a write transaction with one byte of data, which is the number of the register. The transaction is terminated with a restart condition (i.e., without releasing the bus). The I^{2}C master then issues a read transaction to retrieve two bytes of data. Reading one byte is similar except that only one read action is needed in the second transaction. This form of protocol is very common for I^{2}C devices. However, some device may use a stop condition after the first transaction and issue a new start condition for the second transaction.

To verify the operation of the I^{2}C core, our test reads the device signature and the temperature. Following is the relevant information for our test:

- The I^{2}C address of ADT7420 device is configured to be `0x4b` on the Nexys 4 DDR board.
- The ADT7420 signature is stored in register 11 (`0x0b`) and its value is `0xcb`.
- The temperature and status are stored in register 0 (the upper byte) and register 1 (the lower byte). After the 16-bit word is reconstructed, the 13 MSBs are the Celsius temperature reading in signed format.

More detailed usage and configuration information can be found in the ADT7420 data sheet.

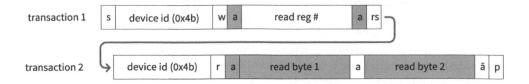

Figure 16.8 ADT7420 read operation.

16.5.2 Test program

A simple test function is derived to access the ADT7420 temperature measurement via the I²C core. The code is shown in Listing 16.5. It checks the ADT7420's signature to verify the existence of the device, retrieves the raw data, and derives the temperature reading.

Listing 16.5 ADT7420 I²C test function (in `main_sampler_test.cpp`)

```
void adt7420_check(I2cCore *adt7420_p, GpoCore *led_p) {
   const uint8_t DEV_ADDR = 0x4b;
   uint8_t wbytes[2], bytes[2];
   uint16_t tmp;
   float tmpC;

   // check part id
   wbytes[0] = 0x0b;
   adt7420_p->write_transaction(DEV_ADDR, wbytes, 1, 1);
   adt7420_p->read_transaction(DEV_ADDR, bytes, 1, 0);
   uart.disp("read ADT7420 id (should be 0xcb): ");
   uart.disp(bytes[0], 16);
   uart.disp("\n\r");
   wbytes[0] = 0x00;
   adt7420_p->write_transaction(DEV_ADDR, wbytes, 1, 1);
   adt7420_p->read_transaction(DEV_ADDR, bytes, 2, 0);
   // conversion
   tmp = (uint16_t) bytes[0];
   tmp = (tmp << 8) + (uint16_t) bytes[1];
   if (tmp & 0x8000) {
      tmp = tmp >> 3;
      tmpC = (float) ((int) tmp - 8192) / 16;
   } else {
      tmp = tmp >> 3;
      tmpC = (float) tmp / 16;
   }
   uart.disp("temperature (C): ");
   uart.disp(tmpC);
   uart.disp("\n\r");
   led_p->write(tmp);
   sleep_ms(1000);
   led_p->write(0);
}
```

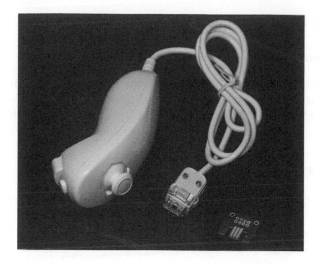

Figure 16.9   Wii Nunchuk and adaptor.

16.6   PROJECT IDEA

The I^2C and SPI buses are the two most common standards for an embedded processor to interface I/O peripherals. As with the SPI, I^2C is used for a wide variety of peripherals, including sensors, ADCs, DACs, LCDs, and EPROM and flash memory. An FPro system can instantiate multiple I^2C cores and thus incorporate any number of desired peripherals into the system.

An interesting device is *Wii Nunchuk*, which is shown in Figure 16.9. It is an accessory for the Wii game console remote but can function as a stand-alone device. Nunchuk contains a three-axis accelerometer, joystick, and three push-buttons. Its internal circuit encodes the measurements into a single serial data stream and transmits the stream via an I^2C port. The signals of the port can be brought out with a simple adaptor, which is shown at the right in Figure 16.9. Nunchuk is a very versatile I/O peripheral and can be used as a general handheld controller.

16.7   BIBLIOGRAPHIC NOTES

I^2C Manual from Philips Semiconductors gives the detailed technical specifications of the I^2C standard and Wikipedia website provides a comprehensive overview. The ADT7420 data sheet from Analog Devices, Inc. contains detailed information on the I^2C timing, command, data transfer protocol, and register usage.

The data format and protocol of Nunchuk are widely available on the internet. The information can be found by searching the terms "Wii" and "I2C."

16.8 SUGGESTED EXPERIMENTS

16.8.1 Thermometer

Use the ADT7420 device to implement a thermometer, in which the temperature reading is shown in the seven-segment LED display. The reading can be either in Celsius (C) or Fahrenheit (F) and the format of the display will look like "18.25C" or "64.85F." Derive the program and verify its operation.

16.8.2 ADT7420 C++ class

The ADT7420 is a versatile device with many features. Consult the data sheet and create an ADT7420 C++ class with the following methods:

- Read device id.
- Perform a software reset.
- Read status.
- Set configuration register.
- Read temperature data.

Write a test program to verify the operation of these methods.

16.8.3 Enhanced I²C core

The I²C core in Section 16.3 processes five basic commands. The read_transaction() and write_transaction() methods use these commands to retrieve or send data. We can enhance the core by implementing the two routines in hardware and improve its performance. For simplicity, we assume that the number of bytes accessed is between one and four bytes and can be accommodated by a 32-bit word. Design the new I²C controller, revise the I²C core and driver, and verify its operation with the ADT7420 device.

16.8.4 Automatic-read ADT7420 wrapper

An automatic reading wrapping circuit retrieves and outputs the temperature data from ADT7420 continuously. An external system can access the data directly without worrying about the I²C protocol. The entity declaration is

```
entity adt7420_wrapper is
   port (
      clk, reset : in  std_logic;
      -- 12-bit temperature reading
      tmp: out std_logic_vector (11 downto 0);
      -- I2C interface connected to ADT7420
      scl : out   std_logic;
      sda : inout std_logic
   );
end adt7420_wrapper;
```

The wrapping circuit can be constructed by an FSMD on top of the I²C controller. The FSMD should follow the ADT7420 read transaction protocol and retrieve the measurement data. Derive the wrapping circuit and verify its operation.

16.8.5   I^{2}C slave controller: part I

An I^{2}C slave controller accepts commands from an I^{2}C master and responds accordingly. With an I^{2}C slave interface on an FPGA board, other prototyping boards can access the FPGA board and exchange information. We want to construct a slave controller following the transaction protocol similar to the one discussed in Section 16.5.1. Follow the procedure in Experiment 15.8.9 to design the slave controller and verify its operation.

16.8.6   I^{2}C slave controller: part II

Repeat the Experiment 15.8.10 but use an "I^{2}C slave wrapping circuit."

16.8.5 I^2C slave controller part 1

CHAPTER 17

PS2 CORE

The PS2 standard is a serial interface commonly used by a mouse or a keyboard. A PS2 device's activities are embedded in a stream of packets and transmitted to the host via two serial lines. In this chapter, we design a PS2 core to transmit and receive data packets and develop driver routines to process the packets and to decode the keyboard or mouse activities.

17.1 INTRODUCTION

The PS2 standard was introduced in IBM's Personal System/2 personal computers. It is a widely supported interface for a keyboard or mouse to communicate with the host, which is usually a processor. As in the I^2C interface, the PS2 port contains two wires for communication purposes. One wire is for the bidirectional data, which is transmitted in a serial stream. The other wire is for the clock information, which specifies when the data is valid and can be retrieved. Unlike the I^2C interface, in which only the master can start a transaction, both the host and PS2 device can initiate the data transfer. A host "listens" and receives data from a PS2 device most of the time but occasionally sends a command to the keyboard or mouse to set certain parameters. In addition, the PS2 device, not the host, generates the clock signal.

Both the data and clock lines are bidirectional. They use open-drain technology, similar to that discussed in Section 16.1.1.

FPGA Prototyping by VHDL Examples 2nd ed., Pong P. Chu.
Copyright © 2017, John Wiley & Sons, Inc.

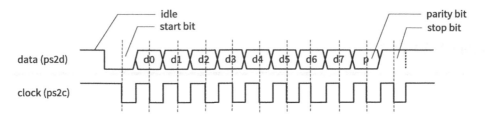

Figure 17.1   PS2-device-to-host timing diagram of a PS2 port.

The information in a PS2 interface is transmitted as an 11-bit "packet" that contains a start bit, eight data bits, an odd parity bit, and a stop bit. Whereas the basic format of the packet is identical for a keyboard and a mouse, the interpretation for the data bits is different. A keyboard data stream contains the scan codes of keys and a mouse data stream contains the movement information and button status. Thus, separate software drivers are needed.

17.1.1   PS2-device-to-host communication protocol and timing

A PS2 device and its host communicate via packets. The basic timing diagram of transmitting a packet from a PS2 device to a host is shown in Figure 17.1, in which the data and clock signals are labeled ps2d and ps2c, respectively.

The data is transmitted in a serial stream. Transmission begins with a start bit, followed by eight data bits and an odd parity bit, and ends with a stop bit. The clock information is carried in a separate clock signal, ps2c. The falling edge of the ps2c signal indicates that the corresponding bit in the ps2d line is valid and can be retrieved. The clock period of the ps2c signal is between 60 and 100 μs (i.e., 10K Hz to 16.7K Hz), and the ps2d signal is stable at least 5 μs before and after the falling edge of the ps2c signal.

17.1.2   Host-to-PS2-device communication protocol and timing

The host-to-PS2-device communication protocol involves bidirectional data exchange. The mouse's data and clock lines are open-drain circuits. For our design purposes, we treat them as tristate lines. The basic timing diagram of transmitting a packet from a host to a PS2 device is shown in Figure 17.2, in which the data and clock signals are labeled ps2d and ps2c. For clarity, the diagram is split into two parts to show which activities are generated by the host (i.e., the FPGA-based controller) and which activities are generated by the device (e.g., a mouse). The basic operation sequence is as follows:

1. The host forces the ps2c line to be 0 for at least 100 μs to inhibit any mouse activity. It can be considered that the host requests to send a packet.
2. The host forces the ps2d line to be 0 and disables the ps2c line (i.e., makes it high impedance). This step can be interpreted as the host sending a start bit.
3. The PS2 device now takes over the ps2c line and is responsible for future PS2 clock signal generation. After sensing the starting bit, the PS2 device generates a 1-to-0 transition.

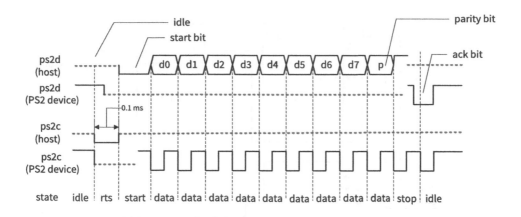

Figure 17.2   Host-to-PS2-device timing diagram of a PS2 port.

4. Once detecting the transition, the host shifts out the least significant data bit over the ps2d line. It holds this value until the PS2 device generates a 1-to-0 transition in the ps2c line, which essentially acknowledges retrieval of the data bit.

5. The host repeats step 4 for the remaining seven data bits and one parity bit.

6. After sending the parity bit, the host disables the ps2d line (i.e., makes it high impedance). The PS2 device now takes over the ps2d line and acknowledges completion of the transmission by asserting the ps2d line to 0. If desired, the host can check this value at the last 1-to-0 transition in the ps2c line to verify that the packet has been transmitted successfully.

17.2   PS2 CONTROLLER

17.2.1   Conceptual design

The top-level diagram of a PS2 controller is shown in Figure 17.3. It consists of the receiving subsystem, the transmitting subsystem, and a FIFO buffer. The tx_idle (for "transmitter idle"), rx_idle (for "receiver idle"), and rx_en (for "receiver enable") signals are used to coordinate the transmitting and receiving operations so that only one type of operation can be performed at a time. The FIFO buffer is inserted after the receiving subsystem to provide some cushion space since a PS2 device may send packets continuously as we move a mouse or type on a keyboard. On the other hand, since the processor is expected to issue commands occasionally and it can control the rate, the transmitting subsystem does not need a buffer.

17.2.2   PS2 receiving subsystem

The basic design of the PS2 receiving subsystem consists of a falling-edge detection circuit, which generates a one-clock-cycle tick at the falling edge of the ps2c signal, and a shift circuit, which shifts in and assembles the serial bits. An FSMD is used to coordinate the overall operation.

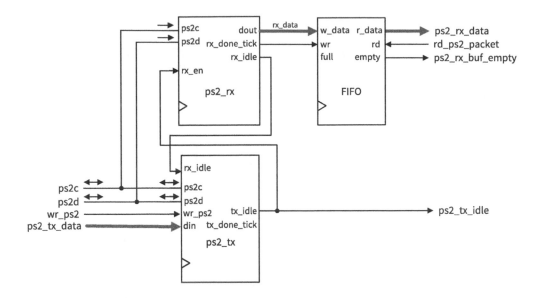

Figure 17.3 Block diagram of a complete PS2 controller.

The edge detection circuit detects the falling edge of the incoming clock signal and generates an enable tick. Because of the potential noise and slow transition, a simple filtering circuit is added to eliminate glitches. Its code is

```
-- register
process (clk, reset)
    filter_reg <= filter_next;
    . . .
end process;

-- 1-bit shifter
filter_next <= ps2c & filter_reg(7 downto 1);
-- "filter"
f_ps2c_next <= '1' when filter_reg="11111111" else
               '0' when filter_reg="00000000" else
               f_ps2c_reg;
```

The circuit is composed of an 8-bit shift register and returns a 1 or 0 when eight consecutive 1's or 0's are received. Any glitch shorter than eight clock cycles will be ignored (i.e., filtered out). The filtered output signal is then fed to the regular falling-edge detection circuit. The cascading registers of the filtering circuit also function as a synchronizer.

The ASMD chart of the receiver is shown in Figure 17.4. The receiver is initially in the idle state. It includes an additional control signal, rx_en, which is used to enable or disable the receiving operation. The purpose of the signal is to coordinate the receiving subsystem operation. After the first falling-edge tick and the rx_en signal are asserted, the FSMD shifts in the start bit and moves to the dps state. In the dps state, ten bits, which include eight data bits, one parity bit, and one stop bit, are sampled at the falling edge of ps2c and the first nine bits are shifted into

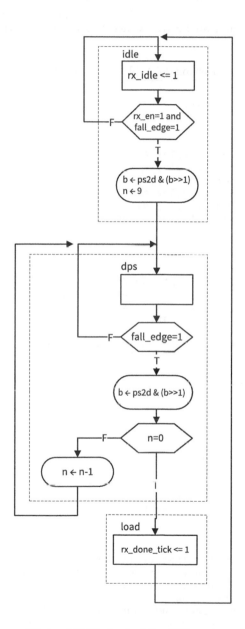

Figure 17.4   ASMD chart of the PS2 port receiver.

the b register. The FSMD then moves to the load state, in which one extra clock cycle is provided to complete the shifting of the stop bit.

There are two output signals. The rx_idle signal indicates whether the receiving subsystem is idle. The rx_done_tick signal is asserted in the load state for one clock cycle to indicate the completion of receiving a packet. The HDL code consists of the filtering circuit and an FSMD, which follows the ASMD chart. It is shown in Listing 17.1.

Listing 17.1   PS2 port receiver

```vhdl
library ieee;
use ieee.std_logic_1164.all;
use ieee.numeric_std.all;
entity ps2rx is
   port(
      clk, reset    : in  std_logic;
      ps2d, ps2c    : in  std_logic;
      rx_en         : in  std_logic;
      rx_done_tick  : out std_logic;
      rx_idle       : out std_logic;
      dout          : out std_logic_vector(7 downto 0)
   );
end ps2rx;

architecture arch of ps2rx is
   type statetype is (idle, dps, load);
   signal state_reg, state_next    : statetype;
   signal filter_reg, filter_next  : std_logic_vector(7 downto 0);
   signal f_ps2c_reg, f_ps2c_next  : std_logic;
   signal b_reg, b_next            : std_logic_vector(10 downto 0);
   signal n_reg, n_next            : unsigned(3 downto 0);
   signal fall_edge                : std_logic;
begin
   --*****************************************************************
   -- filter and falling-edge tick generation for ps2c
   --*****************************************************************
   process(clk, reset)
   begin
      if reset = '1' then
         filter_reg <= (others => '0');
         f_ps2c_reg <= '0';
      elsif (clk'event and clk = '1') then
         filter_reg <= filter_next;
         f_ps2c_reg <= f_ps2c_next;
      end if;
   end process;

   filter_next <= ps2c & filter_reg(7 downto 1);
   f_ps2c_next <= '1' when filter_reg = "11111111" else
                  '0' when filter_reg = "00000000" else
                  f_ps2c_reg;
   fall_edge   <= f_ps2c_reg and (not f_ps2c_next);
   --*****************************************************************
   -- fsmd to extract the 8-bit data
   --*****************************************************************
   -- registers
   process(clk, reset)
   begin
      if reset = '1' then
         state_reg <= idle;
```

```
            n_reg       <= (others => '0');
            b_reg       <= (others => '0');
        elsif (clk'event and clk = '1') then
            state_reg <= state_next;
            n_reg       <= n_next;
            b_reg       <= b_next;
        end if;
    end process;
    -- next-state logic
    process(state_reg, n_reg, b_reg, fall_edge, rx_en, ps2d)
    begin
        rx_idle       <= '0';
        rx_done_tick <= '0';
        state_next   <= state_reg;
        n_next       <= n_reg;
        b_next       <= b_reg;
        case state_reg is
            when idle =>
                rx_idle <= '1';
                if fall_edge = '1' and rx_en = '1' then
                    -- shift in start bit
                    b_next       <= ps2d & b_reg(10 downto 1);
                    n_next       <= "1001";
                    state_next <= dps;
                end if;
            when dps =>    -- 8 data + 1 parity + 1 stop
                if fall_edge = '1' then
                    b_next <= ps2d & b_reg(10 downto 1);
                    if n_reg = 0 then
                        state_next <= load;
                    else
                        n_next <= n_reg - 1;
                    end if;
                end if;
            when load =>
                -- 1 extra clock to complete the last shift
                state_next   <= idle;
                rx_done_tick <= '1';
        end case;
    end process;
    -- output
    dout <= b_reg(8 downto 1); -- data bits
end arch;
```

There is no error detection circuit in the description. A more robust design should check the correctness of the start, parity, and stop bits and include a watch-dog timer to prevent the keyboard or mouse from being locked in an incorrect state.

17.2.3   PS2 transmitting subsystem

Unlike the receiving subsystem, the ps2c and ps2d signals communicate in both directions. A tristate buffer is needed for each signal. The tristate interface is shown in Figure 17.5. The tri_c and tri_d signals are enable signals that control the tristate buffers. When they are asserted, the corresponding ps2c_out and ps2d_out signals will be routed to the output ports.

To design the transmitting subsystem, we can follow the sequence of the preceding protocol to create an ASMD chart, as shown in Figure 17.6. The FSMD

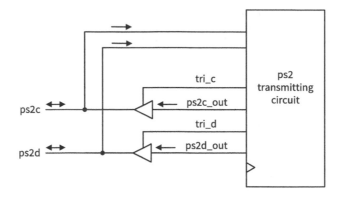

Figure 17.5 Tristate buffers of the PS2 transmission subsystem.

is initially in the `idle` state. To start the transmission, the main system (e.g., a processor) asserts the `wr_ps2` signal and places the data on the `din` bus. The FSMD loads `din`, along with the parity bit, `par` to the `shift_reg` register, loads $1 \cdots 1$ to `c_reg`, and moves to the `waitr` (for "wait receiving") state. In this state, it examines the `rx_idle` signal to determine whether any receiving operation is in progress and waits there until the operation is completed. The FSMD then moves to the `rts` (for "request to send") state. In this state, the `ps2c_out` is set to 0 and the corresponding `tri_c` is asserted to enable the corresponding tristate buffer. The `c_reg` is used as a 13-bit counter to generate an 82-μs delay. The FSMD then moves to the `start` state, in which the PS2 clock line is disabled and the data line is set to 1. The PS2 device now takes over and generates a clock signal over the `ps2c` line. After detecting the falling edge of the `ps2c` signal through the `fall_edge` signal, the FSMD goes to the `data` state and shifts eight data bits and one parity bit. The `n` register is used to keep track of the number of bits shifted. The FSMD then moves to the `stop` state, in which the data line is disabled. It returns to the `idle` state after sensing the last falling edge.

Similar to those of the receiving subsystem, the `tx_idle` signal indicates whether the transmission subsystem is idle and the `tx_done_tick` signal is asserted for one clock cycle when the transmission operation is completed. The code follows the ASMD chart and is shown in Listing 17.2. A filtering circuit is also used to generate the `fall_edge` signal.

Listing 17.2 PS2 port transmitter

```
library ieee;
use ieee.std_logic_1164.all;
use ieee.numeric_std.all;
entity ps2tx is
   port(
      clk, reset   : in     std_logic;
      din          : in     std_logic_vector(7 downto 0);
      wr_ps2       : in     std_logic;
      rx_idle      : in     std_logic;
      ps2d, ps2c   : inout  std_logic;
      tx_idle      : out    std_logic;
      tx_done_tick : out    std_logic
   );
```

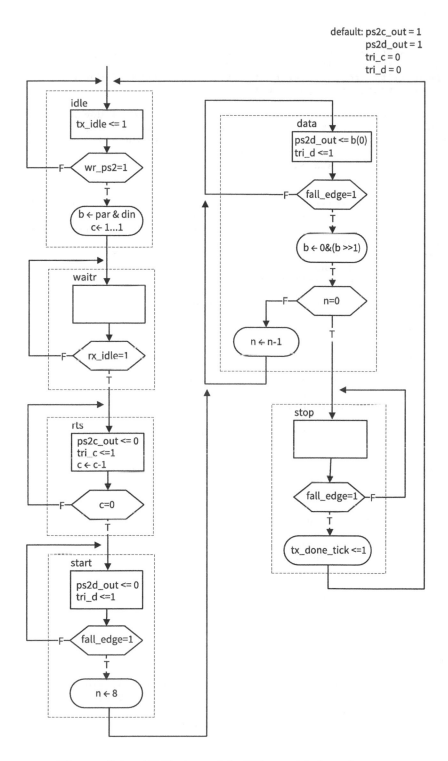

Figure 17.6   ASMD chart of the PS2 transmitting subsystem.

```
end ps2tx;

architecture arch of ps2tx is
    type statetype is (idle, waitr, rts, start, data, stop);
    signal state_reg, state_next   : statetype;
    signal filter_reg, filter_next : std_logic_vector(7 downto 0);
    signal f_ps2c_reg, f_ps2c_next : std_logic;
    signal fall_edge               : std_logic;
    signal b_reg, b_next           : std_logic_vector(8 downto 0);
    signal c_reg, c_next           : unsigned(12 downto 0);
    signal n_reg, n_next           : unsigned(3 downto 0);
    signal par                     : std_logic;
    signal ps2c_out, ps2d_out      : std_logic;
    signal tri_c, tri_d            : std_logic;
begin
    --*****************************************************************
    -- filter and falling edge tick generation for ps2c
    --*****************************************************************
    process(clk, reset)
    begin
        if reset = '1' then
            filter_reg <= (others => '0');
            f_ps2c_reg <= '0';
        elsif (clk'event and clk = '1') then
            filter_reg <= filter_next;
            f_ps2c_reg <= f_ps2c_next;
        end if;
    end process;

    filter_next <= ps2c & filter_reg(7 downto 1);
    f_ps2c_next <= '1' when filter_reg = "11111111" else
                   '0' when filter_reg = "00000000" else
                   f_ps2c_reg;
    fall_edge   <= f_ps2c_reg and (not f_ps2c_next);
    --*****************************************************************
    -- fsmd
    --*****************************************************************
    -- registers
    process(clk, reset)
    begin
        if reset = '1' then
            state_reg <= idle;
            c_reg     <= (others => '0');
            n_reg     <= (others => '0');
            b_reg     <= (others => '0');
        elsif (clk'event and clk = '1') then
            state_reg <= state_next;
            c_reg     <= c_next;
            n_reg     <= n_next;
            b_reg     <= b_next;
        end if;
    end process;
    -- odd parity bit
    par <= not (din(7) xor din(6) xor din(5) xor din(4) xor
                din(3) xor din(2) xor din(1) xor din(0));
    -- next-state logic
    process(state_reg,n_reg,b_reg,c_reg,wr_ps2,din,par,fall_edge,rx_idle)
    begin
        state_next    <= state_reg;
        c_next        <= c_reg;
        n_next        <= n_reg;
```

```vhdl
        b_next        <= b_reg;
        tx_done_tick <= '0';
        ps2c_out      <= '1';
        ps2d_out      <= '1';
        tri_c         <= '0';
        tri_d         <= '0';
        tx_idle       <= '0';
        case state_reg is
            when idle =>
                tx_idle <= '1';
                if wr_ps2 = '1' then
                    b_next       <= par & din;
                    c_next       <= (others => '1');
                    state_next <= waitr;
                end if;
            when waitr =>    -- wait if receiving in progress
                if rx_idle = '1' then
                    state_next <= rts;
                end if;
            when rts =>       -- request to send
                ps2c_out <= '0';
                tri_c    <= '1';
                c_next   <= c_reg - 1;
                if (c_reg = 0) then
                    state_next <= start;
                end if;
            when start =>    -- assert start bit
                ps2d_out <= '0';
                tri_d    <= '1';
                if fall_edge = '1' then
                    n_next      <= "1000";
                    state_next <= data;
                end if;
            when data =>      -- 8 data + 1 parity
                ps2d_out <= b_reg(0);
                tri_d      <= '1';
                if fall_edge = '1' then
                    b_next <= '0' & b_reg(8 downto 1);
                    if n_reg = 0 then
                        state_next <= stop;
                    else
                        n_next <= n_reg - 1;
                    end if;
                end if;
            when stop =>      -- assume floating high for ps2d
                if fall_edge = '1' then
                    state_next    <= idle;
                    tx_done_tick <= '1';
                end if;
        end case;
    end process;
    -- tristate buffers
    ps2c <= ps2c_out when tri_c = '1' else 'Z';
    ps2d <= ps2d_out when tri_d = '1' else 'Z';
end arch;
```

As in the receiving subsystem, there is no error detection circuit in this code.

17.2.4 Complete PS2 system

The top-level HDL code follows the diagram in Figure 17.3 and instantiates the three major components. It is shown in Listing 17.3.

Listing 17.3 Complete PS2 system

```
library ieee;
use ieee.std_logic_1164.all;
use ieee.numeric_std.all;
entity ps2_top is
    generic(W_SIZE : integer := 4);   -- 2^W_SIZE words in FIFO
    port(
        clk, reset        : in    std_logic;
        wr_ps2            : in    std_logic;
        rd_ps2_packet     : in    std_logic;
        ps2_tx_data       : in    std_logic_vector(7 downto 0);
        ps2_rx_data       : out   std_logic_vector(7 downto 0);
        ps2_tx_idle       : out   std_logic;
        ps2_rx_buf_empty  : out   std_logic;
        ps2d, ps2c        : inout std_logic
    );
end ps2_top;

architecture arch of ps2_top is
    signal rx_data      : std_logic_vector(7 downto 0);
    signal rx_done_tick : std_logic;
    signal rx_idle      : std_logic;
    signal tx_idle      : std_logic;
begin
    ps2_tx_unit : entity work.ps2tx(arch)
        port map(
            clk           => clk,
            reset         => reset,
            wr_ps2        => wr_ps2,
            rx_idle       => rx_idle,
            din           => ps2_tx_data,
            ps2d          => ps2d,
            ps2c          => ps2c,
            tx_idle       => tx_idle,
            tx_done_tick  => open
        );
    ps2_rx_unit : entity work.ps2rx(arch)
        port map(
            clk           => clk,
            reset         => reset,
            rx_en         => tx_idle,
            ps2d          => ps2d,
            ps2c          => ps2c,
            rx_idle       => rx_idle,
            rx_done_tick  => rx_done_tick,
            dout          => rx_data
        );
    rx_fifo_unit : entity work.fifo
        generic map(
            DATA_WIDTH => 8,
            ADDR_WIDTH => W_SIZE
        )
        port map(
            clk    => clk,
            reset  => reset,
            rd     => rd_ps2_packet,
```

```
            wr      => rx_done_tick,
            w_data => rx_data,
            empty  => ps2_rx_buf_empty,
            full   => open,
            r_data => ps2_rx_data
        );
    ps2_tx_idle <= tx_idle;
end arch;
```

The generic W specifies the size of the FIFO buffer.

17.3 PS2 CORE DEVELOPMENT

We can follow the procedure in Section 10.3.2 to add a wrapping circuit for the PS2 controller and create an MMIO IP core.

17.3.1 Register map

The processor interacts with the PS2 controller as follows:

- receive (i.e., read) an 8-bit data packet from the PS2 controller's receiving FIFO buffer.
- generate (i.e., write) a pulse to remove a packet from the receiving FIFO buffer.
- issue (i.e., write) an 8-bit command to a PS2 device.
- check (i.e., read) the ps2_rx_buf_empty signal to determine whether a packet is in the receiving FIFO buffer.
- check (i.e., read) the ps2_tx_idle signal to determine whether the transmitting subsystem is available.

Based on these interactions, we can define the PS2 core's register map. For clarity, we separate the read and write operations into different registers. There is one read register. Its address offset and fields are:

- offset 0 (read data and status register)
    - bits 7 to 0: 8-bit received data
    - bit 8: empty status of the receiving FIFO buffer
      bit 9: idle status of the PS2 transmitter

There are two write registers. Their address offsets and fields are:

- offset 1 (write data register)
    - bits 7 to 0: 8-bit transmitted data
- offset 2 (read data removal register)
    - dummy data write to generate a pulse to remove a data byte from the receiving FIFO buffer

17.3.2 Wrapping circuit for the slot interface

Based on the register map and the I/O signals of the PS2 controller, we can derive a wrapping circuit that complies with the slot specification and create the PS2 core. The HDL code of the core is shown in Listing 17.4.

Listing 17.4   PS2 core

```vhdl
library ieee;
use ieee.std_logic_1164.all;
use ieee.numeric_std.all;
entity chu_ps2_core is
    generic(W_SIZE : integer := 8);    -- 2^W_SIZE words in FIFO
    port(
        clk     : in std_logic;
        reset   : in std_logic;
        -- slot interface
        cs      : in std_logic;
        write   : in std_logic;
        read    : in std_logic;
        addr    : in std_logic_vector(4 downto 0);
        rd_data : out std_logic_vector(31 downto 0);
        wr_data : in std_logic_vector(31 downto 0);
        -- external signals
        ps2d    : inout std_logic;
        ps2c    : inout std_logic
    );
end chu_ps2_core;

architecture arch of chu_ps2_core is
    signal ps2_rx_data      : std_logic_vector(7 downto 0);
    signal wr_en, wr_ps2    : std_logic;
    signal rd_fifo          : std_logic;
    signal ps2_rx_buf_empty : std_logic;
    signal ps2_tx_idle      : std_logic;
begin
    -- instantiation
    ps2_unit : entity work.ps2_top
        generic map(W_SIZE => W_SIZE)
        port map(
            clk              => clk,
            reset            => reset,
            ps2d             => ps2d,
            ps2c             => ps2c,
            wr_ps2           => wr_ps2,
            rd_ps2_packet    => rd_fifo,
            ps2_tx_data      => wr_data(7 downto 0),
            ps2_rx_data      => ps2_rx_data,
            ps2_tx_idle      => ps2_tx_idle,
            ps2_rx_buf_empty => ps2_rx_buf_empty
        );
    -- decoding logic
    wr_en  <= '1' when cs='1' and write='1' else '0';
    wr_ps2 <= '1' when wr_en='1' and addr(1 downto 0)="10" else '0';
    rd_fifo <= '1' when wr_en='1' and addr(1 downto 0)="11" else '0';
    -- read data
    rd_data <= x"00000"&"00"& ps2_tx_idle & ps2_rx_buf_empty & ps2_rx_data;
end arch;
```

The wrapping circuit consists of a PS2 controller instance and a write decoding circuit. The FIFO buffer configuration imposes a separate write instruction to perform the removal operation, similar to the UART core discussed in Section 11.3.2

17.4   PS2 DRIVER

PS2 driver consists of two layers of routines. The methods in the lower layer retrieve raw byte data from the incoming stream and issue commands. The methods in the upper layer examine and decode the byte stream and extract and return the keyboard activities or mouse movement information.

17.4.1   Class definition

The class definition of the PS2 core is shown in Listing 17.5.

Listing 17.5   Ps2Core class definition (in ps2_core.h)

```
#ifndef _PS2_H_INCLUDED
#define _PS2_H_INCLUDED

#include "chu_init.h"
class Ps2Core {
public:
   enum {
      RD_DATA_REG = 0,
      PS2_WR_DATA_REG = 2,
      RM_RD_DATA_REG = 3
   };
   enum {
      TX_IDLE_FIELD = 0x00000200,
      RX_EMPT_FIELD = 0x00000100,
      RX_DATA_FIELD = 0x000000ff
   };
   Ps2Core(uint32_t core_base_addr);
   ~Ps2Core();          // not used
   int rx_fifo_empty();
   int tx_idle();
   void tx_byte(uint8_t cmd);
   int rx_byte();
   int init();
   int get_mouse_activity(int *lbtn, int *rbtn, int *xmov, int *ymov);
   int get_kb_ch(char *ch);
private:
   uint32_t base_addr;
};

#endif   // _PS2_H_INCLUDED
```

17.4.2   Lower layer methods

The class implementation of the constructor and the lower layer methods is shown in Listing 17.6.

Listing 17.6   Ps2Core class lower layer method implementation (in ps2_core.cpp)

```
Ps2Core::Ps2Core(uint32_t core_base_addr) {
   base_addr = core_base_addr;
}

Ps2Core::~Ps2Core() { }

int Ps2Core::rx_fifo_empty() {
```

```
   uint32_t rd_word;
   int empty;

   rd_word = io_read(base_addr, RD_DATA_REG);
   empty = (int) (rd_word & RX_EMPT_FIELD) >> 8;
   return (empty);
}

int Ps2Core::tx_idle() {
   uint32_t rd_word;
   int idle;

   rd_word = io_read(base_addr, RD_DATA_REG);
   idle = (int) (rd_word & TX_IDLE_FIELD) >> 9;
   return (idle);
}

void Ps2Core::tx_byte(uint8_t cmd) {
   io_write(base_addr, PS2_WR_DATA_REG, (uint32_t ) cmd);
}

int Ps2Core::rx_byte() {
   uint32_t data;

   if (rx_fifo_empty())   // no data
      return (-1);
   else {
      data = io_read(base_addr, RD_DATA_REG) & RX_DATA_FIELD;
      io_write(base_addr, RM_RD_DATA_REG, 0);
      return ((int) data);
   }
}
```

The receiving FIFO buffer structure of the PS2 core is similar to that of the UART core and thus the corresponding reading methods of the `UartCore` class in Section 11.4 can be applied. The PS2 core does not use a transmitting FIFO buffer and the new methods just check the idle status and write the command byte to the core directly.

17.4.3   PS2 initialization routine

A PS2 device has an internal controller that monitors the device's activities and encodes the activities into a byte stream following a predefined protocol. At power-on, the controller automatically resets the parameters to the default configuration and performs a diagnostic self-test. After passing the test, a PS2 keyboard transmits a 0xaa packet to a host and a PS2 mouse transmits two packets, 0xaa and 0x00, to a host.

While a PS2 host functions as a receiver most of the time, it can send a command to a PS2 device to inquire the status and set certain parameters, such as the typematic rate of a keyboard. The commands are in the form of 8-bit packets. After receiving the command, the PS2 device first transmits a 0xfa acknowledge packet and then performs the designated operation. The detailed commands can be found in the references of the bibliographic section.

The power-on default setting of a PS2 keyboard works properly for most applications and thus we can use the keyboard without ever sending any command. On the other hand, the power-on setting of a PS2 mouse is not adequate. We need

to issue additional commands to configure the mouse in *stream mode*. The init()
method performs the required mouse initialization. In the process, it also resets
the device, verifies the existence of a PS2 device, and identifies the type of the PS2
device (i.e., a keyboard or a mouse). The procedure is

- The host flushes the receiving FIFO buffer.
- The host issues a software reset command, 0xff.
- A device acknowledges with 0xfa and performs diagnostic self-test.
- If the test passes, a keyboard responds with a single packet of 0xaa and a
  mouse responds with two packets of 0xaa and 0x00. The host can identify
  the type of device by examining the response.
- If it is a mouse, the host sends command 0xf4 to enable the stream mode.
- The mouse acknowledges with 0xfa and the initialization completes.

The code of the init() method is shown in Listing 17.7.

Listing 17.7   init() method (in ps2_core.cpp)

```
int Ps2Core::init() {
   int packet;

   /* flush fifo buffer */
   while (!rx_fifo_empty()) {
      rx_byte();
   }
   /* send reset 0xff */
   debug("ps2 reset: write command ", 0, 0);
   tx_byte(0xff);
   sleep_ms(1000);
   /* check 0xfa 0xaa */
   if (rx_byte() != 0xfa){
      return (-1);          // no response or wrong response
   }
   if (rx_byte() != 0xaa){
      return (-1);          // no response or wrong response
   }
   debug("ps2 reset: 0xfa 0xaa valid ", 0, 0);
   /* check whether 0x00 is received */
   packet = rx_byte();
   if (packet == -1) {
      return (1);           // fifo has no more packet, device is keyboard
   }
   if (packet != 0x00) {
      return (-2);          // unknown ps2 device (unlikely)
   }
   /* device is a mouse; set it to stream mode */
   tx_byte(0xf4);
   sleep_ms(100);
   /* check 0xfa (acknowledge) */
   if (rx_byte()!= 0xfa){
      return (-3);          // no response or wrong response
   }
   return (2);              //success
```

It returns a 1 for a keyboard, a 2 for a successfully initiated mouse, and a negative
number for various error conditions.

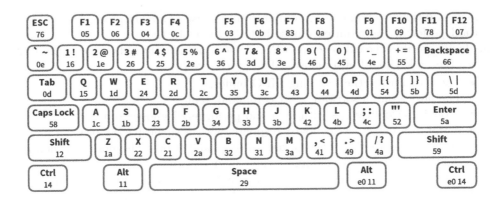

Figure 17.7 Scan code of the PS2 keyboard.

17.4.4 Keyboard routine

Overview of the scan code A keyboard consists of a matrix of keys and an embedded microcontroller that monitors (i.e., scans) the activities of the keys and sends the *scan code* accordingly. Three types of key activities are observed:

- When a key is pressed, the *make code* of the key is transmitted.
- When a key is held down continuously, a condition known as *typematic*, the make code is transmitted repeatedly at a specific rate. By default, a PS2 keyboard transmits the make code about every 100 ms after a key has been held down for 0.5 second.
- When a key is released, the *break code* of the key is transmitted.

The make code of the main part of a PS2 keyboard is shown in Figure 17.7. It is normally one byte wide and is represented by two hexadecimal numbers. For example, the make code of the A key is 0x1c. This code can be conveyed by one packet when transmitted. The make codes of a handful of special-purpose keys, which are known as *extended keys*, can have 2 to 4 bytes. A few of these keys are shown in Figure 17.7. For example, the make code of the right control key (labeled Ctrl) is 0xe0 0x14. Multiple packets are needed for the transmission. The break codes of the regular keys consist of 0xf0 followed by the make code of the key. For example, the break code of the A key is 0xf0 0x1c.

A PS2 keyboard transmits a sequence of codes according to the key activities. For example, when we press and release the A key, the keyboard first transmits its make code and then the break code:

```
0x1c 0xf0 0x1c
```

If we hold the key down for awhile before releasing it, the make code will be transmitted multiple times:

```
0x1c 0x1c 0x1c ... 0x1c 0xf0 0x1c
```

Multiple keys can be pressed at the same time. For example, we can first press the shift key (whose make code is 0x12) and then the A key, and release the A key and then release the shift key. The transmitted code sequence follows the make and break codes of the two keys:

```
0x12 0x1c 0xf0 0x1c 0xf0 0x12
```

The previous sequence is how we normally obtain an uppercase A. Note that there is no special code to distinguish the lowercase and uppercase keys. It is the responsibility of the host device to keep track of whether the shift key is pressed and to determine the case accordingly.

Implementation   The get_kb_ch() method reads a character from a keyboard. Because a PS2 keyboard contains many special-purpose keys and the keys can be pressed and released in an arbitrary combination (such as Ctrl-D and Ctrl-Alt-Del), developing a robust, comprehensive routine is quite involved and beyond the scope of this book. For our purposes, we use the keyboard to obtain "printable" ASCII characters and digits and develop a routine accordingly. Except for the shift keys, no other special-purpose key is processed. The main task of get_kb_ch() is to convert the scan codes to proper characters. Our development ignores the extended scan codes and assumes that these keys are not used.

In C and C++, a character is represented by the 8-bit **char** data type. The representations are based on ASCII codes, which are 7 bits and consist of 128 code words (0x00 to 0x7f). The complete characters and their code words are shown in Table 11.1. There is no clear relationship between the scan codes and ASCII codes. A simple way to do the conversion is to define the mapping in a lookup table. In C, the lookup table can be defined as a one-dimensional constant array with the scan code as the index. The table for the lowercase characters is as follows:

```
const uint8_t SCAN2ASCII_LO_TABLE[128]={
   0,    F9,    0,    F5,   F3,   F1,   F2,   F12,  //00
   0,    F10,   F8,   F6,   F4,   TAB,  '`',  0,    //08
   0,    0,     L_SFT, 0,   L_CTR,'q',  '1',  0,    //10
   0,    0,     'z',  's',  'a',  'w',  '2',  0,    //18
   0,    'c',   'x',  'd',  'e',  '4',  '3',  0,    //20
   0,    ' ',   'v',  'f',  't',  'r',  '5',  0,    //28
   0,    'n',   'b',  'h',  'g',  'y',  '6',  0,    //30
   0,    0,     'm',  'j',  'u',  '7',  '8',  0,    //38
   0,    ',',   'k',  'i',  'o',  '0',  '9',  0,    //40
   0,    '.',   '/',  'l',  ';',  'p',  '-',  0,    //48
   0,    0,     '\'', 0,    '[',  '=',  0,    0,    //50
   CAPS, R_SFT, ENTER,']',  0,    BKSL, 0,    0,    //58
   0,    0,     0,    0,    0,    0,    BKSP, 0,    //60
   0,    '1',   0,    '4',  '7',  0,    0,    0,    //68
   0,    '.',   '2',  '5',  '6',  '8',  ESC,  NUM,  //70
   F11,  '+',   '3',  '-',  '*',  '9',  0,    0     //78
};
```

For example, we can use the expression SCAN2ASCII_LO_TABLE[21] to obtain the corresponding character of scan code 21, which is the letter c. In addition to the normal single-quoted characters, the table contains 0's, which correspond to undefined scan codes, and uppercase constants, which are special ASCII characters and unmapped keys. The special C characters consist of

```
#define TAB    0x09   // tab
#define BKSP   0x08   // backspace
#define ENTER  0x0d   // enter (new line)
#define ESC    0x1b   // escape
#define BKSL   0x5c   // back slash
```

The other uppercase constants correspond to the keys that don't map to ASCII characters, such as the function keys (F1, · · ·, F12) and the control key (Ctrl). We can assign unused 8-bit values (0x80 to 0xff) to these keys and use them for special purposes. For example, we can display the help message when the F1 key is pressed.

A similar table is needed for the uppercase characters as well:

```
const uint8_t SCAN2ASCII_UP_TABLE[128] = {
  0,     F9,     0,      F5,    F3,     F1,    F2,    F12,   //00
  0,     F10,    F8,     F6,    F4,     TAB,   '~',   0,     //08
  0,     0,      L_SFT,  0,     L_CTR,  'Q',   '!',   0,     //10
  0,     0,      'Z',    'S',   'A',    'W',   '@',   0,     //18
  0,     'C',    'X',    'D',   'E',    '$',   '#',   0,     //20
  0,     '\ ',   'V',    'F',   'T',    'R',   '%',   0,     //28
  0,     'N',    'B',    'H',   'G',    'Y',   '^',   0,     //30
  0,     0,      'M',    'J',   'U',    '&',   '*',   0,     //38
  0,     '<',    'K',    'I',   'O',    ')',   '(',   0,     //40
  0,     '>',    '?',    'L',   ':',    'P',   '_',   0,     //48
  0,     0,      '\"',   0,     '{',    '+',   0,     0,     //50
  CAPS,  R_SFT,  ENTER,  '}',   0,      '|',   0,     0,     //58
  0,     0,      0,      0,     0,      0,     BKSP,  0,     //60
  0,     '1',    0,      '4',   '7',    0,     0,     0,     //68
  0,     '.',    '2',    '5',   '6',    '8',   ESC,   NUM,   //70
  F11,   '+',    '3',    '-',   '*',    '9',   0,     0      //78
};
```

The code of the get_kb_ch() method is shown in Listing 17.8.

Listing 17.8   get_kb_ch() method (in ps2_core.cpp)

```cpp
int Ps2Core::get_kb_ch(char *ch) {
  // special   characters
  #define TAB      0x09   // tab
  #define BKSP     0x08   // backspace
  #define ENTER    0x0d   // enter (new line)
  #define ESC      0x1b   // escape
  #define BKSL     0x5c   // back slash
  #define SFT_L    0x59   // left shift
  #define SFT_R    0x12   // right shift

  #define CAPS     0x80
  #define NUM      0x81
  #define CTR_L    0x82
  #define F1       0xf0
  #define F2       0xf1
  . . .

  // keyboard scan code to ascii (lower case)
  const char SCAN2ASCII_LO_TABLE[128] = {...}
  // keyboard scan code to ascii (upper case)
  const char SCAN2ASCII_UP_TABLE[128] = {...};

  static int sft_on = 0;
  uint8_t scode;

  while (1){
    if (rx_fifo_empty())         // no packet
      return(0);
    scode = rx_byte();
    switch (scode){
```

```
      case 0xf0:                    // break code
        while (rx_fifo_empty());    // get next
        scode = rx_byte();
      if (scode==SFT_L || scode==SFT_R)
        sft_on = 0;
        break;
      case SFT_L:                   // shift key make code
      case SFT_R:
        sft_on = 1;
        break;
      default:                      // normal make code
        if (sft_on)
          *ch = SCAN2ASCII_UP_TABLE[scode];
        else
          *ch = SCAN2ASCII_LO_TABLE[scode];
        return(1);
    }   // end switch
  }   // end while
}
```

The routine treats the two shift keys as special cases. It keeps track of whether a shift key is pressed and then uses the lowercase or uppercase lookup table accordingly. A static variable, sft_on, is used for tracking. The routine processes the received packets as follows:

- If it is the break code (i.e., beginning with a 0xf0 packet), remove two packets. If the code is for the shift key, clear sft_on to 0.
- If it is the make code of the shift key, set sft_on to 1.
- If it is the make code of another key, obtain the character value from the proper lookup table and return the character.

Note that this routine does not process other special keys but just returns their designated codes.

17.4.5 Mouse routine

Overview of PS2 mouse protocol   A computer mouse is designed to detect two-dimensional motion on a surface. Its internal circuit measures the relative distance of movement. A standard PS2 mouse reports the x-axis (right/left) and y-axis (up/down) movement and the status of the left button, middle button, and right button. The amount of each movement is recorded in a mouse's internal counter. When the data is transmitted to the host, the counter is cleared to zero and restarts the counting. The content of the counter represents a 9-bit signed integer in which a positive number indicates the right or up movement and a negative number indicates the left or down movement.

The relationship between the physical distances is defined by the mouse's *resolution* parameter. The default value of resolution is four counts per millimeter. When a mouse moves continuously, the data is transmitted at a regular rate. The rate is defined by the mouse's *sampling rate* parameter. The default value of the sampling rate is 100 samples per second. If a mouse moves too fast, the amount of the movement during the sampling period may exceed the maximum range of the counter. The counter is set to the maximum magnitude in the appropriate direction. Two overflow bits are used to indicate the conditions.

Table 17.1   Mouse data packet format

byte 1	y_v	x_v	y_8	x_8	1	m	r	l
byte 2	x_7	x_6	x_5	x_4	x_3	x_2	x_1	x_0
byte 3	y_7	y_6	y_5	y_4	y_3	y_2	y_1	y_0

The mouse reports the movement and button activities in 3 bytes, which are embedded in three PS2 packets. The detailed format of the 3-byte data is shown in Table 17.1. It contains the following information:

- $x_8, \ldots, x_0$: x-axis movement in 2's-complement format
- x_v: x-axis movement overflow
- $y_8, \ldots, y_0$: y-axis movement in 2's-complement format
- y_v: y-axis movement overflow
- l: left button status, which is 1 when the left button is pressed
- r: right button status, which is 1 when the right button is pressed
- m: optional middle button status, which is 1 when the middle button is pressed

During transmission, the byte 1 packet is sent first and the byte 3 packet is sent last.

A mouse has several different operation modes. The most commonly used one is the *stream mode*, in which a mouse sends the movement data when it detects movement or button activity. If the movement is continuous, the data is generated at the designated sampling rate. The init() method discussed in Section 17.4.3 enables this mode.

Implementation   After a mouse is initialized in stream mode, it sends a stream of packets when a movement or button activity is detected. The get_mouse_activity() method extracts and assembles the information. The code is shown in Listing 17.9.

Listing 17.9   get_mouse_activity() method (in **ps2_core.cpp**)

```
int Ps2Core::get_mouse_activity(int *lbtn, int *rbtn, int *xmov, int *ymov)
{
  uint8_t b1, b2, b3;

  uint32_t tmp;

  /* check and retrieve 1st byte */
  if (rx_fifo_empty())
    return (0);                        // no data in rx fifo buffer
  b1 = rx_byte();
  /* wait and retrieve 2nd byte */
  while (rx_fifo_empty());
  b2 = rx_byte();
  /* wait and retrieve 3rd byte */
  while (rx_fifo_empty());
  b3 = rx_byte();
  /* extract button info */
  *lbtn = (int) (b1 & 0x01);      // extract bit 0
  *rbtn = (int) (b1 & 0x02)>>1;   // extract bit 1
  /* extract x movement; manually convert 9-bit 2's comp to int */
  tmp = (uint32_t) b2;
  if (b1 & 0x10)                  // check MSB (sign bit) of x movement
    tmp = tmp | 0xffffff00;       // manual sign-extension if negative
```

```
    *xmov = (int) tmp;                // data conversion
    /* extract y movement; manually convert 9-bit 2's comp to int */
    tmp = (uint32_t) b3;
    if (b1 & 0x20)                    // check MSB (sign bit) of y movement
      tmp = tmp | 0xffffff00;         // manual sign-extension if negative
    *ymov = (int) tmp;                // data conversion
    /* success */
    return(1);
}
```

The routine obtains three packets, extracts the information, and stores it to the proper fields. The sign extension is performed manually to extend the 9-bit movement data to the 32-bit integer data type. It is done by setting 24 MSBs to 1's if the movement is negative (i.e., the MSB of the nine-bit data is 1).

17.5 TEST

The Nexys 4 DDR board does not have a PS2 port. However, it contains an "auxiliary microcontroller" configured as an USB HID (human interface device) host after the FPGA device is programmed. The microcontroller functions as a "USB mouse and keyboard protocol translator." It has an emulated PS2 port and can generate an emulated PS2 data stream from a USB mouse or keyboard attached to the type A USB connector on Nexys 4 DDR board (labeled as J5 and "USB Host"). The Nexys 4 DDR manual states that "only keyboards and mice supporting the Boot HID interface are supported." It appears that some USB devices may not work due to their initialization requirements.

A simple test function is derived to verify the operation of the PS2 core. A USB mouse or keyboard must be connected to the Nexys 4 DDR board in advance. The code is shown in Listing 17.10. It first calls init() to determine the type of PS2 device (i.e., a mouse or a keyboard) and then performs the testing accordingly. If a mouse is attached, the function displays the mouse activities on the UART console in the form of [l,r,x,y], where l and r are left and right button status, and x and y are x-axis and y-axis movement. If a keyboard is attached, the function displays the pressed key. The test function exits after there is no activity for five seconds.

Listing 17.10   PS2 test function (in `main_sampler_test.cpp`)

```
void ps2_check(Ps2Core *ps2_p) {
    int id;
    int lbtn, rbtn, xmov, ymov;
    char ch;
    unsigned long last;

    uart.disp("\n\rPS2 device (1-keyboard / 2-mouse): ");
    id = ps2_p->init();
    uart.disp(id);
    uart.disp("\n\r");
    last = now_ms();
    do {
        if (id == 2) {  // mouse
            if (ps2_p->get_mouse_activity(&lbtn, &rbtn, &xmov, &ymov)) {
                uart.disp("[");
                uart.disp(lbtn);
                uart.disp(", ");
                uart.disp(rbtn);
```

```
                    uart.disp(", ");
                    uart.disp(xmov);
                    uart.disp(", ");
                    uart.disp(ymov);
                    uart.disp("] \r\n");
                    last = now_ms();
              }    // end get_mouse_activity()
        } else {         // keyboard
            if (ps2_p->get_kb_ch(&ch)) {
                    uart.disp(ch);
                    uart.disp(" ");
                    last = now_ms();
            } // end get_kb_ch()
        }    // end id==2
    } while (now_ms() - last < 5000);
    uart.disp("\n\rExit PS2 test \n\r");
}
```

17.6 BIBLIOGRAPHIC NOTES

Three articles, "PS/2 Mouse/Keyboard Protocol," "PS/2 Keyboard Interface," and "PS/2 Mouse Interface," by Adam Chapweske, provide detailed information on the PS2 keyboard and mouse interface. They can be found at the http://www.computer-engineering.org site. *Nexys 4 DDR FPGA Board Reference Manual* contains more information about the USB HID emulation of the PS2 port.

17.7 SUGGESTED EXPERIMENTS

The mouse is used mainly with a graphic video interface. A custom mouse pointer sprite core is developed in Chapter 22. Additional experiments can be found in that chapter.

17.7.1 PS2 receiving subsystem with watchdog timer

There is no error-handling capability in the PS2 receiving subsystem in Section 17.2.2. The potential noise and glitches in the ps2c signal may cause the FSMD to be stuck in an incorrect state. One way to deal with this problem is to add a watchdog timer. The timer is initiated every time the fall_edge_tick signal is asserted in the dps state. The time_out signal is asserted if no subsequent falling edge arrives in the next 20 μs, and the FSMD returns to the idle state. Design the modified receiving subsystem, derive a testbench, and use simulation to verify its operation.

17.7.2 Keyboard-controlled LED flashing circuit

Consider the LED flashing circuit discussed in Experiment 9.10.2. Instead of switches and buttons, we can use the keyboard to send commands:
- Use the P (for "pause") key to pause and resume the flashing operation.
- Use the following key sequence to enter the desired flashing period: F1 and then three digit keys (i.e., 000 to 999).
- All other keys or illegal sequences will be ignored.

Derive the revised program, and verify the operation.

17.7.3   Enhanced keyboard driver routine I

The get_kb_ch() method in Section 17.4.4 processes only the shift keys. Many additional functionalities can be added:

- Use the Caps Lock key to toggle between the lowercase and uppercase modes.
- Use the Caps Lock LED to indicate the status of the Caps Lock key.
- Use the Ctrl key for special functions (e.g., return a special Ctrl-C code when both the Ctrl and C keys are pressed).

Derive a new driver method and verify the operation.

17.7.4   Enhanced keyboard driver routine II

The get_kb_ch() method in Section 17.4.4 covers only the standard scan codes. A current keyboard usually contains extended scan codes for additional keys and these codes can be found in the references of the bibliographic section. Derive a new driver method to cover the extended scan codes and verify the operation.

17.7.5   Remote-mode mouse driver

An alternative to a mouse's stream mode is the *remote mode*, in which the mouse only transmits data packets after receiving a read data command from the host. More detailed information can be found in the references of the bibliographic section. Derive new mouse driver methods using this mode and verify the operation.

17.7.6   Scroll-wheel mouse driver

In addition to the x- and y-axis activities, newer mice can add a third dimension, which corresponds to the movement of the scroll wheel. More detailed information can be found in the references of the bibliographic section. Derive a new mouse driver method for this type of mice and verify the operation.

CHAPTER 18

SOUND I: DDFS CORE

DDFS (*direct digital frequency synthesis*) is a scheme to generate tunable waveforms from a single fixed clock source. The main application of DDFS is in communication systems, usually involving the generation and modulation of high-frequency signals. However, in this book we mainly use it to generate audio frequency signals. In this chapter, we implement this scheme and a one-bit DAC (digital-to-analog converter) to generate an audio frequency sine wave. In the next chapter, we construct a custom circuit to generate the ADSR (attack-decay-sustain-release) amplitude envelope to mimic the sound of various musical instruments.

18.1 INTRODUCTION

Many communication-related applications need to generate a signal of specific frequency and phase. DDFS is a method of producing a tunable digital or analog waveform from a single fixed clock source. The data points of the waveform are first generated in digital format and then converted to analog format with a DAC and a low-pass filter. Because of the digital implementation, DDFS can offer fast switching between output frequencies, fine frequency resolution, and operation over a wide range of frequencies.

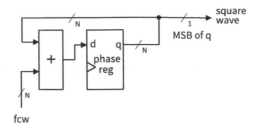

Figure 18.1   Block diagram for synthesizing a digital waveform.

18.2   DESIGN AND IMPLEMENTATION

The DDFS scheme is versatile and flexible. We examine the synthesis and implementation of three types of waveforms:

- Digital waveform, which is a square wave with constant amplitude
- Unmodulated analog sine waveform
- Modulated analog sine waveform, in which the phase, frequency, and amplitude of the output signal can be controlled by another signal

Once understanding the basic implementation, the HDL code can be derived accordingly.

18.2.1   Direct synthesis of a digital waveform

To synthesize a digital waveform, the DDFS scheme requires a register, which is known as the *phase register* or *phase accumulator*, and an adder, as shown in Figure 18.1. The output of the circuit is the MSB of the register. It is a square wave with the designated frequency and the duty cycle is close to 50%. The input is the fcw (for *frequency control word*) signal, which controls the frequency of the output signal. The value of fcw is added to the phase register in every clock cycle.

To explain the operation of this scheme, let us first define the relevant parameters:

- N: the width (i.e., number of bits) of the register and adder
- f_{sys} and t_{sys}: the frequency and period of the system clock
- f_{out} and t_{out}: the frequency and period of the output signal
- M: the value of fcw

This system works as follows. For the phase register, its value starts from 0 and gradually increments to $2^N - 1$ and then wraps around. If we observe the MSB of the register in the process, it starts as 0, changes to 1 when the phase register reaches halfway of $2^N - 1$, and then returns to 0 and repeats when the phase register wraps around. The duration of incrementing from 0 to $2^N - 1$ can be considered as one period of the MSB (i.e., t_{out}). Since M is added to the phase register each time, it requires $\frac{2^N}{M}$ additions to complete one circulation and the corresponding duration is $\frac{2^N}{M} * t_{sys}$; i.e.,

$$t_{out} = \frac{1}{M} * 2^N * t_{sys}$$

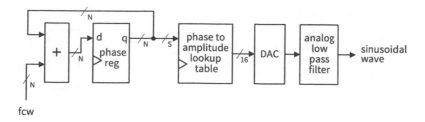

Figure 18.2 Block diagram for synthesizing an analog waveform.

The equation can be rewritten in terms of frequencies:

$$f_{out} = M * \frac{f_{sys}}{2^N}$$

The $\frac{f_{sys}}{2^N}$ term can be considered the "resolution" of a DDFS system. As N increases, finer frequency can be obtained accordingly. The typical width of N is between 24 and 48 bits. Clearly, we can set M to a proper value to obtain the desired frequency:

$$M = \frac{f_{out}}{f_{sys}} * 2^N$$

The value of M must be rounded to a whole integer. The rounding error may introduce a small variation in t_{out} and the effect is known as *jitter*.

18.2.2 Direct synthesis of an unmodulated analog waveform

Assume that the N bits of the phase register are $p_{N-1}p_{N-2}\cdots p_0$. The digital waveform uses the MSB, p_{N-1}, as the output. If we ignore the small jitter, it is a square wave with one half 0 and one half 1, each lasting $\frac{t_{out}}{2}$. The p_{N-1} bit essentially divides t_{out} into two equal regions. The second MSB, p_{N-2}, switches twice faster than p_{N-1}. If we consider the p_{N-1} and p_{N-2} bits together, they divide t_{out} into four equal regions. We can continue the process and divide t_{out} to smaller and smaller regions. The regions are commonly referred to as the *phases* of the period and this is the reason that the register is known as a *phase register*.

We can generate an unmodulated analog waveform by mapping phases to digitized amplitude points and then converting the value to the analog format by a DAC. The conceptual diagram is shown in Figure 18.2. The phase-to-amplitude lookup table performs the mapping and can be implemented by a ROM or RAM. The DAC converts the digitized amplitude value to an analog value and the low-pass filter removes unwanted high-frequency signals. The "shape" of an analog waveform is determined by the values loaded to the lookup table and thus the DDFS scheme can generate any type of analog waveform. The sine waveform is used in most applications.

It is neither practical nor necessary to use all N bits for the lookup table. We usually use 8 to 10 MSBs from the N-bit phase register output. It is labeled S in the diagram. For example, to use an 8-bit (i.e., 2^8 entries) lookup table to implement the sine function, we can divide one period into 256 equally spaced points, obtain the corresponding values, and load them into the lookup table. During the DDFS

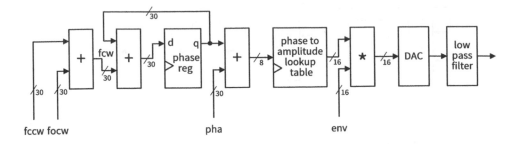

Figure 18.3   Block diagram for synthesizing a modulated analog waveform.

operation, the lookup table is swept every t_{out} seconds and the corresponding output waveform is $\sin(2\pi f_{out}t)$. A larger S value increases the size of the lookup table but puts less constraint on the low-pass filter.

18.2.3   Direct synthesis of a modulated analog waveform

DDFS is used widely in communication applications. A typical communication system involves two types of signals: a low-frequency *message* signal, such as an audio signal, and a high-frequency *carrier* signal to convey the message, such as the radio-frequency sine signal. *Modulation* is the process of modifying the carrier signal in accordance with the message signal. The carrier signal is usually a sine waveform and the message can be used to adjust its amplitude, frequency, phase, or a combination of them. Assume that the carrier signal is $\sin(2\pi ft)$. The modulated signals become the following:

- Amplitude modulation: $A(t) * \sin(2\pi ft)$
- Frequency modulation: $\sin(2\pi(f + \Delta f(t))t)$
- Phase modulation: $\sin(2\pi ft + \Delta p(t))$

The $A(t)$, $\Delta f(t)$, and $\Delta p(t)$ terms are slow time-varying signals that embed the message.

A DDFS system can incorporate the desired modulation scheme by inserting an additional adder or multiplier in its path. We can actually construct an extended DDFS system that supports all three modulation schemes. Instead of $\sin(2\pi ft)$, the extended system generates $A(t) * \sin(2\pi(f + \Delta f(t))t + \Delta p(t))$. The expanded diagram is shown in Figure 18.3. We assume that the $A(t)$, $\Delta f(t)$, and $\Delta p(t)$ terms are pre-processed and converted to proper digitized format:

- fccw: the frequency control word to generate the carrier frequency, f
- focw: the frequency control word to generate the offset frequency, $\Delta f(t)$
- pha: the phase value corresponding to the desired phase offset, $\Delta p(t)$
- env (for envelope): the digitized value of $A(t)$

The bus widths are the ones used in our implementation in Section 18.4.

18.3   FIXED-POINT ARITHMETIC

The amplitude modulation requires multiplication. Ideally, the *floating-point* data type should be used since it provides a large range and good precision. However, its complexity can severely degrade the software performance and complicate the design of a dedicated hardware accelerator. An alternative is to use the *fixed-point* data type, which requires only simple modification over the integer data type.

The fixed-point data type is essentially an integer that is scaled by a specific factor. For explanation purposes, let us first consider the fixed-point data type in the decimal number system. For example, we assume a fixed-point data type with five decimal digits and a scaling factor of 10^{-3}. The value 12.345 can be represented as 12345 (i.e., $12345 * 10^{-3}$). The scaling factor corresponds to the exponent of the floating-point data type. Unlike the floating-point data type, the scaling factor is the same for all entities of the same type and does not change during the computation. This particular data type can be interpreted as a five-digit integer type with an implicit decimal point after the second most significant digit. Thus, the value 12345 becomes 12.345. One way to represent a fixed-point data type is the $\mathbf{Q}m.f$ notation, where m represents the number of integer digits and f represents the number of fractional digits. The previous data type can be represented as the $\mathbf{Q}2.3$ format.

To add or subtract two numbers in the same fixed-point type, it is sufficient to add or subtract the underlying integers and keep their common scaling factor. The result is in the same type, as long as no overflow occurs. In other words, if the two numbers are in the $\mathbf{Q}m.f$ format, the sum will also be in the $\mathbf{Q}m.f$ format. The addition and subtraction thus are actually identical to those in the integer data type except that we assume that there is an implicit decimal point after the mth digit.

To multiply two fixed-point numbers, it suffices to multiply the two underlying integers. Unlike the addition operation, the resulting scaling factor changes. It becomes *the product of two scaling factors of the two numbers*. For example, assume that 12345 and 00025 are in the $\mathbf{Q}2.3$ format, which are interpreted as $12345 * 10^{-3}$ and $00025 * 10^{-3}$ (i.e., 12.345 and 0.025). The five-digit integer multiplication leads to a ten-digit product 0000308625, which is interpreted as $308625 * 10^{-6}$ (i.e., 0.308625). In other words, if the two numbers are in the $\mathbf{Q}m.f$ format, the product will be in the $\mathbf{Q}2m.2f$ format. For a fixed-point operation, it is desirable to have the product in the same $\mathbf{Q}m.f$ format. This can be done by trimming the m most significant digits and f least significant digits. In the previous example, the trimmed product becomes 00308, which is interpreted as $308 * 10^{-3}$ (i.e., 0.308). When using a fixed-point data type, we usually know the range of operation in advance and select a format to ensure that no overflow occurs after multiplication i.e., the m most significant digits in the product are always 0's. Thus, we only need to perform the shift-right operation to remove the f least significant digits.

The main advantage of the fixed-point data type is in its implementation. We can use the same integer arithmetic operations, plus shift operation, for the real numbers. On the other hand, the fixed-point data type operations tend to lose precision even if all operations are within the range. For example, the previous fixed-point multiplication can be rewritten as $12345*10^{-3} * 25*10^{-3} = 308*10^{-3}$, in which the product only has three digits of accuracy. If the floating-point data type is used, the multiplication can be expressed as $12345*10^{-3} * 25*10^{-3} = 30862*10^{-5}$, in which the product maintains five digits of accuracy. If the computation involves a larger number of iterations, the inaccuracy may be accumulated and lead to a much larger error.

In a digital system, the fixed-point data type is based on the binary system. The basic properties are similar to those in the decimal system except that binary bits are used in place of decimal digits. In the rest of the chapter, both software and hardware implementations are based on binary representation.

18.4 DDFS CONSTRUCTION

We can construct the digital portion of a DDFS circuit following the diagram in Figure 18.3. We use the following parameters in the design:

- f_{sys}: 100 MHz, which is the frequency of the onboard clock
- N: 30 bits
- Width of the lookup table address: 8 bits (i.e., 2^8 entries)
- sine wave amplitude resolution (i.e., number of bits of a lookup table entry): 16 bits in signed format

The 30-bit-wide phase register can obtain about 0.1-Hz resolution (i.e., $\frac{50*10^6}{2^{30}}$).

The size of the lookup table is 2^8-by-16 (i.e., 4K bits) and can be implemented by a synchronous ROM discussed in Section 7.4.5. The HDL code is shown in Listing 18.1.

Listing 18.1 Sine lookup table

```
library ieee;
use ieee.std_logic_1164.all;
use ieee.numeric_std.all;
entity sin_rom is
   generic(
      ADDR_WIDTH: integer:=8;
      DATA_WIDTH:integer:=16
   );
   port(
      clk: in std_logic;
      addr_r: in std_logic_vector(ADDR_WIDTH-1 downto 0);
      dout: out std_logic_vector(DATA_WIDTH-1 downto 0)
   );
end sin_rom;

architecture beh_arch of sin_rom is
   type ram_type is array (0 to 2**ADDR_WIDTH-1)
        of std_logic_vector (DATA_WIDTH-1 downto 0);
   -- sine LUT
   -- for symmetry, 0x8000 (i.e., -1) is replaced by 0x8001
   constant SIN_LUT: ram_type:=(    -- 2^8-by-16
      x"0000", x"0324", x"0648", x"096B", x"0C8C", x"0FAB", x"12C8",
      x"15E2", x"18F9", x"1C0C", x"1F1A", x"2224", x"2528", x"2827",
      x"2B1F", x"2E11", x"30FC", x"33DF", x"36BA", x"398D", x"3C57",
      x"3F17", x"41CE", x"447B", x"471D", x"49B4", x"4C40", x"4EC0",
      x"5134", x"539B", x"55F6", x"5843", x"5A82", x"5CB4", x"5ED7",
      x"60EC", x"62F2", x"64E9", x"66D0", x"68A7", x"6A6E", x"6C24",
      x"6DCA", x"6F5F", x"70E3", x"7255", x"73B6", x"7505", x"7642",
      x"776C", x"7885", x"798A", x"7A7D", x"7B5D", x"7C2A", x"7CE4",
      x"7D8A", x"7E1E", x"7E9D", x"7F0A", x"7F62", x"7FA7", x"7FD9",
      x"7FF6", x"7FFF", x"7FF6", x"7FD9", x"7FA7", x"7F62", x"7F0A",
```

```
        x"7E9D",  x"7E1E",  x"7D8A",  x"7CE4",  x"7C2A",  x"7B5D",  x"7A7D",
        x"798A",  x"7885",  x"776C",  x"7642",  x"7505",  x"73B6",  x"7255",
        x"70E3",  x"6F5F",  x"6DCA",  x"6C24",  x"6A6E",  x"68A7",  x"66D0",
        x"64E9",  x"62F2",  x"60EC",  x"5ED7",  x"5CB4",  x"5A82",  x"5843",
        x"55F6",  x"539B",  x"5134",  x"4EC0",  x"4C40",  x"49B4",  x"471D",
        x"447B",  x"41CE",  x"3F17",  x"3C57",  x"398D",  x"36BA",  x"33DF",
        x"30FC",  x"2E11",  x"2B1F",  x"2827",  x"2528",  x"2224",  x"1F1A",
        x"1C0C",  x"18F9",  x"15E2",  x"12C8",  x"0FAB",  x"0C8C",  x"096B",
        x"0648",  x"0324",  x"0000",  x"FCDC",  x"F9B8",  x"F695",  x"F374",
        x"F055",  x"ED38",  x"EA1E",  x"E707",  x"E3F4",  x"E0E6",  x"DDDC",
        x"DAD8",  x"D7D9",  x"D4E1",  x"D1EF",  x"CF04",  x"CC21",  x"C946",
        x"C673",  x"C3A9",  x"C0E9",  x"BE32",  x"BB85",  x"B8E3",  x"B64C",
        x"B3C0",  x"B140",  x"AECC",  x"AC65",  x"AA0A",  x"A7BD",  x"A57E",
        x"A34C",  x"A129",  x"9F14",  x"9D0E",  x"9B17",  x"9930",  x"9759",
        x"9592",  x"93DC",  x"9236",  x"90A1",  x"8F1D",  x"8DAB",  x"8C4A",
        x"8AFB",  x"89BE",  x"8894",  x"877B",  x"8676",  x"8583",  x"84A3",
        x"83D6",  x"831C",  x"8276",  x"81E2",  x"8163",  x"80F6",  x"809E",
        x"8059",  x"8027",  x"800A",  x"8001",  x"800A",  x"8027",  x"8059",
        x"809E",  x"80F6",  x"8163",  x"81E2",  x"8276",  x"831C",  x"83D6",
        x"84A3",  x"8583",  x"8676",  x"877B",  x"8894",  x"89BE",  x"8AFB",
        x"8C4A",  x"8DAB",  x"8F1D",  x"90A1",  x"9236",  x"93DC",  x"9592",
        x"9759",  x"9930",  x"9B17",  x"9D0E",  x"9F14",  x"A129",  x"A34C",
        x"A57E",  x"A7BD",  x"AA0A",  x"AC65",  x"AECC",  x"B140",  x"B3C0",
        x"B64C",  x"B8E3",  x"BB85",  x"BE32",  x"C0E9",  x"C3A9",  x"C673",
        x"C946",  x"CC21",  x"CF04",  x"D1EF",  x"D4E1",  x"D7D9",  x"DAD8",
        x"DDDC",  x"E0E6",  x"E3F4",  x"E707",  x"EA1E",  x"ED38",  x"F055",
        x"F374",  x"F695",  x"F9B8",  x"FCDC");
    signal ram: ram_type := SIN_LUT;
begin
    process(clk)
    begin
        if (clk'event and clk = '1') then
            dout <= ram(to_integer(unsigned(addr_r)));
        end if;
    end process;
end beh_arch;
```

The top-level DDFS system code follows the diagram in Figure 18.3 and is shown in Listing 18.2.

Listing 18.2   DDFS system

```
library ieee;
use ieee.std_logic_1164.all;
use ieee.numeric_std.all;
entity ddfs is
    generic(PW : integer := 30);    -- width of phase accumulator
    port(
        clk         : in  std_logic;
        reset       : in  std_logic;
        fccw        : in  std_logic_vector(PW - 1 downto 0);
        focw        : in  std_logic_vector(PW - 1 downto 0);
        pha         : in  std_logic_vector(PW - 1 downto 0);
        env         : in  std_logic_vector(15 downto 0);
        pcm_out     : out std_logic_vector(15 downto 0);
        pulse_out   : out std_logic
    );
end ddfs;

architecture arch of ddfs is
    signal fcw, pcw       : unsigned(PW - 1 downto 0);
    signal p_reg, p_next  : unsigned(PW - 1 downto 0);
```

```
   signal p2a_raddr      : std_logic_vector(7 downto 0);
   signal amp            : std_logic_vector(15 downto 0);
   signal modu           : signed(31 downto 0);
begin
   -- instantiate sin ROM
   sin_rom_unit : entity work.sin_rom
      port map(
         clk   => clk,
         addr_r => p2a_raddr,
         dout  => amp);
   -- phase register
   process(clk, reset)
   begin
      if reset = '1' then
         p_reg <= (others => '0');
      elsif (clk'event and clk = '1') then
         p_reg <= p_next;
      end if;
   end process;
   -- frequency modulation
   fcw       <= unsigned(fccw) + unsigned(focw);
   -- phase accumulation
   p_next    <= p_reg + fcw;
   -- phase modulation
   pcw       <= p_reg + unsigned(pha);
   -- phase to amplitude mapping address
   p2a_raddr <= std_logic_vector(pcw(PW - 1 downto PW - 8));
   -- amplitude modulation
   modu      <= signed(env) * signed(amp);
   -- modulated and square-wave ouptut
   pcm_out   <= std_logic_vector(modu(29 downto 14));
   pulse_out <= p_reg(PW - 1);
end arch;
```

Note that both the lookup table output (i.e., amp) and env are 16 bits wide. After multiplication, we must trim the 32-bit multiplication result back to 16 bits. This issue can be solved by representing the signals in the fixed-point format. Assume that amp is in the **Q**16.0 format and the range of env is $[-1.0, 1.0]$. To represent the range, we may be tempted to use **Q**1.15. However, the most positive number in this format is $011\cdots11$, which corresponds to $1 - 2^{15}$, and the range is like $[-1.0, 1.0)$. This range is not ideal since it is desirable to include the value of 1.0, which represents no attenuation. A better alternative is the **Q**2.14 format, in which -1.0 and $+1.0$ are represented by $1100\cdots00$ and $0100\cdots00$, respectively. Note that the range of this format is larger than $[-1.0, 1.0]$. To ensure the correct operation, we must artificially limit the env input within this range. The multiplication result, modu, is in the **Q**18.14 format. We can select the proper portion of the modu signal and trim it back to the **Q**16.0 format (i.e., 16-bit signed integer) to match the input format of the DAC.

There are two outputs. The pulse_out signal is the square wave and the pcm_out signal is the digitized sine wave, a format commonly referred to as *PCM* (*pulse code modulation*).

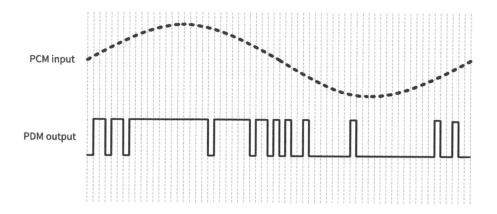

Figure 18.4   Input and output of PDM circuit.

18.5   DAC (DIGITAL-TO-ANALOG CONVERTER)

Converting the PCM signal to a true analog signal requires a *DAC* (*digital-to-analog converter*) and a *low-pass filter* to remove high-frequency noises. We will use a *one-bit delta-sigma DAC*, which generates *PDM* (*pulse density modulation*) output and can be realized by pure digital logic without any analog component. In our configuration, the frequency of the system clock is 100M Hz and the frequency of the generated audio signal is around 20K Hz. The *oversampling* leads to satisfactory result. Furthermore, the delta-sigma scheme performs *noise shaping*, which "pushes" quantization noises into the high-frequency range. This imposes fewer constraints on the subsequent low-pass filter. The theoretical analysis of the delta-sigma modulation is beyond the scope of the book. The remaining section introduces the basic concept and focuses on implementation.

18.5.1   Conceptual design

The one-bit delta-sigma DAC generates PDM output. It is a sequence of pulses that can be at a low-voltage level (i.e., logic '0') or high-voltage level (i.e., logic '1') and can switch in every clock period. Note that the pulses are driven by the system clock, whose frequency is much faster than the frequency of the analog sine audio signal. The *density* of the high-voltage pulses corresponds to the amplitude of the analog signal. The input waveform (which is in the PCM format) and the output PDM waveform are illustrated in Figure 18.4. The gray grid represents one clock period.

The conceptual block diagram of a one-bit delta-sigma DAC is shown in Figure 18.5. It consists of a *delta-sigma* modulation circuit and a one-bit ADC. The *delta* term refers to the subtraction operation and the *sigma* term is originated from the Σ symbol used in mathematics and refers to the summation operation. We assume that the PCM input is in a 16-bit unsigned integer format, whose range is between 0x0000 and 0xffff. The data width is expanded to 17 bits internally. The main part of the system is a *sigma accumulator*, which is composed of an adder and a register. It continuously adds the input PCM data samples. If the accumulation

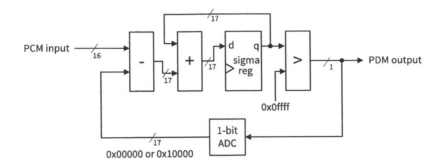

Figure 18.5   Conceptual block diagram of a one-bit delta-sigma DAC.

exceeds the maximum value of 0xffff, the PDM pulse becomes '1' and the amount of 0x10000 is subtracted from the accumulation. The one-bit ADC circuit simply converts logic '0' and logic '1' into two PCM values of 0x00000 and 0x10000. Clearly, more (i.e., "denser") high-voltage pulses will be generated if the PCM amplitude is larger.

18.5.2   HDL implementation

The actual implementation of the one-bit delta-sigma DAC system is simpler than the conceptual block diagram shown in Figure 18.5. This is based on two observations. First, the comparator can be eliminated. Note that the internal data width of the PDM signal is 17 bits wide. If the output of the sigma register is greater than 0x0ffff, it must be in the form of 0x1dddd, where dddd can be any value. Thus, the MSB can be used as the output and no comparator is needed.

Second, the subtractor can be eliminated. The output from the sigma register can be either 0x1dddd or 0x0dddd. If it is 0x1dddd, the MSB of '1' causes the one-bit ADC to output 0x10000. After subtraction, 0x0dddd will be fed back and added to the sigma accumulator. If it is 0x0dddd, the MSB of '0' causes the one-bit ADC to output 0x00000 and 0x0dddd will be fed back and added to the sigma accumulator. In either case, 0x0dddd is used. Thus, we can simply append a 0 to the 16 LSBs and use it as the feedback value.

The corresponding HDL code is shown in Listing 18.3. Note that the output of the DDFS lookup table is converted from a 16-bit signed format to a 17-bit unsigned format. It is first sign-extended to 17 bits and then added with a bias of 0x08000.

Listing 18.3   One-bit delta-sigma DAC

```
library ieee;
use ieee.std_logic_1164.all;
use ieee.numeric_std.all;
entity ds_1bit_dac is
   generic(W : integer := 16);    -- width of input
   port(
      clk     : in  std_logic;
      reset   : in  std_logic;
      pcm_in  : in  std_logic_vector(W - 1 downto 0);
      pdm_out : out std_logic
   );
```

```vhdl
end ds_1bit_dac;

architecture arch of ds_1bit_dac is
   constant BIAS    : unsigned(W downto 0) := ((W-1)=>'1', others=>'0');
   signal pcm_biased : unsigned(W downto 0);
   signal acc_reg    : unsigned(W downto 0);
   signal acc_next   : unsigned(W downto 0);
begin
   -- shift range from [-(2^(W-1)), 2^(W-1)-1] to [0, 2^W-1)]
   pcm_biased <= unsigned(pcm_in(W - 1) & pcm_in) + BIAS;
   -- signal treated as unsigned number in delta-sigma modulation
   acc_next   <= ('0' & acc_reg(W - 1 downto 0)) + pcm_biased;
   process(clk, reset)
   begin
      if reset = '1' then
         acc_reg <= (others => '0');
      elsif (clk'event and clk = '1') then
         acc_reg <= acc_next;
      end if;
   end process;
   pdm_out <= acc_reg(W);
end arch;
```

18.6   DDFS CORE DEVELOPMENT

The main components of the DDFS core are the DDFS system and a one-bit delta-sigma DAC. Its output waveforms are controlled by four input signals, which are fccw, focw, pha, and env. The latter three are considered modulation signals. To make the core more flexible, we include an additional routing control so that the modulation signals can be connected either to core's I/O registers, whose data is written by a processor, or directly to external signal sources. For simplicity, we use a fixed 30-bit phase accumulator (i.e., PW is set to 30).

18.6.1   Register map

The processor interacts with the DDFS system as follows:
- set (i.e., write) the value of fccw.
- set (i.e., write) the value of focw.
- set (i.e., write) the value of pha.
- set (i.e., write) the value of env.
- specify (i.e., write) the sources of modulation signals.

Based on these interactions, we can define the DDFS core's register map. There are five write registers. Their address offsets and fields are:
- offset 0 (fccw register)
  - bits 29 to 0: 30-bit carrier frequency control word
- offset 1 (focw register)
  - bits 29 to 0: 30-bit offset frequency control word
- offset 2 (phase register)
  - bits 29 to 0: 30-bit phase offset word
- offset 3 (envelope register)

  − bits 15 to 0: 16-bit amplitude modulation envelope in **Q**2.14 format
- offset 4 (routing control register)
  − bit 0: source selection for the `env` signal (0 for I/O register and 1 for external signal)
  − bit 1: source selection for the `focw` signal
  − bit 2: source selection for the `pha` signal

18.6.2 Wrapping circuit for the slot interface

Based on the register map and the I/O signals of the DDFS system, we can derive a wrapping circuit that complies to the slot specification and create the DDFS core. The HDL code of the core is shown in Listing 18.4. It consists of an instance of the DDFS system, an instance of a one-bit delta-sigma DAC, five registers, a decoding circuit, and a source routing circuit.

Listing 18.4 DDFS core

```vhdl
library ieee;
use ieee.std_logic_1164.all;
use ieee.numeric_std.all;
entity chu_ddfs_core is
   generic(PW : integer := 30);    -- width of phase accumulator
   port(
      clk         : in  std_logic;
      reset       : in  std_logic;
      -- slot interface
      cs          : in  std_logic;
      write       : in  std_logic;
      read        : in  std_logic;
      addr        : in  std_logic_vector(4 downto 0);
      rd_data     : out std_logic_vector(31 downto 0);
      wr_data     : in  std_logic_vector(31 downto 0);
      -- external modulation input
      focw_ext    : in  std_logic_vector(PW-1 downto 0);
      pha_ext     : in  std_logic_vector(PW-1 downto 0);
      env_ext     : in  std_logic_vector(15 downto 0);
      -- external output
      digital_out : out std_logic;
      pcm_out     : out std_logic_vector(15 downto 0);
      pdm_out     : out std_logic
   );
end chu_ddfs_core;

architecture arch of chu_ddfs_core is
   signal fccw_reg          : std_logic_vector(PW-1 downto 0);
   signal focw_reg, focw    : std_logic_vector(PW-1 downto 0);
   signal pha_reg, pha      : std_logic_vector(PW-1 downto 0);
   signal env_reg, env      : std_logic_vector(15 downto 0);
   signal ctrl_reg          : std_logic_vector(2 downto 0);
   signal wr_en             : std_logic;
   signal wr_fccw, wr_focw  : std_logic;
   signal wr_pha, wr_env    : std_logic;
   signal wr_ctrl           : std_logic;
   signal pcm               : std_logic_vector(15 downto 0);
begin
   -- instantiate ddfs
   ddfs_unit : entity work.ddfs
      generic map(PW => PW)
```

```vhdl
        port map(
            clk        => clk,
            reset      => reset,
            fccw       => fccw_reg,
            focw       => focw,
            pha        => pha,
            env        => env,
            pcm_out    => pcm,
            pulse_out  => digital_out
        );
    dac_unit : entity work.ds_1bit_dac
        generic map(W => 16)
        port map(clk       => clk,
                 reset     => reset,
                 pcm_in    => pcm,
                 pdm_out   => pdm_out
        );
    pcm_out <= pcm;
    -- registers
    process(clk, reset)
    begin
        if reset = '1' then
            fccw_reg <= (others => '0');
            focw_reg <= (others => '0');
            pha_reg  <= (others => '0');
            env_reg  <= x"4000";    -- 1.00
            ctrl_reg <= (others => '0');
        elsif (clk'event and clk = '1') then
            if wr_fccw = '1' then
                fccw_reg <= wr_data(PW-1 downto 0);
            end if;
            if wr_focw = '1' then
                focw_reg <= wr_data(PW-1 downto 0);
            end if;
            if wr_pha = '1' then
                pha_reg <= wr_data(PW-1 downto 0);
            end if;
            if wr_env = '1' then
                env_reg <= wr_data(15 downto 0);
            end if;
            if wr_ctrl = '1' then
                ctrl_reg <= wr_data(2 downto 0);
            end if;
        end if;
    end process;
    --decoding logic
    wr_en   <= '1' when write='1' and cs='1' else '0';
    wr_fccw <= '1' when addr(2 downto 0)="000" and wr_en='1' else '0';
    wr_focw <= '1' when addr(2 downto 0)="001" and wr_en='1' else '0';
    wr_pha  <= '1' when addr(2 downto 0)="010" and wr_en='1' else '0';
    wr_env  <= '1' when addr(2 downto 0)="011" and wr_en='1' else '0';
    wr_ctrl <= '1' when addr(2 downto 0)="100" and wr_en='1' else '0';
    -- input signal routing
    env  <= env_reg when  ctrl_reg(0) = '0' else env_ext;
    focw <= focw_reg when ctrl_reg(1) = '0' else focw_ext;
    pha  <= pha_reg when  ctrl_reg(2) = '0' else pha_ext;
    -- read data
    rd_data <= X"0000" & pcm;
end arch;
```

18.7 DDFS DRIVER

The driver consists of a set of routines to configure the DDFS core's parameters.

18.7.1 Class definition

The class definition of the DDFS core is shown in Listing 18.5.

Listing 18.5 DdfsCore class definition (in `ddfs_core.h`)

```
#ifndef _DDFS_H_INCLUDED
#define _DDFS_H_INCLUDED

#include "chu_io_basic.h"

class DdfsCore {
   enum {
      FCW_REG      = 0,
      FOW_REG      = 1,
      PHA_REG      = 2,
      ENV_REG      = 3,
      SRC_SEL_REG  = 4
   };
   enum {
      PHA_WIDTH = 30 // # bits in ddfs phase register
   };

public:
   DdfsCore(uint32_t core_base_addr);
   ~DdfsCore();                      // not used
   void set_carrier_freq(int freq);
   void set_offset_freq(int freq);
   void set_phase_degree(int phase);
   void set_env(float env);
   void set_fow_source(int channel);
   void set_env_source(int channel);
   void set_pha_source(int channel);
private:
   uint32_t base_addr;
   uint32_t ch_select_reg;
};

#endif   // _DDFS_H_INCLUDED
```

The first `enum` definition uses symbolic names for the four register offsets. The second `enum` definition specifies the width of the DDFS phase accumulator, which is 30 bits wide.

18.7.2 Class implementation

The class implementation of the constructor and the configuration methods is shown in Listing 18.6.

Listing 18.6 DdfsCore class implementation (in `ddfs_core.cpp`)

```
#include "ddfs_core.h"

DdfsCore::DdfsCore(uint32_t core_base_addr){
   base_addr = core_base_addr;
```

```
    // select processor bus
    set_env_source(0);
    set_fow_source(0);
    set_pha_source(0);
    // set note C
    set_carrier_freq(262);
    set_offset_freq(0);
    set_phase_degree(0);
    set_env(1.0);
}

DdfsCore::~DdfsCore(){};
void DdfsCore::set_carrier_freq(int freq){
    uint32_t fcw, p2n;
    float tmp;

    p2n = 1<<PHA_WIDTH;   //2^PHA_WIDTH
    tmp = ((float)p2n) / float(SYS_CLK_FREQ*1000000);
    fcw = uint32_t (freq * tmp);
    io_write(base_addr, FCW_REG, fcw);
    debug("ddfs set_carrier_freq - fcw: ", fcw, 0);
}

void DdfsCore::set_offset_freq(int freq){
    uint32_t fow, p2n;
    float tmp;

    p2n = 1<<PHA_WIDTH;   //2^PHA_WIDTH
    tmp = ((float)p2n) / float(SYS_CLK_FREQ*1000000);
    fow = uint32_t (freq * tmp);
    io_write(base_addr, FOW_REG, fow);
}

void DdfsCore::set_phase_degree(int phase){
    uint32_t pha;

    pha = (SYS_CLK_FREQ*1000000) * phase /360;
    io_write(base_addr, PHA_REG, pha);
}

void DdfsCore::set_env(float env){
    // convert floating point to fixed-point Q2.14 format
    int32_t q216;
    float max_amp;

    max_amp = (float)(0x4000);   // env * 2^15
    q216 = (int32_t) (env*max_amp);
    io_write(base_addr, ENV_REG, q216 & 0x0000ffff);
    debug("ddfs set_carrier_freq - env: ", q216, 0);
}

void DdfsCore::set_fow_source(int channel){
    int ch = 0;

    if (channel==1)
        ch=1;
    bit_write(ch_select_reg, 1, ch);
    io_write(base_addr, SRC_SEL_REG, ch_select_reg);
}

void DdfsCore::set_env_source(int channel){
```

```
    int ch = 0;

    if (channel==1)
      ch=1;
    bit_write(ch_select_reg, 0, ch);
    io_write(base_addr, SRC_SEL_REG, ch_select_reg);
}

void DdfsCore::set_pha_source(int channel){
    int ch = 0;

    if (channel==1)
      ch=1;
    bit_write(ch_select_reg, 2, ch);
    io_write(base_addr, SRC_SEL_REG, ch_select_reg);
}
```

The `DdfsCore()` constructor saves the base address, selects the core's registers as the modulation sources, and sets the frequency to 262 Hz.

The `set_carrier_freq()` and `set_offset_freq()` methods calculate carrier and offset frequency control words and write them to respective registers. They use the formula, $M = \frac{f_{out}}{f_{sys}} * 2^N$, to calculate the values. The `set_phase_degree()` method calculates the phase control word and writes it to the register. The input phase offset is represented in degrees. Since 2^N steps in the phase register represent one period, which is 360 degrees, the amount to be added to the phase register is $\frac{offset}{360} * 2^N$. The `set_env()` method calculates the amplitude of the envelope and writes it to the register. The amplitude is represented as a real number between -1.0 and $+1.0$. The method converts it into a **Q**2.14 format.

The `set_fow_source()` method selects the source of the frequency modulation signal. It sets the corresponding bit in the `ch_select_reg` variable and writes the variable to the register. If the `channel` parameter is 1, an external signal is used as the source. If it is 0, the core's `focw_reg` register is the source. This setting allows the software to generate the modulation signal by writing to the core's register. The `set_env_source()` and `set_pha_source()` methods are similar and select sources for amplitude modulation and phase modulation, respectively.

18.8   TESTING

The main application of DDFS is in communication systems, usually involving the generation and modulation of high-frequency signals. However, in this book we mainly use it to generate audio frequency signals. The Nexys 4 DDR board does not contain a DAC but includes an active fourth-order filter, whose output is connected to a mono audio jack. It can be used in conjunction with the PDM output of the one-bit DAC for the audio frequency signal. We can plug in an earphone or a powered speaker to listen to the sound generated by the DDFS system. Note that in the sampler FPro system an ADSR core is also instantiated and its output is connected to amplitude modulation input of the DDFS core.

For a board without a low-pass filter, a simple passive RC filter can be constructed on a breadboard. The design is discussed in a Xilinx application note in the bibliographic section.

The test routine demonstrates the basic operation of the DDFS system. Additional sound tests are performed in conjunction with the ADSR envelope generators in the next chapter. The code is shown in Listing 18.7. The test produces a single tone at 262 Hz ("middle C"), attenuates the volume via amplitude modulation, and generates a continuously sweeping siren sound via frequency modulation.

Listing 18.7   DDFS test function (in `main_sampler_test.cpp`)

```
void ddfs_check(DdfsCore *ddfs_p, GpoCore *led_p) {
   int i, j;
   float env;

   ddfs_p->set_env_source(0);  // select envelope source
   ddfs_p->set_env(0.0);    // set volume
   sleep_ms(500);
   ddfs_p->set_env(1.0);    // set volume
   ddfs_p->set_carrier_freq(262);
   sleep_ms(2000);
   ddfs_p->set_env(0.0);    // set volume
   sleep_ms(2000);
   // volume control (attenuation)
   ddfs_p->set_env(0.0);    // set volume
   env = 1.0;
   for (i = 0; i < 1000; i++) {
      ddfs_p->set_env(env);
      sleep_ms(10);
      env = env / 1.0109;  //1.0109**1024=2**16
   }
   // frequency modulation 635-912 siren sound
   ddfs_p->set_env(1.0);    // set volume
   ddfs_p->set_carrier_freq(635);
   for (i = 0; i < 5; i++) {          // 10 cycles
      for (j = 0; j < 30; j++) {      // sweep 30 steps
         ddfs_p->set_offset_freq(j * 10); // 10 Hz increment
         sleep_ms(25);
      } // end j loop
   } // end i loop
   ddfs_p->set_offset_freq(0);
   ddfs_p->set_env(0.0);    // set volume
   sleep_ms(1000);
}
```

It will be helpful to have an oscilloscope to observe the output signal.

18.9   BIBLIOGRAPHIC NOTES

The DDFS circuit is a key component is today's communication systems. *Direct Digital Synthesizers: Theory, Design and Applications* by J. Vankka provides detailed coverage of this subject. Basic concepts behind delta-sigma modulation and PDM can be found on the Wikipedia website. The Xilinx manual *DS588, XPS Delta-Sigma Digital-to-Analog Converter (DAC),* includes a discussion of the design of low-pass RC filter.

18.10   SUGGESTED EXPERIMENTS

DDFS can be used to construct a music synthesizer and to generate special sound effects. These experiments are more interesting when implemented with the ADSR envelope generator and are listed in the next chapter. The experiments in this section focus mainly on the DDFS system itself.

18.10.1   Quadrature phase carrier generation

Many communication schemes require an additional 90-degree out-of-phase signal, known as the *quadrature* component. In other words, the $\sin(2\pi ft)$ and $\cos(2\pi ft)$ waveforms must be generated at the same time. Expand the DDFS circuit to generate both signals at the same time. Note that the FPGA's internal memory module supports dual-port operation and thus two lookup operations can be done by using the same memory module.

18.10.2   Reduced-size phase-to-amplitude lookup table

The size of the lookup table can grow large when high-resolution output is needed. However, it can be reduced to one quarter of the original size by taking advantage of the symmetry of the sine function. We only need to include data points in the first quadrant (i.e., between 0 and $\frac{\pi}{2}$) and derive the rest of the data points using the following equations:

$$\sin(x) = \quad \sin(\pi - x) \quad \text{if } \tfrac{\pi}{2} < x \leq \pi$$
$$\sin(x) = \quad -\sin(x - \pi) \quad \text{if } \pi < x < 2\pi$$

Design the new DDFS circuit using this approach, derive the HDL code, and verify its operation.

18.10.3   Additive harmonic synthesis

A harmonic is a signal whose frequency is an integer multiple of the fundamental frequency. For example, if the fundamental frequency is f, its harmonics are $2f$, $3f$, $4f$, $\cdots$. One scheme to generate synthesized music is to add attenuated harmonics to the original signal. Expand the DDFS IP core to allow the addition of three harmonics. The integer multiple and attenuation level of each harmonic can be controlled individually. Create the necessary software driver and verify its operation.

18.10.4   Simple function generator

Modify the DDFS system to generate a square wave, triangular wave, ramp wave, and sine wave. Use two additional bits in the control register as a selection signal to route the desired signal to output. Extend the DDFS core to include this feature, modify the driver, and use an oscilloscope to verify its operation.

18.10.5   Arbitrary waveform generator

To generate a sine waveform, the DDFS system uses a phase-to-amplitude lookup table for the sine function, as discussed in Section 18.4. Other waveforms can be generated if the values of a different (i.e., "arbitrary") function are stored in the table. One way to write the lookup table is to add an additional field in the control register. After this field is asserted for one clock cycle, the subsequent 256 write operations will be used to write the lookup table sequentially. Extend the DDFS core, modify the driver, and use an oscilloscope to verify its operation.

18.10.6   Sample-based synthesis

After completing Experiment 18.10.5, we can use the enhanced core to perform sample-based synthesis. This is a scheme to create better sound by recording a sample from a real instrument and storing the waveform in the phase-to-amplitude lookup table. Use a microphone and the XADC core to record a note of an instrument (e.g., a harmonica), extract data points from one cycle, and store the results to the lookup table. Derive the software and verify its operation.

CHAPTER 19

SOUND II: ADSR CORE

The DDFS core can generate an audio-frequency tone. The unmodulated tone is plain and uninteresting. One method to produce better sound is to modulate the sine wave with an ADSR (attack-decay-sustain-release) amplitude envelope, which is the scheme used in a music synthesizer. In the chapter, we construct a custom core to generate the ADSR envelope and use it as the amplitude modulation input of the DDFS core.

19.1 INTRODUCTION

A musical instrument creates a direct acoustic sound. A music synthesizer imitates an instrument by producing electronic signals and playing them through a speaker. The unmodulated DDFS circuit can be used to generate the basic tone. Additional schemes, such as adding harmonic components and performing frequency modulation, can produce more interesting effects.

Applying an ADSR amplitude envelope is the most commonly used scheme and the foundation of a *music synthesizer*. The scheme is based on observation that when a real musical instrument produces a note, the loudness changes over time. It rises quickly from zero and then steadily decays. To model the effect, we can multiply the constant tone by a loudness *ADSR envelope*, which contains the *attack*, *decay*, *sustain*, and *release* segments. An ADSR envelope is shown in Figure 19.1. The contour of the envelope corresponds to pressing and releasing a key of a musical instrument, such as a piano. When a key is pressed, the loudness quickly rises

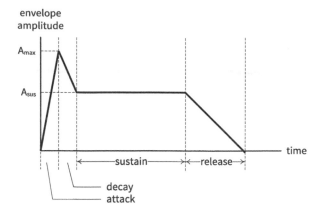

Figure 19.1   ADSR envelope for a piano.

to the maximum (attack segment), then falls fast (decay segment) to a rather constant level (sustain segment), which continues until the key is released. The sound then quickly fades away (release segment). We can imitate the sounds of different instruments by adjusting the levels and lengths of various segments.

The ADSR envelope can be specified by two amplitude parameters and four timing parameters:

- Maximum level: the maximal amplitude the envelope can reach, which is labeled A_{max} in Figure 19.1
- Sustain level: the amplitude of of the sustain segment, which is labeled A_{sus} in Figure 19.1
- Attack time: the time interval of the attack segment
- Decay time: the time interval of the decay segment
- Sustain time: the time interval of the sustain segment
- Release time: the time interval of the release segment

To obtain better resolution in our design, we set the maximum level as a constant and represent the sustain level as a percentage of the maximum level. Thus, there are only five parameters. In software, the four timing parameters are defined in terms of milliseconds and the sustain level as a percentage in the floating-point format.

In terms of communication discussed in Section 18.2.3, the original tone can be considered as the carrier signal and the ADSR envelope can be considered as the message. Applying the ADSR envelope essentially performs the amplitude modulation over the original tone.

19.2   ADSR ENVELOPE GENERATOR

An *ADSR envelope generator* is a circuit that generates an amplitude envelope following the ADSR model. Its output can be connected to a DDFS as an external amplitude modulation signal. The generator can be constructed by an FSMD with an amplitude counter and a sustain time counter.

The ADSR circuit models the sound generation of a musical instrument. Its output does not run continuously. A single envelope is generated at a time and the generation is controlled by a trigger signal. We consider two types of triggering modes. The *automatic mode* uses a trigger signal as a "start" tick. When it is asserted, the circuit generates the predefined envelope. The *real-time mode* uses the trigger signal to mimic the operation of a key on the piano keyboard. The ADSR circuit monitors the trigger signal continuously. It initiates the operation when the trigger signal is asserted (i.e., key is pressed) and stays on the sustain segment until the trigger signal is deasserted (i.e., key is released). The automatic mode is used for design in this section and the real-time mode is left as an exercise in Experiment 19.8.6.

19.2.1 Conceptual FSMD design

We develop an FSMD to construct an ADSR envelope generator. The FSM uses the `attack`, `decay`, `sustain`, and `release` states to represent the four segments of the ADSR envelope. The data paths contain an amplitude counter and a sustain time counter.

The amplitude counter is used to generate the envelope amplitude. It is also used implicitly to determine the time spent in attack, decay, and release segments. In each segment, it increments or decrements a specific amount every clock cycle. The exact value can be calculated from the ADSR timing and amplitude parameters. Assume that the desired attack time and the system clock period are t_{attack} and t_{sys}, and the maximum amplitude is A_{max}. There are t_{attack}/t_{sys} clock cycles in the attack segment. For the counter to reach A_{max} from 0 in t_{attack}, the incrementing amount in each clock cycle must be $\frac{A_{max}-0}{t_{attack}/t_{sys}}$. The pre-calculated value becomes an external input signal. In the `attack` state, the FSMD increments the amplitude counter with this amount every clock cycle. When the counter reaches A_{max}, the FSMD moves to the `decay` state.

The decrementing amounts in the decay and release segments can be calculated in a similar fashion. Assume that the desired decay time and the release time are t_{decay} and $t_{release}$, and the sustain amplitude is A_{sus}. The decrementing amount in the decay segment is $\frac{A_{max}-A_{sus}}{t_{sustain}/t_{sys}}$ and the decrementing amount in the release segment $\frac{A_{sus}-0}{t_{release}/t_{sys}}$. The FSMD decrements the counter in these states and transits to a new state when the counter reaches the threshold.

The amplitude is constant in the sustain segment and its interval is controlled by the sustain time. There are $t_{sustain}/t_{sys}$ clock cycles in this segment. A separate sustain time counter is used for this purpose. The FSMD decrements the counter in the `sustain` state and moves to the `release` state when the counter reaches 0.

19.2.2 ASMD chart

The ASMD chart of the ADSR envelope generator is shown in Figure 19.2. In addition to the `attack`, `decay`, `sustain`, and `release` states discussed earlier, the `idle` state is for the idle condition and the `launch` state is to facilitate the re-triggering. The `start` signal is the trigger to initiate the envelope generation. After it is asserted, the FSMD usually goes through the entire envelope generation process. However, if the `start` signal is asserted before completion (i.e., re-triggered), the FSMD aborts the current operation and starts a new one. Thus, the `start` signal

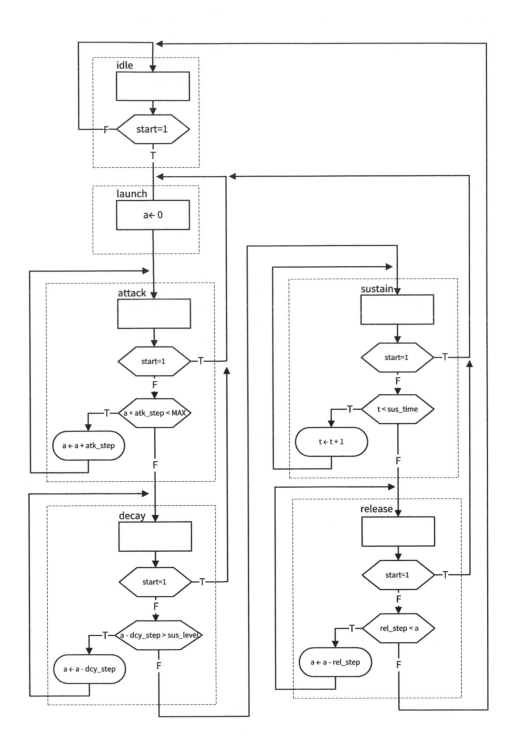

Figure 19.2   ASMD chart of the ADSR circuit.

is checked in every state and the FSMD moves to the `launch` state immediately if it is asserted.

The two data registers, represented by `a` and `t`, are used for the amplitude counter and sustain time counter. The timing parameters are converted into incrementing and decrementing steps, as discussed in the previous subsection. They appear as external input signals, `atk_step`, `dcy_step`, `rel_step`, and `sus_level`, in the ADSR circuit. After `start` is asserted, the FSMD moves to the `launch` state and then the `attack` state, in which `a` is incremented with `atk_step` until it reaches `MAX`. The FSMD transits to the `decay` state, in which `a` is decremented with `dcy_step` until it reaches the designated sustain amplitude `sus_level`. The FSMD then moves to the `sustain` state, in which `t` is incremented until reaching `sus_time`. The FSMD then reaches the `release` state, in which `a` is decremented with `rel_step` until it becomes 0, and eventually returns to the `idle` state.

19.2.3 HDL implementation

The HDL code of the ADSR envelope generator can be derived following the ASMD chart, as shown in Listing 19.1.

Listing 19.1 ADSR envelope generator

```
library ieee;
use ieee.std_logic_1164.all;
use ieee.numeric_std.all;
entity adsr is
   port(
      clk        : in   std_logic;
      reset      : in   std_logic;
      start      : in   std_logic;
      atk_step   : in   std_logic_vector(31 downto 0);
      dcy_step   : in   std_logic_vector(31 downto 0);
      sus_level  : in   std_logic_vector(31 downto 0);
      sus_time   : in   std_logic_vector(31 downto 0);
      rel_step   : in   std_logic_vector(31 downto 0);
      env        : out  std_logic_vector(15 downto 0);
      adsr_idle  : out  std_logic
   );
end adsr;

architecture arch of adsr is
   type state_type is (idle, launch, attack, decay, sustain, release);
   constant MAX    : unsigned(31 downto 0) := x"80000000";
   constant BYPASS : std_logic_vector(31 downto 0):=(others => '1');
   constant ZERO   : std_logic_vector(31 downto 0):=(others => '0');
   signal state_reg      : state_type;
   signal state_next     : state_type;
   signal a_reg, a_next  : unsigned(31 downto 0);
   signal t_reg, t_next  : unsigned(31 downto 0);
   signal env_i          : unsigned(31 downto 0);
   signal a_atk, a_dcy   : unsigned(31 downto 0);

begin
   -- fsmd state and data registers
   process(clk, reset)
   begin
      if reset = '1' then
         state_reg <= idle;
         a_reg     <= (others => '0');
```

```vhdl
        t_reg        <= (others => '0');
    elsif (clk'event and clk = '1') then
        state_reg <= state_next;
        a_reg        <= a_next;
        t_reg        <= t_next;
    end if;
end process;
-- atttack value calculation
a_atk <= a_reg + unsigned(atk_step);
a_dcy <= a_reg - unsigned(dcy_step);
-- next-state logic and data path logic
process(state_reg, a_reg, start, sus_level, rel_step,
        sus_time, t_reg, a_atk, a_dcy)
begin
    state_next <= state_reg;
    a_next        <= a_reg;
    adsr_idle    <= '0';
    t_next        <= t_reg;
    case state_reg is
        when idle =>
            adsr_idle <= '1';
            if (start = '1') then
                state_next <= launch;
            end if;
        when launch =>
            state_next <= attack;
            a_next        <= (others => '0');
        when attack =>
            if (start = '1') then
                state_next <= launch;
            else
                if (a_atk < MAX) then
                    a_next <= a_atk;
                else
                    state_next <= decay;
                end if;
            end if;
        when decay =>
            if (start = '1') then
                state_next <= launch;
            else
                if (a_dcy > unsigned(sus_level)) then
                    a_next <= a_dcy;
                else
                    a_next        <= unsigned(sus_level);
                    state_next <= sustain;
                    t_next        <= (others => '0');
                end if;
            end if;
        when sustain =>
            if (start = '1') then
                state_next <= launch;
            else
                if (t_reg <= unsigned(sus_time)) then
                    t_next <= t_reg + 1;
                else
                    state_next <= release;
                end if;
            end if;

        when release =>
```

```
              if (start = '1') then
                 state_next <= launch;
              else
                 if (a_reg > unsigned(rel_step)) then
                    a_next <= a_reg - unsigned(rel_step);
                 else
                    state_next <= idle;
                 end if;
              end if;
      end case;
   end process;
   -- special cases
   env_i <= MAX when atk_step=BYPASS else
            (others => '0') when atk_step = ZERO else
            a_reg;
   env <= '0' & std_logic_vector(env_i(31 downto 17));
end arch;
```

We use the 32-bit unsigned representation for the amplitude computation and define the maximum level, MAX, to be 0x80000000. It can be interpreted as the unsigned **Q**1.31 format and its range is between +0.0 and +1.0. The internal registers and input step signals are represented in this format. The output of the envelope generator is expected to be connected to the DDFS circuit's amplitude modulation input, which is in signed **Q**2.14 format. The last statement in the program does the format conversion.

In addition to the normal envelope generation, we include special *bypass* and *mute* operations. The former bypasses the envelope generation circuit completely by setting the envelope amplitude to the maximum value. The latter turns off the sound immediately by setting the envelope amplitude to 0. The two special operations are conveyed by all 1's and all 0's patterns in the atk_step input signal. A special output checking code segment examines the two patterns and sets the output signal to special values when the patterns are detected.

19.3 ADSR CORE DEVELOPMENT

The ADSR core is composed of the ADSR envelope generator and a wrapping circuit.

19.3.1 Register map

The processor interacts with the ADSR envelope generator as follows:
- generate (i.e., write) a pulse to start the generation.
- set (i.e., write) the value of atk_step.
- set (i.e., write) the value of dcy_step.
- set (i.e., write) the value of sus_time.
- set (i.e., write) the value of rel_step.
- set (i.e., write) the value of sus_level.
- check (i.e., read) the idle status and retrieve the envelope amplitude data.

Based on these interactions, we can define the ADSR core's register map. There are six write registers. Their address offsets and fields are:
- offset 0 (start register)

> − dummy data write to generate a pulse to start the envelope generation
- offset 1 (attack step register)
  - bits 31 to 0: 32-bit attack step value
- offset 2 (decay step register)
  - bits 31 to 0: 32-bit decay step value
- offset 3 (sustain time register)
  - bits 31 to 0: 32-bit sustain time
- offset 4 (release step register)
  - bits 31 to 0: 32-bit release step value
- offset 5 (sustain level register)
  - bits 31 to 0: sustain level

There is one read register. Its address offset and fields are:

- offset 8 (read data and status register)
  - bits 31 to 16: 16-bit envelope data
  - bit 0: idle status

19.3.2 Wrapped ADSR circuit

Based on the register map and the I/O signals of the ADSR envelope generator, we can derive a wrapping circuit that complies to the slot specification and create the ADSR core. The HDL code of the core is shown in Listing 19.2. It consists of an instance of the ADSR generator, five registers, and a decoding circuit.

Listing 19.2   ADSR core

```
library ieee;
use ieee.std_logic_1164.all;
use ieee.numeric_std.all;
entity chu_adsr_core is
   port(
      clk      : in  std_logic;
      reset    : in  std_logic;
      -- slot interface
      cs       : in  std_logic;
      write    : in  std_logic;
      read     : in  std_logic;
      addr     : in  std_logic_vector(4 downto 0);
      rd_data  : out std_logic_vector(31 downto 0);
      wr_data  : in  std_logic_vector(31 downto 0);
      -- external signal
      adsr_env : out std_logic_vector(15 downto 0)
   );
end chu_adsr_core;

architecture arch of chu_adsr_core is
   signal atk_step_reg   : std_logic_vector(31 downto 0);
   signal dcy_step_reg   : std_logic_vector(31 downto 0);
   signal sus_level_reg  : std_logic_vector(31 downto 0);
   signal rel_step_reg   : std_logic_vector(31 downto 0);
   signal sus_time_reg   : std_logic_vector(31 downto 0);
   signal idle           : std_logic;
   signal wr_en          : std_logic;
   signal wr_atk, wr_dcy : std_logic;
   signal wr_sus_lvl     : std_logic;
```

```vhdl
    signal wr_rel          : std_logic;
    signal wr_start        : std_logic;
    signal wr_sus_time     : std_logic;

begin
    -- instantiate adsr generator
    adsr_unit : entity work.adsr
        port map(
            clk          => clk,
            reset        => reset,
            start        => wr_start,
            atk_step     => atk_step_reg,
            dcy_step     => dcy_step_reg,
            sus_level    => sus_level_reg,
            sus_time     => sus_time_reg,
            rel_step     => rel_step_reg,
            env          => adsr_env,
            adsr_idle    => idle
        );
    -- registers
    process(clk, reset)
    begin
        if reset = '1' then
            atk_step_reg   <= (others => '0');
            dcy_step_reg   <= (others => '0');
            sus_level_reg  <= (others => '0');
            rel_step_reg   <= (others => '0');
            sus_time_reg   <= (others => '0');
        elsif (clk'event and clk = '1') then
            if wr_atk = '1' then
                atk_step_reg <= wr_data;
            end if;
            if wr_dcy = '1' then
                dcy_step_reg <= wr_data;
            end if;
            if wr_sus_lvl = '1' then
                sus_level_reg <= wr_data;
            end if;
            if wr_rel = '1' then
                rel_step_reg <= wr_data;
            end if;
            if wr_sus_time = '1' then
                sus_time_reg <= wr_data;
            end if;
        end if;
    end process;
    -- decoding logic
    wr_en        <= '1' when write = '1' and cs = '1' else '0';
    wr_start     <= '1' when addr(2 downto 0)="000" and wr_en='1' else '0';
    wr_atk       <= '1' when addr(2 downto 0)="001" and wr_en='1' else '0';
    wr_dcy       <= '1' when addr(2 downto 0)="010" and wr_en='1' else '0';
    wr_sus_time  <= '1' when addr(2 downto 0)="011" and wr_en='1' else '0';
    wr_rel       <= '1' when addr(2 downto 0)="100" and wr_en='1' else '0';
    wr_sus_lvl   <= '1' when addr(2 downto 0)="101" and wr_en='1' else '0';
    -- read data
    rd_data      <= x"0000000" & "000" & idle;
end arch;
```

19.4 ADSR DRIVER

The ADSR driver is composed of two sets of methods. The first set controls the basic operation and configures the ADSR envelope parameters. The second set generates a music note with a specific frequency and duration. It includes the calc_note_freq() and play_note() methods.

19.4.1 Class definition

The class definition of the ADSR core is shown in Listing 19.3.

Listing 19.3   `AdsrCore` class definition (in `adsr_core.h`)

```
#ifndef _ADSR_H_INCLUDED
#define _ADSR_H_INCLUDED

#include "chu_init.h"
#include "ddfs_core.h"

class AdsrCore {
public:
    enum {
        START_REG = 0,
        ATK_REG   = 1,
        DCY_REG   = 2,
        SUS_REG   = 3,
        REL_REG   = 4,
        SUS_LEVEL_REG = 5
    };
    enum {
        MAX = 0x7fffffff,
        BYPASS_PATTERN = 0xffffffff,
        STOP_PATTERN   = 0x00000000
    };

    AdsrCore(uint32_t adsr_base_addr, DdfsCore *ddfs);
    ~AdsrCore();                     // not used
    void init();
    int idle();
    void start();
    void abort();
    void bypass();
    void select_env(int n);
    int calc_note_freq(int oct, int ni);
    void play_note(int note, int oct, int dur);
private:
    uint32_t base_addr;
    /* current envelope parameters  */
    int ams, dms, sms, rms;
    float slevel;
    /* DDFS instance */
    DdfsCore *_ddfs;
    /* method */
    void write_adsr_reg();
};

#endif   // _ADSR_H_INCLUDED
```

The first **enum** definition uses symbolic names for the six register offsets. The second **enum** definition lists the constant patterns used in the ADSR controller. To reduce errors of interpreting 0x80000000 as a negative number, we use 0x7fffffff for the maximum amplitude instead. The relevant ADSR parameters are kept in the private section. Note that the attack time (**ams**), decay time (**dms**), sustain time (**sms**), and release time (**rms**) are represented in terms of millisecond, and the sustain level are represented as a fraction (between 0.0 and 1.0) of the maximum amplitude.

Since the ADSR generator's output is fed to a DDFS circuit, the **AdsrCore()** constructor includes a parameter that is a pointer to a DDFS instance. From pure C++ programming's point of view, the **AdsrCore** class should be derived from the **DdfsCore** class and inherits its members. This implies that an **AdsrCore** instance implicitly controls the physical DDFS core and maintains its own "system state." If a separate **DdfsCore** instance is created, it controls the same physical DDFS core and may lead to conflicting access. To maintain a clear view of the instantiated physical cores, we intentionally avoid deriving the **AdsrCore** class from the **DdfsCore** class.

19.4.2 Configuration methods

The class implementation of the constructor and the configuration methods is shown in Listing 19.4.

Listing 19.4 **AdsrCore** configuration methods (in **adsr_core.cpp**)

```
#include "adsr_core.h"

AdsrCore::AdsrCore(uint32_t adsr_base_addr, DdfsCore *ddfs) {
    base_addr = adsr_base_addr;
    _ddfs = ddfs;
    init();
    select_env(1);
}
AdsrCore::~AdsrCore() {
}      // not used

void AdsrCore::init() {
    _ddfs->set_env_source(1);   //select external env source (i.e., adsr)
    _ddfs->set_fow_source(0);
    _ddfs->set_pha_source(0);
    // set note C
    _ddfs->set_carrier_freq(262);
    _ddfs->set_offset_freq(0);
    _ddfs->set_phase_degree(0);
}

int AdsrCore::idle() {
    int idle_bit;

    idle_bit = (int) io_read(base_addr, 0) & 0x00000001;
    return (idle_bit);
}

void AdsrCore::start() {
    // write a dummy data to generate a start pulse
    io_write(base_addr, START_REG, 0);
}
```

```cpp
void AdsrCore::abort() {
   // write 0 to attack register
   io_write(base_addr, ATK_REG, (uint32_t )STOP_PATTERN);
}

void AdsrCore::bypass() {
   ams = BYPASS_PATTERN;
   io_write(base_addr, ATK_REG, (uint32_t )BYPASS_PATTERN);
}

void AdsrCore::set_env(int attack_ms, int decay_ms, int sustain_ms,
                       int release_ms, float sus_level) {
   ams = attack_ms;
   dms = decay_ms;
   sms = sustain_ms;
   rms = release_ms;
   slevel = sus_level;
   write_adsr_reg();
}

void AdsrCore::select_env(int n) {
   switch (n) {
   case 1:
      set_env(100, 50, 100, 50, 0.9);
      break;
   case 2:
      set_env(10, 50, 100, 100, 0.9);
      break;
   default:
      set_env(10, 200, 100, 100, 0.1);
      break;
   }
   return;
}

void AdsrCore::write_adsr_reg() {
   uint32_t nc, step, sus_abs;
   //# clocks per ms = 0.001 / (1/(SYS_CLK_FREQ*1000000))
   const uint32_t clks = SYS_CLK_FREQ * 1000;

   if (ams == BYPASS_PATTERN) {
      io_write(base_addr, ATK_REG, (uint32_t )BYPASS_PATTERN);
      return;
   }
   if (ams == STOP_PATTERN) {
      io_write(base_addr, ATK_REG, (uint32_t )STOP_PATTERN);
      return;
   }

   // convert sustain level in absolute value
   sus_abs = (unsigned int) MAX * slevel;
   io_write(base_addr, SUS_LEVEL_REG, (uint32_t )sus_abs);
   // convert attack time (in ms) into envelope increment step
   nc = ams * clks;
   step = MAX / nc;              // increment step
   if (step == 0)
      step = 1;
   io_write(base_addr, ATK_REG, (uint32_t )step);
   debug("adsr set - sus_level/atk_step: ", sus_abs, step);
   // convert decay time (in ms) into envelope decrement step
```

```
  nc = dms * clks;
  step = (MAX - sus_abs) / nc;
  if (step == 0)
     step = 1;
  io_write(base_addr, DCY_REG, (uint32_t )step);
  // convert sustain time (in ms) into #clocks
  nc = sms * clks;
  io_write(base_addr, SUS_REG, (uint32_t )nc);
  debug("adsr set - sus_time/dcy_step: ", nc, step);
  // convert release time (in ms) into envelope decrement step
  nc = rms * clks;
  step = sus_abs / nc;
  if (step == 0)
     step = 1;
  io_write(base_addr, REL_REG, (uint32_t )step);
}
```

The `AdsrCore()` constructor saves the base address and the `DdfsCore` instance and calls the initial method, `init()`, which sets up the DDFS circuit and then selects the external input as its amplitude modulation source.

The `set_env()` method inputs and stores various envelope parameters, which are expressed in terms of milliseconds, and calls `write_adsr_reg()`. The private `write_adsr_reg()` method converts the timing parameters from milliseconds into increment and decrement steps discussed in Section 19.2.1 and writes them into the ADSR core's registers. The `uint32_t` type is used for calculation and it is interpreted as **Q**1.31.

The `idle()` method checks whether the ADSR is idle; i.e., whether the envelope generation is completed. The `start()` method writes the start register to generate a pulse to initiate the operation. The `abort()` method writes the predefined abort pattern to the attack time register to turn off the amplitude. The `bypass()` method writes the predefined bypass pattern to the attack time register to bypass the envelope generation and set the amplitude to the maximum value. The `select_env()` method selects a predefined envelope.

19.4.3 `calc_note_freq()` method

A music *note* specifies the *pitch* of a sound. For our purposes, we treat the *pitch* the same as the *frequency*. The `calc_note_freq()` method calculates the frequency of a note at a specific *octave*. There are 12 notes in an *octave*, represented by C, C$^\sharp$, D, D$^\sharp$, E, F, F$^\sharp$, G, G$^\sharp$, A, A$^\sharp$, and B. The frequencies from octave 0 to octave 8 are summarized in Table 19.1.

There is a simple relationship between two successive notes. If the frequencies of two successive notes are f_i and f_{i+1}, then

$$f_{i+1} = 2^{\frac{1}{12}} * f_i$$

The notes are standardized around the A note of octave 4 (A4), which is 440.0 Hz. The frequencies of other notes are then derived accordingly. The previous equation indicates that a frequency is doubled after one octave, i.e.,

$$f_{i+12} = (2^{\frac{1}{12}})^{12} * f_i = 2 * f_i$$

The frequency of a specific note can be obtained by applying the first formula. However, since the direct calculation involves expensive floating-point operations,

Table 19.1   Note frequencies of nine octaves

	oct 0	oct 1	oct 2	oct 3	oct 4	oct 5	oct 6	oct 7	oct 8
C	16.4	32.7	65.4	130.8	261.6	523.3	1046.5	2093.0	4186.0
C♯	17.3	34.7	69.3	138.6	277.2	554.4	1108.7	2217.5	4434.9
D	18.4	36.7	73.4	146.8	293.7	587.3	1174.7	2349.3	4698.6
D♯	19.5	38.9	77.8	155.6	311.1	622.3	1244.5	2489.0	4978.0
E	20.6	41.2	82.4	164.8	329.6	659.3	1318.5	2637.0	5274.0
F	21.8	43.7	87.3	174.6	349.2	698.5	1396.9	2793.8	5587.7
F♯	23.1	46.3	92.5	185.0	370.0	740.0	1480.0	2960.0	5919.9
G	24.5	49.0	98.0	196.0	392.0	784.0	1568.0	3136.0	6271.9
G♯	26.0	51.9	103.8	207.7	415.3	830.6	1661.2	3322.4	6644.9
A	27.5	55.0	110.0	220.0	440.0	880.0	1760.0	3520.0	7040.0
A♯	29.1	58.3	116.5	233.1	466.2	932.3	1864.7	3729.3	7458.6
B	30.9	61.7	123.5	246.9	493.9	987.8	1975.5	3951.1	7902.1

we use an alternative lookup table scheme. This scheme is based on the second formula. Note that if the frequency of a note in octave 0 is f_0, the frequency in octave i becomes $2^i * f_0$. We can store the precalculated frequencies of octave 0 in a table and obtain the frequencies in other octaves by multiplying the octave-0 frequency with 2^i. The multiplication operation corresponds to shift f_0 to left i positions. The code is shown in Listing 19.5.

Listing 19.5   `calc_note_freq()` method (in `adsr_core.cpp`)

```
int AdsrCore::calc_note_freq(int oct, int ni)
{
  // frequency table for octave 0
  const float NOTES[]={
  16.3516,    //   0  C
  17.3239,    //   1  C#
  18.3541,    //   2  D
  19.4454,    //   3  D#
  20.6017,    //   4  E
  21.8268,    //   5  F
  23.1247,    //   6  F#
  24.4997,    //   7  G
  25.9565,    //   8  G#
  27.5000,    //   9  A
  29.1352,    //  10  A#
  30.8677     //  11  B
  };
  int freq;

  freq = (unsigned int) NOTES[ni] * (1<<oct);
  return(freq);
}
```

19.4.4   `play_note()` method

This method plays a note for a specific interval. In music, a note can also include additional information to express the relative *duration* of a sound. A *quarter note* is considered as the basic unit. Its duration is twice the duration of an *eighth note* and four times the duration of a *sixteenth note* and is half the duration of a *half note* and a quarter the duration of a *whole note*. The *tempo* of a music determines the exact time interval of a quarter note. It is usually defined in terms of *BPM* (*beats per minute*). For example, 120 BPM specifies that the interval of a quarter note is 0.5 second (500 milliseconds). The time intervals of other notes can be determined accordingly. For example, the time intervals of a whole note and eighth note are 2 seconds and 0.25 second, respectively.

One way to obtain the desired time interval is to adjust the sustain time of the amplitude envelope. Let the desired interval be T and the sustain time becomes

$$t_{sustain} = T - (t_{attack} + t_{decay} + t_{release})$$

The code is shown in Listing 19.6.

Listing 19.6 play_note() method (in `adsr_core.cpp`)

```
void AdsrCore::play_note(int note, int oct, int dur) {
   int sus_tmp;
   int freq;

   freq = calc_note_freq(oct, note);
   _ddfs->set_carrier_freq(freq);

   sus_tmp = dur - (ams + dms + rms);
   if (sus_tmp <= 0) {
      // sustain time must be greater than 0
      sus_tmp = 10;
   }
   set_env(ams, dms, sus_tmp, rms, slevel);
   // start envelope
   io_write(base_addr, START_REG, 0);
```

19.5 TESTING

The physical setup for testing is the same as the DDFS core in Section 18.8. The test routine demonstrates the basic operation of the ADSR core. The code is shown in Listing 19.7. The test first plays the seven primary notes in octave 3 in unmodulated format (i.e., no envelope) and then plays the same notes in octaves 3, 4, and 5 with the predefined envelope, which is selected by slide switches.

Listing 19.7 ADSR test function (in `main_sampler_test.cpp`)

```
void adsr_check(AdsrCore *adsr_p, GpoCore *led_p, GpiCore *sw_p) {
   const int melody[] = {0, 2, 4, 5, 7, 9, 11 };
   int i, oct;

   adsr_p->init();
   // no adsr envelope and  play one octave
   adsr_p->bypass();
   for (i = 0; i < 7; i++) {
      led_p->write(bit(i));
      adsr_p->play_note(melody[i], 3, 500);
      sleep_ms(500);
   }
   adsr_p->abort();
```

```
sleep_ms(1000);
// set and enable adsr envelope
// play 4 octaves
adsr_p->select_env(sw_p->read());
for (oct = 3; oct < 6; oct++) {
    for (i = 0; i < 7; i++) {
        led_p->write(bit(i));
        adsr_p->play_note(melody[i], oct, 500);
        sleep_ms(500);
    }
}
led_p->write(0);
// test duration
sleep_ms(1000);
for (i = 0; i < 4; i++) {
    adsr_p->play_note(0, 4, 500 * i);
    sleep_ms(500 * i + 1000);
}
```

19.6   PROJECT IDEA

One application of ADSR and DDFS system is to play a predefined *melody*, which is a sequence of music notes. One simple representation of a melody is the *RTTTL* (ring tone text transfer language) format, which was developed for cell phone's ring tone. The RTTTL string for "Jingle Bell" looks like:

```
JingleBell:d=8,o=5,b=112:32p,a,a,4a,a,a,4a,a,c6,f.,16g,2a,
a#,a#,a#.,16a#,a#,a,a.,16a,a,g,g,a,4g,4c6
```

It is divided into *name, default value,* and *data* sections, separated by a ":" character.

The name section is a string representing the name of the melody, as in `JingleBell`. The default value section sets the default values of the melody and includes three parts:

- Duration of note (as in d=8): which can be 1, 2, 4, 8, 16, and 32, representing whole note, half note, etc.
- Octave (o=5): which can be 4, 5, 6, and 7.
- Beats (b=112): which is the tempo in terms of beats per minutes.

The data sections contains a series of "note elements" separated by a "," character. The element is in "duration-note-octave-dot" form. The note portion can be p, a, a#, b, $\cdots$, g, and g#, representing pause (i.e., silence), A, $A^\sharp$, B, $\cdots$, G, and $G^\sharp$, respectively. The dot portion is a "(.)" character appended in the end. If it exists, the duration of the basic note is increased by half of its original value. The duration and octave portions are optional and, if not present, the default values will be applied. Following are a few examples (assume that the default values of "Jingle Bell" melody are used):

- a#: $A^\sharp$ in octave 5 with a duration of an eighth note
- 4a#: $A^\sharp$ in octave 5 with a duration of a quarter note
- a#6: $A^\sharp$ in octave 6 with a duration of an eighth note
- 4a#6: $A^\sharp$ in octave 6 with a duration of a quarter note
- 4a#6.: $A^\sharp$ in octave 6 with a duration of one and half quarter note
- 4p: pause for duration of a quarter note

A program can parse the RTTTL string, extract the note and duration, and play the melody. It is left as an exercise in Experiment 19.8.1

19.7 BIBLIOGRAPHIC NOTES

Basic concepts behind modulation and the ADSR envelope can be found on the Wikipedia website. *The Theory and Technique of Electronic Music* by M. Puckette discusses various techniques and their mathematical foundations of music synthesis. The book is available online and can be found by searching the title.

19.8 SUGGESTED EXPERIMENTS

19.8.1 RTTTL music player

We want to develop a software program that can play the RTTTL melody discussed in Section 19.6. The program should parse the RTTTL string and call the play_note() method. Derive the code and verify its operation.

19.8.2 ADSR envelope testing

We want to develop a software program that can take ADSR parameters from the UART console and listen to its effect. Derive the code and verify its operation.

19.8.3 Pushbutton piano

Pushbutton switches can be used as piano key to play synthesized music. Use the pushbutton and slide switches as follows:
- Five pushbutton switches on the Nexys 4 DDR board for five notes
- Three slide switches for the octave
- Three slide switches for the note duration
- Two slide switches for the envelope

Note that a button cannot control the duration because the ADSR circuit constructed in Section 19.2.3 only supports the automatic mode. Derive the software code and verify its operation.

19.8.4 Keyboard piano

Repeat Experiment 19.8.3 but use a PS2 keyboard as piano keyboard. Assign the PS2 keys for notes, octaves, note durations, and ADSR envelopes. Derive software and verify its operation.

19.8.5 Keyboard recorder

For the keyboard piano in Experiment 19.8.4, we want to add an additional "recording feature" that records the notes and octaves played in an interval. Use one pushbutton switch to start and stop the recording session and another pushbutton switch to play back the stored information.

19.8.6 Real-time mode ADSR generator

The real-time mode mimics the operation of a key on the piano keyboard, as discussed in Section 19.2. The `trigger` signal controls the envelope initiation and the sustain time interval. The ADSR circuit initiates the operation when `trigger` is asserted. After progressing through the attack and decay intervals, the circuit stays on the sustain segment until `trigger` is deasserted. Revise the original ADSR generator, modify the ADSR core and driver, and verify its operation.

19.8.7 Real-time mode button piano

Repeat Experiment 19.8.3 with the real-time mode ADSR generator designed in Experiment 19.8.6. Use the pushbuttons to control the sustain interval. Derive software and verify its operation.

19.8.8 Merged DDFS and ADSR core

Merge the DDFS circuit and ADSR circuit into a single MMIO core. Derive the core and driver and verify its operation.

19.8.9 ADSR core with an automatic play FIFO buffer

A piece of music is represented by a sequence of notes and their durations. When playing the music, the main program must keep track of the duration of each note and initiate the following note in proper time. To reduce the overhead in software, the ADSR core can add a FIFO buffer and a FIFO controller. The buffer stores the notes and duration of each note and the controller plays back the music. The main program only needs to fill the FIFO buffer. Derive the new core and driver and verify the operation.

19.8.10 ADSR core for frequency modulation

While the ADSR scheme is mainly applied for amplitude modulation, it can also be used for frequency modulation. Modify the sampler FPro system to include a second ADSR core and connect its output to the frequency modulation input of the DDFS core. Derive the system and test its effect.

PART IV

EMBEDDED SOC III: VIDEO CORES

CHAPTER 20

INTRODUCTION TO THE VIDEO SYSTEM

A video monitor displays a visual image, which is composed of *pixels* arranged as a two-dimensional grid. Since a pixel can only retain color for a short amount of time, it must be refreshed at a constant rate. A video system generates and processes the pixel data stream. This chapter demonstrates the construction of a basic video controller and introduces the concept of stream interface.

20.1 INTRODUCTION TO A VIDEO DISPLAY

20.1.1 Conceptual video display

A computer display screen is composed of *pixels*. The pixels are arranged as a two-dimensional grid, as shown in Figure 20.1. The *resolution* defines the number of distinct pixels in each dimension, usually quoted as *width-by-height*, as in 640-by-480. One horizontal row is referred to as a *line* and the entire screen is referred to as a *frame*. Note that the coordinate of the vertical axis increases downward.

The video display is based on an *RGB color model*, in which a color is sum of the three primary colors — *red*, *green*, and *blue*. A wide range of visible colors can be obtained by adjusting the intensities of the primary colors. A pixel of a color display actually contains three *subpixels*, corresponding to the three primary colors. The *color depth* specifies the number of bits used for a color representation. If one bit is used for each primary color (which means a color can be ether turned on or turned off), the color depth becomes three bits and eight (i.e., 2^3) distinctive colors

FPGA Prototyping by VHDL Examples 2nd ed., Pong P. Chu.
Copyright © 2017, John Wiley & Sons, Inc.

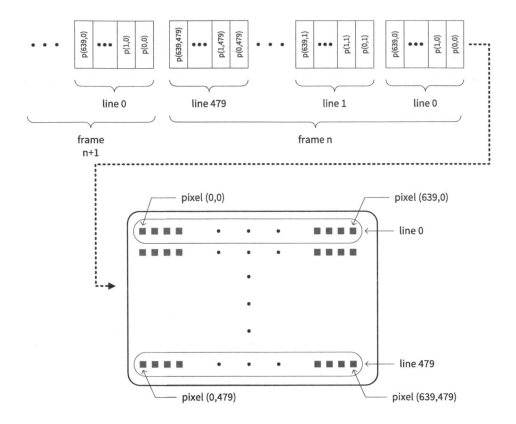

Figure 20.1 Display screen.

can be displayed. Multiple bits can be used to specify the intensity of each primary color. For example, the *true color* mode has a color depth of 24 bits, in which eight bits are used for each primary color. More than 16 million (i.e., 2^{24}) distinctive colors can be displayed. On the other hand, *monochrome* mode uses a single bit for three pixels and the bit is used to turn on or off the three primary colors at the same time. Only two colors, black (all off) and white (all on), can be displayed.

A pixel can retain information for a very short amount of time and must be constantly *refreshed*. The data is fed to the display serially, pixel by pixel and line by line for the entire frame, as shown in Figure 20.1. To obtain flicker-free operation, the screen needs to be refreshed around 60 times per second, a *refreshing rate* of 60 Hz. This implies that a frame of data is transmitted to the display every $\frac{1}{60}$ second.

20.1.2 VGA interface

VGA (Video Graphics Array) is a display hardware introduced with the IBM PS/2 PC in 1980s and is widely supported by PC graphics hardware and monitors. The term VGA is not precisely defined. It may refer to the 15-pin D-subminiature connector, the analog interface standard, or the 640-by-480 resolution. If clarification is needed, we use the terms *VGA port*, *VGA interface*, and *VGA resolution* to emphasize the specific aspect.

Although graphic hardware has improved considerably over time, VGA is considered the lowest common denominator and supported by most graphic cards and monitors. In this book, we use VGA for the video subsystem development. Furthermore, we assume that color depth is 12 bits (i.e., four bits for each color). The selection is based on the configuration of Nexys 4 DDR and Basys 3 boards.

Although VGA is an old standard, the same design principles and concepts can be applied to systems with higher resolutions and larger color depths. The factors that affect the "scaling" are discussed in Section 20.7.

20.2 STREAM INTERFACE

20.2.1 Random-access interface versus stream interface

In the FPro bus, the data transfer between the processor and I/O cores is based on a memory-mapped I/O scheme. This scheme constitutes a *random-access interface*. The term, *random access*, as in *random-access memory (RAM)*, means that a data item can be accessed in the same amount of time regardless the physical location of the data. In a transfer, the *address line* specifies the location and the *control line* indicates the type of transaction (e.g., read or write). The data is then transferred via the *data line* from or to the designated location accordingly.

The data transfer between a video source and a sink (such as the monitor) is very different. The pixel data is arranged in a predefined format and transmitted in a steady, continuous flow. This type of configuration constitutes a *stream interface*. The stream interface is intended to provide a high-speed unidirectional data link between two fixed points. It contains a data line and optional flow-control signals to regulate the data rate. The address line is not needed since the data is transmitted as a large batch in a predefined format. The control lines are not needed since the transfer is one direction and the link is between two fixed entities. In addition to video, the stream interface is used in audio processing, communication links, high-speed serial links, etc.

20.2.2 Flow control of the stream interface

A key task of a stream interface is to regulate the flow rate between the *source* (the subsystem that produces data) and the *sink* (the subsystem that "consumes" and processes the data). We consider three basic schemes of flow control in the book:

- Centralized flow control
- Distributed flow control with back pressure
- Distributed flow control with a FIFO buffer

The centralized flow control scheme uses a master circuit to coordinate the entire operation and there is only a stream data signal between a source and a sink, as shown in the top of Figure 20.2. The master circuit checks the status of subsystems and controls the rate of data generation.

The distributed flow control scheme uses a simple handshake protocol to exchange information between a source and a sink, as shown in the middle of Figure 20.2. A pair of signals, valid and ready, is included in the interface. The sink

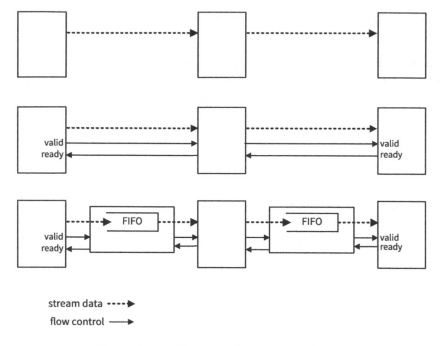

stream data ‒‒‒‒▶

flow control ⟶

Figure 20.2   Flow control conceptual diagrams.

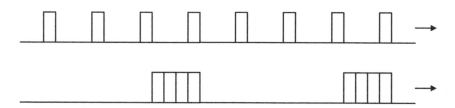

Figure 20.3   Periodic data streams versus burst data streams.

asserts the ready signal when it is ready to accept new data and the source asserts the valid signal when a new data item becomes available. The sink can apply *back pressure* via the handshake to slow down the source data generation.

In a stream interface, the sink's "consumption rate" must be larger or equal to the source's "production rate" to prevent data overflow. However, a source sometimes generates data in a "burst" and its peak rate may exceed the sink's capability. For example, consider an interface between a fast source and a slow sink. The source is running a 40M-Hz clock and can generate data packets at a peak rate of 40M packets per second. On the other hand, the sink is running a 10M-Hz clock and can process 10M packets per second. The source could reduce the rate to match the speed of the sink. The two possible scenarios are shown Figure 20.3. In the first scenario, the source generates packets in a periodic manner, one packet every four clocks, as shown in the top of the diagram. In the second scenario, it generates a four-packet burst and then becomes idle for 12 clocks, as shown in the bottom of the diagram. Although the average rate is still 10M packets per second, the peak rate reaches 40M packets per second.

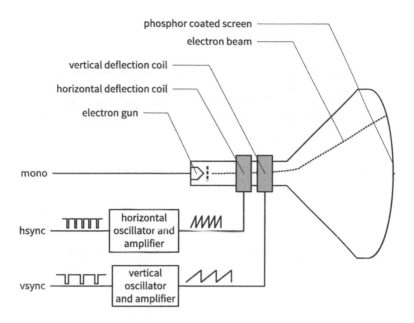

Figure 20.4   Conceptual diagram of a CRT monitor.

Many types of sources, such as an SDRAM, involve a large delay or overhead to obtain the first data item, so generating a burst of data is more efficient and sometimes necessary. To accommodate this type of source, a FIFO buffer can be added as a cushion, as shown in the bottom of Figure 20.2. For example, a four-word FIFO buffer can be used to accommodate the data burst discussed earlier. The empty and full status signals can be used to produce the handshake signals.

20.3   VGA SYNCHRONIZATION

Video data is transmitted in a steady, continuous stream. The term *VGA synchronization* refers to the mechanism to match the onset of a display's internal scanning with the beginning of a frame and the beginning of a line. As stated in Section 20.1, the 640-by-480 VGA resolution is used for the video development in this book. The example and explanation in this section are based on this resolution.

20.3.1   Basic operation of a CRT monitor

The VGA standard was designed for CRT (cathode ray tube) monitors. Although the CRT technology is no longer in common use, most modern LCD (liquid crystal display) monitors still include a VGA port. Studying the basic operation of a CRT monitor can help us understand the concept of video synchronization. A conceptual sketch of a monochrome CRT monitor is shown in Figure 20.4. The electron gun (cathode) generates a focused electron beam, which traverses a vacuum tube and eventually hits the phosphorescent screen. Light is emitted at the instant that electrons hit a phosphor dot on the screen. The intensity of the electron beam and the brightness of the dot are determined by the voltage level of the external video

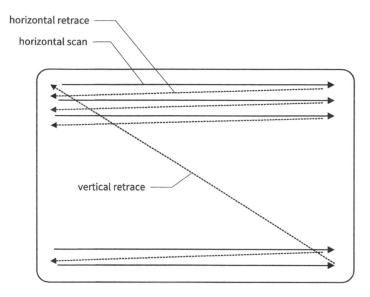

Figure 20.5 CRT scanning pattern.

input signal, labeled `mono` in Figure 20.4. The `mono` signal is an analog signal whose voltage level is between 0 and 0.7 V.

A vertical deflection coil and a horizontal deflection coil outside the tube produce magnetic fields to control how the electron beam travels and to determine where on the screen the electrons hit. The electron beam traverses (i.e., scans) the screen systematically in a fixed pattern, from left to right and from top to bottom, as shown in Figure 20.5.

The monitor's internal oscillators and amplifiers generate sawtooth waveforms to control the two deflection coils. In a horizontal scan, the electron beam moves from the left edge to the right edge as the voltage applied to the horizontal deflection coil gradually increases. After reaching the right edge, the beam returns rapidly to the left edge (i.e., *retraces*) when the voltage changes to 0. The relationship between the sawtooth waveform and the scan is shown in Figure 20.6. An external horizontal synchronization signal, `hsync`, controls the generation of the sawtooth and thus determines the required time to traverse (i.e., scan) a row. The `hsync` is a digital signal and its "1" and "0" periods correspond to the rising and falling ramps of the sawtooth waveform, as shown in Figure 20.6. The vertical scan is performed in a similar way and its operation is controlled by an external vertical synchronization signal, `vsync`.

The operation of a color CRT is similar except that it has three electron beams, which are projected to the red, green, and blue phosphor dots on the screen. The three dots are combined to form a pixel. We can adjust the voltage levels of the three video input signals to obtain the desired pixel color.

20.3.2 Horizontal synchronization

A detailed timing diagram of one horizontal scan and its relationship to the monitor screen are shown in Figure 20.6. Note that the screen of a CRT monitor usually

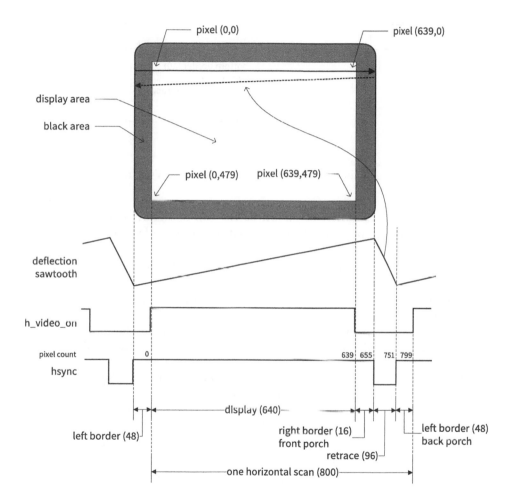

Figure 20.6   Timing diagram of a horizontal scan (with VGA resolution).

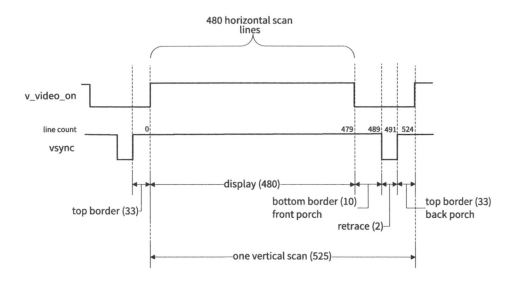

Figure 20.7   Timing diagram of a vertical scan (with VGA resolution).

includes a small black border, as shown at the top of Figure 20.6. The middle rectangle is the visible portion. The coordinates of the top-left and bottom-right corners are (0, 0) and (639, 479), respectively.

A period of the `hsync` signal contains 800 pixels and can be divided into four regions:

- *Display*: region where the pixels are actually displayed on the screen. The length of this region is 640 pixels.
- *Retrace*: region in which the electron beams return to the left edge. The video signal should be disabled (i.e., black) and the length of this region is 96 pixels.
- *Right border*: region that forms the right border of the display region. It is also known as the *front porch* (i.e., porch before retrace). The video signal should be disabled and the length of this region is 16 pixels.
- *Left border*: region that forms the left border of the display region. It is also known as the *back porch* (i.e., porch after retrace). The video signal should be disabled and the length of this region is 48 pixels.

Note that the lengths of the right and left borders may vary for different brands of monitors.

20.3.3   Vertical synchronization

During the vertical scan, the electron beams move gradually from top to bottom and then return to the top. This corresponds to the time required to refresh the entire screen. The format of the `vsync` signal is similar to that of the `hsync` signal, as shown in Figure 20.7. The time unit of the movement is represented in terms of horizontal scan lines. A period of the `vsync` signal is 525 lines and can be divided into four regions:

- *Display*: region where the horizontal lines are actually displayed on the screen. The length of this region is 480 lines.
- *Retrace*: region that the electron beams return to the top of the screen. The video signal should be disabled and the length of this region is 2 lines.
- *Bottom border*: region that forms the bottom border of the display region. It is also known as the *front porch* (i.e., porch before retrace). The video signal should be disabled and the length of this region is 10 lines.
- *Top border*: region that forms the top border of the display region. It is also known as the *back porch* (i.e., porch after retrace). The video signal should be disabled and the length of this region is 33 lines.

As in the horizontal scan, the lengths of the top and bottom borders may vary for different brands of monitors.

20.3.4 Pixel clock rate

The pixel data needs to be fed to the VGA monitor at a specific rate to obtain the desired resolution and refresh rate. The pixel data is transmitted according to a *pixel clock*. Its rate is determined by three parameters:

- *p*: the number of pixels in a horizontal scan line. For 640-by-480 VGA resolution, it is

$$p = 800 \, \frac{pixels}{line}$$

- *l*: the number of lines in a screen (i.e., a vertical scan). For 640-by-480 VGA resolution, it is

$$l = 525 \, \frac{lines}{screen}$$

- *s*: the number of screens per second. For flicker-free operation, we can set it to

$$s = 60 \, \frac{screens}{second}$$

The *s* parameter specifies how fast the screen should be refreshed. For a human eye, the refresh rate must be at least 30 screens per second to make the motion appear to be continuous. To reduce flickering, the monitor usually has a much higher rate, such as the 60 screens per second specification above. The pixel clock rate can be calculated by the three parameters:

$$\text{pixel clock rate} = p * l * s \approx 25M \, \frac{clocks}{second}$$

The pixel data rate is lower than the clock rate because no pixel is generated in the retrace and porch intervals. The data rate is

$$\text{pixel data rate} = 640 * 480 * 60 \approx 18.4M \, \frac{pixels}{second}$$

The pixel clock rate and data rate for other resolutions and refresh rates can be calculated in a similar fashion. Clearly, they increase as the resolution and refresh rate grow.

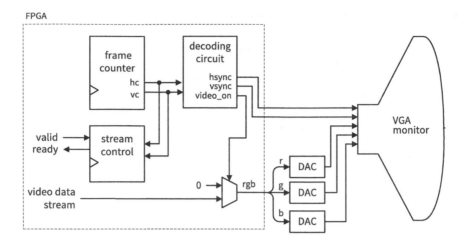

Figure 20.8 Conceptual block diagram of a VGA synchronization circuit.

20.3.5 VGA synchronization circuit

The simplified block diagram of VGA synchronization circuit and VGA monitor is shown in Figure 20.8. The VGA port has five active signals, including the horizontal and vertical synchronization signals, hsync and vsync, and three analog video signals for the red, green, and blue beams. Three DACs convert the video streams to the desired analog levels. The VGA synchronization circuit generates the hsync and vsync signals and synchronizes the incoming video data stream with these signals. It contains a *frame counter*, a *decoding circuit*, and an optional *stream interface control circuit* (for distributed flow control). This section describes the design and construction of the frame counter and the decoding circuit and Chapter 21 discusses the stream interface circuit.

Frame counter A *frame counter* is composed of a *horizontal counter* and a *vertical counter*. The counting patterns resemble the horizontal scan and vertical scan of a screen. The horizontal counter is a mod-800 counter and the counts are marked on the top of the hsync signal in Figure 20.6. We intentionally start the counting from the beginning of the display region. This allows us to use the counter output as the horizontal (x-axis) coordinate. This output constitutes the hcount signal.

The vertical counter is a mod-525 counter, which increments once when the horizontal counter reaches the end of line (i.e., 799). The counts are marked on the top of the vsync signal in Figure 20.7. Again, we intentionally start counting from the beginning of the display region. This allows us to use the counter output as the vertical (y-axis) coordinate. This output constitutes the vcount signal.

The HDL code of the frame counter is shown in Listing 20.1. To make it more flexible, we include an enable signal, inc, to enable or pause the counting, and two status signals, frame_start and frame_end, to indicate the beginning and end of a frame.

Listing 20.1 Frame counter

```
library ieee;
use ieee.std_logic_1164.all;
```

```vhdl
use ieee.numeric_std.all;
entity frame_counter is
    generic(
        HMAX : integer := 640;    -- max horizontal count
        VMAX : integer := 480     -- max vertical count
    );
    port(
        clk, reset   : in  std_logic;
        inc          : in  std_logic;
        sync_clr     : in  std_logic;
        hcount       : out std_logic_vector(10 downto 0);
        vcount       : out std_logic_vector(10 downto 0);
        frame_start  : out std_logic;
        frame_end    : out std_logic
    );
end frame_counter;

architecture arch of frame_counter is
    signal hc_reg, hc_next : unsigned(10 downto 0);
    signal vc_reg, vc_next : unsigned(10 downto 0);
begin
    -- horizontal and vertical pixel counters
    -- registers
    process(clk, reset)
    begin
        if reset = '1' then
            vc_reg <= (others => '0');
            hc_reg <= (others => '0');
        elsif (clk'event and clk = '1') then
            if (sync_clr = '1') then
                vc_reg <= (others => '0');
                hc_reg <= (others => '0');
            else
                vc_reg <= vc_next;
                hc_reg <= hc_next;
            end if;
        end if;
    end process;
    -- next-state logic of horizontal counter
    process(hc_reg, inc)
    begin
        if (inc = '1') then
            if hc_reg = (HMAX - 1) then
                hc_next <= (others => '0');
            else
                hc_next <= hc_reg + 1;
            end if;
        else
            hc_next <= hc_reg;
        end if;
    end process;
    -- next-state logic of vertical counter
    process(vc_reg, hc_reg, inc)
    begin
        if (inc = '1') and hc_reg = (HMAX - 1) then
            if vc_reg = (VMAX - 1) then
                vc_next <= (others => '0');
            else
                vc_next <= vc_reg + 1;
            end if;
        else
```

```
            vc_next <= vc_reg;
       end if;
   end process;
   -- output
   hcount        <= std_logic_vector(hc_reg);
   vcount        <= std_logic_vector(vc_reg);
   frame_start <= '1' when vc_reg = 0 and hc_reg = 0 else '0';
   frame_end   <=
       '1' when vc_reg = (VMAX - 1) and hc_reg = (HMAX - 1) else '0';
end arch;
```

The widths of the two counters should depend on the values of **VMAX** and **HMAX** and be specified by generics. To reduce the clutter in code, we set them to 11 bits, which should be large enough for most resolutions.

Decoding circuit   A decoding circuit generates the **hsync** and **vsync** from the counters. The **hsync** signal goes low when the horizontal counter's output is between 656 and 751, as shown in Figure 20.6. The **vsync** signal goes low when the vertical counter's output is 490 or 491, as shown in Figure 20.7.

In addition to the two signals, the decoding circuit generates a **video_on** signal to indicate whether the current horizontal and vertical coordinates are in the displayable region. The signal turns off the video data (i.e., outputs 0 to make the display black) when the scan retraces or is within the border areas. It is asserted only when the horizontal count is smaller than 640 and the vertical count is smaller than 480, as shown in Figures 20.6 and 20.7.

The HDL code of a demo synchronization circuit is shown in Listing 20.2. We assume that the centralized flow control is used and thus no stream interface control circuit is included.

Listing 20.2   Demo synchronization circuit

```
library ieee;
use ieee.std_logic_1164.all;
use ieee.numeric_std.all;
entity vga_sync_demo is
   generic(CD : integer := 12);      -- color depth
      port(
      clk, reset : in  std_logic;
      -- stream input
      vga_si_rgb : in  std_logic_vector(CD-1 downto 0);
      -- to vga monitor
      hsync   : out std_logic;
      vsync   : out std_logic;
      rgb     : out std_logic_vector(CD-1 downto 0);
      -- frame counter output
      hc, vc : out std_logic_vector(10 downto 0)
   );
end vga_sync_demo;

architecture arch of vga_sync_demo is
   -- vga 640-by-480 sync parameters
   constant HD : integer := 640;    -- horizontal display area
   constant HF : integer := 16;     -- h. front porch
   constant HB : integer := 48;     -- h. back porch
   constant HR : integer := 96;     -- h. retrace
   constant HT : integer := HD+HF+HB+HR; -- horizontal total (800)
   constant VD : integer := 480;    -- vertical display area
   constant VF : integer := 10;     -- v. front porch
```

```vhdl
    constant VB : integer := 33;    -- v. back porch
    constant VR : integer := 2;     -- v. retrace
    constant VT : integer := VD+VF+VB+VR; -- vertical total (525)
    -- sync counetr and signals
    signal x, y       : unsigned(10 downto 0);
    signal hcount     : std_logic_vector(10 downto 0);
    signal vcount     : std_logic_vector(10 downto 0);
    signal hsync_i    : std_logic;
    signal vsync_i    : std_logic;
    signal video_on_i : std_logic;
    signal q_reg      : unsigned(1 downto 0);
    signal tick_25M   : std_logic;
begin
    -- 25M Hz pixel tick generation
    process(clk)
    begin
        if (clk'event and clk = '1') then
            q_reg <= q_reg + 1;
        end if;
    end process;
    tick_25M <= '1' when q_reg = "11" else '0';
    -- instantiate frame counter
    counter_unit : entity work.frame_counter
        generic map(
            HMAX => HT,
            VMAX => VT
        )
        port map(
            clk         => clk,
            reset       => reset,
            sync_clr    => '0',
            hcount      => hcount,
            vcount      => vcount,
            inc         => tick_25M,
            frame_start => open,
            frame_end   => open
        );
    x  <= unsigned(hcount);
    y  <= unsigned(vcount);
    hc <= hcount;
    vc <= vcount;
    -- horizontal sync decoding
    hsync_i <= '0' when (x>=(HD+HF)) and (x<=(HD+HF+HR-1)) else '1';
    -- vertical sync decoding
    vsync_i <= '0' when (y>=(VD+VF)) and (y<=(VD+VF+VR-1)) else '1';
    -- display on/off
    video_on_i <= '1' when (x < HD) and (y < VD) else '0';
    -- buffered output to vga monitor
    process(clk)
    begin
        if (clk'event and clk = '1') then
            vsync <= vsync_i;
            hsync <= hsync_i;
            if (video_on_i = '1') then
                rgb <= vga_si_rgb;
            else
                rgb <= (others => '0');    -- black when display off
            end if;
        end if;
    end process;
end arch;
```

Figure 20.9   Bar testing screen.

The code instantiates a frame counter and uses the counter's outputs for decoding. The decoded signals and stream data are fed to a register to remove potential glitches. The VGA resolution imposes a pixel clock rate of 25M Hz, as discussed in Section 20.3.4. Since a system clock is 100M Hz, we can construct a two-bit mod-4 counter to obtain a 25M-Hz tick signal to control the increment of the frame counter. The circuit also outputs the frame counter's values for reference.

Note that this clock-divider approach does not work in general since the system clock rate is usually not a multiple of the pixel clock rate. A better, more robust alternative is discussed in Chapter 21.

A simple test can be performed by connecting 12 switches to the `vga_si_rgb` port, which leads to a video stream of constant values. It generates a screen of a single color based on the switch values.

20.4 BAR TEST-PATTERN GENERATOR

We consider two types of modules in our stream video system — *pixel generation circuit* and *pixel transformation circuit*. A pixel generation circuit is a video source that generates a graphic frame to be displayed on the screen. A pixel transformation circuit performs certain operation on the pixel stream. We use a bar test-pattern generator circuit to demonstrate the construction of a pixel generation circuit in this section.

The bar test-pattern generator generates a screen with bars of available colors and bars of different shades of gray, as shown in Figure 20.9. The screen is divided into three regions. The top region demonstrates the *grayscale*, in which the intensities of red, green, and blue colors are identical. For a color depth of 12 bits, four bits is used for each color and thus 16 (i.e., 2^4) shades of gray can be generated.

The second region shows the primary colors and their combinations, which are red, green, blue, black (all off), cyan (green plus blue), yellow (red plus green), magenta (red plus blue), and white (all on). The bottom region displays a "rainbow color spectrum," similar to that in Figure 13.5.

The inputs of the generator are x and y, representing the x- and y-coordinates of the current pixel. Based on the values of coordinates, a color is produced accordingly. The y signal is used to determine the region. In regions 1 and 2, the MSBs of x set the shades or primary colors and implicitly form the bar. In region 3, the MSBs of x define the "segments" of Figure 13.5 and the LSBs derive the "ramps" of the three colors. The HDL code of the generator is shown in Listing 20.3.

Listing 20.3 Bar test-pattern generator

```vhdl
library ieee;
use ieee.std_logic_1164.all;
use ieee.numeric_std.all;
entity bar_demo is
   port(
      x, y    : in  std_logic_vector(10 downto 0);
      bar_rgb : out std_logic_vector(11 downto 0)
   );
end bar_demo;

architecture arch of bar_demo is
   -- intermediate counter value
   signal up, down : std_logic_vector(3 downto 0);
   -- misc signals
   signal r, g, b  : std_logic_vector(3 downto 0);
   -- delay line
begin
   -- pixel data generation
   up   <= x(6 downto 3);
   down <= not x(6 downto 3);   -- "not" reverses the binary sequence
   process(x, y, up, down)
   begin
      -- 16 shades of gray
      if unsigned(y) < 128 then
         r <= x(8 downto 5);
         g <= x(8 downto 5);
         b <= x(8 downto 5);
      -- 8 prime color with 50% intensity
      elsif unsigned(y) < 256 then
         r <= x(8) & x(8) & "00";
         g <= x(7) & x(7) & "00";
         b <= x(6) & x(6) & "00";
      -- a continuous color spectrum
      else
         case x(9 downto 7) is
            when "000" =>
               r <= (others => '1');
               g <= up;
               b <= "0000";
            when "001" =>
               r <= down;
               g <= "1111";
               b <= "0000";
            when "010" =>
               r <= "0000";
               g <= "1111";
               b <= up;
            when "011" =>
               r <= "0000";
               g <= down;
               b <= "1111";
            when "100" =>
               r <= up;
               g <= "0000";
               b <= "1111";
            when "101" =>
               r <= "1111";
               g <= "0000";
               b <= down;
            when others =>
```

```
                    r <= "1111";
                    g <= "1111";
                    b <= "1111";
             end case;
         end if;
      end process;
      bar_rgb <= r & g & b;
end arch;
```

If x and y are connected to the horizontal count and vertical count of a frame counter, the circuit generates a video pixel stream resembling that in Figure 20.1.

20.5 COLOR-TO-GRAYSCALE CONVERSION CIRCUIT

Video and image processing consist of many levels. A *pixel transformation circuit* performs *pixel-level* conversions and transformations, such as color space conversion, brightness adjustment, and edge detection. In this section, we use a color-to-grayscale conversion circuit for demonstration.

A video display is driven by red, green and blue signals and thus each color pixel is represented by a triple of r, g, and b, which corresponds to the intensities of the respective colors. On the other hand, the grayscale represents the overall light intensity or brightness of a pixel. Three color-to-grayscale conversion methods are suggested by GIMP (GNU image manipulation program):

- Lightness method: $gray = 0.5 * \max(r, g, b) + 0.5 * \min(r, g, b)$
- Average method: $gray = 0.33 * r + 0.33 * g + 0.33 * b$
- Luminosity method: $gray = 0.21 * r + 0.72 * g + 0.07 * b$

The lightness method takes the average of the most prominent and least prominent colors and the average method takes the average of three colors. The luminosity method uses a weighted average to account for human perception since human eyes are more sensitive to green and less sensitive to blue. The same formula is also used to obtain the Y component in the $YCrCb$ color space, which is used for video transmission and storage.

We use the luminosity method to construct a color-to-grayscale circuit that converts a color image to a "black-and-white" (i.e., grayscale) image. The HDL of the conversion circuit is shown in Listing 20.4.

Listing 20.4   Color-to-grayscale conversion circuit

```
library ieee;
use ieee.std_logic_1164.all;
use ieee.numeric_std.all;

entity rgb2gray is
   port(
      color_rgb : in std_logic_vector(11 downto 0);
      gray_rgb  : out std_logic_vector(11 downto 0)
   );
end rgb2gray;

architecture arch of rgb2gray is
   signal r, g, b : unsigned(3 downto 0);
   signal gray    : std_logic_vector(3 downto 0);
   signal gray12  : unsigned(11 downto 0);
   constant RW : unsigned(7 downto 0) := x"35"; --weight for red
```

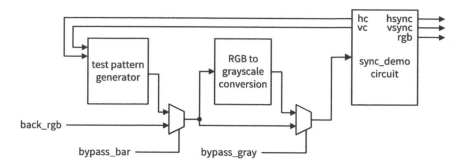

Figure 20.10   Block diagram of a demo video system.

```
   constant GW : unsigned(7 downto 0) := x"b8"; --weight for green
   constant BW : unsigned(7 downto 0) := x"12"; --weight for blue
begin
   r <= unsigned(color_rgb(11 downto 8));
   g <= unsigned(color_rgb(7 downto 4));
   b <= unsigned(color_rgb(3 downto 0));
   gray12 <= r*RW + g*GW + b*BW;
   gray <= std_logic_vector(gray12(11 downto 8));
   gray_rgb <= gray & gray & gray;
end arch;
```

The code uses fixed-point arithmetic operation. We assume that the r, g, and b signals are in a **Q4.0** format and the constant weights are in a **Q0.8** format. The resulting grayscale signal, gray12, is in a **Q4.8** format, and then trimmed to a **Q4.0** gray signal.

20.6   DEMO VIDEO SYSTEM

We create a simple system to demonstrate the basic operation of video stream processing and generation. The top-level diagram is shown in Figure 20.10. It is composed of a test-pattern generator, a color-to-grayscale conversion circuit, a video synchronization circuit, and two multiplexers. The first multiplexer selects the background color or the test-pattern generator's output. It can also be interpreted as whether to bypass the generator. The second multiplexer selects the color or grayscale test screen. It can also be interpreted as whether to bypass the conversion circuit.

Since both the test-pattern generator and the color-to-grayscale conversion circuit are combinational circuits, the pixel stream is produced and consumed at the same time. The stream interface implicitly uses the centralized flow control scheme discussed in Section 20.2. The frame counter of the sync_demo circuit drives both the generator and synchronization circuit and can be considered as the "centralized controller."

The HDL follows the block diagram and is shown in Listing 20.5. The bypass signals and background colors are connected to the discrete switches for physical testing.

Listing 20.5 Demo video system

```vhdl
library ieee;
use ieee.std_logic_1164.all;
use ieee.numeric_std.all;
entity vga_demo is
   generic(
      CD : integer := 12   -- color depth
   );
   port(
      clk   : in  std_logic;
      -- switch control
      sw    : in  std_logic_vector(13 downto 0);
      -- to vga monitor
      hsync : out std_logic;
      vsync : out std_logic;
      rgb   : out std_logic_vector(CD - 1 downto 0)
   );
end vga_demo;

architecture arch of vga_demo is
   signal hc, vc        : std_logic_vector(10 downto 0);
   signal bar_rgb       : std_logic_vector(CD - 1 downto 0);
   signal back_rgb      : std_logic_vector(CD - 1 downto 0);
   signal gray_rgb      : std_logic_vector(CD - 1 downto 0);
   signal color_rgb     : std_logic_vector(CD - 1 downto 0);
   signal vga_rgb       : std_logic_vector(CD - 1 downto 0);
   signal bypass_bar    : std_logic;
   signal bypass_gray   : std_logic;
begin
   -- use switches to set background color
   back_rgb     <= sw(13 downto 2);
   bypass_bar   <= sw(1);
   bypass_gray  <= sw(0);
   -- instantiate bar generator
   bar_demo_unit : entity work.bar_demo
      port map(
         x       => hc,
         y       => vc,
         bar_rgb => bar_rgb
      );
   -- instantiate color-to-gray conversion circuit
   c2g_unit : entity work.rgb2gray
      port map(
         color_rgb => color_rgb,
         gray_rgb  => gray_rgb
      );
   -- instantiate video synchronization circuit
   sync_unit : entity work.vga_sync_demo
      generic map(CD => CD)
      port map(
         clk        => clk,
         reset      => '0',
         vga_si_rgb => vga_rgb,
         hsync      => hsync,
         vsync      => vsync,
         rgb        => rgb,
         hc         => hc,
         vc         => vc
      );
   -- video source selection mux #1
   color_rgb <= back_rgb when bypass_bar = '1' else bar_rgb;
```

Table 20.1   Main characteristics of video display standards

Standard	Resolution	Pixels	Clock rate	Frame buffer	Bandwidth
VGA	640-by-480	0.31 M	25M Hz	0.93 MB	55.3 MB/s
SVGA	800-by-600	0.48 M	40M Hz	1.44 MB	86.4 MB/s
XGA	1024-by-768	0.79 M	65M Hz	2.36 MB	141.3 MB/s
SXGA	1280-by-1024	1.31 M	108M Hz	3.93 MB	235.8 MB/s
UXGA	1600-by-1200	1.92 M	162M Hz	5.76 MB	345.6 MB/s
Full HD	1920-by-1080	2.07 M	165M Hz	6.22 MB	373.2 MB/s

```
   — video source selection mux #0
   vga_rgb <= color_rgb when bypass_gray = '1' else gray_rgb;
end arch;
```

20.7   ADVANCED VIDEO STANDARDS

The VGA is the most basic video standard. After its introduction, the resolution and performance of video display improved continuously. Several key standards, including SVGA (super VGA), XGA (extended graphics array), SXGA (super XGA), UXGA (Ultra XGA), and full HD (high-definition), and their main characteristics are listed in Table 20.1.

The resolution is the key parameter. The other aspects are derived from the resolution. A refresh rate of 60 frames per second is used for calculation. The "pixels" column shows the total number of pixels in a frame. The "clock rate" column specifies the pixel clock rate. The values are obtained following the procedure in Section 20.3.4. Since the screen is refreshed at the same rate, a higher resolution leads to a higher clock rate.

The "frame buffer" column shows the required memory space to store a frame. It is determined by

$$buffer\_size = pixel\_per\_frame * color\_depth/8$$

The "data bandwidth" column indicates to the pixel data processed or transmitted per second. It is derived as follows

$$bandwidth = frame\_per\_second * pixel\_per\_frame * color\_depth/8$$

In the table, the "true-color" format, which has a color depth of 24 bits, is used for calculation for the last two columns.

In addition to resolution, newer standards, such as HDMI and DVI, use digital interface to transmit pixel data to the video display. Conceptually, the digital interface replaces three DACs in Figure 20.8 with three *SerDes* (serializer/deserializer) blocks. The blocks convert the parallel data to serial data, perform 8b/10b line encoding, and then transmit the data through three high-speed serial links.

A high-performance memory system is needed to support the bandwidth and capacity imposed by the advanced video standards. Implementing this type of system is very involved, as discussed in Section 7.5. Furthermore, the memory controller and interface circuit use proprietary IP cores and the implementation is usually device and board dependent. Since the goal of the book is to develop portable codes and illustrate the key hardware concepts, we use the basic VGA standard in Part IV. However, the same concepts and design practices can be "scaled" and applied to a high-performance video system as well.

20.8 BIBLIOGRAPHIC NOTES

The VGA standard is the most commonly used analog video interface. The Wikipedia website provides detailed technical specifications and lists the available resolutions. Image processing is quite complex and involves many sophisticated algorithms. *Design for Embedded Image Processing on FPGAs* by D. G. Bailey discusses the basic concepts and algorithms and demonstrates how to use FPGA to realize these algorithms.

20.9 SUGGESTED EXPERIMENTS

20.9.1 Horizontal bar test-pattern generator

The test-pattern generator discussed in Section 20.4 generates a screen composed of three horizontal regions of vertical bars. Design an alternative screen that is composed of three vertical regions of horizontal bars and verify its operation.

20.9.2 Color channel selection circuit

A color channel selection circuit separates a color channel (which can be red, green, or blue) from the pixel data and blocks the other two colors (by making them 0's). Incorporate this circuit into the demo video system in Section 20.6, and verify its operation.

20.9.3 Enhanced color-to-grayscale conversion circuit

The color-to-grayscale conversion circuit discussed in Section 20.5 is based on the luminosity method. We wish to expand the circuit to include the two other methods. A selection signal should be included to select the desired method. Design the new circuit, incorporate this circuit into the demo video system in Section 20.6, and verify its operation.

20.9.4 Square test-pattern generator: part I

A square test-pattern generator produces a screen as follows:
- A square lies on the middle of the screen.
- The side of the square can be 16, 32, 64, or 128 pixels and can be selected by switches.
- The color of the square can be set by switches.
- The background displays the *complement color*. The complement color of (r, g, b) is $(1111 - r, 1111 - g, 1111 - b)$.

Design the new circuit, replace the bar test-pattern generator in the demo video system in Section 20.6, and verify its operation.

20.9.5   Square test-pattern generator: part II

In Experiment 20.9.4, the color of the square is set by switches. We can enhance the test-pattern generator by circulating a spectrum of colors through the square and its complementary background. This can be done by a circuit that generates the colors sequentially from "rainbow spectrum" of Figure 13.5. Design the new circuit, incorporate it into the demo video system in Section 20.6, and verify its operation.

20.9.6   Square test-pattern generator: part III

In Experiment 20.9.4, the location of the square is fixed. We can enhance the test-pattern generator by animating the square. The square normally moves on a straight line. When reaching the boundary of the screen, it bounces back like a ball. The angle of reflection should be incremented each time to avoid repeated path. Design the new circuit, incorporate it into the demo video system in Section 20.6, and verify its operation.

20.9.7   Square test-pattern generator: part IV

We can combine the features in Experiments 20.9.5 and 20.9.6. In the new generator, the square will advanced to the next color in the rainbow spectrum when it hits a boundary. Design the new circuit, incorporate it into the demo video system in Section 20.6, and verify its operation.

CHAPTER 21

FPRO VIDEO SUBSYSTEM

The ad hoc video subsystem in Chapter 20 demonstrates the basic concepts of video stream processing. This chapter introduces the video IP core architecture used in this book and illustrates the clock-domain-crossing techniques to accommodate different pixel clock rates. The systematic approach helps us construct a more versatile and robust subsystem and encourages the IP reuse.

21.1 ORGANIZATION OF THE VIDEO SUBSYSTEM

The video subsystem in Chapter 20 is constructed in an ad hoc manner. We introduce a more systematic approach in the chapter and create a video subsystem that can be integrated into the FPro bus. The framework defines the general architecture and interface of an *FPro video IP core* and provides an easy way to integrate and reuse the cores to construct a custom embedded video subsystem.

21.1.1 Overview

The top-level FPro system diagram in Figure 8.2 is repeated in Figure 21.1. The bottom portion is the video subsystem. The main functionality of the subsystem is to generate, process, and display pixel data. It contains a collection of *FPro video IP cores* and a *video controller*.

An FPro video core inputs a pixel stream, performs computation or adds new contents to the stream, and then outputs the processed pixel stream. The video

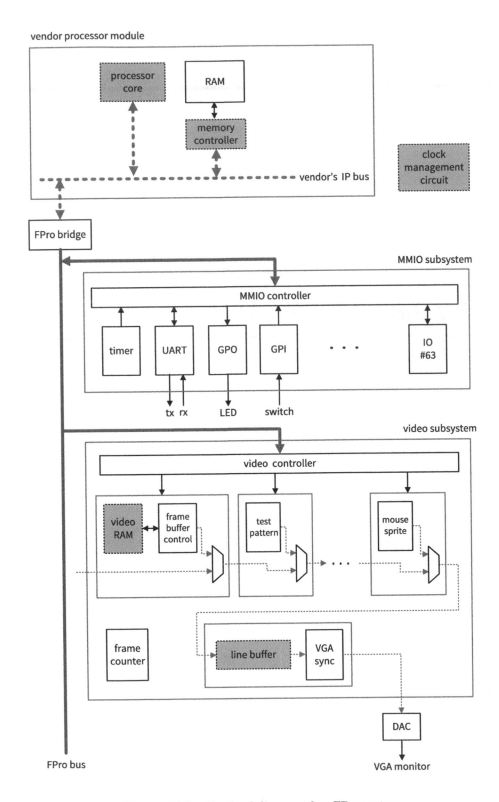

Figure 21.1 Top-level diagram of an FPro system

cores are arranged as a *cascading chain* and the pixel data travels through the chain stage by stage. The cores in the left are video sources that generate full-screen pixel data and stream it to the cores in the following stages. The cores in the middle perform certain transformations on the pixels or augment the frame with additional graphics. The stream is eventually routed to the video synchronization circuit, which is the rightmost core in the chain, and displayed on a monitor.

The cascading chain of video cores forms a dedicated path for the pixel data stream. To maintain a stable image on the monitor, the cores in the cascading chain must generate or process the pixel data at a constant data rate, which is 18.4M pixels per second for the VGA resolution, as discussed in Section 20.3.4. The nature of the dedicated stream data path is very different from the traffic flow in the MMIO subsystem, in which an MMIO I/O core exchanges data with the processor via the FPro bus sporadically.

21.1.2 Video controller

A video core may contain a collection of control registers and maintain a local memory to store image patterns. While the processor does not interact with the pixel stream directly, it writes the video core's registers to configure its operation and updates its local memory with new patterns. The functionality of the *video controller* of the video subsystem is similar to the MMIO controller of the MMIO subsystem discussed in Section 10.5. It decodes the address and writes the data to the designated video core. Since a video core rarely connects to an external stimulus, it does not need to transmit data to a processor. Therefore, the controller does not implement the read functionality.

Recall that the MMIO subsystem can accommodate up to 64 (i.e., 2^6) I/O cores, each with 32 (i.e., 2^5) registers. Thus, the subsystem takes a 2^{11}-word memory space. Because a video core may contain a local memory for image pattern storage, it needs more memory space. On the other hand, the needed functionalities of the video subsystem are less diverse and thus the number of video cores in the subsystem is smaller. In addition to normal cores, the video subsystem includes a *frame buffer core*, which stores the pixel data of one screen (i.e., a frame). The memory space of the frame buffer core must be large enough to meet the storage requirement. Based on these criteria, we can allocate the video subsystem memory space as follows:

- Eight (i.e., 2^3) slots for video cores, each with 16K (i.e., 2^{14}) words.
- One frame buffer core with 1M (i.e., 2^{20}) words.

The FPro system uses a 22-bit word space (i.e., 24-bit byte space) for the two subsystems. Its memory map (in terms of word) is:

- `0x xxxx xxxx xmmm mmma aaaa`: 64 MMIO I/O cores, each with a 2^5-word address space
- `10 xxxv vvaa aaaa aaaa aaaa`: 8 video cores, each with a 2^{14}-word address space
- `11 aaaa aaaa aaaa aaaa aaaa`: 1 frame buffer core with a 2^{20}-word address space

In the expression, x represents a "don't-care" bit, a represents a local address bit, m represents a slot id bit of an MMIO core, and v represents a slot id bits of a video core. The map indicates the following:

- MSB is used to distinguish the MMIO subsystem and video subsystem (0 for the MMIO subsystem and 1 for the video subsystem).
- In the video subsystem (in which MSB is 1), the second MSB is used to distinguish normal video cores and the frame buffer core (0 for a normal video core and 1 for the frame buffer core).

The physical implementation of memory decoding of an FPro system is distributed in three components. The decoding of the MSB is done in the FPro bridge of Figure 21.1, which generates the `fp_mmio_cs` and `fp_video_cs` signals to enable the designated subsystem, as shown in Listing 10.7. The decoding within the MMIO subsystem is done by MMIO controller and its HDL code is shown in Listing 10.5. The decoding within the video subsystem is done by the video controller. Its detailed implementation is discussed in the next subsection.

The memory allocation is aimed to simplify the address decoding and data routing. A large fraction of the space is not used. However, the 24-bit byte memory space only counts for $\frac{1}{256}$ (i.e., $\frac{2^{24}}{2^{32}}$) of processor's 32-bit memory space. It does not cause any serious issue in an embedded system.

21.1.3   HDL of the video controller

The design and implementation of the video controller are similar to those of the MMIO controller. There are two modifications. First, since the video subsystem does not output data to the processor, no read signal or read data bus is needed. Second, because the memory mapping of the frame buffer core is different from the normal video slots, it is treated as a special case with a separate write interface and decoding logic.

As for the MMIO controller, two new data types are defined in the `chu_io_map` package to facilitate the video core decoding:

```
type slot_2d_video_data_type is array (7 downto 0) of
     std_logic_vector(31 downto 0);
type slot_2d_video_reg_type is array (7 downto 0) of
     std_logic_vector(13 downto 0);
```

The HDL code of the video controller is shown in Listing 21.1.

Listing 21.1   Video controller

```
library ieee;
use ieee.std_logic_1164.all;
use ieee.numeric_std.all;
use work.chu_io_map.all;

entity chu_video_controller is
   port(
      -- FPro bus
      video_cs          : in  std_logic;
      video_wr          : in  std_logic;
      video_addr        : in  std_logic_vector(20 downto 0);
      video_wr_data     : in  std_logic_vector(31 downto 0);
      -- MM frame buffer interface
      frame_cs          : out std_logic;
      frame_wr          : out std_logic;
      frame_addr        : out std_logic_vector(19 downto 0);
      frame_wr_data     : out std_logic_vector(31 downto 0);
      -- MM video core interface
```

```
      slot_cs_array         : out std_logic_vector(7 downto 0);
      slot_mem_wr_array     : out std_logic_vector(7 downto 0);
      slot_reg_addr_array   : out slot_2d_video_reg_type;
      slot_wr_data_array    : out slot_2d_video_data_type
   );
end chu_video_controller;

architecture arch of chu_video_controller is
   alias slot_addr : std_logic_vector(2 downto 0) is video_addr(16 downto 14);
   alias reg_addr  : std_logic_vector(13 downto 0) is video_addr(13 downto 0);
   signal slot_cs  : std_logic;

begin
   -- address decoding
   frame_cs <= '1' when video_cs = '1' and video_addr(20) = '1' else '0';
   slot_cs  <= '1' when video_cs = '1' and video_addr(20) = '0' else '0';
   process(slot_addr, slot_cs)
   begin
      slot_cs_array <= (others => '0');
      if slot_cs = '1' then
         slot_cs_array(to_integer(unsigned(slot_addr))) <= '1';
      end if;
   end process;
   -- to frame buffer
   frame_addr       <= video_addr(19 downto 0);
   frame_wr         <= video_wr;
   frame_wr_data    <= video_wr_data;
   -- to normal video cores
   slot_mem_wr_array   <= (others => video_wr);
   slot_wr_data_array  <= (others => video_wr_data);
   slot_reg_addr_array <= (others => reg_addr);
end arch;
```

21.2 FPRO VIDEO IP CORE

An FPro video IP core uses the similar architecture and utilizes the same interface.
The core can be used as the basic building block of a video subsystem. It simplifies
the integration and enhances the modularity and reuse.

21.2.1 Basic functionality

An FPro video core inputs a pixel stream, performs computation or adds new
contents to the stream, and then outputs the processed pixel stream. Based on the
functionality, it can be either a *pixel transformation core* or a *pixel generation core*.

A pixel transformation core performs certain computation on the incoming pixel
stream and transforms it into a new format. The color-to-grayscale conversion
operation discussed in Section 20.5 is a good example. The simplified conceptual
diagram is shown in Figure 21.2(a). The transformed pixel stream and the original
pixel stream are connected to a multiplexer in the output stage. The multiplexer
selects and routes the desired stream to the next core in the chain. It provides a
bypassing mechanism to skip the transformation when desirable.

A pixel generation core functions as a video source and generates a new pixel
stream. The stream is *blended* with the incoming stream to form a new stream. The
test pattern generation operation discussed in Section 20.4 is a good example. The

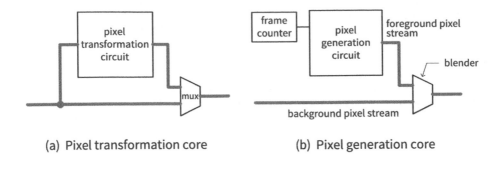

(a) Pixel transformation core   (b) Pixel generation core

Figure 21.2   Two types of FPro video core.

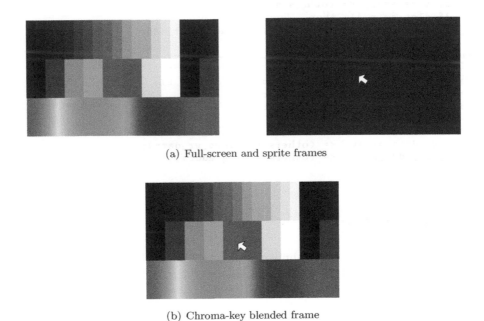

(a) Full-screen and sprite frames

(b) Chroma-key blended frame

Figure 21.3   Chroma-key demonstration.

simplified conceptual diagram is shown in Figure 21.2(b). It includes a *blending circuit* in the output stage. The core can produces pixel data for the entire screen or a small bit-mapped image, known as a *sprite*. A representative full-screen frame (the test pattern) and a sprite frame (a mouse pointer) are shown in Figure 21.3(a).

21.2.2   Blending operation

Blending is the process to combine two video frames into a single frame. To facilitate the discussion, we can treat the two frames as two vertical layers. The top layer and bottom layer are known as a *foreground* frame and a *background* frame, respectively. An example is shown in Figure 21.3(a), in which the foreground frame is a mouse pointer sprite and the background frame is the test-pattern screen. There are several commonly used blending schemes:

- multiplexing (also known as binary alpha blending)
- α blending (alpha blending)
- chroma-key blending
- chroma-key α blending

In the FPro video subsystem, a pixel is composed with three color channels. Without loss of generality, we use f and b to represent a color of the foreground pixel and a color of the background pixel, respectively, and use r for the blended result.

The *multiplexing scheme* treats the frames as two independent video sources and selects one source and routes it to the output, just like a normal multiplexer. To be incorporated into the blending framework, it can be mathematically expressed as

$$r = \alpha f + (1 - \alpha)b, \text{ where } \alpha = 0 \text{ or } \alpha = 1$$

The α can only assume a value of 0 or 1 in this scheme (and thus it is also known as *binary alpha blending*). The foreground frame is selected if it is 1 and the background frame is selected if it is 0. There is actually no blending or mixing between the two frames.

The α *blending scheme* combines the pixels from two frames and sets the relative "weight" of each frame. Its formula is

$$r = \alpha f + (1.0 - \alpha)b, \text{ where } 0.0 \leq \alpha \leq 1.0$$

Note that α now is a continuous value between 0.0 and 1.0. The α parameter can be interpreted as the "opacity level" of the foreground frame, which is the opposite of the "transparency level" of the background frame. When it is 1.0, the foreground frame is completely opaque and the background frame does not show at all. As its value decreases, the background frame starts to show. The background frame becomes more dominant as α continues to decrease and eventually becomes completely transparent as α reaches 0.0.

The *chroma-key blending scheme* is a technique to display a selective region of foreground frame on top of a background frame. In this scheme, a special pre-selected color is defined as the *chroma key*. The chroma key is used as the "background color" of a foreground frame. When two frames are blended, the pixels with the chroma key of the foreground frame are replaced by the corresponding pixels of the background frame. A simple example is demonstrated in Figure 21.3, in which the black color is defined as the chroma key. The foreground frame contains a simple mouse pointer in black background, as shown in the right panel of Figure 21.3(a). When it is blended with the background frame, only the pointer maintains its pixels and the black pixels are replaced with pixels from of the background frame, as shown in Figure 21.3(b). The scheme can be applied multiple times to form a multi-layered image. Let us assume that the chroma key is C_K, then the chroma-key scheme can be mathematically expressed as

$$r = \begin{cases} f & \text{if } f \neq C_K \\ b & \text{if } f = C_K \end{cases}$$

In video production, the green color is frequently used for the chroma key and this technique is also referred to as the *green-screen* scheme.

It is possible to combine the chroma-key blending and α-blending schemes together to make the selective region of the foreground frame semi-opaque. The equation becomes

$$r = \begin{cases} \alpha f + (1.0 - \alpha)b & \text{if } f \neq C_K \\ b & \text{if } f = C_K \end{cases}$$

21.2.3   Core architecture

The basic sketches of the video cores are shown in Figure 21.2. We define a more elaborated core architecture in this section. Because in this book we are mainly interested in the development of the pixel generation cores, the discussion is focused on this type of core. However, the concepts can be applied to the pixel transformation core as well.

Based on the nature of processing, an FPro video core can use a global or local frame counter. The block diagrams of the two basic types are shown in Figure 21.4. The main components are

- Pixel generation circuit
- Blender
- Video slot wrapping circuit
- Optional local frame-counter and FIFO buffers

The blender merges the original pixel stream and the processed pixel stream. It is a circuit incorporating a blending algorithm similar to those discussed in Section 21.2.2. In addition to the blending operation, the circuit always incorporates the multiplexing functionality, which provides a *bypass mechanism* to skip the current operation. To maintain consistency, we include a one-bit *bypass control register* at the word location of 0x2000 in the wrapping interface of the video core.

The pixel data stream does not connect to the FPro's I/O bus. However, the processor may need to write the video core to configure the pixel processing circuit or to control the blending operation. The FPro slot wrapping circuit is the interface circuit to the FPro video controller. It consists of a collection of registers and a decoding circuit and its design is similar to the write interface circuit of the MMIO core discussed in Section 10.2.

The pixel generation circuit gets the x- and y-coordinates as inputs and produces a pixel accordingly. The coordinates are generated by a frame counter. The video core can utilize an external frame counter or have a dedicated frame counter residing within the module. The first option is shown in Figure 21.4(a). It allows multiple cores to share the same frame counter and simplifies the stream interface between cores. The simple centralized flow control scheme discussed in Section 20.2.2 can be used for this option. Most video cores discussed in this book utilize this option.

Sometimes it is not possible to synchronize all core operations. The second option includes a local, independent frame counter within the core, as shown in Figure 21.4(b). The lack of "global" information requires more sophisticated stream interface, such as the distributed schemes discussed in Section 20.2.2. This scheme leads to an additional input FIFO buffer and a local frame counter. The video synchronization core discussed in Section 21.4 uses this option.

A complete video subsystem can be constructed by cascading the stream interfaces of individual video cores, as shown in the bottom of Figure 21.1. Each core

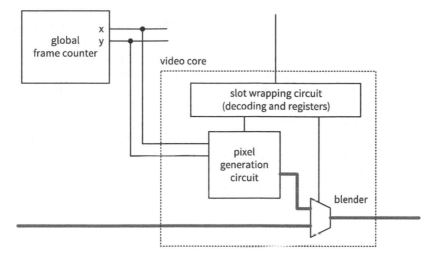

(a) Core with a global frame counter

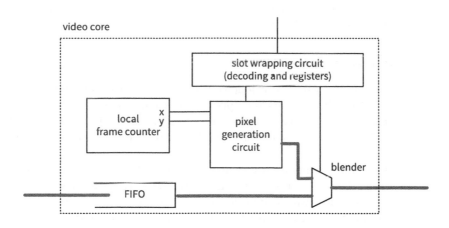

(b) Core with a local frame counter

Figure 21.4   Block diagram of an FPro video core.

performs an additional transformation or adds an additional overlay on top of the current frame.

21.2.4 Alternative core partition

An FPro video core is constructed with a pixel processing circuit, a blender, a slot wrapping circuit, and an optional FIFO buffer. The architecture imposes a single stream input and a single stream output. This leads to a video subsystem of cascading cores, as shown in the bottom of Figure 21.1. The straightforward configuration simplifies the coding and reduces potential errors. However, this configuration may occasionally be too restricted and limits the potential for parallel operations, such as routing a video source to several pixel processing chains concurrently.

An alternative partition scheme is to reduce the "granularity" of the cores. We can define the pixel processing circuit, the blender, and the FIFO buffer as a core, each with its own slot wrapping circuit. Although this partition scheme is more flexible, there will be many more components and stream data routing in the system level. This is doable with schematic entry. However, since the FPro system is derived with textual HDL codes, the fine partition makes the HDL code development tedious and error prone. Thus, this alternative scheme is not used in this book.

21.3 EXAMPLE VIDEO CORES

The ad hoc video subsystem in Chapter 20 contains a color-to-grayscale conversion circuit and a bar test-pattern generator. The following subsections demonstrate the procedure to construct FPro video cores from the original circuits.

21.3.1 Bar test-pattern generator core

The bar test-pattern generator core follows the basic structure of Figure 21.4(a) and is composed of the bar test-pattern pixel generation circuit, a blender, and a slot wrapping circuit. It produces a full screen of pixels shown in the left panel of Figure 21.3(a). A global frame counter is used to supply the x- and y-coordinate signals.

The test-pattern pixel generation circuit is similar to the one discussed in Section 20.4. However, the video subsystem constructed in Section 21.5 assumes that a pixel generation core takes two clock cycles to produce a pixel data after obtaining the x- and y-coordinate signals from the global frame counter. A two-stage delay line is added to accommodate the requirement and the revised HDL code is shown in Listing 21.2.

Listing 21.2 Bar test-pattern pixel generation circuit

```
library ieee;
use ieee.std_logic_1164.all;
use ieee.numeric_std.all;
entity bar_src is
    port(
        clk     : in   std_logic;
        x, y    : in   std_logic_vector(10 downto 0);
        -- stream out
```

```vhdl
      bar_rgb : out std_logic_vector(11 downto 0)
   );
end bar_src;

architecture arch of bar_src is
   -- intermediate counter value
   signal up, down   : std_logic_vector(3 downto 0);
   -- misc signals
   signal r, g, b    : std_logic_vector(3 downto 0);
   -- delay line
   signal rgb        : std_logic_vector(11 downto 0);
   signal reg_d1_reg : std_logic_vector(11 downto 0);
   signal reg_d2_reg : std_logic_vector(11 downto 0);
begin
   -- pixel data generation
   up   <= x(6 downto 3);
   down <= not x(6 downto 3);   -- reverse the binary sequence
   process(x, y, up, down)
   begin
      -- 16 shades of gray
      if unsigned(y) < 128 then
         r <= x(8 downto 5);
         g <= x(8 downto 5);
         b <= x(8 downto 5);
      -- 8 prime color with 50% intensity
      elsif unsigned(y) < 256 then
         r <= x(8) & x(8) & "00";
         g <= x(7) & x(7) & "00";
         b <= x(6) & x(6) & "00";
      -- a continuous "rainbow" color spectrum
      else
         case x(9 downto 7) is
            when "000" =>
               r <= (others => '1');
               g <= up;
               b <= "0000";
            when "001" =>
               r <= down;
               g <= "1111";
               b <= "0000";
            when "010" =>
               r <= "0000";
               g <= "1111";
               b <= up;
            when "011" =>
               r <= "0000";
               g <= down;
               b <= "1111";
            when "100" =>
               r <= up;
               g <= "0000";
               b <= "1111";
            when "101" =>
               r <= "1111";
               g <= "0000";
               b <= down;
            when others =>
               r <= "1111";
               g <= "1111";
               b <= "1111";
         end case;
```

```
            end if;
      end process;
      rgb <= r & g & b;
      -- 2-stage delay line
      process(clk)
      begin
         if (clk'event and clk = '1') then
            reg_d1_reg <= rgb;
            reg_d2_reg <= reg_d1_reg;
         end if;
      end process;
      bar_rgb <= reg_d2_reg;
end arch;
```

The test-pattern pixel generation circuit produces a full-screen frame. It functions as an independent video source and its output is not likely to be mixed with the incoming background pixel data stream. Thus, the blender of the core should function as a switch that selects one of the two video sources and should be implemented as a multiplexer.

The operation of the test-pattern pixel generation circuit is fixed and does not involve any control signal. However, a one-bit selection signal is needed to choose the source of the multiplexer. The slot wrapping circuit contains a one-bit register and a decoding logic to write the register.

The HDL code follows the diagram in Figure 21.4(a) and is shown in Listing 21.3.

Listing 21.3   Bar test-pattern generator core

```
library ieee;
use ieee.std_logic_1164.all;
entity chu_vga_bar_core is
   port(
      clk      : in   std_logic;
      reset    : in   std_logic;
      -- frame counter
      x, y     : in   std_logic_vector(10 downto 0);
      -- video slot interface
      cs       : in   std_logic;
      write    : in   std_logic;
      addr     : in   std_logic_vector(13 downto 0);
      wr_data  : in   std_logic_vector(31 downto 0);
      -- stream interface
      si_rgb   : in   std_logic_vector(11 downto 0);
      so_rgb   : out  std_logic_vector(11 downto 0)
   );
end chu_vga_bar_core;

architecture arch of chu_vga_bar_core is
   signal wr_en       : std_logic;
   signal bypass_reg  : std_logic;
   signal bar_rgb     : std_logic_vector(11 downto 0);
begin
   -- instantiate bar generator
   bar_src_unit : entity work.bar_src
      port map(
         clk     => clk,
         x       => x,
         y       => y,
         bar_rgb => bar_rgb);
   -- register
   process(clk, reset)
```

```
      begin
         if reset = '1' then
            bypass_reg <= '0';
         elsif (clk'event and clk = '1') then
            if wr_en = '1' then
               bypass_reg <= wr_data(0);
            end if;
         end if;
      end process;
      -- decoding
      wr_en  <= '1' when write = '1' and cs = '1' else '0';
      -- blending: bypass mux
      so_rgb <= si_rgb when bypass_reg = '1' else bar_rgb;
end arch;
```

The si_rgb and so_rgb ports are for the pixel streams. We use the si_ and so_ prefixes to represent the stream input and and stream output, respectively.

Since there is only one register within the core, the address decoding of the bypass control register can be omitted; i.e., the Boolean expression addr=10···0 is not needed in the condition of the statement:

```
      wr_en  <= '1' when write = '1' and cs = '1' else '0';
```

21.3.2 Color-to-grayscale conversion core

The color-to-grayscale conversion is a pixel transformation core. It is composed of the conversion circuit, a blender (which is a two-to-one multiplexer), and a slot wrapping circuit. The core performs conversion on the incoming stream pixel by pixel and does not need a frame counter to supply the coordinates. The circuit described in Listing 20.4 can be instantiated in the core.

Similar to the bar test-pattern generator core, the operation of the color-to-grayscale circuit does not involve any control signal and the slot wrapping circuit is just an interface with a one-bit control register. The HDL code is shown in Listing 21.4.

Listing 21.4   Color-to-grayscale conversion core

```
library ieee;
use ieee.std_logic_1164.all;
use ieee.numeric_std.all;

entity chu_rgb2gray_core is
   port(
      clk     : in   std_logic;
      reset   : in   std_logic;
      -- video slot interface
      cs      : in   std_logic;
      write   : in   std_logic;
      addr    : in   std_logic_vector(13 downto 0);
      wr_data : in   std_logic_vector(31 downto 0);
      -- stream interface
      si_rgb  : in   std_logic_vector(11 downto 0);
      so_rgb  : out  std_logic_vector(11 downto 0)
   );
end chu_rgb2gray_core;

architecture arch of chu_rgb2gray_core is
```

```
   signal wr_en      : std_logic;
   signal bypass_reg : std_logic;
   signal gray_rgb   : std_logic_vector(11 downto 0);
begin
   -- instantiate rgb-to-grayscale conversion circuit
   rgb2gray_unit : entity work.rgb2gray
      port map(
         color_rgb => si_rgb,
         gray_rgb  => gray_rgb
      );
   -- register
   process(clk, reset)
   begin
      if reset = '1' then
         bypass_reg <= '1';
      elsif (clk'event and clk = '1') then
         if wr_en = '1' then
            bypass_reg <= wr_data(0);
         end if;
      end if;
   end process;
   -- decoding
   wr_en  <= '1' when write = '1' and cs = '1' else '0';
   -- blending: bypass mux
   so_rgb <= si_rgb when bypass_reg = '1' else gray_rgb;
end arch;
```

21.3.3 "Dummy" core

A *dummy core* is a "placeholder" in the video cascading chain. It simply connects the output stream to the input stream and is used to facilitate the system development. The HDL code is shown in Listing 21.5.

Listing 21.5 Dummy core

```
library ieee;
use ieee.std_logic_1164.all;
use ieee.numeric_std.all;

entity chu_vga_dummy_core is
   port(
      clk      : in  std_logic;
      reset    : in  std_logic;
      -- MM video interface
      cs       : in  std_logic;
      write    : in  std_logic;
      addr     : in  std_logic_vector(13 downto 0);
      wr_data  : in  std_logic_vector(31 downto 0);
      -- stream interface
      si_rgb   : in  std_logic_vector(11 downto 0);
      so_rgb   : out std_logic_vector(11 downto 0)
   );
end chu_vga_dummy_core;

architecture arch of chu_vga_dummy_core is
begin
   so_rgb <= si_rgb;
end arch;
```

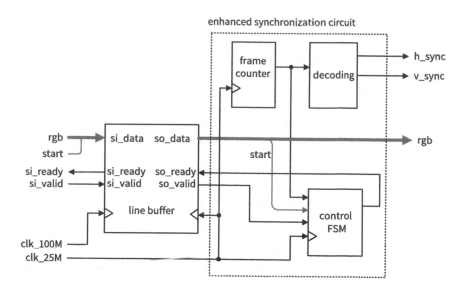

Figure 21.5   Conceptual diagram of the video synchronization core.

21.4   FPRO VIDEO SYNCHRONIZATION CORE

A video synchronization circuit supplies the pixels to a monitor at a rate specified by the pixel clock. The pixel clock rate depends on the resolution and refresh rate of a display standard, as shown in the "clock rate" column of Table 20.1.

In the ad hoc video system of Chapter 20, the system clock rate (100M Hz) is exactly four times that of the VGA pixel clock rate (25M Hz) and the synchronization circuit uses a mod-4 counter to obtain a 25M-Hz tick signal. However, since the pixel clock rate is independent of the system clock rate, this is just a coincidence and the scheme cannot be used for other scenarios.

In a general setting, a video subsystem should be able to accommodate various display resolution standards. It needs a flexible and robust video synchronization core to reliably transmit the video data stream from the system clock domain to the pixel clock domain. The most general and effective method is to use the FIFO-based distributed flow control, the third scheme discussed in Section 20.2.2.

We develop a more general *video synchronization core* for the FPro system. The conceptual diagram is shown in Figure 21.5. It contains a *line buffer* and an *enhanced synchronization circuit*.

Because the pixel generation cores and video synchronization circuit reside in different clock domains, they use their own frame counters. The synchronization circuit does not have access to the counter in other clock domain and thus cannot use it to identify the beginning of a frame. One scheme to resolve the issue is to transmit an additional start signal along with the pixel data. The start signal is used as a flag and asserted only in the beginning of a frame, when both horizontal count and vertical count of the frame counter are 0's. The pixel generation circuit produces this signal and transmits it along with the normal pixel data.

21.4.1   Line buffer

The FIFO buffer serves as a cushion between a source and a sink. In a video system, the buffer is usually set to be large enough to accommodate the pixel data of one horizontal line. This provides additional flexibility for the source to generate pixel data during the horizontal retrace interval. For example, in a VGA system, there are 800 pixel clock cycles in the horizontal line but only 640 clock cycles are used for data pixels. If the FIFO buffer can accommodate more than 640 words, the pixel generation circuit can use the extra 160 clock cycles to fill the buffer. Because of this, the buffer is referred to as a *line buffer*. To meet the specification of the VGA line buffer, the width of the FIFO should be 13 bits, which is composed of 12-bit color data and one-bit `start` signal, and the depth has to be at least 10 bits (i.e., $\lceil \log_2 640 \rceil$ bits), which corresponds to 2^{10} words.

The FIFO buffer involves two different clock signals. The write portion is driven by the system clock and the read portion is driven by the pixel clock. Because of the potential metastability, the internal construction of a dual-clock FIFO buffer is different from that of a normal single-clock FIFO buffer discussed in Section 7.3. The design is beyond the scope of this book and additional information can be found in the bibliographic section. The Xilinx Artix device can configure its BRAM as a dual-clock FIFO buffer and a single 18 Kb BRAM module can accommodate the required 2^{10}-by-13 capacity. We use this module for our purposes. The FIFO macro core, `fifo_dualclock_macro`, can be instantiated directly in the HDL code and the procedure is outlined in Appendix A.4.1. It uses generics of the instance to specify the desired configuration. The HDL code in Listing 21.6 shows a "wrapped" instance for the line buffer.

Listing 21.6   Xilinx BRAM-based dual-clock FIFO buffer

```
library ieee;
use ieee.std_logic_1164.all;
-- Xilinx macro libraries
library unisim;
use unisim.vcomponents.all;
library unimacro;
use unimacro.vcomponents.all;

entity bram_fifo_fpro is
   generic(
      DW : integer := 14  -- # data width (bits per word; 10-18)
   );
   port(
      reset        : in  std_logic;
      -- read port
      clk_rd       : in  std_logic;      -- read clock
      empty        : out std_logic;      -- read port empty
      almost_empty : out std_logic;      -- read port almost empty
      rd_ack       : in  std_logic;      -- read acknowledge
      rd_data      : out std_logic_vector(DW-1 downto 0); -- read data
      -- write port
      clk_wr       : in  std_logic;      -- write clock
      full         : out std_logic;      -- write port full
      almost_full  : out std_logic;      -- write port almost full
      wr_en        : in  std_logic;      -- write enable
      wr_data      : in  std_logic_vector(DW-1 downto 0); -- write data
      -- occupancy of fifo
      rdcount      : out std_logic_vector(9 downto 0); -- read count
```

```
        wrcount         : out std_logic_vector(9 downto 0)  — write count
    );
end bram_fifo_fpro;

architecture wrapper_arch of bram_fifo_fpro is
—signal rdcount, wrcount: std_logic_vector(9 downto 0);
begin
    — instantiate macro
    bram_fifo_unit : FIFO_DUALCLOCK_MACRO
        generic map(
            device                => "7SERIES",  — target device: "7SERIES"
            almost_full_offset    => x"0080",    — almost full threshold
            almost_empty_offset   => x"0080",    — almost empty threshold
            data_width            => DW,
            fifo_size             => "18Kb",     — BRAM: "18Kb" or "36Kb"
            first_word_fall_through => true
        )
        port map(
            rst         => reset,
            — read port
            rdclk       => clk_rd,     — read clock
            do          => rd_data,    — read data out
            rden        => rd_ack,     — remove word from head
            empty       => empty,      — fifo empty
            almostempty => almost_empty,
            rdcount     => rdcount,
            rderr       => open,       — read error
            — write port
            wrclk       => clk_wr,     — write clock
            di          => wr_data,    — write data in
            wren        => wr_en,      — write enable
            full        => full,       — fifo full
            almostfull  => almost_full,
            wrcount     => wrcount,
            wrerr       => open        — write error
        );
end wrapper_arch;
```

The main input and output ports of the core are similar to those of Listing 7.6. However, it includes output ports for word counts and additional statuses. The almost_full and almost_empty signals are asserted after the numbers of words in the FIFO buffer reach the thresholds defined in the generic mapping. For example, almost_full is asserted when there are less than 128 (i.e., x"0080") words left.

The distributed stream interface consists of a pair of signals, ready and valid, to exchange control information between the source and sink. The line buffer functions as a sink for the pixel processing circuits and as a source for the enhanced synchronization circuit. The HDL code of the line buffer is shown in Listing 21.7. It instantiates the dual-clock FIFO buffer and maps the FIFO buffer's status and control signals to the stream interface's signals.

Listing 21.7   Line buffer

```
library ieee;
use ieee.std_logic_1164.all;
use ieee.numeric_std.all;
entity line_buffer is
    generic(
        CD : integer := 12   — color depth
    );
```

```vhdl
   port(
      reset               : in   std_logic;
      clk_stream_in   : in   std_logic;
      clk_stream_out  : in   std_logic;
      -- stream in (sink)
      si_data           : in   std_logic_vector(CD downto 0);
      si_valid          : in   std_logic;
      si_ready          : out std_logic;
      -- stream out (source)
      so_data           : out std_logic_vector(CD downto 0);
      so_valid          : out std_logic;
      so_reday          : in   std_logic
   );
end line_buffer;

architecture str_arch of line_buffer is
   constant DW        : integer := CD + 1; -- colors+start
   signal almost_full : std_logic;
   signal empty       : std_logic;
   signal fifo_wr_en  : std_logic;
   signal fifo_rd_ack : std_logic;
   signal rdcount     : std_logic_vector(9 downto 0);
   signal wrcount     : std_logic_vector(9 downto 0);
begin
   -- instantiate dual-clock fifo
   fifo_unit : entity work.bram_fifo_fpro
      generic map(DW => DW)
      port map(
         reset         => reset,
         -- read port
         clk_rd        => clk_stream_out,
         rd_data       => so_data,
         rd_ack        => fifo_rd_ack,
         empty         => empty,
         almost_empty  => open,
         rdcount       => rdcount,
         -- write port
         clk_wr        => clk_stream_in,
         wr_data       => si_data,
         wr_en         => fifo_wr_en,
         full          => open,
         almost_full   => almost_full,
         wrcount       => wrcount
      );
   -- stream interface signals
   fifo_wr_en <= si_valid;
   si_ready    <= not almost_full;
   so_valid    <= not empty;
   fifo_rd_ack <= so_reday;
end str_arch;
```

The si_ready signal is asserted when the FIFO buffer has space to accept new pixel data and is connected to the inverted almost_full signal of the FIFO buffer. It functions as a request signal for the pixel generation cores to produce new pixels. The si_valid signal is connected to the write enable signal of the FIFO buffer. When a new pixel is available, the rightmost pixel generation core in the cascading chain can assert si_valid to store the data to the buffer.

The so_valid signal is asserted when the FIFO buffer has data available (i.e., is not empty) and is connected to the inverted empty of the FIFO buffer signal. The

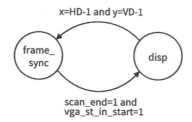

x=HD-1 and y=VD-1

scan_end=1 and
vga_st_in_start=1

Figure 21.6   Simplified synchronization FSM state diagram.

so_ready signal is connected to the read acknowledge signal of the FIFO buffer
to remove data from the head of the FIFO buffer. The enhanced synchronization
asserts this signal after retrieving the current pixel data.

21.4.2   Enhanced video synchronization circuit

The enhanced video synchronization circuit generates the horizontal and vertical
synchronization signals and retrieves the pixel data from the line buffer. It uses the
start flag to synchronize the pixel stream with its own frame counter. Its block
diagram is shown in the dotted box of Figure 21.5. The circuit contains a frame
counter, a decoding circuit to generate the horizontal and vertical synchronization
signals, and a *frame synchronization FSM*. The first two are similar to those in
Section 20.3.5.

The frame synchronization FSM synchronizes the beginning of a frame with the
frame counter. Its simplified state diagram is shown in Figure 21.6. The FSM has
two states. In the frame_sync state, the FSM waits for the beginning of a new frame.
It uses the scan_end signal of the frame counter to monitor the end of the current
frame. When the current frame is completed, the initial pixel data of a new frame
should already be in the line buffer, a condition reflected by the assertion of the
vga_st_in_start signal. The FSM then moves to the disp state to retrieve and
display the pixel data for this frame. However, a frame may be corrupted or late
and then the vga_st_in_start signal is not asserted. In this case, the FSM stays on
the frame_sync state and keeps on reading the line buffer until the vga_st_in_start
signal becomes 1. The operation flushes the corrupted frame from the line buffer
and re-synchronizes the next frame.

In the disp state, the FSM retrieves the pixel data from the line buffer in each
pixel clock cycle when the scan is in the "displayable" area, in which the video_on
signal is asserted. After the last displayable line is retrieved, the FSM returns to
the frame_sync state and waits for the completion of the vertical retrace to start a
new frame. Note that the FSM assumes that the valid pixel data can be provided
on time. The frame counter cannot be paused even when the pixel data is not
ready (i.e., the line buffer is empty). If the video sources fail to supply pixel data
on time, some frame fragments will be left in the line buffer. These late data will
be flushed out later in the frame_sync state.

The frame synchronization FSM makes the overall video system more robust.
Even though the system loses some data (due to occasional mismatched speed,
buffer overflow, etc.) in one frame, this will not affect the subsequent frames.

The complete HDL code of the enhanced synchronization circuit is shown in Listing 21.8.

Listing 21.8 Enhanced synchronization circuit

```vhdl
library ieee;
use ieee.std_logic_1164.all;
use ieee.numeric_std.all;
entity vga_sync is
   generic(
      CD : integer := 12   -- color depth
   );
   port(
      clk, reset    : in  std_logic;
      -- stream interface
      vga_si_data  : in  std_logic_vector(CD downto 0); -- color+start
      vga_si_valid : in  std_logic;
      vga_si_ready : out std_logic;
      -- to vga monitor
      hsync        : out std_logic;
      vsync        : out std_logic;
      rgb          : out std_logic_vector(CD - 1 downto 0)
   );
end vga_sync;

architecture arch of vga_sync is
   -- vga 640-by-480 sync parameters
   constant HD : integer := 640; -- horizontal display area
   constant HF : integer := 16; -- h. front porch
   constant HB : integer := 48; -- h. back porch
   constant HR : integer := 96; -- h. retrace
   constant HT : integer := HD + HF + HB + HR; -- horizontal total (800)
   constant VD : integer := 480; -- vertical display area
   constant VF : integer := 10; -- v. front porch
   constant VB : integer := 33; -- v. back porch
   constant VR : integer := 2; -- v. retrace
   constant VT : integer := VD + VF + VB + VR; -- vertical total (525)
   -- stream input
   alias vga_st_in_start : std_logic is vga_si_data(0);
   alias vga_st_in_color : std_logic_vector(CD - 1 downto 0) is
         vga_si_data(CD downto 1);
   -- sync counter and signals
   signal x, y       : unsigned(10 downto 0);
   signal hc, vc     : std_logic_vector(10 downto 0);
   signal scan_end   : std_logic;
   signal hsync_i    : std_logic;
   signal vsync_i    : std_logic;
   signal video_on_i : std_logic;
   -- fsm
   type state_type is (frame_sync, disp);
   signal state_reg, state_next : state_type;
begin
   --*****************************************************************
   -- instantiate frame counter
   --*****************************************************************
   counter_unit : entity work.frame_counter
      generic map(
         HMAX => HT,
         VMAX => VT
      )
      port map(
         clk           => clk,
```

```vhdl
        reset       => reset,
        sync_clr    => '0',
        hcount      => hc,
        vcount      => vc,
        inc         => '1',
        frame_start => open,
        frame_end   => scan_end
    );
x <= unsigned(hc);
y <= unsigned(vc);

--*********************************************************************
-- horizontal and vertical sync
--*********************************************************************
hsync_i <= '0' when (x >= (HD+HF)) and (x <= (HD+HF+HR-1)) else '1';
vsync_i <= '0' when (y >= (VD+VF)) and (y <= (VD+VF+VR-1)) else '1';
video_on_i <= '1' when (x < HD) and (y < VD) else '0';

--*********************************************************************
-- buffered output to vga monitor
--*********************************************************************
process(clk)
begin
    if (clk'event and clk = '1') then
        vsync <= vsync_i;
        hsync <= hsync_i;
        if (video_on_i = '1') then
            rgb <= vga_st in_color;
        else
            rgb <= (others => '0');     -- black when display off
        end if;
    end if;
end process;

--*********************************************************************
-- FSM to synchronize data for each frame
--*********************************************************************
-- state register
process(clk, reset)
begin
    if (reset = '1') then
        state_reg <= frame_sync;
    elsif (clk'event and clk = '1') then
        state_reg <= state_next;
    end if;
end process;
-- next-state/output logic
process(state_reg, x, y, vga_st_in_start, video_on_i, scan_end)
begin
    state_next   <= state_reg;
    vga_si_ready <= '0';
    case state_reg is
        when frame_sync =>
            -- wait for end of current scan (end of v/h retrace)
            if scan_end = '1' then
                if vga_st_in_start = '1' then
                    state_next <= disp;
                else
                    state_next <= frame_sync;
                end if;
            end if;
```

```
                    —— flush out partial frame fragment
                    —— (due to corruption or incorrectly formed long packet)
                    if vga_st_in_start = '0' then
                        vga_si_ready <= '1';
                    end if;
                when disp =>
                    —— resync when reaching end of the displayable data
                    if ((x = HD - 1) and (y = VD - 1)) then
                        state_next <= frame_sync;
                    end if;
                    if video_on_i = '1' then
                        vga_si_ready <= '1';
                    end if;
            end case;
        end process;
    end arch;
```

21.4.3 HDL code

The FPro video synchronization core combines the line buffer and the enhanced synchronization circuit. The HDL code is shown in Listing 21.9.

Listing 21.9 FPro video synchronization and buffer core

```
library ieee;
use ieee.std_logic_1164.all;
entity chu_vga_sync_core is
    generic(CD : integer := 12);
    port(
        clk_sys  : in   std_logic;
        clk_25M  : in   std_logic;
        reset    : in   std_logic;
        —— MM interface
        cs       : in   std_logic;
        write    : in   std_logic;
        addr     : in   std_logic_vector(13 downto 0);
        wr_data  : in   std_logic_vector(31 downto 0);
        ——
        si_data  : in   std_logic_vector(CD downto 0);
        si_valid : in   std_logic;
        si_ready : out std_logic;
        —— to vga monitor
        hsync    : out std_logic;
        vsync    : out std_logic;
        rgb      : out std_logic_vector(CD - 1 downto 0)
    );
end chu_vga_sync_core;

architecture str_arch of chu_vga_sync_core is
    signal line_so_data  : std_logic_vector(CD downto 0);
    signal line_so_valid : std_logic;
    signal vga_si_ready  : std_logic;
begin
    —— instantiate line buffer
    line_unit : entity work.line_buffer
        generic map(CD => CD)
        port map(
            reset           => reset,
            clk_stream_in   => clk_sys,
            clk_stream_out  => clk_25M,
```

```
           si_data          => si_data,
           si_valid         => si_valid,
           si_ready         => si_ready,
           so_data          => line_so_data,
           so_valid         => line_so_valid,
           so_reday         => vga_si_ready
      );
   -- instantiate vga controller
   vga_unit : entity work.vga_sync
      generic map(CD => CD)
      port map(
           clk              => clk_25M,
           reset            => reset,
           vga_si_data   => line_so_data,
           vga_si_valid  => line_so_valid,
           vga_si_ready  => vga_si_ready,
           hsync            => hsync,
           vsync            => vsync,
           rgb              => rgb
      );
end str_arch;
```

21.5 DAISY VIDEO SUBSYSTEM

The ad hoc video subsystem in Chapter 20 can be reconstructed with the three video cores introduced in Sections 21.3 and 21.4. However, for completeness, a more comprehensive subsystem is created to include the cores developed in the next three chapters as well. We call it *daisy video subsystem* since the video cores are arranged as a daisy chain.

21.5.1 Subsystem overview

The daisy video subsystem incorporates all video cores developed in this chapter and subsequent chapters. The main components are as follows:

- Video decoding logic
- Global frame counter
- Frame buffer core
- Eight normal FPro video cores, including two "dummy" cores

The video cores in the daisy video subsystem form a cascading chain where the pixel stream is generated, processed, augmented, and eventually displayed on a monitor, as shown in the bottom portion of Figure 21.1. The cores in the left, such as the frame buffer and the bar test-pattern generator, are the video sources and generate a full-screen frame. The cores in the middle, such as the color-to-grayscale conversion circuit, process and modify the pixels. The cores in the right, such as the on-screen text display circuit and the mouse pointer circuits, augment the frame with overlays of auxiliary sprites. The rightmost one is the video synchronization core for the monitor.

To simplify the stream interface, we use a common global frame counter for the pixel generation cores in the daisy subsystem. After the valid x- and y-coordinate signals are sampled, a pixel generation core outputs the valid pixel data after two clock cycles. The two-clock delay is needed to accommodate the internal memory

Table 21.1 Slot assignment of the daisy video subsystem

Slot	Symbolic name	Description
0	V0_SYNC0	synchronization core
1	V1_MOUSE1	mouse sprite core
2	V2_OSD2	OSD (on screen display) core
3	V3_GHOST3	ghost sprite core
4	V4_USER4	dummy core
5	V5_USER5	dummy core
6	V6_GRAY6	color-to-grayscale core
7	V7_BAR7	bar test-pattern generator core

access of some cores. For a simple core, such as the bar test-pattern generator discussed in Section 21.3, additional registers should be inserted to satisfy the two-clock delay requirement. Note that the synchronization core contains a local frame counter driven by the pixel clock. Its stream interface utilizes a FIFO buffer and handshake mechanism and is different from other pixel generation and transformation cores.

The slot assignments of the video cores and their corresponding symbolic names are shown in Table 21.1. The names are added to the chu_io_map.vhd file and the chu_io_map.h file. Note that the frame buffer core is decoded separately and does not occupy a slot.

Unlike the MMIO cores of an MMIO subsystem, in which each one can be treated as an independent entity, the order and arrangement of video cores in the cascading chain are important. Misconnecting a single core can affect, and even fail, the entire video subsystem. To reduce potential errors, we use all eight slots in the HDL code and systematically connect the pixel streams among the cores. If a slot is not physically used, a dummy core is inserted as a placeholder.

21.5.2 Interface to the video synchronization core

The stream interface of the video synchronization core includes the si_ready and si_valid signals for the handshake and is expected to have an additional start bit in the pixel data stream to mark the beginning of a frame. It is not compatible with other video cores of the daisy video subsystem.

We design an ad hoc circuit to connect the two types of interfaces. The first part of the circuit is to use the si_ready signal to trigger a new pixel generation and to assert the si_valid signal to write the generated pixel to the FIFO buffer. The si_ready is connected to the inc (incremental) signal of the global and a two-stage delay line (i.e., two cascading registers). Recall that the si_ready signal is asserted when there is enough FIFO space for new pixels, as specified in Listing 21.7. The assertion increments the global frame counter to generate new coordinates, which in turn trigger the operation of the pixel generation cores. The new blended pixel becomes available after two clock cycles. At this point, the asserted signal is also looped back via the two-stage delay line and activates the si_valid signal, whose action writes the new pixel into the FIFO buffer. In summary, because of the deterministic nature of the pixel generation cores, the handshake mechanism is automatically completed through a delayed feedback loop.

The second part of the circuit is to insert a `start` bit to the pixel stream. This bit can be obtained from global frame counter's `frame_start` status signal, which is asserted when the x- and y-coordinates are (0, 0). Because the pixel generation requires two clock cycles, this signal is delayed via a two-stage delay line as well.

21.5.3 HDL code

The code of the daisy video subsystem is shown in Listing 21.10. The normal video cores are instantiated based on the slot assignment in Table 21.1 and the frame buffer is placed on the leftmost location of the cascading chain. Two two-stage delay lines are created to implement the start bit and the flow-control scheme.

Listing 21.10   Daisy video subsystem

```vhdl
library ieee;
use ieee.std_logic_1164.all;
use work.chu_io_map.all;

entity video_sys_daisy is
   generic(
      CD               : integer := 12;   -- color depth
      VRAM DATA WIDTH : integer := 9      -- frame buffer data width
   );
   port(
      clk_sys        : in   std_logic;
      clk_25M        : in   std_logic;
      reset_sys      : in   std_logic;
      -- generic bus interface
      video_cs       : in   std_logic;
      video_wr       : in   std_logic;
      video_addr     : in   std_logic_vector(20 downto 0);
      video_wr_data  : in   std_logic_vector(31 downto 0);
      -- to vga monitor
      vsync, hsync   : out  std_logic;
      rgb            : out  std_logic_vector(11 downto 0)
   );
end video_sys_daisy;

architecture arch of video_sys_daisy is
   constant KEY_COLOR : std_logic_vector(CD-1 downto 0):=(others=>'0');
   -- video data stream
   signal frame_rgb8         : std_logic_vector(CD - 1 downto 0);
   signal bar_rgb7           : std_logic_vector(CD - 1 downto 0);
   signal gray_rgb6          : std_logic_vector(CD - 1 downto 0);
   signal user5_rgb5         : std_logic_vector(CD - 1 downto 0);
   signal user4_rgb4         : std_logic_vector(CD - 1 downto 0);
   signal ghost_rgb3         : std_logic_vector(CD - 1 downto 0);
   signal osd_rgb2           : std_logic_vector(CD - 1 downto 0);
   signal mouse_rgb1         : std_logic_vector(CD - 1 downto 0);
   signal line_data_in       : std_logic_vector(CD downto 0);
   -- frame counter
   signal inc                : std_logic;
   signal x, y               : std_logic_vector(10 downto 0);
   signal frame_start        : std_logic;
   -- delay line
   signal frame_start_d1_reg : std_logic;
   signal frame_start_d2_reg : std_logic;
```

```vhdl
   signal inc_d1_reg           : std_logic;
   signal inc_d2_reg           : std_logic;
   -- frame interface
   signal frame_wr, frame_cs   : std_logic;
   signal frame_addr           : std_logic_vector(19 downto 0);
   signal frame_wr_data        : std_logic_vector(31 downto 0);
   signal slot_cs_array        : std_logic_vector(7 downto 0);
   signal slot_mem_wr_array    : std_logic_vector(7 downto 0);
   signal slot_reg_addr_array  : slot_2d_video_reg_type;
   signal slot_wr_data_array   : slot_2d_data_type;
begin
   --****************************************************************
   -- 2-stage delay line for start signal
   --****************************************************************
   process(clk_sys)
   begin
      if (clk_sys'event and clk_sys = '1') then
         frame_start_d1_reg <= frame_start;
         frame_start_d2_reg <= frame_start_d1_reg;
         inc_d1_reg         <= inc;
         inc_d2_reg         <= inc_d1_reg;
      end if;
   end process;
   --****************************************************************
   -- instantiate global frame counter
   --****************************************************************
   counter_unit : entity work.frame_counter
      generic map(
         HMAX => 640,
         VMAX => 480)
      port map(
         clk         => clk_sys,
         reset       => reset_sys,
         inc         => inc,
         sync_clr    => '0',
         hcount      => x,
         vcount      => y,
         frame_start => frame_start,
         frame_end   => open);
   --****************************************************************
   -- instantiate video controller
   --****************************************************************
   ctrl_unit : entity work.chu_video_controller
      port map(
         video_cs            => video_cs,
         video_wr            => video_wr,
         video_addr          => video_addr,
         video_wr_data       => video_wr_data,
         frame_cs            => frame_cs,
         frame_wr            => frame_wr,
         frame_addr          => frame_addr,
         frame_wr_data       => frame_wr_data,
         slot_cs_array       => slot_cs_array,
         slot_mem_wr_array   => slot_mem_wr_array,
         slot_reg_addr_array => slot_reg_addr_array,
         slot_wr_data_array  => slot_wr_data_array
      );
   --****************************************************************
   -- frame buffer and video cores
   --****************************************************************
   -- instantiate frame buffer
```

```
frame_unit : entity work.chu_frame_buffer_core
    generic map(
        CD => CD,
        DW => VRAM_DATA_WIDTH
    )
    port map(
        clk     => clk_sys,
        reset   => reset_sys,
        x       => x,
        y       => y,
        cs      => frame_cs,
        write   => frame_wr,
        addr    => frame_addr,
        wr_data => video_wr_data,
        si_rgb  => x"008",          -- blue screen
        so_rgb  => frame_rgb8
    );
-- instantiate bar generator
v7_bar_unit : entity work.chu_vga_bar_core
    port map(
        clk     => clk_sys,
        reset   => reset_sys,
        x       => x,
        y       => y,
        cs      => slot_cs_array(V7_BAR),
        write   => slot_mem_wr_array(V7_BAR),
        addr    => slot_reg_addr_array(V7_BAR),
        wr_data => slot_wr_data_array(V7_BAR),
        si_rgb  => frame_rgb8,
        so_rgb  => bar_rgb7);
-- instantiate rgb-to-gray generator
v6_gray_unit : entity work.chu_rgb2gray_core
    port map(
        clk     => clk_sys,
        reset   => reset_sys,
        cs      => slot_cs_array(V6_GRAY),
        write   => slot_mem_wr_array(V6_GRAY),
        addr    => slot_reg_addr_array(V6_GRAY),
        wr_data => slot_wr_data_array(V6_GRAY),
        si_rgb  => bar_rgb7,
        so_rgb  => gray_rgb6);
-- instantiate dummy unit (placeholder for 1st user defined unit)
v5_user_unit : entity work.chu_vga_dummy_core
    port map(
        clk     => clk_sys,
        reset   => reset_sys,
        cs      => slot_cs_array(V5_USER5),
        write   => slot_mem_wr_array(V5_USER5),
        addr    => slot_reg_addr_array(V5_USER5),
        wr_data => slot_wr_data_array(V5_USER5),
        si_rgb  => gray_rgb6,
        so_rgb  => user5_rgb5
    );
-- instantiate dummy unit (placeholder for 2nd user defined unit)
v4_user_unit : entity work.chu_vga_dummy_core
    port map(
        clk     => clk_sys,
        reset   => reset_sys,
        cs      => slot_cs_array(V4_USER4),
        write   => slot_mem_wr_array(V4_USER4),
        addr    => slot_reg_addr_array(V4_USER4),
```

```vhdl
         wr_data => slot_wr_data_array(V4_USER4),
         si_rgb  => user5_rgb5,
         so_rgb  => user4_rgb4
      );
   -- instantiate ghost sprite
   v3_ghost_unit : entity work.chu_vga_slot_ghost_core
      generic map(
         CD         => CD,
         ADDR_WIDTH => 10,
         KEY_COLOR  => KEY_COLOR
      )
      port map(
         clk     => clk_sys,
         reset   => reset_sys,
         x       => x,
         y       => y,
         cs      => slot_cs_array(V3_GHOST),
         write   => slot_mem_wr_array(V3_GHOST),
         addr    => slot_reg_addr_array(V3_GHOST),
         wr_data => slot_wr_data_array(V3_GHOST),
         si_rgb  => user4_rgb4,
         so_rgb  => ghost_rgb3
      );
   -- instantiate osd
   v2_osd_unit : entity work.chu_vga_osd_core
      generic map(
         CD         => CD,
         KEY_COLOR  => KEY_COLOR
      )
      port map(
         clk     => clk_sys,
         reset   => reset_sys,
         x       => x,
         y       => y,
         cs      => slot_cs_array(V2_OSD),
         write   => slot_mem_wr_array(V2_OSD),
         addr    => slot_reg_addr_array(V2_OSD),
         wr_data => slot_wr_data_array(V2_OSD),
         si_rgb  => ghost_rgb3,         --bar_rgb,
         so_rgb  => osd_rgb2
      );
   -- instantiate mouse sprite
   v1_mouse_unit : entity work.chu_vga_slot_mouse_core
      generic map(
         CD         => CD,
         ADDR_WIDTH => 10,
         KEY_COLOR  => KEY_COLOR
      )
      port map(
         clk     => clk_sys,
         reset   => reset_sys,
         x       => x,
         y       => y,
         cs      => slot_cs_array(V1_MOUSE),
         write   => slot_mem_wr_array(V1_MOUSE),
         addr    => slot_reg_addr_array(V1_MOUSE),
         wr_data => slot_wr_data_array(V1_MOUSE),
         si_rgb  => osd_rgb2,
         so_rgb  => mouse_rgb1
      );
   -- merge start bit to rgb data stream
```

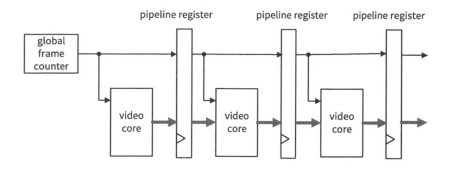

Figure 21.7   Conceptual pipelined video subsystem.

```
line_data_in <= mouse_rgb1 & frame_start_d2_reg;
— instantiate sync_core
v0_vga_sync_unit : entity work.chu_vga_sync_core
   generic map(CD => CD)
   port map(
      clk_sys    => clk_sys,
      reset      => reset_sys,
      clk_25M    => clk_25M,
      cs         => slot_cs_array(V0_SYNC),
      write      => slot_mem_wr_array(V0_SYNC),
      addr       => slot_reg_addr_array(V0_SYNC),
      wr_data    => slot_wr_data_array(V0_SYNC),
      si_data    => line_data_in,
      si_valid   => inc_d2_reg,
      si_ready   => inc,
      hsync      => hsync,
      vsync      => vsync,
      rgb        => rgb
   );
end arch;
```

21.5.4   Timing and performance considerations

In the daisy video subsystem, all cores are driven by the same 100M Hz clock. Although the cores are arranged as a chain, the operations are performed concurrently. After the global frame counter produces new coordinates, the pixel generation circuit of every core outputs a new pixel after two clock cycles. These pixels may be further processed by the pixel transformation cores and then are combined and mixed via the cascading blending network.

The cascading blending network contains no register and thus is a pure combinational circuit. The cascading chain with complex blenders may lead to a large propagation delay and cause timing violation. Since the blenders of daisy video subsystem's cores are quite simple, the overall propagation delay is tolerable.

For a video subsystem with complex blenders or more video cores, the long propagation delay of the cascading chain may be an issue. One way to solve the problem is to insert a register in each cascading stage and form a pipeline, as illustrated in Figure 21.7. Note that registers are also needed to delay the frame counter's output in each stage and to expand the delay line of the synchronization core's si_ready signal.

Another way to remedy the problem is to put a local frame counter in each core and to utilize a FIFO buffer for stream interface, similar to that in the synchronization core. This approach makes the core a completely independent entity but complicates the stream interface. It is overkill for the simple cores.

21.6 VANILLA DAISY FPRO SYSTEM

The daisy video subsystem can be incorporated into the vanilla FPro system of Part II or into the sampler FPro system of Part III. We use the former in the remaining book and call it the *vanilla daisy FPro system*. A similar *sampler daisy FPro system* is also constructed and can be found in the companion website.

21.6.1 Clock management core

The video synchronization core of the video subsystem needs a separate pixel clock signal. Modern FPGA devices contain special proprietary macro cells for the clock management. We instantiate a Xilinx Artix clock management core in the system level for this purpose, as shown in the top right portion of Figure 21.1.

The clock management core can be instantiated as an IP core and the procedure is outlined in Appendix A.4.5. We configure the core to generate 100M-Hz, 25M-Hz, 40M-Hz, and 67M-Hz clock signals. The 100M-Hz clock signal is for the system clock and the 25M-Hz clock signal is for the VGA pixel clock. The other two clocks are used as pixel clocks for the SVGA and XGA systems in the homework experiments. Although the rate of the 100M-Hz output clock signal is identical to the rate of the input clock signal, the clock management core generated signal can compensate for the internal delay and has a smaller clock skew. The internal clock management circuit is constructed with a closed feedback loop and may need a small amount of time to obtain the stable output initially. Its status signal, locked, indicates whether the output is stabilized (i.e., locked). The entity declaration of the generated HDL code of the core is shown in Listing 21.11.

Listing 21.11   Xilinx clock management circuit

```
module mmcm_fpro (
 // clock in port
  input           clk_in_100M,
  // clock out ports
  output          clk_100M,
  output          clk_25M,
  output          clk_40M,
  output          clk_67M,
  // status and control signals
  input           reset,
  output          locked
);

// instantiate customized mmcm macro core
mmcm_fpro_clk_wiz inst (
  .clk_in_100M(clk_in_100M),
  // Clock out ports
  .clk_100M(clk_100M),
```

```
  .clk_25M(clk_25M),
  .clk_40M(clk_40M),
  .clk_67M(clk_67M),
  .reset(reset),
  .locked(locked)
 );
endmodule
```

This is a Verilog file since the software used at the time of writing (Vivado v2016 and clock Wizard v5.3) does not generate a VHDL file. Since Vivado supports mixed-language synthesis, this module can still be instantiated in a VHDL file, as shown in Listing 21.12. This file and its companion `mmcm_fpro_clk_wiz.v` file can be imported directly to the project.

21.6.2   Updated `chu_io_map` VHDL package

To facilitate the video subsystem, the `chu_io_map` VHDL package is expanded to include a new data type for the video subsystem controller and the symbolic constants of the video slots. The added statements are:

```
type slot_2d_video_reg_type is array (7 downto 0) of
     std_logic_vector(13 downto 0);
constant V0_SYNC  : integer := 0;
constant V1_MOUSE : integer := 1;
constant V2_OSD   : integer := 2;
constant V3_GHOST : integer := 3;
constant V4_USER4 : integer := 4;
constant V5_USER5 : integer := 5;
constant V6_GRAY  : integer := 6;
constant V7_BAR   : integer := 7;
```

21.6.3   HDL code

The code of the vanilla daisy FPro system is shown in Listing 21.12. It expands the vanilla FPro system of Listing 10.8 by adding the daisy video subsystem and a clock management core. The instantiated clock management core generates a 25M-Hz clock signal, which is connected to the VGA pixel clock, and a 100M-Hz clock signal, which is used as a system clock to drive all other logic. The system `reset` signal is now connected to the negated `locked` signal so that the system is kept in the "reset state" until the stable clock signals are obtained.

Listing 21.12   Vanilla daisy FPro system

```
library ieee;
use ieee.std_logic_1164.all;
use work.chu_io_map.all;
entity mcs_top_vanilla_daisy is
   generic(BRIDGE_BASE : std_logic_vector(31 downto 0) := x"C0000000");
   port(
      clk     : in  std_logic;
      reset_n : in  std_logic;
      -- switches and LEDs
      sw      : in  std_logic_vector(15 downto 0);
      led     : out std_logic_vector(15 downto 0);
      -- uart
```

```vhdl
      rx          : in  std_logic;
      tx          : out std_logic;
      -- to vga monitor
      hsync       : out std_logic;
      vsync       : out std_logic;
      rgb         : out std_logic_vector(11 downto 0)
   );
end mcs_top_vanilla_daisy;

architecture arch of mcs_top_vanilla_daisy is
   component cpu
      port(
         clk                : in  std_logic;
         reset              : in  std_logic;
         io_addr_strobe     : out std_logic;
         io_read_strobe     : out std_logic;
         io_write_strobe    : out std_logic;
         io_address         : out std_logic_vector(31 downto 0);
         io_byte_enable     : out std_logic_vector(3 downto 0);
         io_write_data      : out std_logic_vector(31 downto 0);
         io_read_data       : in  std_logic_vector(31 downto 0);
         io_ready           : in  std_logic
      );
   end component;
   component mmcm_fpro
      port(
         clk_in_100M : in  std_logic;
         clk_100M    : out std_logic;
         clk_25M     : out std_logic;
         clk_40M     : out std_logic;
         clk_67M     : out std_logic;
         reset       : in  std_logic;
         locked      : out std_logic
      );
   end component;
   signal clk_25M, clk_100M : std_logic;
   signal locked, reset_sys : std_logic;
   -- MCS IO bus
   signal io_addr_strobe     : std_logic;
   signal io_read_strobe     : std_logic;
   signal io_write_strobe    : std_logic;
   signal io_byte_enable     : std_logic_vector(3 downto 0);
   signal io_address         : std_logic_vector(31 downto 0);
   signal io_write_data      : std_logic_vector(31 downto 0);
   signal io_read_data       : std_logic_vector(31 downto 0);
   signal io_ready           : std_logic;
   -- fpro bus
   signal fp_mmio_cs         : std_logic;
   signal fp_wr              : std_logic;
   signal fp_rd              : std_logic;
   signal fp_addr            : std_logic_vector(20 downto 0);
   signal fp_wr_data         : std_logic_vector(31 downto 0);
   signal fp_rd_data         : std_logic_vector(31 downto 0);
   signal fp_video_cs        : std_logic;

begin
   -- clock and reset
   reset_sys <= (not locked) or (not reset_n);
   -- instantiate clock management unit
   clk_mmcm_unit : mmcm_fpro
      port map(
```

```
        —— Clock in ports
        clk_in_100M => clk,
        —— Clock out ports
        clk_100M     => clk_100M,
        clk_25M      => clk_25M,
        clk_40M      => open,
        clk_67M      => open,
        —— Status and control signals
        reset        => '0',
        locked       => locked
    );
—— instantiate microBlaze MCS
mcs_0 : cpu
    port map(
        clk              => clk_100M,
        reset            => reset_sys,
        io_addr_strobe   => io_addr_strobe,
        io_read_strobe   => io_read_strobe,
        io_write_strobe  => io_write_strobe,
        io_byte_enable   => io_byte_enable,
        io_address       => io_address,
        io_write_data    => io_write_data,
        io_read_data     => io_read_data,
        io_ready         => io_ready
    );
—— instantiate bridge
bridge_unit : entity work.chu_mcs_bridge
    generic map(BRG_BASE => BRIDGE_BASE)
    port map(
        io_addr_strobe   => io_addr_strobe,
        io_read_strobe   => io_read_strobe,
        io_write_strobe  => io_write_strobe,
        io_byte_enable   => io_byte_enable,
        io_address       => io_address,
        io_write_data    => io_write_data,
        io_read_data     => io_read_data,
        io_ready         => io_ready,
        fp_video_cs      => fp_video_cs,
        fp_mmio_cs       => fp_mmio_cs,
        fp_wr            => fp_wr,
        fp_rd            => fp_rd,
        fp_addr          => fp_addr,
        fp_wr_data       -> fp_wr_data,
        fp_rd_data       => fp_rd_data
    );
—— instantiate vanilla MMIO subsystem
mmio_sys_unit : entity work.mmio_sys_vanilla
    generic map(
        N_LED => 16,
        N_SW  => 16
    )
    port map(
        clk              => clk_100M,
        reset            => reset_sys,
        mmio_cs          => fp_mmio_cs,
        mmio_wr          => fp_wr,
        mmio_rd          => fp_rd,
        mmio_addr        => fp_addr,
        mmio_wr_data     => fp_wr_data,
        mmio_rd_data     => fp_rd_data,
        sw               => sw,
```

```
        led             => led,
        rx              => rx,
        tx              => tx
    );
  -- instantiate daisy video subsystem
  video_sys_unit : entity work.video_sys_daisy
    generic map(
        CD              => 12,
        VRAM_DATA_WIDTH => 9)
    port map(
        clk_sys         => clk_100M,
        clk_25M         => clk_25M,
        reset_sys       => reset_sys,
        video_cs        => fp_video_cs,
        video_wr        => fp_wr,
        video_addr      => fp_addr,
        video_wr_data   => fp_wr_data,
        vsync           => vsync,
        hsync           => hsync,
        rgb             => rgb
    );
end arch;
```

21.7   VIDEO DRIVER AND TESTING PROGRAM

We follow the same principle and practice discussed in Section 9.6.2 to develop video core drivers.

21.7.1   Updated chu_io_map.h and chu_io_rw.h files

The chu_io_map.h and chu_io_rw.h files are expanded to facilitate the video subsystem. Symbolic constant definitions for the frame buffer's base address and the video slot assignments are added to the chu_io_map.h file:

```
#define FRAME_OFFSET  0x00c00000
#define FRAME_BASE    BRIDGE_BASE+FRAME_OFFSET
#define V0_SYNC       0
#define V1_MOUSE      1
#define V2_OSD        2
#define V3_GHOST      3
#define V4_USER4      4
#define V5_USER5      5
#define V6_GRAY       6
#define V7_BAR        7
```

The get_slot_addr() macro in the chu_io_rw.h file calculates the base address of an MMIO slot. A similar macro, get_video_slot_addr(), is added to the file to calculate the base address of a video slot. It is defined as

```
#define get_video_slot_addr(base, vslot) \
        ((uint32_t)((base) + 0x00800000 + (vslot)*16384*4))
```

The value of 0x00800000 represents the 21st bit of the word address (i.e., the 23rd bit of the byte address) assigned to identify the video subsystem, as discussed in Section 21.1.2, and the value of 16384 corresponds to the 2^{14}-word memory space of each video core.

21.7.2 GPV core driver

Since the main data flow of a video core is through the stream interface, its bus transaction is usually simpler than that of an MIMO core. We define a *GPV* (for *general-purpose video core*) class for the general video cores. It contains methods to write a register and to bypass the core, which is done by writing register 0x2000, as discussed in Section 21.2.3. The class definition and implementation of the GPV core are shown in Listings 21.13 and Listings 21.14, respectively.

Listing 21.13   GpvCore class definition (in `vga_core.h`)

```
class GpvCore {
public:
   enum {
      BYPASS_REG = 0x2000,
   };
   /* methods */
   GpvCore(uint32_t core_base_addr);
   ~GpvCore();                       // not used
   void wr_mem(int addr, uint32_t color);
   void bypass(int by);
private:
   uint32_t base_addr;
};
```

Listing 21.14   GpvCore class implementation (in `vga_core.cpp`)

```
GpvCore::GpvCore(uint32_t core_base_addr) {
   base_addr = core_base_addr;
}
GpvCore::~GpvCore() {}

void GpvCore::wr_mem(int addr, uint32_t data) {
   io_write(base_addr, addr, data);
}

void GpvCore::bypass(int by) {
   io_write(base_addr, BYPASS_REG, (uint32_t ) by);
}
```

The `GpvCore` class can be used for the bar test-pattern generation core and color-to-grayscale conversion core in this chapter. Because of the simplicity of the video core drivers, we group the class definitions of all drivers into a single `vga_core.h` file and group the class implementations into a single `vga_core.cpp` file.

21.7.3 Testing program

We follow the same procedure discussed in Section 12.4.4 to develop testing software. A test function is derived for each video core to verify the basic operation of the hardware implementation and driver. To maintain portability, the function utilizes only the four basic MMIO cores in the vanilla FPro system.

The testing function can be incorporated into the main test program as follows:

- Include the core driver header file.
- Develop the test function.
- Create an instance.
- Call the function in `main()`.

The test program for the video cores of the vanilla daisy FPro system is shown in Listing 21.15. Several test functions involve the video cores in the later chapters and their bodies can be found in the respective chapters. The basic outline of the main program is:

- Flash LEDs to signal the start of the video demonstration.
- Bypass all video cores.
- Demonstrate the basic operations of video cores one at a time.
- Enter the "overlay test," in which the desired overlays are selected by the switches.

Listing 21.15   Vanilla daisy FPro video test program (in `main_video_test.cpp`)

```
#include "chu_init.h"
#include "gpio_cores.h"
#include "vga_core.h"

void test_start(GpoCore *led_p) {
   int i;

   for (i = 0; i < 20; i++) {
      led_p->write(0xff00);
      sleep_ms(50);
      led_p->write(0x0000);
      sleep_ms(50);
   }
}

void bar_check(GpvCore *bar_p) {
   bar_p->bypass(0);
   sleep_ms(3000);
}

void gray_check(GpvCore *gray_p) {
   gray_p->bypass(0);
   sleep_ms(3000);
   gray_p->bypass(1);
}

void ghost_check(SpriteCore *ghost_p) {...}
void mouse_check(SpriteCore *mouse_p) {...}
void osd_check(OsdCore *osd_p) {...}
void frame_check(FrameCore *frame_p) {...}

// external core instantiation
GpoCore led(get_slot_addr(BRIDGE_BASE, S2_LED));
GpiCore sw(get_slot_addr(BRIDGE_BASE, S3_SW));
FrameCore frame(FRAME_BASE);
GpvCore bar(get_sprite_addr(BRIDGE_BASE, V7_BAR));
GpvCore gray(get_sprite_addr(BRIDGE_BASE, V6_GRAY));
SpriteCore ghost(get_sprite_addr(BRIDGE_BASE, V3_GHOST), 1024);
SpriteCore mouse(get_sprite_addr(BRIDGE_BASE, V1_MOUSE), 1024);
OsdCore osd(get_sprite_addr(BRIDGE_BASE, V2_OSD));

int main() {
   while (1) {
      test_start(&led);
      // bypass all cores; blue screen
      frame.bypass(1);
      bar.bypass(1);
      gray.bypass(1);
```

```
ghost.bypass(1);
osd.bypass(1);
mouse.bypass(1);
sleep_ms(3000);

// enable cores one by one
frame_check(&frame);
bar_check(&bar);
gray_check(&gray);
ghost_check(&ghost);
osd_check(&osd);
mouse_check(&mouse);
while (sw.read(0)) {
    // test composition with different overlays if sw(0) is 1
    mouse.bypass(sw.read(1));
    osd.bypass(sw.read(2));
    ghost.bypass(sw.read(3));
    gray.bypass(sw.read(6));
    bar.bypass(sw.read(7));
    frame.bypass(sw.read(8));
    // set osd parameters
    osd.set_color(0x0f0, sw.read(9));
    // set ghost sprite parameters
    ghost.wr_ctrl(sw.read() >> 11);
}
    }
}
```

21.8 BIBLIOGRAPHIC NOTES

Designing a dual-clock FIFO involves several clock-domain-crossing issues. The author's other book, *RTL Hardware Design with VHDL*, explains the issues in detail and discusses the design techniques. Xilinx's user guide UG472, *7 Series FPGAs Clocking Resources User Guide*, and user guide UG473, *7 Series FPGAs Memory Resources User Guide*, provide detailed information on the clock management macro cell and the BRAM based FIFO macro core.

21.9 SUGGESTED EXPERIMENTS

21.9.1 Color channel selection core

We can convert the color channel selection circuit of Experiment 20.9.2 to an FPro video core by adding a wrapping circuit and a blender. Develop the core, attach it to a user slot of the daisy video subsystem, derive the software driver to select the channel, and develop a test function to verify its operation.

21.9.2 Enhanced color-to-grayscale conversion core

We can convert the enhanced color-to-grayscale conversion circuit of Experiment 20.9.3 to an FPro video core by adding a wrapping circuit and a blender. Develop the core, replace the original core of the daisy video subsystem, derive the software driver to select the intended conversion algorithm, and develop a test function to verify its operation.

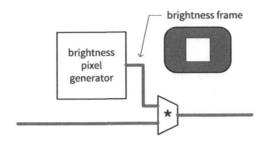

Figure 21.8   Conceptual diagram of the highlight core.

21.9.3   Square test-pattern generator core

We can convert the square test pattern generator circuit of Experiment 20.9.4 to an FPro video core by adding a wrapping circuit and a blender. Design the core, attach it to a user slot of the daisy video subsystem, derive the software driver to set the color and square size, and develop a test function to verify its operation.

21.9.4   Alpha blending circuit

The alpha blending circuit is discussed in Section 21.2.2. We can assume that the r, g, and b signals are in a **Q**4.0 format and α is in a **Q**0.8 format and perform multiplication, similar to that in Section 20.5. Design the alpha blending circuit, use it in the bar test-pattern generator core, derive the software driver to set the α value, and develop a test function to verify its operation.

21.9.5   "Highlight" core

A "highlight" core maintains the brightness of the central portion of a frame but dims its surroundings. The conceptual diagram is shown in Figure 21.8. The brightness generator circuit is similar to the square test pattern generator of Experiment 20.9.4. Instead of a color, the brightness generator outputs a "brightness level," β, where $0.0 \leq \beta \leq 1.0$. The input video stream, p, and the brightness stream, β, are mixed by a *multiplication blender*, which performs multiplication operation and output $\beta * p$. We can set the pixel brightness within the center square to 1.0 and the brightness in the surrounding area to a smaller value. This will "dim" the surrounding area and highlight the center portion. Design the core, attach it to a user slot of the daisy video subsystem, derive the software driver to set the square size and brightness level, and develop a test function to verify its operation.

21.9.6   SVGA synchronization core

In the SVGA standard, the screen resolution is 800 by 600 pixels and the pixel clock rate is 40M Hz, as shown in Table 20.1. Modify the video synchronization core, and the global frame counter of the vanilla daisy FPro system to facilitate the new standard. Derive the new system and verify its operation.

21.9.7   Configurable video synchronization core

Instead of just generating fixed clock frequencies, the Artix clock management core can be configured to change the output frequencies on the fly. It is done via the *DRP interface*, which is similar to the XADC macro cell discussed in Section 12.2.2. We can use this core to develop an enhanced video synchronization core, which can be configured by software to accommodate different video standards (VGA, SVGA, XVGA, etc.). The basic design procedure is as follows:

- Instantiate a dedicated clock management core inside the synchronization core.
- Derive a wrapping circuit to write clock management core's registers.
- Consult the Xilinx user guide to learn the procedure to set the output frequency.
- Revise the frame counter so that its "counting boundaries" can be dynamically adjusted.

Derive the code for the new core and verify its operation.

21.9.8   Pipelined video subsystem

Follow the discussion in Section 21.5.4 to add pipeline registers for the video cores of the daisy video subsystem. Modify the cores, reconstruct the subsystem, and verify its operation.

CHAPTER 22

SPRITE CORE

In computer graphics, a *sprite* is a small bitmap image that is integrated into a larger scene. A sprite can be realized by hardware or software. In this chapter, we describe the construction of a sprite video core that can create a still or an animated sprite and place it on top of a full-screen background frame.

22.1 INTRODUCTION

A *bitmap* defines a rectangular display area and the color for each pixel (i.e., "bit") in the display area. A *sprite* is a "small" bitmap, whose display area is a tiny fraction of the frame. A 12-pixel-by-20-pixel mouse pointer sprite is shown in Figure 22.1. One main application of the sprites is to integrate small graphic objects into a larger scene, such as creating an animated character on an otherwise static screen. Another application is to use them as *tiles* to construct a full-screen frame.

The sprites can be implemented by software or custom hardware. In the FPro video subsystem, we adopt the latter approach and construct custom video cores to generate sprites. The core's output constitutes a new layer and the sprites are overlaid over the background frame via the chroma-key blending scheme discussed in Section 21.2.2.

We demonstrate the general sprite construction with a mouse pointer core and an animated Pac-Man "ghost" character core in this chapter and illustrate the tile-based design with the on-screen text display in Chapter 23.

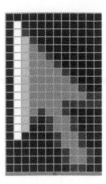

Figure 22.1 Mouse pointer sprite.

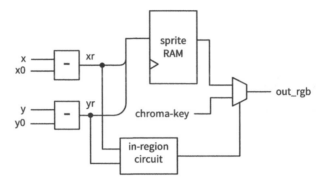

Figure 22.2 Block diagram of a sprite generation circuit.

22.2 BASIC DESIGN

The *sprite core* is a pixel generation video core. It is a video source and generates the entire frame. However, the sprite only occupies a small area and the remaining portion of the frame is filled with the chroma-key color. A representative screen of a mouse pointer is shown in the right panel of Figure 21.3(a). When it is mixed with a full-screen background frame (shown in the left) with chroma-key blending, the resulting frame shows just the sprite on top of the background frame, as demonstrated in Figure 21.3(b). Note that the black color is used as the chroma key in the daisy video subsystem.

The top-level conceptual diagram of the sprite pixel generation circuit is shown in Figure 22.2. It contains a sprite RAM, an "in-region" comparison circuit, two subtractors to calculate the relative addresses, and a multiplexer.

22.2.1 Sprite RAM

The sprite RAM stores the pixel data of the bitmap. The simple arrangement is to make the data width of a memory word identical to the color depth of a pixel and use one address for each pixel.

Since the RAM module is constructed as a one-dimensional array, the two-dimensional pixel coordinates must be converted to a one-dimensional memory

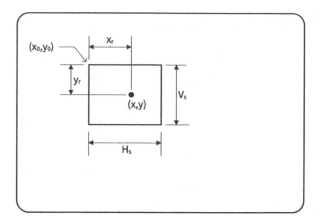

Figure 22.3   In-region comparison screen.

address. Let the pixel coordinates be (x_r, y_r) and let the dimension of the bitmap be H_S-pixels by V_S-pixels. We use the *row-major order* method, in which consecutive pixels of the rows are contiguous in memory, to do the conversion. The formula is

$$address = y_r * H_s + x_r$$

The calculation involves two arithmetic operations and infers a rather complex circuit with an adder and a multiplier, .

To simplify the processing, we restrict the dimensions of sprite to be a power of 2. Let H_s and V_s be 2^h and 2^v, respectively. The previous formula can be rewritten as

$$address = y_r * H_s + x_r = y_r * 2^h + x_r = y_r \ll h + x_r$$

where $\ll$ is the shift-left operator. In hardware implementation, it can be realized by a simple concatenation operation:

```
address <= yr & xr;
```

This arrangement can be summarized as follows. A 2^h-by-2^v sprite infers a 2^{v+h}-word RAM module, in which the h LSBs of the RAM address corresponds to a pixel's horizontal (x-axis) coordinate and the v MSBs of the RAM address corresponds to a pixel's vertical (y-axis) coordinate.

22.2.2   In-region comparison circuit

The in-region comparison circuit can be best explained by examining the sprite screen in Figure 22.3. The small rectangle represents a sprite. Its placement is specified by (x_0, y_0), the location of sprite's origin point (the top-left corner). The relevant parameters in the diagram are:

- x, y: current x- and y-coordinates of the scan point (which are generated by the frame counter)
- x_0, y_0: the x- and y-coordinates of the top-left corner of the sprite

- x_r, y_r: the "relative coordinates" defined as $x - x_0$ and $y - y_0$, representing the x- and y-axis differences between the current scan point and the sprite's origin
- H_s, V_s: the horizontal and vertical dimensions of the sprite

Note that x and y are driven by the 100M-Hz system clock but the x_0 and y_0 vary slowly and even can be static.

In the pixel generation circuit, the frame counter continues updating x and y. When x and y lie within the sprite region, the sprite pixels are routed to the output. In this condition, x_r and y_r correspond to the column number and row number of the bitmap and they can be used to form the RAM address and retrieve the pixel from the sprite RAM. On the other hand, when x and y lie outside the sprite region, the chroma-key color will be generated.

The illustration in Figure 22.3 shows that the condition for x and y within the sprite region is

$$0 \leq x_r < H_s \text{ and } 0 \leq y_r < V_s$$

The in-region comparison circuit implements this condition. Its output controls the multiplexer to select either the sprite RAM's readout or the chroma key, as shown in Figure 22.2.

22.3 MOUSE POINTER CORE

A *mouse pointer core* generates a mouse pointer overlay that can be placed on top of a full-screen frame. It is a custom sprite core and its construction follows the basic design approach outlined in Section 22.2.

22.3.1 Pointer sprite RAM

A 12-by-20 mouse pointer bitmap is shown in Figure 22.1. To make the core more flexible and facilitate future expansion, a larger 32-by-32 (2^5-by-2^5) RAM is selected for construction. The unused portion can be filled with the chroma-key color. The RAM can accommodate a sprite up to 2^{10} pixels. Since the 12-bit color is used in the FPro video subsystem, we create a 2^{10}-by-12 RAM module for this purpose.

Although a ROM is adequate for a pure mouse pointer application, we add an additional write port so that the bitmap can be updated by software later when needed. Since the memory module is implemented by Artix's internal BRAMs, no additional resource is required for the additional port. The HDL code follows the simple dual-port RAM template discussed in Section 7.4.3 and is shown in Listing 22.1.

Listing 22.1   Pointer sprite RAM

```
library ieee;
use ieee.std_logic_1164.all;
use ieee.numeric_std.all;
entity mouse_ram_lut is
   generic(
      ADDR_WIDTH: integer : =10;
      DATA_WIDTH:integer   : =12
   );
   port(
```

```vhdl
      clk    : in  std_logic;
      we     : in  std_logic;
      addr_w : in  std_logic_vector(ADDR_WIDTH-1 downto 0);
      addr_r : in  std_logic_vector(ADDR_WIDTH-1 downto 0);
      din    : in  std_logic_vector(DATA_WIDTH-1 downto 0);
      dout   : out std_logic_vector(DATA_WIDTH-1 downto 0)
   );
end mouse_ram_lut;

architecture beh_arch of mouse_ram_lut is
   type ram_type is array (0 to 2**ADDR_WIDTH-1)
       of std_logic_vector (DATA_WIDTH-1 downto 0);
   -- pointer pattern
   constant INIT_LUT: ram_type :=
   (
      x"f00",
         . . .
         . . .
   );
   signal ram: ram_type:=INIT_LUT;
begin
   process(clk)
   begin
      if (clk'event and clk = '1') then
         if (we = '1') then
            ram(to_integer(unsigned(addr_w))) <= din;
         end if;
         dout <= ram(to_integer(unsigned(addr_r)));
      end if;
   end process;
end beh_arch;
```

The initial pointer bitmap is defined as a 2^{10}-element constant array, INIT_LUT, and assigned as the initial value of the BRAM. These values will be loaded into the BRAM when the FPGA device is programmed. The module functions as a ROM if no software initiated write operation is performed.

22.3.2 Pixel generation circuit

The code for the mouse pointer pixel generation circuit follows the block diagram of Figure 22.2 and is shown in Listing 22.2.

Listing 22.2 Mouse pixel generation circuit

```vhdl
library ieee;
use ieee.std_logic_1164.all;
use ieee.numeric_std.all;
entity mouse_src is
   generic(
      CD        : integer := 12;
      ADDR      : integer := 10;
      KEY_COLOR : std_logic_vector(11 downto 0) := (others => '0')
   );
   port(
      clk       : std_logic;
      -- frame counter input
      x, y      : in  std_logic_vector(10 downto 0);
      -- origin of sprite
      x0, y0    : in  std_logic_vector(10 downto 0);
      -- sprite ram write
```

```
      we        : in   std_logic;
      addr_w    : in   std_logic_vector(9 downto 0);
      pixel_in  : in   std_logic_vector(CD - 1 downto 0);
      -- pixel output
      mouse_rgb : out  std_logic_vector(CD - 1 downto 0)
   );
end mouse_src;

architecture arch of mouse_src is
   -- sprite size
   constant H_SIZE        : integer := 32; -- horizontal size of sprite
   constant V_SIZE        : integer := 32; -- vertical size of sprite
   signal xr              : signed(11 downto 0); -- relative x position
   signal yr              : signed(11 downto 0); -- relative y position
   signal in_region       : std_logic;
   signal addr_r          : std_logic_vector(9 downto 0);
   signal full_rgb        : std_logic_vector(CD - 1 downto 0);
   signal out_rgb         : std_logic_vector(CD - 1 downto 0);
   signal out_rgb_d1_reg  : std_logic_vector(CD - 1 downto 0);
begin
   -- instantiate sprite RAM
   sprite_ram_unit : entity work.mouse_ram_lut
      generic map(
         ADDR_WIDTH => ADDR,
         DATA_WIDTH => CD
      )
      port map(
         clk    => clk,
         we     => we,
         addr_w => addr_w,
         din    => pixel_in,
         addr_r => addr_r,
         dout   => full_rgb
      );
   addr_r <= std_logic_vector(yr(4 downto 0) & xr(4 downto 0));
   -- relative coordinate calculation
   xr <= signed('0' & x) - signed('0' & x0);
   yr <= signed('0' & y) - signed('0' & y0);
   -- in-region comparison and multiplexing
   in_region <= '1' when (0 <= xr) and (xr < H_SIZE) and
                         (0 <= yr) and (yr < V_SIZE) else
                '0';
   out_rgb   <= full_rgb when in_region = '1' else KEY_COLOR;
   -- delay line (one clock)
   process(clk)
   begin
      if (clk'event and clk = '1') then
         out_rgb_d1_reg <= out_rgb;
      end if;
   end process;
   mouse_rgb <= out_rgb_d1_reg;
end arch;
```

There is one subtle point of the relative coordinate calculation. The relative coordinates are defined as $x - x_0$ and $y - y_0$, which can lead to negative results. The code converts the relevant signals to the signed format and expands them by one bit. Note that the in_region signal is not asserted when xr or yr is negative.

The RAM read operation introduces a one-clock delay for the full_rgb signal. Since the daisy video subsystem imposes a two-clock delay for the pixel generation operation, as discussed in Section 21.5.1, an artificial one-clock delay line is added for the output pixel stream.

22.3.3 Top-level design

The top-level of the mouse pointer core consists of the mouse pointer pixel generation circuit, the slot wrapping circuit, and the blending circuit.

Register map The processor interacts with the mouse pointer core as follows:

- set (i.e., write) the value of bypass register.
- specify (i.e., write) the location (i.e., x_0, y_0) of the mouse pointer sprite.
- update (i.e., write) the sprite RAM bitmap.

Recall that each video core occupies a 14-bit word address space. We use x to represent a "don't-care" bit and a to represent a local address bit. The address offsets and fields are:

- address offset 0x xxaa aaaa aaaa (sprite RAM)
  - bits 11 to 0: 12-bit pixel color
- address offset 1x xxxx xxxx xx00 (bypass register)
  - bit 0: bypass bit
- address offset 1x xxxx xxxx xx01 (x_0 register)
  - bits 9 to 0: x-coordinate of the mouse pointer sprite origin
- address offset 1x xxxx xxxx xx10 (y_0 register)
  - bits 9 to 0: y-coordinate of the mouse pointer sprite origin

HDL code The top-level HDL code instantiates the mouse pixel generator circuit, provides a write slot interface, and implements the blender. The blender consists of two stages. The first stage performs the chroma-key blending and the second stage uses a multiplexer for bypassing. The HDL code of the core is shown in Listing 22.3.

Listing 22.3   Mouse pointer core

```vhdl
library ieee;
use ieee.std_logic_1164.all;
use ieee.numeric_std.all;
entity chu_vga_slot_mouse_core is
   generic(
      CD          : integer                          := 12;
      ADDR_WIDTH : integer                          := 10;
      KEY_COLOR  : std_logic_vector(11 downto 0) := (others => '0')
   );
   port(
      clk      : in  std_logic;
      reset    : in  std_logic;
      -- frame counter
      x, y     : in  std_logic_vector(10 downto 0);
      -- video slot interface
      cs       : in  std_logic;
      write    : in  std_logic;
      addr     : in  std_logic_vector(13 downto 0);
      wr_data  : in  std_logic_vector(31 downto 0);
      -- stream interface
      si_rgb   : in  std_logic_vector(CD - 1 downto 0);
      so_rgb   : out std_logic_vector(CD - 1 downto 0)
```

```vhdl
    );
end chu_vga_slot_mouse_core;

architecture arch of chu_vga_slot_mouse_core is
    signal wr_en       : std_logic;
    signal wr_ram      : std_logic;
    signal wr_reg      : std_logic;
    signal wr_bypass   : std_logic;
    signal wr_x0       : std_logic;
    signal wr_y0       : std_logic;
    signal x0_reg      : std_logic_vector(10 downto 0);
    signal y0_reg      : std_logic_vector(10 downto 0);
    signal bypass_reg  : std_logic;
    signal mouse_rgb   : std_logic_vector(CD - 1 downto 0);
    signal chrom_rgb   : std_logic_vector(CD - 1 downto 0);
begin
    -- instantiate sprite generator
    slot_unit : entity work.mouse_src
        generic map(
            CD        => 12,
            KEY_COLOR => (others => '0')
        )
        port map(
            clk       => clk,
            x         => x,
            y         => y,
            x0        => x0_reg,
            y0        => y0_reg,
            we        => wr_ram,
            addr_w    => addr(ADDR_WIDTH - 1 downto 0),
            pixel_in  => wr_data(CD - 1 downto 0),
            mouse_rgb => mouse_rgb
        );
    -- registers and decoding
    process(clk, reset)
    begin
        if reset = '1' then
            x0_reg <= (others => '0');
            x0_reg <= (others => '0');
        elsif (clk'event and clk = '1') then
            if wr_x0 = '1' then
                x0_reg <= wr_data(10 downto 0);
            end if;
            if wr_y0 = '1' then
                y0_reg <= wr_data(10 downto 0);
            end if;
            if wr_bypass = '1' then
                bypass_reg <= wr_data(0);
            end if;
        end if;
    end process;
    wr_en      <= '1' when write = '1' and cs = '1' else '0';
    wr_ram     <= '1' when addr(13) = '0' and wr_en = '1' else '0';
    wr_reg     <= '1' when addr(13) = '1' and wr_en = '1' else '0';
    wr_bypass  <= '1' when wr_reg='1' and addr(1 downto 0)="00" else '0';
    wr_x0      <= '1' when wr_reg='1' and addr(1 downto 0)="01" else '0';
    wr_y0      <= '1' when wr_reg='1' and addr(1 downto 0)="10" else '0';
    -- chroma-key blending and multiplexing
    chrom_rgb <= mouse_rgb when mouse_rgb /= KEY_COLOR else si_rgb;
    so_rgb    <= si_rgb when bypass_reg = '1' else chrom_rgb;
end arch;
```

Figure 22.4   Ghost sprites.

22.4   "GHOST" CHARACTER CORE

The sprite scheme was used widely in early computer games. We create a sprite core for the "ghost" character of the Pac-Man game to demonstrate several other sprite-related techniques, including access of multiple images, animation, and palette lookup.

22.4.1   Multiple images and animation

The mouse video core displays a single static bitmap on the designated screen area. A sprite core can maintain multiple images and select one of them for display. For example, a set of four ghost images is shown in Figure 22.4. One of the four images can be selected and loaded. The image set is referred to as a *sprite sheet*.

Animation is the process to make the illusion of motion by rapidly presenting a sequence of slightly modified images. For example, the ghost images in Figure 22.4 have similar body shape but different eye positions and "skirts." When animated, a ghost appears to look around and run.

The *frame rate* specifies the speed with which the images are to be presented and is defined as the number of images (i.e., frames) per second. The exact frame rate depends on the nature of images, sizes, etc. Ten frames per second is a good starting point and can be adjusted afterward accordingly.

The animation can be achieved by software or by hardware. The software-based approach just needs a general sprite core, similar to the mouse pointer core described in Section 22.3. It stores the sprite sheet in the computer's main memory and uses software to load the designated image to the core at a specific time interval. The hardware-based approach creates an enhanced sprite core that has an expanded

Table 22.1 Palette coding of the ghost images

Code	12-bit color	Description
00	0x000 (black)	chroma key
01	0x111 (dark gray)	dark portion of eye
10	0xf00 (red)	body
11	0xfff (white)	white portion of eye

RAM storage for all images of the sprite sheet and a timing circuit to load the images automatically. We use the latter approach in this section.

22.4.2 Overview of the palette scheme

In computer graphics, a *palette* is defined as a set of colors that can be simultaneously displayed. In a given image, the palette may only use a subset of all the available colors. Properly encoding the palette subset can save memory space.

For example, there are 2^{12} possible colors for a video system with a color depth of 12 bits. To use all available colors, each pixel in an image consumes a 12-bit word. However, an image seldom contains all 2^{12} possible colors. For example, the ghost images in Figure 22.4 only uses four colors, which are black (used as the chroma key), red (for the body), white (for the white portion of the eye), and dark gray (for the dark portion of the eye). It is a very small subset of the 2^{12} colors. A two-bit code can be used to represent the four colors, as shown in Table 22.1. A *palette circuit* translates a code word into the desired color. It can be implemented as a lookup table or a combinational circuit.

The palette scheme can significantly reduce the memory usage and is an important technique for a system with limited memory. For example, the four ghost images contain 2^{10} pixels. For the 12-bit color representation, it requires a 2^{10}-by-12 RAM, which is 1536 bytes (i.e., $2^{10}*12$ bits). With the palette scheme, it only needs a 2^{10}-by-2 RAM and a 2^2-by-12 ROM lookup table, totaling 262 bytes (i.e., $2^{10}*2 + 2^2*12$ bits).

22.4.3 Ghost sprite RAM and the palette circuit

Four ghost images are shown in the sprite sheet of Figure 22.4. Each image is a 16-by-16 (2^4-by-2^4) sprite and contains 2^8 pixels. For the 12-bit color depth, it can be stored in a 2^8-by-12 RAM module. To accommodate the four images, a 2^{10}-by-12 ($4*2^8$-by-12) RAM module is needed. The two MSBs are used as the *sprite id* to identify the individual image. To save memory, we use a palette circuit discussed in the previous subsection. It requires a 2^{10}-by-2 RAM to store the encoded images and a 2^2-by-12 ROM lookup table.

The HDL code of the ghost sprite RAM is similar to that of the pointer RAM in Listing 22.1 and is shown in Listing 22.4. The RAM's data width is reduced from 12 to 2 and the INIT_LUT constant contains the encoded images.

Listing 22.4   Ghost sprite RAM

```vhdl
-- with four 16-by-16 Pac-Man ghost sprite
library ieee;
use ieee.std_logic_1164.all;
use ieee.numeric_std.all;
entity ghost_ram_lut is
   generic(
      ADDR_WIDTH : integer := 10;
      DATA_WIDTH : integer := 2
   );
   port(
      clk    : in   std_logic;
      we     : in   std_logic;
      addr_w : in   std_logic_vector(ADDR_WIDTH - 1 downto 0);
      addr_r : in   std_logic_vector(ADDR_WIDTH - 1 downto 0);
      din    : in   std_logic_vector(DATA_WIDTH - 1 downto 0);
      dout   : out  std_logic_vector(DATA_WIDTH - 1 downto 0)
   );
end ghost_ram_lut;

architecture beh_arch of ghost_ram_lut is
   type ram_type is array (0 to 2 ** ADDR_WIDTH - 1) of
     std_logic_vector(DATA_WIDTH - 1 downto 0);
   -- sprite LUT
   constant INIT_LUT : ram_type :=
   (
      "00",
      "00",
      . . .
   );
   signal ram : ram_type := INIT_LUT;
begin
   process(clk)
   begin
      if (clk'event and clk = '1') then
         if (we = '1') then
            ram(to_integer(unsigned(addr_w))) <- din;
         end if;
         dout <= ram(to_integer(unsigned(addr_r)));
      end if;
   end process;
end beh_arch;
```

The palette circuit is a lookup table and can be implemented by a ROM module. However, since the ghost palette contains only four code words, it can be easily realized by a selected signal assignment statement. Assume that the sprite RAM readout is connected to the plt_code signal. The HDL segment is

```vhdl
with plt_code select
   full_rgb <=
      x"000" when "00",      -- chroma key
      x"111" when "01",      -- dark gray
      x"f00" when "10",      -- red
      x"fff" when others;    -- white
```

In the Pac-Man game, there are four ghosts. They have the same body shapes but different colors, which are red, pink, orange, and cyan. The color variation can be realized by replacing the original red body color with a selectable color. This can be implemented by a multiplexer controlled by the `gc_color_sel` color selection signal. The revised code segment becomes

```
with gc_color_sel select
   ghost_rgb <=
      x"f00" when "00",        -- red
      x"f8b" when "01",        -- pink
      x"fa0" when "10",        -- orange
      x"0ff" when others;      -- cyan

with plt_code select
   full_rgb <=
      x"000"    when "00",     -- chroma key
      x"111"    when "01",     -- dark gray
      ghost_rgb when "10",     -- body color
      x"fff"    when others;   -- white
```

22.4.4 Animation timing circuit

The animation timing circuit is used to load a sequence of sprites at a designated rate. For a sprite RAM module with multiple images, the MSBs of the address are used as an id to identify a specific image. The previous read address statement in Section 22.2.1

```
address = yr(v-1 downto 0) & xr(h-1 downto 0);
```

is expanded to include `sid` (for "sprite id") bits and becomes

```
address = sid & yr(v-1 downto 0) & xr(h-1 downto 0);
```

While the `yr` and `xr` signals are updated continuously through the frame counter, the `sid` signal is controlled by the animation timing circuit.

The animation timing circuit constitutes a *frame-tick counter* and a *sprite id counter*. The frame-tick counter generates a one-clock *animation tick* based on the designated sprite animation frame rate. If the frame rate is f_{sprite}, a new image must be loaded every $\frac{1}{f_{sprite}}$ second. With the system clock frequency of f_{sys}, there are $\frac{f_{sys}}{f_{sprite}}$ system clocks in the interval and a mod-$\frac{f_{sys}}{f_{sprite}}$ counter can be derived accordingly.

An alternative design is to use the global frame counter's output, x and y, which can be decoded to generate a 60-Hz tick (the VGA frame refresh rate). A mod-$\frac{60}{f_{sprite}}$ counter can be derived and its size is much smaller than the previous approach.

The sprite id counter keeps track of the sprite id value. For a sequence of four ghost sprites, it can be implemented by a two-bit binary counter whose enable signal is connected to the animation tick signal.

22.4.5 Pixel generation circuit

The ghost pixel generation circuit generates a frame with a 16-by-16 sprite that can be overlaid over a full-screen background frame. It follows the basic approach of the mouse pixel generation circuit but incorporates the palette circuit and animation timing circuit.

A five-bit control signal, `ctrl`, is included to demonstrate the circuit's various capabilities. The first two MSBs specify the color of the ghost. The next MSB sets the operation mode. In the *still mode*, one of the four images is selected for display. In the *animation mode*, the four images are retrieved in a round-robin fashion automatically. If the still mode is used, the two LSBs select the desired still image. The code is shown in Listing 22.5.

Listing 22.5   Ghost pixel generation circuit

```vhdl
library ieee;
use ieee.std_logic_1164.all;
use ieee.numeric_std.all;
entity ghost_src is
   generic(
      CD        : integer                          := 12;
      ADDR      : integer                          := 10;
      KEY_COLOR : std_logic_vector(11 downto 0) := (others => '0')
   );
   port(
      clk      : std_logic;
      -- frame counter input
      x, y     : in  std_logic_vector(10 downto 0);
      -- orgin of sprite
      x0, y0   : in  std_logic_vector(10 downto 0);
      -- sprite control
      ctrl     : in  std_logic_vector(4 downto 0);
      -- sprite ram write
      we       : in  std_logic;
      addr_w   : in  std_logic_vector(9 downto 0);
      pixel_in : in  std_logic_vector(1 downto 0);
      sprite_rgb : out std_logic_vector(CD - 1 downto 0)
   );
end ghost_src;

architecture arch of ghost_src is
   constant H_SIZE    : integer := 16;   -- horizontal size of sprite
   constant V_SIZE    : integer := 16;   -- vertical size of sprite
   signal addr_r      : std_logic_vector(9 downto 0);
   signal sid         : std_logic_vector(1 downto 0); -- sprite id
   signal xr          : signed(11 downto 0); -- sprite x position
   signal yr          : signed(11 downto 0); -- sprite y position
   signal in_region   : std_logic;
   signal plt_code    : std_logic_vector(1 downto 0);
   signal frame_tick  : std_logic;
   signal ani_tick    : std_logic;
   signal c_reg       : unsigned(3 downto 0);
   signal c_next      : unsigned(3 downto 0);
   signal ani_reg     : unsigned(1 downto 0);
   signal ani_next    : unsigned(1 downto 0);
   signal x_d1_reg    : std_logic_vector(10 downto 0);
   signal full_rgb    : std_logic_vector(CD - 1 downto 0);
   signal ghost_rgb   : std_logic_vector(CD - 1 downto 0);
   signal out_rgb     : std_logic_vector(CD - 1 downto 0);
   signal out_d1_reg  : std_logic_vector(CD - 1 downto 0);
   alias gc_color_sel : std_logic_vector(1 downto 0) is ctrl(4 downto 3);
   alias auto         : std_logic is ctrl(2);
   alias gc_id_sel    : std_logic_vector(1 downto 0) is ctrl(1 downto 0);
begin
   --*****************************************************************
   -- sprite RAM
   --*****************************************************************
```

```
— instantiate sprite RAM
slot_ram_unit : entity work.ghost_ram_lut
    generic map(
        ADDR_WIDTH => ADDR,
        DATA_WIDTH => 2
    )
    port map(
        clk    => clk,
        we     => we,
        addr_w => addr_w,
        din    => pixel_in,
        addr_r => addr_r,
        dout   => plt_code
    );
addr_r <= sid & std_logic_vector(yr(3 downto 0) & xr(3 downto 0));
—-*****************************************************************
— ghost color control
—-*****************************************************************
— ghost color selection
with gc_color_sel select
    ghost_rgb <=
        x"f00" when "00",    — red
        x"f8b" when "01",    — pink
        x"fa0" when "10",    — orange
        x"0ff" when others;  — cyan
— palette table
with plt_code select
    full_rgb <=
        x"000" when "00",    — chroma key
        x"111" when "01",    — dark gray
        ghost_rgb when "10", — ghost body color
        x"fff" when others;  — white
—-*****************************************************************
— in—region circuit
—-*****************************************************************
— relative coordinate calculation
xr  <= signed('0' & x) - signed('0' & x0);
yr  <= signed('0' & y) - signed('0' & y0);
— in—region comparison and multiplexing
in_region <= '1' when (0 <= xr) and (xr < H_SIZE) and
                      (0 <= yr) and (yr < V_SIZE) else
             '0';
out_rgb   <= full_rgb when in_region = '1' else KEY_COLOR;
—-*****************************************************************
— animation timing control
—-*****************************************************************
— counters
process(clk)
begin
    if (clk'event and clk = '1') then
        x_d1_reg <= x;
        c_reg <= c_next;
        ani_reg <= ani_next;
    end if;
end process;
c_next <=
    (others => '0') when frame_tick='1' and c_reg=9 else
    c_reg +1 when frame_tick='1' else
    c_reg;
ani_next <= ani_reg + 1 when ani_tick='1' else ani_reg;
— 60—Hz tick from frame counter
```

```
      frame_tick <= '1' when signed(x_d1_reg) = 0 and
                            signed(x) = 1 and signed(y) = 0 else
                    '0';
      — sprite  animation  id  tick
      ani_tick   <= '1' when frame_tick = '1' and c_reg = 0 else '0';
      — sprite  id  selection
      sid <= std_logic_vector(ani_reg) when auto = '1' else gc_id_sel;
      —*****************************************************************
      — delay  line  (one  clock)
      —*****************************************************************
      process(clk)
      begin
         if (clk'event and clk = '1') then
            out_d1_reg <= out_rgb;
         end if;
      end process;
      sprite_rgb <= out_d1_reg;
end arch;
```

The animation timing circuit uses the frame counter to generate the 60-Hz tick. The tick, `frame_tick`, is asserted at the transition edge of $(0, 0)$ to $(0, 1)$, as shown in the following conditional expression

$$\texttt{(signed(x\_d1\_reg)=0)} \textbf{ and } \texttt{(signed(x)=1)} \textbf{ and } \texttt{(signed(y)=0)}$$

The Boolean condition indicates that current `y` is 0, current `x` is 1, and the previous `x` (i.e., `x_d1_reg`) is 0. The `frame_tick` signal is used to trigger a mod-10 counter and to generate 6-Hz `ani_tick` signal, which triggers the two-bit sprite animation id counter and leads to the animation speed of six frames per second.

22.4.6   Top-level design

The top-level of the ghost pointer core consists of the ghost pointer pixel generation circuit, the slot wrapping circuit, and the blending circuit.

Register map   The interaction between the processor and the ghost core is similar to that of the mouse pointer core and thus the mouse pointer core's register map in Section 22.3.3 can be used for the ghost core. However, in the ghost core, the processor can issue a command to configure the ghost color and the animation mode. This leads to one additional register, whose address offset and fields are

- address offset 1x xxxx xxxx xx11 (control register)
  - bits 4 to 3: selection of ghost color
  - bit 2: selection of mode (1 for the animation mode and 0 for the still mode)
  - bits 1 to 0: selection of a specific sprite in still mode

HDL code   The top-level HDL code instantiates the ghost pixel generator circuit, provides a write slot interface, and implements the blender. The blender consists of two stages. It is similar to the mouse pointer core code in Listing 22.3 and is shown in Listing 22.6.

Listing 22.6   Ghost core

```vhdl
library ieee;
use ieee.std_logic_1164.all;
use ieee.numeric_std.all;
entity chu_vga_slot_ghost_core is
   generic(
      CD          : integer := 12;
      ADDR_WIDTH : integer := 10;
      KEY_COLOR  : std_logic_vector(11 downto 0) := (others => '0')
   );
   port(
      clk      : in  std_logic;
      reset    : in  std_logic;
      -- frame counter
      x        : in  std_logic_vector(10 downto 0);
      y        : in  std_logic_vector(10 downto 0);
      -- video slot interface
      cs       : in  std_logic;
      write    : in  std_logic;
      addr     : in  std_logic_vector(13 downto 0);
      wr_data  : in  std_logic_vector(31 downto 0);
      -- stream interface
      si_rgb   : in  std_logic_vector(CD - 1 downto 0);
      so_rgb   : out std_logic_vector(CD - 1 downto 0)
   );
end chu_vga_slot_ghost_core;

architecture arch of chu_vga_slot_ghost_core is
   signal wr_en          : std_logic;
   signal wr_ram         : std_logic;
   signal wr_reg         : std_logic;
   signal wr_ctrl        : std_logic;
   signal wr_bypass      : std_logic;
   signal wr_x0, wr_y0   : std_logic;
   signal bypass_reg     : std_logic;
   signal x0_reg         : std_logic_vector(10 downto 0);
   signal y0_reg         : std_logic_vector(10 downto 0);
   signal ctrl_reg       : std_logic_vector(4 downto 0);
   signal sprite_rgb     : std_logic_vector(CD - 1 downto 0);
   signal chrom_rgb      : std_logic_vector(CD - 1 downto 0);
begin
   -- instantiate sprite generator
   sprite_unit : entity work.ghost_src
      generic map(
         CD        => 12,
         KEY_COLOR => (others => '0')
      )
      port map(
         clk        => clk,
         x          => x,
         y          => y,
         x0         => x0_reg,
         y0         => y0_reg,
         we         => wr_ram,
         ctrl       => ctrl_reg,
         addr_w     => addr(ADDR_WIDTH - 1 downto 0),
         pixel_in   => wr_data(1 downto 0),
         sprite_rgb => sprite_rgb
      );
   -- registers and decoding
   process(clk, reset)
```

```vhdl
   begin
      if reset = '1' then
         bypass_reg <= '0';
         x0_reg      <= (others => '0');
         y0_reg      <= (others => '0');
         ctrl_reg    <= "00100";          -- red animation
      elsif (clk'event and clk = '1') then
         if wr_bypass = '1' then
            bypass_reg <= wr_data(0);
         end if;
         if wr_x0 = '1' then
            x0_reg <= wr_data(10 downto 0);
         end if;
         if wr_y0 = '1' then
            y0_reg <= wr_data(10 downto 0);
         end if;
         if wr_ctrl = '1' then
            ctrl_reg <= wr_data(4 downto 0);
         end if;
      end if;
   end process;
   wr_en     <= '1' when write = '1' and cs = '1' else '0';
   wr_ram    <= '1' when addr(13) = '0' and wr_en = '1' else '0';
   wr_reg    <= '1' when addr(13) = '1' and wr_en = '1' else '0';
   wr_bypass <= '1' when wr_reg='1' and addr(1 downto 0)="00" else '0';
   wr_x0     <= '1' when wr_reg='1' and addr(1 downto 0)="01" else '0';
   wr_y0     <= '1' when wr_reg='1' and addr(1 downto 0)="10" else '0';
   wr_ctrl   <= '1' when wr_reg='1' and addr(1 downto 0)="11" else '0';
   -- chroma-key blending and multiplexing
   chrom_rgb <= sprite_rgb when sprite_rgb /= KEY_COLOR else si_rgb;
   so_rgb    <= si_rgb when bypass_reg = '1' else chrom_rgb;
end arch;
```

22.5 SPRITE CORE DRIVER AND TESTING PROGRAM

The mouse pointer core and ghost core have similar organization and share the similar register map. We use a single driver for both cores.

22.5.1 Sprite core driver

We define a `SpriteCore` class for the sprite cores. The class definition is similar to that of the `GpvCore` class. In addition to `wr_mem()` and `bypass()`, it adds two more methods. The `move_xy()` method moves the sprite to a specific location and the `wr_ctr()` method writes the control register. The class definition and implementation are shown in Listings 22.7 and 22.8, respectively.

Listing 22.7 `SpriteCore` class definition (in `vga_core.h`)

```cpp
class SpriteCore {
public:
   // register map
   enum {
      BYPASS_REG = 0x2000,
      X_REG = 0x2001,
      Y_REG = 0x2002,
      SPRITE_CTRL_REG = 0x2003
   };
```

```
    // chroma key
    enum {
        KEY_COLOR = 0
    };
    /* methods */
    SpriteCore(uint32_t core_base_addr, int size);
    ~SpriteCore();                     // not used
    void wr_mem(int addr, uint32_t color);
    void move_xy(int x, int y);
    void wr_ctrl(int32_t cmd);
    void bypass(int by);
private:
    uint32_t base_addr;
    int size;       // sprite memory size
};
```

<hr>

Listing 22.8   SpriteCore class implementation (in **vga_core.cpp**)

<hr>

```
SpriteCore::SpriteCore(uint32_t core_base_addr, int sprite_size) {
    base_addr = core_base_addr;
    size = sprite_size;
}
SpriteCore::~SpriteCore() {}

void SpriteCore::wr_mem(int addr, uint32_t color) {
    io_write(base_addr, addr, color);
}

void SpriteCore::bypass(int by) {
    io_write(base_addr, BYPASS_REG, (uint32_t ) by);
}

void SpriteCore::move_xy(int x, int y) {
    io_write(base_addr, X_REG, x);
    io_write(base_addr, Y_REG, y);
    return;
}

void SpriteCore::wr_ctrl(int32_t cmd) {
    io_write(base_addr, SPRITE_CTRL_REG, cmd);
}
```

<hr>

Note that the address offset for the sprite RAM is 0, as shown in the register map of Section 22.3.3. The wr_mem() method can update a pixel in the sprite RAM and thus can be used to load a new bitmap when needed.

22.5.2 Testing program

The overall testing program for the vanilla daisy FPro system is discussed in Section 21.7.3. The mouse_check() function verifies the basic functionalities of the mouse pointer core and the code is shown in Listing 22.9.

Listing 22.9   Mouse pointer sprite test function (in **main_video_test.cpp**)

<hr>

```
void mouse_check(SpriteCore *mouse_p) {
    int x, y;

    mouse_p->bypass(0);
    // clear top and bottom lines
```

```
for (int i = 0; i < 32; i++) {
   mouse_p->wr_mem(i, 0);
   mouse_p->wr_mem(31 * 32 + i, 0);
}

// slowly move mouse pointer
x = 0;
y = 0;
for (int i = 0; i < 80; i++) {
   mouse_p->move_xy(x, y);
   sleep_ms(50);
   x = x + 4;
   y = y + 3;
}
sleep_ms(3000);
// load top and bottom rows
for (int i = 0; i < 32; i++) {
   sleep_ms(20);
   mouse_p->wr_mem(i, 0x00f);
   mouse_p->wr_mem(31 * 32 + i, 0xf00);
}
sleep_ms(3000);
}
```

The middle segment slowly moves the mouse pointer from the top-left corner to the center. The beginning and ending segments load new pixel data to the top row and bottom row of the sprite. The latter draws a blue line and a red line and the former segment clears the color lines by writing the chroma-key color.

The ghost_check() function verifies the basic functionalities of the ghost core and the code is shown in Listing 22.10.

Listing 22.10   Ghost sprite test function (in main_video_test.cpp)

```
void ghost_check(SpriteCore *ghost_p) {
   int x, y;

   // slowly move mouse pointer
   ghost_p->bypass(0);
   ghost_p->wr_ctrl(0x1c);   //animation; blue ghost
   x = 0;
   y = 100;
   for (int i = 0; i < 156; i++) {
      ghost_p->move_xy(x, y);
      sleep_ms(100);
      x = x + 4;
      if (i == 80) {
         // change to red ghost half way
         ghost_p->wr_ctrl(0x04);
      }
   }
   sleep_ms(3000);
}
```

The program slowly moves the animated ghost character from left to right and changes its color halfway. After demonstration, the program includes a statement to set the parameters of the control register with the five rightmost slide switches (i.e., sw(15 downto 11)):

```
ghost.wr_ctrl(sw.read() >> 11);
```

It can be used to show the animation mode, select one out of four ghost sprites, and set the ghost color.

22.6   BIBLIOGRAPHIC NOTES

The sprite- and tile-based design were used widely in early video games and the concepts are still used in today's computer graphics. Additional materials can be found on the Wikipedia website by searching the keywords "sprite" and "tile-based video game." The website also provides interesting information on the "chroma key" scheme and the "Pac-Man" game.

22.7   SUGGESTED EXPERIMENTS

22.7.1   Mouse pointer control with a PS2 core

The mouse pointer icon is best controlled by a physical mouse device. Expand the vanilla daisy FPro system to include a PS2 core, attach a physical mouse to the port, derive software to control the mouse pointer movement, and verify its operation.

22.7.2   Emulated ghost core

The mouse pointer core sprite RAM can be updated and thus can accommodate any 32-by-32 sprite. We can use it to emulate the ghost core and its animation capability by updating the sprite RAM content with software. Derive a software program to mimic the functionality of `test_ghost()` in Listing 22.10.

22.7.3   Palette circuit for the mouse pointer sprite

The mouse pointer sprite in Figure 22.1 only uses four colors (white, light gray, dark gray, and a black chroma key). The colors can be represented by a two-bit code set. Redesign the mouse pointer core with a new two-bit wide sprite RAM and a palette circuit and verify its operation.

22.7.4   Sprite scaling circuit

A sprite can be "magnified" by enlarging the screen pixels. For example, the 12-by-20 mouse pointer sprite in Figure 22.1 can be scaled to a 24-by-40 sprite by enlarging the original bitmap four times (i.e., expanding one pixel to four pixels). We want to design a new mouse pixel generation circuit that can generate the original pointer and the magnified pointer. An additional one-bit control signal will be used as the selection signal. Develop the new core, update the software driver, and verify its operation.

22.7.5   Portrait mode display

In the *portrait mode*, a monitor is turned 90 degree and the image is displayed in a "vertical" screen. We can enhance the mouse pointer core to support both

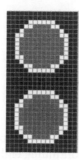

Figure 22.5   LED sprites.

landscape mode (normal "horizontal" screen) and portrait mode and use a one-bit control signal to select the mode. Develop the new core, update the software driver, and verify its operation.

22.7.6   Multiple object generation

It is possible to place multiple sprite objects on screen simultaneously. This can be achieved by duplicating the in-region circuit and relevant registers and adding "glue logic" to do the multiplexing and control. We want to enhance the ghost core to display two ghost objects at the same time. The two ghosts should function independently, each specified with its own location, color, and animation mode. Develop the new core, update the software driver, and verify its operation.

22.7.7   Animation speed control

The animation of the ghost core is controlled by the special timing circuit discussed in Section 22.4.4. The speed of animation (i.e., frame rate) is fixed. We want to enhance the core to make the frame rate configurable. The range of the rate is from 1 frame per second to 30 frames per second. An additional five-bit control signal will be used to set the rate. Develop the new core, update the software driver, and verify its operation.

22.7.8   Imitated blinking LED: part I

We want to mimic the operation of a discrete LED on screen. The two 16-by-16 sprites in Figure 22.5 represents the "off" and "on" states of a discrete LED. The LED can be in one of the following states:
- On
- Off
- Automatically blinking at a high rate (around 15 Hz)
- Automatically blinking at a slow rate (around 5 Hz)

Develop a new LED core and the software driver and verify its operation.

22.7.9 Imitated blinking LED: part II

Repeat Experiment 22.7.8 to mimic a tri-color LED. The core should have an additional 12-bit control signal to set the color of LED. Develop the LED core and the software driver and verify its operation.

22.7.10 Imitated blinking LED: part III

Instead of showing a single LED, we want to display an LED array on screen. The object should be a 128-by-16 image containing eight individual LEDs. The LEDs are controlled and operated independently. Extend the LED core in Experiment 22.7.8, develop the software driver, and verify its operation.

CHAPTER 23

ON-SCREEN-DISPLAY CORE

A sprite occupies a small rectangular space and can be treated as a *tile*. Tiles can be laid out adjacent to one another in a predefined grid to form a larger image or even the entire screen. In this chapter, we use a tile-based scheme to construct an OSD (on-screen-display) core, which adds an overlay of textual display to the background frame.

23.1 INTRODUCTION TO TILE GRAPHICS

In Chapter 22, a sprite is used as a small "floating" object, such as a mouse pointer icon, overlaid on top of a full-screen background frame. Placing the sprite on the designated location involves an in-region circuit to calculate and compare the relative pixel addresses, as discussed in Section 22.2.2. Placing multiple sprites on the screen simultaneously can be achieved by using duplicated in-region circuits, as suggested in Experiment 22.7.6. Because of the complexity of the in-region circuit, this approach is not feasible for a large number of sprites.

A *tile-based graphic* uses the sprites in a different way. It treats a sprite as a rectangular *tile* and a screen as a predefined grid of slots. Instead of being put in any coordinates, a sprite can only be placed in a slot. A tile-based system usually contains a collection of similarly sized sprites, referred to as a *tile set* or a *sprite sheet*. A full screen can be constructed by filling the slots with the desired sprites, just like covering a wall with tiles.

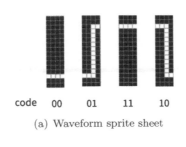

(a) Waveform sprite sheet

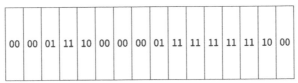

(b) Waveform tiles with codes

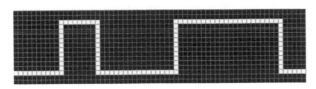

(c) Waveform screen

Figure 23.1 Tile-based waveform screen.

We begin with an example to demonstrate the concept of the tile-based graphics system. Assume that we wish to display a digitized waveform in a 64-pixel-by-16-pixel screen. A representative screen is shown in Figure 23.1(c). Close observation shows that a waveform can be constructed with four basic patterns, which are "steady-0," "0-to-1 edge," "1-to-0 edge," and "steady-1." A tile-based system can be constructed as follows:

- Define a 4-pixel-by-16-pixel area as a tile.
- Create a tile set of four sprites for the four patterns and assign each tile a two-bit code, as shown in Figure 23.1(a).
- Treat the 64-pixel-by-16-pixel screen as a 16-tile-by-1-tile (i.e., $\frac{64}{4}$-by-$\frac{16}{16}$) screen, as shown in Figure 23.1(b).
- Enter the sprite codes to the slots in the tile screen, which is shown inside the tile of Figure 23.1(b).
- Draw the screen by replacing the codes with actual sprites, which leads to the screen in Figure 23.1(c).

Two RAM modules are needed to facilitate the construction. The first module is a sprite RAM, which stores the four bitmap images of the tile set, and its organization is similar to that of the ghost sprites discussed in Section 22.4.3. Since one-bit color is adequate for this application, a $4*4*16$-by-1 RAM module can be used to accommodate the pixel data. The second module is a *tile RAM*, which stores the sprite codes. The four patterns can be encoded with a two-bit code, as defined in the bottom of Figure 23.1(a). The 16-tile-by-1-tile grid requires a $16*1$-by-2

RAM module for storage. The content of the tile RAM of the waveform shown in Figure 23.1(b) is

```
00,00,01,11,10,00,00,00,00,01,11,11,11,11,11,10,00
```

A main advantage of the tile-based scheme is its efficient use of memory. In the previous demonstration, the direct bitmap implementation needs 1024 (i.e., $64*16$) bits of memory. The tile-based system only needs 288 bits, which include 256 bits (i.e., $4*4*16*1$) for the sprite RAM and 32 bits (i.e., $16*1*2$) for the tile RAM. The saving becomes much more significant for a more realistic scenario. For example, a 640-pixel-by-64-pixel four-waveform screen can be converted into a 160-tile-by-4-tile screen. The required memory spaces for the bitmap and tile-based graphic are 41,024 bits (i.e., $640*64*1$) and 1,536 bits (i.e., $4*4*16*1 + 160*4*2$), respectively. The tile-based graphics were used widely in early computer games because of scarcity of hardware resources. It can be an effective scheme for simple embedded systems that only require a primitive display.

23.2 BASIC OSD DESIGN

An OSD (on-screen-display) core generates a textual overlay. The overlay itself constitutes a "text-mode" display. It can be used to implement a console window for a command-line interface, a simple menu (e.g., the configuration menu of a DVD player), or supplemental text information (such as the subtitle and one-line stock ticker banner).

23.2.1 Text-mode display

Text mode is a display mode in which the screen is treated as a uniform grid of *character tiles*. The character is selected from a predefined *character set*, which is generally based on the ASCII codes of Table 11.1. The set includes digits, uppercase and lowercase letters, and punctuation symbols. Note that ASCII codes 00_{16} to $1F_{16}$ (the first column in the ASCII table) are reserved for the nonprintable control characters. They can be used to implement special graphic symbols. For example, the 06_{16} code will generate a spade pattern, ♠, on the screen.

The patterns of the character set constitute the *font*. A variety of fonts is available. We choose a font similar to the one used in the early IBM PC. This font contains 128 characters and uses an 8-pixel-by-16-pixel tile for a character sprite. A representative sprite, the uppercase letter A, is shown in Figure 23.2(a). With a 640-by-480 VGA screen, 80 (i.e., $\frac{640}{8}$) tiles can be fitted into a horizontal line and 30 (i.e., $\frac{480}{16}$) tiles can be fitted into a vertical line.

In summary, the text-mode display can be implemented by a tile-based system and its main characteristics are as follows:

- color depth: one bit
- tile size: 8-pixel by 16-pixel (i.e., 8 columns and 16 rows)
- screen size in terms of tiles: 80-tile by 30-tile (i.e., $\frac{640}{8}$ by $\frac{480}{16}$)
- tile set size: 128 sprites for digits, letters, punctuation symbols, and graphics symbols

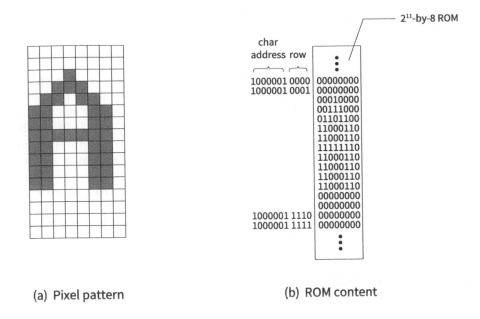

(a) Pixel pattern   (b) ROM content

Figure 23.2   Font pattern for the letter A.

Many fonts extend the character set to 256 characters and use the additional characters to define a set of graphical symbols for drawing boxes and imitating widgets found in GUI (graphic user interface).

23.2.2   Font ROM

The tile set can be stored in a sprite RAM. Since the font rarely changes, a ROM module can be used. This sprite memory module is frequently referred to as *font ROM*.

In Chapter 22, we set the RAM's data width to correspond to the color depth. For example, the 32-by-32 mouse pointer sprite in Section 22.3.1 uses 12-bit color and the memory module is configured as a $32*32$-by-12 RAM. If this scheme is adopted, a single character sprite infers an $8*16$-by-1 RAM. However, to simplify the representation, we use a 16-by-8 (i.e., 2^4-by-8) module instead. This allows us to mirror font ROM data with the bitmap, as shown in Figure 23.2(b).

The font includes 128 (i.e., 2^7) characters and can be accommodated by a 2^{11}-by-8 (i.e., 2^7*2^4-by-8) ROM module. In this ROM, the seven MSBs of the 11-bit address are the sprite id, which is the ASCII code of the corresponding character, and the four LSBs of the address are used to identify the row within a character sprite. The address and ROM content for the letter "A" are shown in Figure 23.2(b).

23.2.3   Tile RAM

The text-mode VGA screen can be treated as a 80-tile-by-30-tile display, in which each tile slot contains the character's ASCII code. This leads to a $80*30$-by-7 RAM module. However, since the on-chip memory is constructed from Xilinx FPGA's

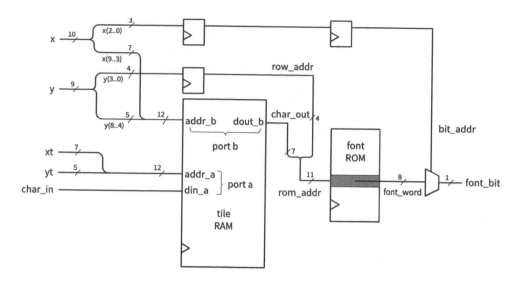

Figure 23.3   Conceptual block diagram of an OSD pixel generation circuit.

2K (2^{11}) byte BRAMs, we use a $128*32$-by-8 (i.e., 2^{12}-byte) module for the tile RAM.

The BRAMs need to be configured as a dual-port RAM. The first port is for the pixel generation, in which the tile data (i.e., the ASCII code) is retrieved (i.e., read) continuously. The second port is for the video controller interface, in which the processor writes the ASCII character to the desired tile location.

23.2.4 Basic organization

The top-level conceptual diagram of the OSD pixel generation circuit is shown in Figure 23.3. The circuit contains a font ROM, a tile RAM, and an 8-to-1 multiplexer. Its operation is done in stages:

- The frame counter generates pixel's x- and y-coordinates, which are the 10-bit x signal and the 9-bit y signal.
- The upper bits, x(9 downto 3) and y(8 downto 4), correspond to the x- and y-coordinates of the current tile. They are concatenated to form the 12-bit tile RAM address and connected to the address input, addr_b, of tile RAM's read port (port b in the diagram). The readout data, char_out, is the character's ASCII code stored in this location.
- The retrieved 7-bit ASCII code becomes the seven MSBs of the font ROM and is used to identify the location the specific character sprite. It is concatenated with the four LSBs of the screen's y-axis coordinate to form the 11-bit font ROM address, rom_addr. The readout data, font_word, corresponds to an 8-pixel row of the sprite.
- The three LSBs of the screen's x-axis coordinate, bit_addr, specify the desired pixel location and an 8-to-1 multiplexer routes the pixel, font_bit, to the output.

Recall that the BRAM's operation is synchronous, as discussed in Section 7.1.1, and the read operation introduces a one-clock delay. Thus, it takes two clock cycles to obtain the pixel data from the cascading tile RAM and font ROM. The x and y signals need to be passed through the delay line, which is implemented by the registers in Figure 23.3, to match the data flow.

The second port (port a in the diagram) of the tile RAM is for an external circuit to write an ASCII character to a specific tile location. The seven-bit x-coordinate and 5-bit y-coordinate, xt and yt, specify the tile location. They are concatenated to form the 12-bit tile RAM address and connected to the address input, addr_a. The character's ASCII code, char_in, is connected to the data input, din_a.

23.3   OSD CORE

An *OSD core* generates a text overlay on top of a background frame. It follows the basic organization discussed in Section 23.2.4.

23.3.1   Font ROM

The font ROM contains 128 character sprites and is implemented by a 2^{11}-by-8 memory module, The seven MSBs of the 11-bit address are the character's ASCII code and the four LSBs are the sprite's row number, as shown in Figure 23.2(b).

The HDL code follows the template discussed in Section 7.4.5 and is shown in Listing 23.1.

Listing 23.1   Font ROM

```
library ieee;
use ieee.std_logic_1164.all;
use ieee.numeric_std.all;
entity font_rom is
   port(
      clk: in std_logic;
      addr: in std_logic_vector(10 downto 0);
      data: out std_logic_vector(7 downto 0)
   );
end font_rom;

architecture arch of font_rom is
   constant ADDR_WIDTH: integer:=11;
   constant DATA_WIDTH: integer:=8;
   signal addr_reg: std_logic_vector(ADDR_WIDTH-1 downto 0);
   type rom_type is array (0 to 2**ADDR_WIDTH-1)
        of std_logic_vector(DATA_WIDTH-1 downto 0);
   -- ROM definition
   constant ROM: rom_type:=(      -- 2^11-by-8
   -- code x00 (null character)
   "00000000", -- 0
   "00000000", -- 1
   "00000000", -- 2
   "00000000", -- 3
   "00000000", -- 4
   "00000000", -- 5
   "00000000", -- 6
   "00000000", -- 7
   "00000000", -- 8
```

```
"00000000", — 9
"00000000", — a
"00000000", — b
"00000000", — c
"00000000", — d
"00000000", — e
"00000000", — f
— code x01 (smiling face)
"00000000", — 0
"00000000", — 1
"01111110", — 2   ******
"10000001", — 3   *      *
"10100101", — 4   *  *   *  *
"10000001", — 5   *      *
"10000001", — 6   *      *
"10111101", — 7   *  ****  *
"10011001", — 8   *   **   *
"10000001", — 9   *      *
"10000001", — a   *      *
"01111110", — b   ******
"00000000", — c
"00000000", — d
"00000000", — e
"00000000", — f
. . .
```

The content of the font ROM is defined as a 2^{11}-element constant array and mirrors the pixel patterns, as shown in Figure 23.2(b). These values will be loaded into the BRAM when the FPGA device is programmed.

23.3.2   Pixel generation circuit

The OSD pixel generation circuit generates a 80-by-30 text-mode frame. To make it more versatile, the following features are added:

- A palette circuit to set the text's foreground and background colors.
- A control bit to reverse the foreground color and background color of a character.
- A mechanism to control the text display area.

The first feature controls the color of the text. The OSD circuit uses one-bit color and thus can only display a foreground color and a background color. A palette circuit can map them into two 12-bit colors, such as green on black or white on blue. It is also possible to make the background transparent by assigning the chroma key as the background color.

The second feature reverses the foreground color and background color, which can be used to highlight a specific portion of the text. The condition can be specified by incorporating one additional control bit to the data word of the tile RAM. Recall that the tile RAM is constructed with eight-bit data but only stores seven-bit ASCII code, as discussed in Section 23.2.2. The extra bit can be used for this purpose.

The third feature controls the text display area. Instead of covering the entire frame with text, some applications only display the text in a limited area. For example, we may just need a small command console window or a one-line status message. Our design accomplishes this by defining a specific "transparency" ASCII code, 0x00 (the "null" character), for this purpose. There is no bitmap associated with this code. When this code is retrieved, the circuit does not read data from the font ROM but generates the chroma key for the entire tile. With the subsequent chroma-key blender, the corresponding tile becomes transparent. Note that the null character is different from the "blank space" character. The latter character's ASCII code is 0x20 and the corresponding tile displays the background color.

The HDL code for the OSD pixel generation circuit is shown in Listing 23.2. The naming and connection of the signals follows the block diagram of Figure 23.3. The code also adds logic to implement the additional features.

Listing 23.2   OSD pixel generation circuit

```vhdl
library ieee;
use ieee.std_logic_1164.all;
use ieee.numeric_std.all;
entity osd_src is
   generic(
      CD         : integer                               := 12;
      KEY_COLOR : std_logic_vector(11 downto 0) := (others => '0')
   );
   port(
      clk        : std_logic;
      x, y       : in  std_logic_vector(10 downto 0);
      -- tile ram write port
      xt         : in  std_logic_vector(6 downto 0);
      yt         : in  std_logic_vector(4 downto 0);
      ch_in      : in  std_logic_vector(7 downto 0); -- char data
      we_ch      : in  std_logic;          -- char write enable
      -- foreground/background color of char tile
      front_rgb : in  std_logic_vector(CD - 1 downto 0);
      back_rgb  : in  std_logic_vector(CD - 1 downto 0);
      -- stream
      osd_rgb    : out std_logic_vector(CD - 1 downto 0)
   );
end osd_src;

architecture arch of osd_src is
   constant NULL_CHAR  : std_logic_vector(6 downto 0):= (others=>'0');
   -- font ROM
   signal char_addr   : std_logic_vector(6 downto 0);
   signal rom_addr    : std_logic_vector(10 downto 0);
   signal row_addr    : std_logic_vector(3 downto 0);
   signal bit_addr    : std_logic_vector(2 downto 0);
   signal font_word   : std_logic_vector(7 downto 0);
   -- char tile RAM
   signal addr_a      : std_logic_vector(11 downto 0);
   signal addr_b      : std_logic_vector(11 downto 0);
   signal ch_ram_out  : std_logic_vector(7 downto 0);
   signal ch_d1_reg   : std_logic_vector(7 downto 0);
   -- delay line
   signal x_delay1_reg : std_logic_vector(2 downto 0);
   signal x_delay2_reg : std_logic_vector(2 downto 0);
   signal y_delay1_reg : std_logic_vector(3 downto 0);
   -- other signals
   signal font_bit    : std_logic;
   signal f_rgb       : std_logic_vector(CD - 1 downto 0);
   signal b_rgb       : std_logic_vector(CD - 1 downto 0);
   signal p_rgb       : std_logic_vector(CD - 1 downto 0);
   signal rev_bit     : std_logic;

begin
```

```vhdl
-- *****************************************************************
-- instantiation
-- *****************************************************************
-- instantiate font ROM
font_unit : entity work.font_rom
   port map(
       clk  => clk,
       addr => rom_addr,
       data => font_word
   );
-- instantiate dual port tile RAM (2^12-by-8)
text_ram_unit : entity work.sync_rw_port_ram
   generic map(ADDR_WIDTH => 12, DATA_WIDTH => 8)
   port map(
       clk     => clk,
       -- write from main system
       we      => we_ch,
       addr_w  => addr_w,
       din     => ch_in,
       -- read to vga
       addr_r  => addr_r,
       dout    => ch_ram_out
   );
-- tile RAM write
addr_a <= yt & xt;

-- *****************************************************************
-- delay-line registers
-- *****************************************************************
process(clk)
begin
   if (clk'event and clk = '1') then
       y_delay1_reg <= y(3 downto 0);
       x_delay1_reg <= x(2 downto 0);
       x_delay2_reg <= x_delay1_reg;
       ch_d1_reg    <= ch_ram_out;
   end if;
end process;

-- *****************************************************************
-- pixel data read
-- *****************************************************************
-- tile RAM address
addr_b    <= y(8 downto 4) & x(9 downto 3);
char_addr <= ch_ram_out(6 downto 0); -- 7 LSBs (ascii code)
-- font ROM
row_addr  <= y_delay1_reg;
rom_addr  <= char_addr & row_addr;
-- select a bit
bit_addr  <= x_delay2_reg;
font_bit  <= font_word(to_integer(unsigned(not bit_addr)));

-- *****************************************************************
-- pixel color control
-- *****************************************************************
-- reverse color control
rev_bit <= ch_d1_reg(7);
f_rgb   <= front_rgb when rev_bit = '0' else back_rgb;
b_rgb   <= back_rgb when rev_bit = '0' else front_rgb;
-- palette circuit
p_rgb   <= f_rgb when font_bit = '1' else b_rgb;
```

```
    — transparency control
  osd_rgb <= KEY_COLOR when ch_d1_reg(6 downto 0)=NULL_CHAR else p_rgb;
end arch;
```

To match the delays introduced by the tile RAM and font ROM, the delayed x, y, and `char_out` signals are used accordingly. The routing of the `font_bit` signal is done by a multiplexer, coded as an array with a dynamic index:

```
  font_bit <= font_word(to_integer(unsigned(not bit_addr)));
```

Note that a row (i.e., a word) in the font ROM is defined in descending order, as in (7 `downto` 0). Since the screen's x-coordinate is defined in an ascending fashion, in which the number increases from left to right, the order of the retrieved bits must be reversed. This is achieved by the **not** operator in the expression.

The last portion of the code controls the transparency and color reversal and sets the final text's foreground and background colors.

23.3.3 Top-level design

The top-level of the OSD core consists of the OSD pointer pixel generation circuit, the slot wrapping circuit, and the blending circuit.

Register map The processor interacts with the OSD core as follows:
- set (i.e., write) the value of bypass register.
- specify (i.e., write) the foreground and background colors.
- write a character to the designated tile RAM location.

Recall that each video core occupies a 14-bit word address space. We use x to represent a "don't-care" bit and a to represent a local address bit. The address offsets and fields are
- address offset 0x aaaa aaaa aaaa (address of the tile RAM, which is the concatenation of tile's y- and x-coordinates)
  - bit 7: 1-bit foreground and background color reversal control
  - bits 6 to 0: 7-bit ASCII code of the character
- address offset 1x xxxx xxxx xx00 (bypass register)
  - bit 0: bypass bit
- address offset 1x xxxx xxxx xx01 (foreground color register)
  - bits 11 to 0: 12-bit foreground color
- address offset 1x xxxx xxxx xx10 (background color register)
  - bits 11 to 0: 12-bit background color

HDL code The top-level HDL code instantiates the OSD pixel generator circuit, provides a write slot interface, and implements the blender. The blender consists of two stages. The first stage performs chroma-key blending and the second stage uses a multiplexer for bypassing. The HDL code of the core is shown in Listing 23.3.

Listing 23.3 OSD core

```
library ieee;
use ieee.std_logic_1164.all;
use ieee.numeric_std.all;
entity chu_vga_osd_core is
```

```vhdl
   generic(
      CD          : integer                             := 12;
      KEY_COLOR : std_logic_vector(11 downto 0) := (others => '0')
   );
   port(
      clk      : in  std_logic;
      reset    : in  std_logic;
      -- frame counter
      x        : in  std_logic_vector(10 downto 0);
      y        : in  std_logic_vector(10 downto 0);
      -- video slot interface
      cs       : in  std_logic;
      write    : in  std_logic;
      addr     : in  std_logic_vector(13 downto 0);
      wr_data : in  std_logic_vector(31 downto 0);
      -- stream interface
      si_rgb  : in  std_logic_vector(CD - 1 downto 0);
      so_rgb  : out std_logic_vector(CD - 1 downto 0)
   );
end chu_vga_osd_core;

architecture arch of chu_vga_osd_core is
   signal wr_en        : std_logic;
   signal wr_reg       : std_logic;
   signal wr_bypass    : std_logic;
   signal wr_fg_color  : std_logic;
   signal wr_bg_color  : std_logic;
   signal wr_char_ram  : std_logic;
   signal fg_color_reg : std_logic_vector(CD - 1 downto 0);
   signal bg_color_reg : std_logic_vector(CD - 1 downto 0);
   signal bypass_reg   : std_logic;
   signal osd_rgb      : std_logic_vector(CD - 1 downto 0);
begin
   -- instantiate osd generator
   osd_src_unit : entity work.osd_src
      generic map(CD => CD)
      port map(
         clk        => clk,
         x          => x,
         y          => y,
         xt         => addr(6 downto 0),
         yt         => addr(11 downto 7),
         ch_in      => wr_data(7 downto 0),
         we_ch      => wr_char_ram,
         front_rgb => fg_color_reg,
         back_rgb  => bg_color_reg,
         osd_rgb    => osd_rgb);
   -- registers
   process(clk, reset)
   begin
      if reset = '1' then
         fg_color_reg <= (others => '1');
         bg_color_reg <= (others => '0');
      elsif (clk'event and clk = '1') then
         if wr_fg_color = '1' then
            fg_color_reg <= wr_data(CD - 1 downto 0);
         end if;
         if wr_bg_color = '1' then
            bg_color_reg <= wr_data(CD - 1 downto 0);
         end if;
         if wr_bypass = '1' then
```

```
          bypass_reg <= wr_data(0);
      end if;
   end if;
end process;
-- decoding logic
wr_en         <= '1' when write = '1' and cs = '1' else '0';
wr_char_ram <= '1' when addr(13) = '0' and wr_en = '1' else '0';
wr_reg        <= '1' when addr(13) = '1' and wr_en = '1' else '0';
wr_bypass   <= '1' when wr_reg='1' and addr(1 downto 0)="00" else '0';
wr_fg_color <= '1' when wr_reg='1' and addr(1 downto 0)="01" else '0';
wr_bg_color <= '1' when wr_reg='1' and addr(1 downto 0)="10" else '0';
-- chroma-key blending and multiplexing
so_rgb <= si_rgb when bypass_reg='1' or osd_rgb=KEY_COLOR else osd_rgb;
end arch;
```

Memory usage   The OSD core can generate a full text-mode screen. The core only requires 6 KB of memory, which includes a 2-KB font ROM module and a 4-KB tile RAM module. The memory requirement is much smaller than that of a frame buffer outlined in Table 20.1. A font can be extended to include additional graphical symbols to draw boxes and GUI widgets. This allows us to construct a simple GUI with limited hardware resources.

23.4   OSD CORE DRIVER AND TESTING PROGRAM

The OSD core adds a layer of text on a frame. Its driver and basic testing program are discussed in this section.

23.4.1   OSD core driver

We define an `OsdCore` class for the OSD core. The class definition is shown in Listings 23.4.

Listing 23.4   `OsdCore` class definition (in `vga_core.h`)

```
class OsdCore {
public:
   /* register map */
   enum {
      BYPASS_REG = 0x2000,
      FG_CLR_REG = 0x2001,
      BG_CLR_REG = 0x2002
   };
   /* chroma key color and constants */
   enum {
      CHROMA_KEY_COLOR = 0,
      NULL_CHAR = 0x00,      // flag for transparent char tile
      CHAR_X_MAX = 80,       // 80 char per row
      CHAR_Y_MAX = 30        // 30 char per column
   };
   /* methods */
   OsdCore(uint32_t core_base_addr);
   ~OsdCore();                      // not used
   void bypass(int by);
   void set_color(uint32_t fg_color, uint32_t bg_color);
   void wr_char(uint8_t x, uint8_t y, char ch, int reverse = 0);
   void clr_screen();
```

```
private:
   uint32_t base_addr;
};
```

The `wr_char()` method writes an ASCII character to the specific location in normal or reversed mode. The `set_color()` method sets the foreground and background colors of the text. The `clr_screen()` method is a simple utility function that clears the screen by writing the null character to all locations.

The class implementation is shown in Listings 23.5.

Listing 23.5 `OsdCore` class implementation (in `vga_core.cpp`)

```
OsdCore::OsdCore(uint32_t core_base_addr) {
   base_addr = core_base_addr;
   set_color(0x0f0, CHROMA_KEY_COLOR);
}

OsdCore::~OsdCore() {}

void OsdCore::set_color(uint32_t fg_color, uint32_t bg_color) {
   io_write(base_addr, FG_CLR_REG, fg_color);
   io_write(base_addr, BG_CLR_REG, bg_color);
}

void OsdCore::wr_char(uint8_t x, uint8_t y, char ch, int reverse) {
   uint32_t ch_offset;
   uint32_t data;

   ch_offset = (y << 7) + (x & 0x07f);    // concatenation of y and x
   if (reverse == 1)
      data = (uint32_t)(ch | 0x80);
   else
      data = (uint32_t) ch;
   io_write(base_addr, ch_offset, data);
   return;
}

void OsdCore::clr_screen() {
   int x, y;

   for (x = 0; x < CHAR_X_MAX; x++)
      for (y = 0; y < CHAR_Y_MAX; y++) {
         wr_char(x, y, NULL_CHAR);
      }
   return;
}

void OsdCore::bypass(int by) {
   io_write(base_addr, BYPASS_REG, (uint32_t ) by);
}
```

Note that the default argument is used for the `reverse`parameter of the `wr_ch()` method. If it is not specified, a normal (i.e., non-reversed) character is displayed.

23.4.2 Testing program

The overall testing program for the vanilla daisy FPro system is discussed in Section 21.7.3. The `osd_check()` function verifies the basic operation and the code is shown in Listing 23.6.

Listing 23.6 OSD test function (in `main_video_test.cpp`)

```
void osd_check(OsdCore *osd_p) {
   osd_p->set_color(0x0f0, 0x001); // dark grey / green
   osd_p->bypass(0);
   osd_p->clr_screen();
   for (int i = 0; i < 64; i++) {
      osd_p->wr_char(8 + i, 20, i);
      osd_p->wr_char(8 + i, 21, 64 + i, 1);
      sleep_ms(100);
   }
   sleep_ms(3000);
}
```

The function first clears the screen and then displays the 128 characters in two separate lines, one in normal display and one in reverse display.

In the "overlay test" portion of Listing 21.15, the background color of tile is controlled by a switch, as in

```
osd.set_color(0x0f0, sw.read(9));
```

The color can be 0x001, which is close to black, or 0x000, which is the chroma-key color. In the latter case, the background of the tiles becomes transparent without blocking the background frame.

23.5 BIBLIOGRAPHIC NOTES

The sprite- and tile-based designs were used widely in early video games and the concept is still used in today's computer graphics. Additional materials can be found in the Wikipedia website by searching the keywords of "sprite" and "tile-based video game."

The font used in this chapter is similar to the one used in the early IBM PC, which is sometimes referred to as "IBM code page 437." The detailed information can be found in the Wikipedia website, including the 128 additional symbolic and graphics characters of the extended set.

23.6 SUGGESTED EXPERIMENTS

23.6.1 Rotating banner

A rotating banner on the monitor screen moves a line of text from right to left and then wraps around. Let the text on the banner be "Hello, FPGA World." Derive the software and verify its operation.

23.6.2 Text console

The OSD core can be used to create a console to display text, similar to a serial-port terminal window on a PC. The detailed specifications of the console are as follows:

- A 40-tile-by-20-tile area is used for display.
- A cursor is used to indicate the current location.
- The screen starts a new line when a "carriage return" ASCII code ($0d_{16}$) is received.

- A line wraps around (i.e., starts a new line) after 40 characters.
- When the cursor reaches the bottom of the screen (i.e., the last line), the first line will be discarded and all other lines move up (i.e., scroll up) one position.

Create a function similar to the `disp()` function discussed in Section 11.4.4 and verify its operation.

23.6.3   Underline for the cursor

The character bitmap has several blank rows, as shown in Figure 23.2. They can be used to accommodate an *underline* for the character. This feature can be implemented by adding an additional "underline" bit to the character data and expanding the OSD core to generate the line as needed, similar to the processing of the reverse bit. The width of the character data will be expanded from eight bits to nine bits accordingly. Develop the new core, update the software driver, and verify its operation.

23.6.4   Portrait-mode display

In the *portrait mode*, a monitor is turned 90 degrees and the text is displayed in a "vertical" screen. We can enhance the OSD core to support both *landscape mode* (normal "horizontal" screen) and portrait mode and use a one-bit control signal to select the mode. Develop the new core, update the software driver, and verify its operation.

23.6.5   Font scaling circuit: part I

A font can be "magnified" by enlarging the screen pixels. The 8-by-16 font can be scaled to a 16-by-32 font by enlarging the original bitmap four times (i.e., expanding one pixel to four pixels). The original 80-tile-by-30-tile text screen will be transformed to a 40-tile-by-15-tile screen accordingly. We want to design a new OSD pixel generation circuit that can accommodate both fonts and to use a one-bit control signal to select the font. Note that the core can generate either the normal font or the magnified font, but not two fonts at the same time. Develop the new core, update the software driver, and verify its operation.

23.6.6   Font scaling circuit: part II

The core of Experiment 23.6.5 can only display one font at a time. One way to overcome the problem is to create two frames, one for the normal font and one for the magnified font. The enhanced core can have two tile RAMs for the two frames and use the chroma-key blending to merge them. The application software should be carefully designed to avoid any overlapping text. Develop the new core, update the software driver, and verify its operation.

23.6.7   Extended font

The extended IBM font includes additional 128 characters, which includes special symbols and widgets to create a simple GUI screen. The information of the detailed patterns can be found in the bibliographic section. Develop a new OSD core to cover the extended font, update the software driver, and verify its operation.

23.6.8   Tile-based ghost core

The ghost core discussed in Section 22.4 can place the ghost bitmap in any location within the screen. This is achieved by a rather complicated in-region circuit. An alternative is to limit the placement. We can treat the 16-by-16 ghost sprite as a tile and the frame as a 40-tile-by-30-tile grid. The ghost sprite can only be placed in one of the 1200 tile slots. This approach can greatly simplify the in-region circuit. Derive the new core, modify the `ghost_check()` function, and compare the animation effect of the original and alternative design.

CHAPTER 24

VGA FRAME BUFFER CORE

A *frame buffer* is a special memory module that holds the bitmap of a complete frame. It provides a temporary storage in which the pixel data is transferred or processed. In this chapter, we use FPGA's internal memory to construct a frame buffer for the VGA system.

24.1 OVERVIEW

In a video system, the monitor cannot maintain data and thus the pixel stream must be fed continuously at the designated pixel rate. This implies that the video sources and processing circuits must match this rate. A general and versatile scheme to ease the constraint is to use a *frame buffer* to hold the pixel data of the entire frame.

Conceptually, a frame buffer is a memory module in which each pixel can be independently retrieved and updated. The module has two access ports. The first port is connected to a *read pipe*, which is a data-retrieving circuit that reads the memory continuously at the pixel rate and outputs a steady pixel data stream. The second port is usually connected to a processor that generates or updates the content of the frame by writing the pixel data to the designated locations. Other types of devices, such as a video camera, can be connected as well.

This arrangement separates the data-generating rate and data-retrieving rate. Since a copy of frame data is stored in the frame buffer and retrieved at pixel rate, the display can always maintain a stable image. A processor can update the frame at a relatively slower rate or just revise a small area of the frame. In fact, the processor can be completely idle and does not generate any new pixel data. A still screen will be displayed.

While the basic concept is straightforward, the complexity of a frame buffer depends on the buffer size and the data rate. The size and rate can vary significantly, as demonstrated in Table 20.1. In this chapter, we use FPGA's internal memory modules to demonstrate the construction and operation of a frame buffer. The FPGA's internal memory modules are intended for small buffers and lookup tables and their capacity is limited. In a more realistic scenario, external memory devices, particularly SDRAM (synchronous dynamic RAM), should be used for massive storage. A vendor's *memory controller core* discussed in Section 7.5 is needed for this purpose.

24.2 FRAME BUFFER CORE

A *frame buffer core* contains a memory buffer and interface circuits to retrieve and update the pixel data. The main design is around the construction of a large dual-port memory module using FPGA's internal memory. The following subsections discuss the issues and show the implementation.

24.2.1 FPGA memory consideration

Modern FPGA devices contain internal memory modules. The module is referred to as the BRAM in a Xilinx device, as discussed in Section 7.1. The numbers of BRAM modules and total usable bits are summarized in Table 2.1.

The 640-by-480 VGA frame contains about 310K (i.e., $640*480$) pixels. Thus, the frame buffer needs $310K*1$ bits (i.e., 38 KB) for the 1-bit color and $310K*12$ bits (i.e., 461 KB) for the 12-bit color. To save memory space, we truncate the color depth from 12 bits to 9 bits in our frame buffer design and use a palette circuit to convert the readout back to 12 bits. This leads to a more manageable size of $310K*9$ bits (i.e., 345 KB).

The Nexys 4 DDR board contains an Artix XC7A100T device, which has 540 KB of usable internal memory. It is large enough to accommodate the 345-KB memory of frame buffer, the 128-KB memory of the MicroBlaze MCS, and other buffers and lookup tables.

In comparison, the Basys 3 board contains a smaller Artix XC7A35T device, which has 200-KB internal memory. Its capacity is too small to accommodate a 9-bit color frame buffer. One possible option is to reduce the color depth to 1 bit in conjunction with a programmable 2-entry palette table.

24.2.2 Video memory module

The 9-bit VGA frame buffer needs a 310K-by-9 dual-port memory module. In the HDL template of Listing 7.7, the size of the inferred memory module is determined indirectly from the width of the address bus. This implies that the number of the data words must be a power of two. The 310K data words need a 19-bit (i.e., $\lceil \log_2 310K \rceil$) memory address and thus a 512K-by-9 (i.e., 2^{19}-by-9) memory module is inferred. Clearly, more than 200 KB of memory will be wasted.

A close observation shows that a 320K-by-9 module can be constructed by combining a 256K-by-9 (i.e., 2^{18}-by-9) module and a 64K-by-9 (i.e., 2^{16}-by-9) module. It is larger than the 310K requirement and all its capacity is utilized. The HDL code is shown in Listing 24.1. It instantiates two modules and uses the MSB of the address to perform the decoding.

Listing 24.1   320K-by-9 memory module

```
library ieee;
use ieee.std_logic_1164.all;
use ieee.numeric_std.all;
entity vga_ram is
   generic(
      DW : integer := 9
   );
   port(
      clk    : in  std_logic;
      we     : in  std_logic;
      addr_w : in  std_logic_vector(18 downto 0);
      data_w : in  std_logic_vector(DW - 1 downto 0);
      addr_r : in  std_logic_vector(18 downto 0);
      data_r : out std_logic_vector(DW - 1 downto 0)
   );
end vga_ram;

architecture arch of vga_ram is
   signal data_r_256k  : std_logic_vector(DW - 1 downto 0);
   signal data_r_64k   : std_logic_vector(DW - 1 downto 0);
   signal we_256k      : std_logic;
   signal we_64k       : std_logic;
begin
   -- instantiate 256K RAM
   ram_256k_unit : entity work.sync_rw_port_ram(beh_arch)
      generic map(
         DATA_WIDTH => DW,
         ADDR_WIDTH => 18)
      port map(
         clk    => clk,
         we     => we_256k,
         addr_w => addr_w(17 downto 0),
         din    => data_w,
         addr_r => addr_r(17 downto 0),
         dout   => data_r_256k
      );
   -- instantiate 64K RAM
   ram_64k_unit : entity work.sync_rw_port_ram(beh_arch)
      generic map(
         DATA_WIDTH => DW,
         ADDR_WIDTH => 16)
      port map(
         clk    => clk,
         we     => we_64k,
         addr_w => addr_w(15 downto 0),
         din    => data_w,
         addr_r => addr_r(15 downto 0),
         dout   => data_r_64k
      );
   -- read data multiplexing
   data_r <= data_r_64k when addr_r(18) = '1' else data_r_256k;
   -- write decoding
   we_256k <= we and (not addr_w(18));
```

```
   we_64k  <= we and addr_w(18);
end arch;
```

24.2.3 Address translation

The VGA's pixel location is represented by two-dimensional coordinates and needs to be converted to a one-dimensional memory address. This issue is discussed in the construction of the sprite RAM and the tile RAM in Sections 22.2.1 and 23.2.3. Both conversions are achieved by concatenating the y-coordinate and the x-coordinate to form a memory address. However, this simple scheme leaves some unused "holes" in the address space if the size of the row is not of a power of 2. While the wasted memory is negligible in a small tile RAM, it becomes very significant for the frame buffer. Instead, we need to use the exact *row-major order* formula to do the address translation. Let the two-dimensional coordinates be (x, y). For the 640-by-480 resolution, the formula to obtain the one-dimensional address is

$$addr = 640*y + x$$

The $*$ operator implies an expensive hardware multiplier. Closer examination shows that
$$addr = 640*y + x = 512*y + 128*y + x = y \ll 9 + y \ll 7 + x$$

where $\ll$ is the shift left operator (recall that in the binary system $y \ll n$ corresponds to $y * 2^n$). This formula can be implemented by an *address translation circuit* and the HDL statement is

```
  addr <= '0' & y(8 downto 0) & "000000000"  +
             "000" & y(8 downto 0) & "0000000") + x;
```

This scheme essentially replaces the multiplier with an adder.

24.2.4 Pixel generation circuit

The frame buffer core can be treated as video source that generates a pixel data stream. The top-level diagram of its pixel generation circuit is shown in Figure 24.1. It is composed of a dual-port RAM module, a read pipe, and a palette circuit. The read pipe just performs address translation. It does not contain any buffer or control logic since the BRAM is synchronous and returns the data in one clock cycle.

The palette circuit maps and converts the retrieved pixel data to 12-bit color format. We treat our frame buffer core as a 9-bit color system and thus the palette circuit performs a 9-bit-to-12-bit conversion. The HDL code is shown in Listing 24.2.

Listing 24.2   9-bit-to-12-bit palette circuit

```
library ieee;
use ieee.std_logic_1164.all;
entity frame_palette_9 is
   port(
      color_in  : in  std_logic_vector(8 downto 0);
      color_out : out std_logic_vector(11 downto 0)
   );
end frame_palette_9;
```

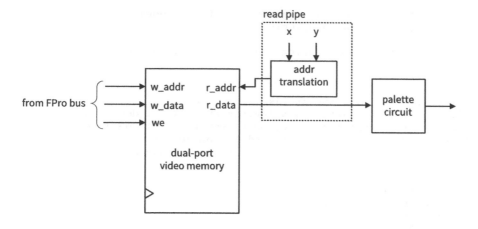

Figure 24.1   Top-level diagram of frame buffer pixel generation circuit.

```
architecture arch of frame_palette_9 is
   alias r_in    : std_logic_vector(2 downto 0) is color_in(8 downto 6);
   alias g_in    : std_logic_vector(2 downto 0) is color_in(5 downto 3);
   alias b_in    : std_logic_vector(2 downto 0) is color_in(2 downto 0);
   signal r_out  : std_logic_vector(3 downto 0);
   signal g_out  : std_logic_vector(3 downto 0);
   signal b_out  : std_logic_vector(3 downto 0);
begin
   r_out       <= r_in & r_in(2);
   g_out       <= g_in & g_in(2);
   b_out       <= b_in & b_in(2);
   color_out <= r_out & g_out & b_out;
```

Because of the limitation of FPGA's internal memory capacity, the data width of the frame buffer may be smaller. The design intentionally codes the palette circuit as a separate component so that it can be substituted later to accommodate a different data width. The _9 suffix indicates that the data width of the frame buffer is 9 bits.

The code for the frame buffer pixel generation circuit is shown in Listing 24.3. It follows the block diagram of Figure 24.1.

Listing 24.3   Frame buffer pixel generation circuit

```
library ieee;
use ieee.std_logic_1164.all;
use ieee.numeric_std.all;
entity frame_src is
   generic(
      CD : integer := 12;              -- color depth
      DW : integer := 9                -- video RAM data width
   );
   port(
      clk            : std_logic;
      -- frame counter coordinate
      x, y           : in  std_logic_vector(10 downto 0);
      -- write interface
      addr_pix       : in  std_logic_vector(18 downto 0);
```

```
        wr_data_pix : in   std_logic_vector(DW - 1 downto 0);
        write_pix   : in   std_logic;
        -- stream
        frame_rgb   : out std_logic_vector(CD - 1 downto 0)
    );
end frame_src;

architecture arch of frame_src is
    signal ram_rd_out_data : std_logic_vector(DW - 1 downto 0);
    signal converted_color : std_logic_vector(CD - 1 downto 0);
    signal y_offset        : unsigned(18 downto 0);
    signal r_addr          : std_logic_vector(18 downto 0);
begin
    -- instantiate video ram
    vram_unit : entity work.vga_ram
        generic map(DW => DW)
        port map(
            clk   => clk,
            -- write port (to processor)
            we    => write_pix,
            addr_w => addr_pix(18 downto 0),
            data_w => wr_data_pix(DW - 1 downto 0),
            -- read port (to read pipe)
            addr_r => r_addr,
            data_r => ram_rd_out_data
        );
    -- instantiate palette circuit
    pallete_unit : entity work.frame_palette_9
        port map(
            color_in  => ram_rd_out_data,
            color_out => converted_color);
    -- read address = 640*y + x = 512*y + 128*y + x
    y_offset <= unsigned('0' & y(8 downto 0) & "000000000") +
                unsigned("000" & y(8 downto 0) & "0000000");
    r_addr   <= std_logic_vector(y_offset + unsigned(x));
    -- 1 clock delay line
    process(clk)
    begin
        if (clk'event and clk = '1') then
            frame_rgb <= converted_color;
        end if;
    end process;
end arch;
```

Note that the memory access takes one clock cycle. To make the core conform to the two-clock delay imposed by the daisy video subsystem, a register is inserted to the output to add an additional one-clock delay.

24.3 REGISTER MAP

The processor interacts with the frame buffer core as follows:

- set (i.e., write) the value of bypass register.
- write a pixel to the designated RAM location.

Because of its large memory space, the frame buffer core is separated from other cores and assigned a 20-bit address space, as discussed in Section 21.1.2. We use x to represent a "don't-care" bit and a to represent a local address bit, whose range is from 0 to $640*480-1$. The address offsets and fields are:

- address offset `xaaa aaaa aaaa aaaa aaaa` (address of the RAM, with a valid value between 0 and $640*480-1$)
    - bits 8 to 0: 9-bit color data
- address offset `1111 1111 1111 1111 1111` (bypass register)
    - bit 0: bypass bit

24.3.1 Top-level HDL code

The top-level HDL code instantiates the frame buffer pixel generator circuit, provides a write slot interface, and implements the blender. The HDL code of the core is shown in Listing 24.4.

Listing 24.4   Frame buffer core

```
library ieee;
use ieee.std_logic_1164.all;
use ieee.numeric_std.all;
entity chu_frame_buffer_core is
   generic(
      CD : integer := 12;   -- color depth
      DW : integer := 9     -- frame buffer RAM data width
   );
   port(
      clk     : in  std_logic;
      reset   : in  std_logic;
      x       : in  std_logic_vector(10 downto 0);
      y       : in  std_logic_vector(10 downto 0);
      -- video interface
      cs      : in  std_logic;
      write   : in  std_logic;
      addr    : in  std_logic_vector(19 downto 0);
      wr_data : in  std_logic_vector(31 downto 0);
      -- stream interface
      si_rgb  : in  std_logic_vector(CD - 1 downto 0);
      so_rgb  : out std_logic_vector(CD - 1 downto 0)
   );
end chu_frame_buffer_core;

architecture arch of chu_frame_buffer_core is
   signal wr_pix     : std_logic;
   signal wr_en      : std_logic;
   signal wr_bypass  : std_logic;
   signal bypass_reg : std_logic;
   signal frame_rgb  : std_logic_vector(CD - 1 downto 0);
begin
   -- instantiate pixel generation circuit
   frame_gen_unit : entity work.frame_src
      generic map(
         CD => CD,
         DW => DW
      )
      port map(
         clk          => clk,
         x            => x,
         y            => y,
         addr_pix     => addr(18 downto 0),
         wr_data_pix  => wr_data(DW - 1 downto 0),
         write_pix    => wr_pix,
         frame_rgb    => frame_rgb
```

```vhdl
    );
   -- register
   process(clk, reset)
   begin
      if reset = '1' then
         bypass_reg <= '0';
      elsif (clk'event and clk = '1') then
         if wr_bypass = '1' then
            bypass_reg <= wr_data(0);
         end if;
      end if;
   end process;
   -- decoding logic
   wr_en      <= '1' when write = '1' and cs = '1' else '0';
   wr_bypass  <= '1' when addr = x"fffff" and wr_en = '1' else '0';
   wr_pix     <= '1' when addr /= x"fffff" and wr_en = '1' else '0';
   -- stream blending: mux
   so_rgb     <= si_rgb when bypass_reg = '1' else frame_rgb;
end arch;
```

24.4   DRIVER AND THE TESTING PROGRAM

The frame buffer core provides storage for the entire screen. Its driver and basic testing program are discussed in this section.

24.4.1   Frame buffer core driver

We define a `FrameCore` class for the frame buffer core. The class definition is shown in Listing 24.5.

Listing 24.5   `FrameCore` class definition (in `vga_core.h`)

```cpp
class FrameCore {
public:
   /* register map */
   enum {
      BYPASS_REG = 0x7ffff
   };
   /* video resolution */
   enum {
      HMAX = 640,   // horizontal   resolution
      VMAX = 480    // vertical resolution
   };
   /* methods */
   FrameCore(uint32_t frame_base_addr);
   ~FrameCore();               // not used
   void wr_pix(int x, int y, int color);
   void clr_screen(int color);
   void plot_line(int x1, int y1, int x2, int y2, int color);
   void bypass(int by);
private:
   uint32_t base_addr;
   void swap(int &a, int &b);
};
```

The `wr_pix()` method writes a pixel color data to a specific location. The `clr_screen()` method is a simple utility function that clears the screen by writing a specific color to all locations. The class also includes a method, `plot_line()`, to demonstrate the concept of *geometrical modeling.*

The implementation of basic methods is shown in Listing 24.6. The implementation of the `plot_line()` method is discussed in the next subsection.

Listing 24.6 `FrameCore` class implementation (in `vga_core.cpp`)

```
FrameCore::FrameCore(uint32_t frame_base_addr) {
   base_addr = frame_base_addr;
}

FrameCore::~FrameCore() {}

void FrameCore::wr_pix(int x, int y, int color) {
   uint32_t pix_offset;

   pix_offset = 640 * y + x;
   io_write(base_addr, pix_offset, color);
   return;
}

void FrameCore::clr_screen(int color) {
   int x, y;

   for (x = 0; x < HMAX; x++)
      for (y = 0; y < VMAX; y++) {
         wr_pix(x, y, color);
      }
   return;
}

void FrameCore::bypass(int by) {
   int pass = 0;

   if (by == 1) {    //bypass only by is 1
      pass = 1;
   }
   io_write(base_addr, BYPASS_REG, (uint32_t ) pass);
}
```

24.4.2 Geometrical modeling

A *geometrical model* is generated by the mathematical description of the object, which sometimes is referred to as a *vector graphic.* For example, we can obtain a line segment from two given points, (x_1, y_1) and (x_2, y_2), by generating a series of pixels based on the equation

$$\frac{y - y_1}{x - x_1} = \frac{y_2 - y_1}{x_2 - x_1}$$

We can obtain the pixel coordinates by using the x as the independent variable and then calculating y:

$$y = \frac{y_2 - y_1}{x_2 - x_1} * (x - x_1) + y_1$$

This can be realized by a C segment:

```
slope=(float)(y2-y1) / (float)(x2-x1);
```

```
for(x=x1; x!=x2; x=x+1){
   y = slope*(x-x1) + y1;
   wr_pix(x, y, color);
}
```

The straightforward code suffers several problems. It uses expensive floating-point operation and the generated line appears as disconnected if the slope is steep. A better alternative is to use the *Bresenham algorithm,* which uses integer arithmetic exclusively and takes into consideration a discrete screen. The `plot_line()` method is based on this algorithm and its implementation is shown in Listing 24.7. The computer graphics and geometric modeling themselves are a separate discipline and involve many sophisticated techniques and algorithms. Deriving even a simple set of driver routines is beyond the scope of the book. The line plotting function just gives us a taste of this type of program and additional information can be found in the bibliographic section.

Listing 24.7   `plot_line()` method (in `vga_core.cpp`)

```
void FrameCore::plot_line(int x0, int y0, int x1, int y1, int color) {
   int dx, dy;
   int err, ystep, steep;

   if (x0 > x1) {
      swap(x0, x1);
      swap(y0, y1); }
   // slope is high
   steep = (abs(y1 - y0) > abs(x1 - x0)) ? 1 : 0;
   if (steep) {
      swap(x0, y0);
      swap(x1, y1); }
   dx = x1 - x0;
   dy = abs(y1 - y0);
   err = dx / 2;
   if (y0 < y1) {
      ystep = 1;
   } else {
      ystep = -1;
   }
   for (; x0 <= x1; x0++) {
      if (steep) {
         wr_pix(y0, x0, color);
      } else {
         wr_pix(x0, y0, color);
      }
      err = err - dy;
      if (err < 0) {
         y0 = y0 + ystep;
         err = err + dx;
      }
   }
}

void FrameCore::swap(int &a, int &b) {
   int tmp;

   tmp = a;
   a = b;
   b = tmp;
}
```

Figure 24.2   OV7670 camera module.

24.4.3 Testing program

The overall testing program for the vanilla daisy FPro system is discussed in Section 21.7.3. The `frame_check()` function verifies the basic operation and the code is shown in Listing 24.8.

Listing 24.8   Frame buffer test function (in `main_video_test.cpp`)

```
void frame_check(FrameCore *frame_p) {
    int x, y, color;

    frame_p->bypass(0);
    for (int i = 0; i < 10; i++) {
        frame_p->clr_screen(0x008);   // dark green
        for (int j = 0; j < 20; j++) {
            x = rand() % 640;
            y = rand() % 480;
            color = rand() % 512;
            frame_p->plot_line(400, 200, x, y, color);
        }
        sleep_ms(300);
    }
    sleep_ms(3000);
}
```

The inner for-loop draws 20 random lines originating from the coordinates $(400, 200)$ and the outer for-loop repeats the action ten times.

24.5 PROJECT IDEAS

A video camera contains an image sensor and converts the image to a pixel stream. Because of the high data rate, most camera modules require a special high-speed I/O interface and thus cannot be used with introductory FPGA prototyping boards. However, there are a few inexpensive modules available. They are based on an OmniVison OV7670 device and can be connected to normal I/O pins. A module is shown in Figure 24.2. The OV7670 device can generate full-frame or sub-sampled images in a wide range of formats and can be configured via an I^{2}C-like serial interface. At the highest performance, it can output 16-bit color VGA data at a rate of 30 frames per second.

The simplified conceptual diagram of the device is shown in the left of Figure 24.3. The device generates the relevant timing signals and outputs the pixel

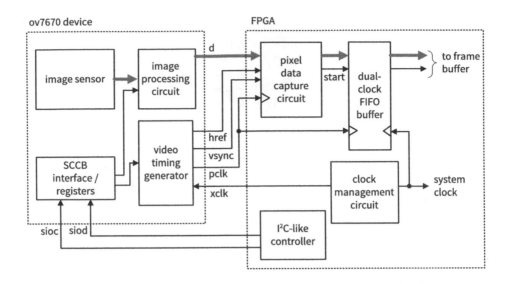

Figure 24.3 Block diagram of OV7670 and camera core

data via the 8-bit d signal. The 16-bit data is divided into two packets and transmitted in two consecutive clock cycles. The output timing signals consists of pclk (pixel clock), href (horizontal sync), and vsync (vertical sync). These signals are derived from xclk, which is the input reference clock signal and can run at a maximum rate of 48M Hz.

The configuration is done through the OmniVision SCCB (serial camera control bus) bus, which is similar to the I^{2}C bus. The sioc and siod signals correspond to the I^{2}C's clock and data signals.

A video core can be developed to integrate the camera module into the FPro video framework. The conceptual diagram is shown in the right of Figure 24.3. It is composed of four major parts. The clock management circuit generates the 48-MHz xclk signal. The I^{2}C-like controller sends the configuration commands via the SCCB bus. The data capture circuit samples the data bus, converts the packets into the 9-bit color format, and decodes the timing signals to generate the start signal, which is asserted at the beginning of a frame. The dual-clock FIFO is used as a buffer for the clock-domain-crossing interface since the data capture circuit is driven by the camera-generated pclk clock.

Because the camera's frame rate (30 frames per second) is much slower than the VGA's frame rate (60 frames per second rate), the core's output pixel stream cannot be connected to the FPro stream interface directly. It must be fed to the frame buffer.

24.6 BIBLIOGRAPHIC NOTES

A wide variety of references are available in the area of computer graphics. These books are based either on an existing API (such as OpenGL or DirectX) or primitive routines. A text, *Fundamentals of Computer Graphics, fourth ed.* by P. Shirley and

S. Marschner, provides a comprehensive coverage of general computer graphics. A graphics library, *Adafruit_GFX*, developed by Adafruit Industries contains many basic functions and is intended for a simple embedded system. The `plot_line()` method is based on a function in the *Adafruit_GFX* library.

24.7 SUGGESTED EXPERIMENTS

24.7.1 Virtual prototyping board panel

We can derive a virtual graphic panel that mirrors the condition of the FPGA prototyping board. The panel should consist of the same numbers of slide switches, pushbutton switches, discrete LEDs, and seven-segment LED displays. Derive the software program and verify its operation.

24.7.2 Virtual analog wall clock

We wish to implement an analog wall clock on the VGA monitor. The clock should have rotating hour, minute, and second hands. Derive the code and verify its operation.

24.7.3 Geometrical model functions

In Section 24.4.1, only the line plotting function is provided. Many additional geometrical modeling routines can be added:

- Function to draw a square or rectangle
- Function to draw a polygon
- Function to draw a circle or oval
- Function to fill a closed shape with a specific color

Derive these functions and verify their operations.

24.7.4 Simulated "Etch A Sketch" toy

We can implement a simulated "Etch A Sketch" toy with a mouse and VGA monitor. It functions as follows:

- The mouse pointer can be moved to the desired location.
- Whenever the left button is pressed, the system records the mouse movement and shows the trace on the monitor.
- When the right button is pressed, the system erases the screen.

Create an FPro system with the PS2 core, derive the software code, and verify its operation.

24.7.5 Frame buffer core with 3-bit color depth

The frame buffer memory requirement can be eased by reducing the color depth. We want to redesign the frame buffer core for the three-bit color format. The palette circuit can be redesigned as a 2^3-to-12 lookup table. The processor treats the table as a small memory module and can update the entries. Thus, eight 12-bit colors can be displayed simultaneously. Develop the new core, update the software driver, and verify its operation.

24.7.6 Frame buffer core with 1-bit color depth

Repeat the Experiment 24.7.5 with one-bit color format.

24.7.7 QVGA frame buffer core

The QVGA (quarter VGA) has a 320-by-240 resolution and its display area is one quarter of VGA. We can design a QVGA frame buffer core that outputs a VGA frame. One quarter of the VGA frame is the QVGA display and the other three quarters are filled with the chroma-key color. The QVGA display can be placed in any location within the VGA frame. Develop the new core, update the software driver, and verify its operation.

24.7.8 Line drawing hardware accelerator

A *graphic card* migrates certain graphic operations to hardware to speed the operation. We can accelerate the line plotting operation (i.e., the plot_line() method) by implementing the algorithm in hardware. The accelerator will accept a command composed of the starting point, the end point, and the color. It then generates the pixels and updates the frame buffer. Study the Bresenham algorithm, derive the custom hardware and incorporate it into the frame buffer core, update the software driver, and verify its operation.

24.7.9 Bidirectional frame buffer access: part I

The daisy video subsystem only supports bus write operation; i.e., the processor cannot read data from video core's registers or local RAM. Some applications may need to retrieve data from a video core. To overcome the problem, the application software can maintain a copy of data mirroring the content of the core's RAM content. However, this approach is not feasible for the frame buffer core because of the size of its memory. To incorporate the read functionality in frame buffer core, the bridge and video controller need to be modified to support the bidirectional traffic. Develop the new core, and modify the daisy video system as needed, update the software driver, and verify the system operation.

24.7.10 Bidirectional frame buffer access: part II

With the bidirectional frame buffer, the hardware overlay operation can be simulated by software. Consider a mouse pointer layer. The software can use a smaller buffer to store the background pixels under the mouse bitmap. We can move the mouse pointer icon on top of the background frame as follows:

- Restore the background pixels by writing back the buffered pixels to the frame buffer.
- Read the area of the new destination from the frame buffer and store the pixels to the software buffer.

- Write the mouse pointer bitmap to the destination area of the frame buffer. Derive the software and verify its operation.

PART V

EPILOGUE

CHAPTER 25

WHAT'S NEXT

The primary focus of the book is on the digital system development at register-transfer level. We discuss the design principles and practices through a series of projects and introduce the basic concepts of SoC and software-hardware co-design along the way.

However, today's FPGA devices are very sophisticated and capable. The capability of a high-end FPGA device rivals the entry-level ASIC devices and its development goes beyond register-transfer level design. Following are the main areas to continue:

- EDA (electronic design automation) tool
- IP-centric development flow
- High-level development tool
- Embedded OS
- Full VHDL language
- Testbench and verification

The Nexys 4 DDR board contains a large Artix device and necessary memory device and network ports. It can be used for the first four areas.

EDA tool

Developing a digital system is a complex task and is aided by a collection of EDA tools for synthesis, placement and routing, timing analysis, etc. In Vivado Design Suite, these tools are integrated into a single GUI framework. Our development basically uses the default setting. Studying the user guides and manuals for these tools can help us better understand the development procedure, exercise more control over the processes, and derive more efficient implementation.

IP-centric development flow

Since the book is about learning digital system design and developing portable HDL codes, we deliberately avoid vendor-specific IP cores and design our own cores from scratch. In real-world practice, it is estimated that the pre-designed IP cores accounts for 70% to 90% of the gate count of a complex digital system.

Learning the following topics helps us embrace the IP-centric development flow in Xilinx platform:

- *AXI interface.* The AXI interface is used widely in SoC design and is the interconnect structure used in the Xilinx IP platform. The IP core developed for the FPro bus and even the entire subsystem can be converted to AXI interface and integrated into the Xilinx IP platform.
- *MicroBlaze processor.* The MicroBlaze MCS used in the FPro system is a simple, pre-configured MicroBlaze system, as discussed in Section 10.6.1. The full MicroBlaze processor supports the AXI interface and can incorporate a high-performance memory controller and the full range of IP cores.
- *MIG (memory interface generator) utility and DDR memory controller.* The MIG utility can be used to generate and configure a DDR memory controller to access the external DDR SDRAM device on the Nexsys 4 DDR board. The memory module then can be used as the main memory of the MicorBlaze processor. In addition, it can be shared as the frame buffer through proper partition.
- *IP Integrator utility.* The IP-based system construction in Vivado Design Suite is done by the IP Integrator utility. It is a GUI used to derive system-level schematics.

High-level development tool

Hardware designs can be created at different levels of abstraction. The book is mainly at the *register-transfer* (RT) level, in which a digital system is described in terms of data transfer and manipulation between registers (like an FSMD). *High-level synthesis* (HLS), sometimes referred to as electronic system-level (ESL) synthesis, focuses on a more abstract level. It describes the hardware's "algorithmic behavior" using C-like language constructs and generates RT-level HDL codes. The HLS tools have gradually improved and become more accessible. Vivado Design Suite includes the Vivado HLS package for this purpose.

Embedded OS

The FPro system uses a bare metal model for software development. With the full MicroBlaze processor and adequate external memory, a full-fledge OS, such as Linux, can be installed. The OS enables us to use the more complex I/O peripherals, such as the network and USB controllers, which require sophisticated drivers and protocol stacks. The OS also has better support for task scheduling and file systems.

Xilinx supports the PetaLinux tools, which help developers to configure, build, and deploy Linux to Xilinx FPGA based systems.

Full VHDL language

VHDL is a complex language. It is intended to model hardware behavior at various levels and to facilitate the verification process. Since the book focuses on design, it covers only a small synthesizable subset of VHDL. It is necessary to learn the complete language to perform simulation and verification tasks.

Testbench and verification

Verification is a key part of the development process. In a large digital system, the effort and time spent on verification is comparable, if not exceeding, the effort and time spent on design. One key technique is to develop comprehensive testbenches to simulate and verify the design in a host computer. Another technique is to check the system's properties via assertions. We should learn these skills after having a good comprehension of the full VHDL language.

REFERENCES

1. P. J. Ashenden, *The Designer's Guide to VHDL*, 3rd ed., Morgan Kaufmann, 2008.

2. D. G. Bailey, *Design for Embedded Image Processing on FPGAs*, Wiley-IEEE Press, 2011.

3. M. Barr, *Programming Embedded Systems in C and C++*, 2nd ed., O'Reilly Media, 2006.

4. L. Bening and H. D. Foster, *Principles of Verifiable RTL Design*, 2nd ed., Springer-Verlag, 2001.

5. J. Bergeron, *Writing Testbenches: Functional Verification of HDL Models*, Springer-Verlag, 2003.

6. A. Chapweske, "PS/2 Mouse/Keyboard Protocol," http://www.computer-engineering.org.

7. A. Chapweske, "PS/2 Keyboard Interface," http://www.computer-engineering.org.

8. A. Chapweske, "PS/2 Mouse Interface," http://www.computer-engineering.org.

9. P. P. Chu, *RTL Hardware Design Using VHDL: Coding for Efficiency, Portability, and Scalability*, Wiley-IEEE Press, 2006.

10. P. P. Chu, *Embedded SoPC Design with Nios II Processor and VHDL Examples*, Wiley, 2011.

11. P. P. Chu, *Embedded SoPC Design with Nios II Processor and Verilog Examples*, Wiley, 2012.

12. M. D. Ciletti, *Advanced Digital Design with the Verilog HDL*, 2nd ed., Prentice Hall, 2010.

13. M. D. Ciletti, *Starter's Guide to Verilog 2001*, Prentice Hall, 2003.

14. Digilent, *Nexys 4 DDR FPGA Board Reference Manual*.

15. N. Dutt and S. Pasricha, *On-Chip Communication Architectures: System on Chip Interconnect*, Morgan Kaufmann, 2008.

16. A. Feldman, *Designing Arcade Computer Game Graphics*, Wordware Publishing, 2000.

17. D. D. Gajski, *Principles of Digital Design*, Prentice Hall, 1997.

18. IEEE, *IEEE Standard for Verilog Hardware Description Language (IEEE Std 1364-2005)*, Institute of Electrical and Electronics Engineers, 2006.

19. IEEE, *IEEE Standard VHDL Language Reference Manual (IEEE Std 1076-2008)*, Institute of Electrical and Electronics Engineers, 2009.

20. iRobot, *iRobot Create Open Interface*.

21. B. Jacob et al., *Memory Systems: Cache, DRAM, Disk*, Morgan Kaufmann, 2007.

22. L. Di Jasio, *Programming 32-bit Microcontrollers in C: Exploring the PIC32*, Newnes, 2008.

23. T. Karvinen et al., *Sensors: A Hands-On Primer for Monitoring the Real World with Arduino and Raspberry Pi*, Maker Media, 2014.

24. R. H. Katz and G. Borriello, *Contemporary Logic Design*, 2nd ed., Prentice Hall, 2004.

25. M. Keating and P. Bricaud, *Methodology Manual for System-on-a-Chip Designs*, 3rd ed., Springer-Verlag, 2002.

26. B. W. Kernighan and D. M. Ritchie, *C Programming Language*, 2nd ed., Prentice Hall, 1988.

27. J. J. Labrosse, *Embedded Systems Building Blocks*, 2nd ed., CMP, 1999.

28. C. M. Maxfield, *The Design Warrior's Guide to FPGAs*, Newnes, 2004.

29. S. Meyers, *Effective C++*, Addison-Wesley Professional, 2005.

30. NXP Semiconductor, *I^2C-Bus Specification and User Manual*.

31. J. Nurmi, *Processor Design: System-on-Chip Computing for ASICs and FPGAs*, Springer, 2007.

32. S. Palnitkar, *Verilog HDL*, 2nd ed., Prentice Hall, 2003.

33. D. A. Patterson and J. L. Hennessy, *Computer Organization and Design: The Hardware/Software Interface*, 5th ed., Morgan Kaufmann, 2013.

34. M. Puckette, *The Theory and Technique of Electronic Music*, World Scientific Publishing, 2007.

35. J. M. Rabaey, *Digital Integrated Circuits*, 2nd ed., Prentice Hall, 2002.

36. C. Reas and B. Fry, *Processing: A Programming Handbook for Visual Designers and Artists*, MIT Press, 2014.

37. P. R. Schaumont, *A Practical Introduction to Hardware/Software Codesign*, Springer, 2010.

38. P. Shirley and S. Marschner, *Fundamentals of Computer Graphics, 2nd ed.*, A K Peters, 2009.

39. W. Stallings, *Computer Organization and Architecture, 10th ed.*, Pearson, 2015.

40. J. Tyler, *App Inventor for Android: Build Your Own Apps*, Wiley, 2011.

41. F. Vahid and T. D. Givargis, *Embedded System Design: A Unified Hardware/Software Introduction*, Wiley, 2001.

42. J. Vankka, *Direct Digital Synthesizers: Theory, Design and Applications*, Springer, 2001.

43. J. F. Wakerly, *Digital Design: Principles and Practices*, Prentice Hall, 2002.

44. W. Wolf, *FPGA-Based System Design*, Prentice Hall, 2004.

45. W. Wolf, *Computers as Components: Principles of Embedded Computing System Design*, 2nd ed., Morgan Kaufmann, 2008.

46. Xilinx, *DS180 7 Series FPGAs Data Sheet: Overview*.

47. Xilinx, *UG472 7 Series FPGAs Clocking Resources User Guide*.

48. Xilinx, *UG473 7 Series FPGAs Memory Resources User Guide*.

49. Xilinx, *UG474 7 Series FPGAs Configurable Logic Block User Guide*.

50. Xilinx, *UG480 7 Series FPGAs and Zynq-7000 All Programmable SoC XADC Dual 12-Bit 1 MSPS Analog-to-Digital Converter User Guide*.

51. Xilinx, *UG586 7 Series FPGAs Memory Interface Solutions*.

52. Xilinx, *UG888 Vivado Design Suite Tutorial: Design Flows Overview*,

53. Xilinx, *UG901 Vivado Design Suite User Guide: Synthesis*.

54. Xilinx, *UG936 Vivado Design Suite Tutorial: Programming and Debugging*.

55. Xilinx, *UG937 Vivado Design Suite Tutorial: Logic Simulation*.

56. Xilinx, *UG986 Vivado Design Suite Tutorial: Implementation*.

57. Xilinx, *PG059 Block Memory Generator*.

58. Xilinx, *PG063 Distributed Memory Generator*.

59. Xilinx, *PG111 I/O Module*.

60. Xilinx, *PG116 MicroBlaze Micro Controller System*.

61. Xilinx, *PG150 UltraScale Architecture-Based FPGAs Memory IP*.

APPENDIX A

TUTORIALS

We use Xilinx *Vivado Design Suite* and *SDK* for hardware and software development, respectively. These packages consist of a comprehensive array of tools and are very complex. The detailed discussion of their use is beyond the scope of the book. The appendix gives a brief overview of Vivado Design Suite and SDK and presents short tutorials to illustrate the basic flows of hardware development, simulation, Xilinx IP core instantiation, and software development. The purpose of the tutorials is to "jump-start" the process. The readers then can refer to Xilinx's manuals and user guides for detailed information.

As FPGA's capability and capacity continue to grow, the EDA (electronic design automation) tools evolve at a similar pace. The software is updated and patched almost on a quarterly basis. The tutorials are based on Vivado WebPack version 2016. While the basic development flow is expected to remain the same, some details, such as the menu items and icon arrangement, may differ in the new versions.

A.1 OVERVIEW OF THE XILINX VIVADO IDE

Vivado Design Suite provides comprehensive development facilities for Xilinx's advanced FPGA devices and the Vivado IDE is a graphical interface for users to access EDA tools and to manage design sources, configuration, and results.

A typical Vivado GUI window is shown in Figure A.1. It is divided into three

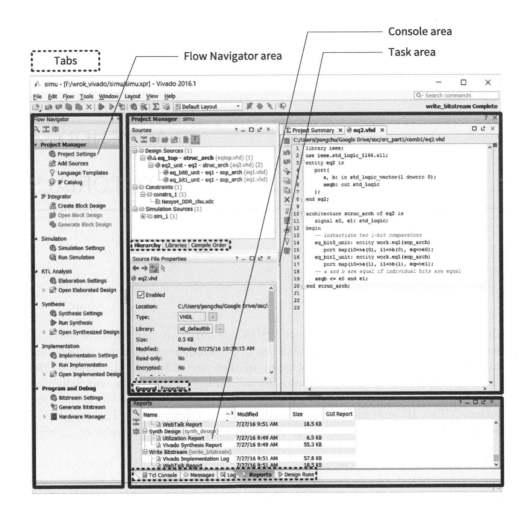

Figure A.1   Typical Vivado window.

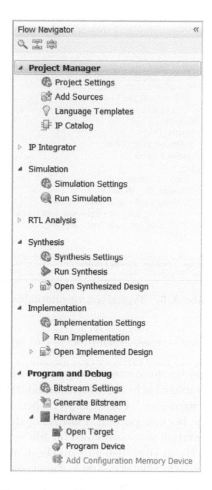

Figure A.2   Flow Navigator subwindow.

major areas:

- *Flow Navigator area.* It contains one subwindow that shows the flow of major development processes.
- *Task area.* It contains several subwindows. The types and layouts of subwindows depend on the process selected in the Flow Navigator. The task area associated with the Project Manager process is shown in Figure A.1.
- *Console area.* It accepts Tcl commands and displays status and error messages, reports, logs, etc.

Each subwindow may be resized, moved, docked, or undocked. The default layout can be restored by selecting the Layout ≻ Reset Layout menu. Note that a subwindow may contain multiple pages. The tabs at the bottom are used to select the desired page. The details are discussed in the following subsections.

Flow Navigator subwindow The Flow Navigator subwindow shows the major development processes and their subprocesses, as shown in Figure A.2. The RTL Analysis, Synthesis, Implementation, and Program and Debug processes resemble the left branch in Figure 2.3 and the Simulation process resembles the right branch. A

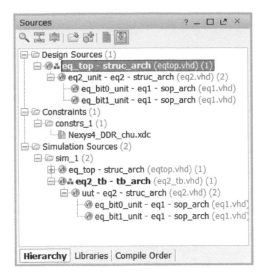

Figure A.3   Typical source subwindow.

process contains several subprocesses. A subprocess can be launched by clicking on the corresponding icon.

The development invokes the processes in a specific order. However, Flow Navigator incorporates a "auto make" scheme, which automatically runs the processes necessary to get to the desired step. For example, when we initiate the Generate Bitstream subprocess in the bottom portion, Flow Navigator automatically invokes the Synthesis and Implementation processes since bit file generation depends on the implementation result, which, in turn, depends on the synthesis result.

The Vivado GUI is not fixed. The available features, such as menu items, icons, and types of documents, are adjusted according to the current process.

Task area for Project Manager   This book mainly uses Project Manager for development and its window as shown in Figure A.1. This is also the default setting when Vivado is launched. The area contains three subwindows:

- Sources subwindow
- Source File Property subwindow
- Workplace subwindow

The Sources subwindow provides automated source file management. A typical subwindow is shown in Figure A.3.

The sources can be displayed in a specific *view*, selected by the tabs at the bottom. We mainly use the Hierarchy view for the basic hardware development.

The Hierarchy view organizes sources into three folders, which are Design Sources, Constraints, and Simulation Sources, and displays the sources according to the internal design hierarchy. The hierarchy updates automatically when new sources are added to the project. The top-level module is identified by boldface and an icon, as in the eq_top module. The synthesis and implementation processes are performed for this particular module. We can manually set the top-level module by right-clicking a module and select Set as Top in the pop-up menu.

The Source File Property subwindow shows the relevant information of the high-lighted source item in the Sources subwindow, such as the file location and size.

The Workplace subwindow contains multiple documents (such as HDL code, report, and schematic) for viewing and editing. Its content and layout depends on the current process. A document is invoked when we select an item in the Sources subwindow or activate certain processes from the Flow Navigator subwindow. During the hardware development, we mainly use it to edit the HDL files and to check the design summary and various reports. During simulation, we use it to observe the simulated waveform.

Console subwindow The Console subwindow accepts Tcl commands, shows the progress of various processes, and displays status and error messages, reports, logs, etc. We mainly use the Message tab, which displays errors, warnings, and information messages. It helps us identify and locate any problems in the HDL code.

A.2 SHORT TUTORIAL ON VIVADO HARDWARE DEVELOPMENT

In the Vivado IDE, the simplified hardware development flow consists of the following major steps:

1. Create a design project.
2. Add or create Xilinx IP core instances.
3. Add or create HDL design codes.
4. Add a constraint file.
5. Perform synthesis, implementation, and bitstream generation.
6. Program an FPGA device.

We use the 2-bit comparator discussed in Chapter 1 for the tutorial. The codes are repeated in Listings A.1, A.2, and A.3.

Listing A.1 Gate-level implementation of a 1-bit comparator

```
library ieee;
use ieee.std_logic_1164.all;
entity eq1 is
   port(
      i0, i1 : in std_logic;
      eq     : out std_logic
   );
end eq1;

architecture sop_arch of eq1 is
   signal p0, p1: std_logic;
begin
   -- sum of two product terms
   eq <= p0 or p1;
   -- product terms
   p0 <= (not i0) and (not i1);
   p1 <= i0 and i1;
end sop_arch;
```

Listing A.2 Structural description of a 2-bit comparator

```
library ieee;
use ieee.std_logic_1164.all;
entity eq2 is
```

```vhdl
   port(
      a, b : in std_logic_vector(1 downto 0);
      aeqb : out std_logic
   );
end eq2;

architecture struc_arch of eq2 is
   signal e0, e1: std_logic;
begin
   -- instantiate two 1-bit comparators
   eq_bit0_unit: entity work.eq1(sop_arch)
      port map(
         i0 => a(0),
         i1 => b(0),
         eq => e0
      );
   eq_bit1_unit: entity work.eq1(sop_arch)
      port map(
         i0 => a(1),
         i1 => b(1),
         eq => e1
      );
   -- a and b are equal if individual bits are equal
   aeqb <= e0 and e1;
end struc_arch;
```

Listing A.3   Top-level wrapping circuit

```vhdl
library ieee;
use ieee.std_logic_1164.all;
entity eq_top is
   port(
      sw  : in std_logic_vector(3 downto 0);   -- 4 switches
      led : out std_logic_vector(0 downto 0)   -- 1 LED
   );
end eq_top;

architecture struc_arch of eq_top is
begin
   -- instantiate 2-bit comparator
   eq2_unit: entity work.eq2(struc_arch)
      port map(
         a => sw(3 downto 2),
         b => sw(1 downto 0),
         aeqb => led(0)
      );
end struc_arch;
```

A.2.1   Create a design project

A new Vivado project can be created as follows:

1. Select Vivado from the Windows start menu or click on the Vivado icon.
2. In the Vivado startup window, click on the Create New Project icon. The New Project window appears.
3. Enter the project name as eq2 and the desired directory location and click Next.

4. In the Project Type dialog, select RTL project and check the Do not specify sources at this time box. Click Next. We will add the files later using Project Manager, which is more flexible and provides more control.

5. In the part selection dialog, click the Parts tab in select field to specify the target FPGA device. For the Nexys 4 DDR board, select the following:

- Product Category: All
- Family: Artix-7
- Sub-Family: Artix-7
- Package: csg324
- Speed: -1
- Part: xc7a100tcsg324-1

For other boards, this information can be found in the FPGA board manual or by checking the marking on the top of the FPGA chip. After selection, click Next and then Finish to complete the creation. The main Vivado window, similar to that in Figure A.1, appears.

Note that there is a Boards tab in the select field and the Nexys 4 DDR board may exist. However, the information specified in the board configuration is not compatible with this book's codes and thus should not be used.

The device and language information can be changed later by invoking the Project Settings subprocess in the Flow Navigator subwindow.

A.2.2 Add or create Xilinx IP core instances

Xilinx provides a comprehensive collection of IP cores. Since the focus of the book is on digital hardware design, we develop most circuits in HDL from scratch. Only a few Xilinx IP cores are used and the creation of the core instances is discussed in Section A.4. Procedure to add existing IP instance files to a project is similar to adding existing HDL files. No Xilinx IP core instance is used in this tutorial.

A.2.3 Add or create HDL design files

After a project is created, we can add existing HDL files to the project or create new HDL files. The procedure to add existing HDL files is as follows:

1. In the Flow Navigator subwindow, expand Project Manager and then select Add Sources. A dialog appears.

2. Select the Add or create design sources button and click Next to proceed to the next dialog.

3. In this dialog window, press the "+" sign in middle of the dialog window. A small window pops up with three items: Add files.., Add directories.., and Create Files...

4. Click the Add files.. item and navigate to the location. Select the three comparator files and add them to the list. Alternatively, if the three files are in the same folder, click Add directories.. to add the directory..

5. After including all the needed files, click the Finish button. The files will be analyzed and imported to the project and displayed hierarchically in the Design Sources of the Sources subwindow, as shown in Figure A.3.

Alternatively, the three HDL files can be constructed from scratch. The procedure to create a new HDL file is as follows:

1. Follow the first three steps as before.
2. Click the Create files.. item and a Create Source File dialog appears.
3. Set the File type: field to VHDL, enter the file name, and click OK to close the dialog.
4. Click Finish to close the Add Sources dialog. A Define Module dialog appears.
5. This dialog allows us to enter port names and the architecture name. These names are embedded in the HDL code later. Click OK and the file is added to the Sources subwindow.
6. Click on the file and it appears in the Workplace subwindow. Enter the HDL code and then save the file.

A.2.4 Add a constraint file

Constraints are certain conditions imposed on the synthesis and implementation processes. For our purposes, the main constraints are the pin assignments of top-level I/O ports and the system clock rate. During the implementation process, an I/O signal of the top-level module must be mapped to a physical pin of the FPGA device. Since the peripherals' I/O signals are already permanently connected to the designated FPGA's pins on the prototyping board, we must ensure that the signals are mapped to the corresponding pins. The other type of constraint is about timing, which specifies the clock frequency of the board's oscillator.

The constraint information is presented in *XDC* (*Xilinx Design Constraints*) format, which is based on the industry-standard *SDC* (*Synopsys Design Constraints*) format. For example, the following statement specifies that signal sw[0], the LSB of the switch input, is mapped to pin J15 with the LVCMOS33 (low-voltage CMOS 3.3 volt) standard. Note that the # sign is used for a comment and the text after it is ignored.

```
# set I/O pin and electrical standard for signal sw[0]
set_property -dict {PACKAGE_PIN J15 IOSTANDARD LVCMOS33}
                    [get_ports {sw[0]}];
```

The constraint information is stored in file with an extension of .xdc and can be edited by a normal text editor.

The nexys4_ddr_chu.xdc constraint file is used for all designs of the book. The file is tailored for the Nexys 4 DDR board and includes the constraints for the pin assignments and the system clock rate. It is recommended to use the same I/O port names in the top-level module. If necessary, a top-level wrapping circuit can be used to re-map the I/O signals, as in Listing A.3. The code essentially maps the "logical" port names of the comparator to the physical signals on the prototyping board.

The procedure to add the constraint file to a project is similar to that of adding a design file:

1. In the Flow Navigator subwindow, expand Project Manager and then select Add Sources. A dialog appears.
2. Select the Add or create constraints button and click Next to proceed to the next dialog.
3. In this dialog window, press the "+" sign.
4. Click Add files.. and navigate to the location. Select the nexys4_ddr_chu.xdc file and check the copy constraints files into project box.

5. Click the Finish button. The file will be imported to the project and displayed under the Constraints folder of the Sources subwindow.

6. Click on the file and it appears in the Workplace subwindow. Comment out the constraints associated with the unused I/O signals and then save the file.

The last step can be skipped. However, this is not recommended since the unused pin assignments in the constraint file will lead to a larger number of warning messages.

A.2.5   Perform synthesis, implementation, and bitstream generation

Realizing a design consists of three cascading processes:

- Synthesis
- Implementation
- Bitstream generation

The processes can be invoked from the Flow Navigator subwindow seqentially. The procedure is as follows:

1. Make sure that the desired module is designated as the top-level module (with boldface and an icon to the left).

2. In the Flow Navigator subwindow, expand Synthesis and then select Run synthesis

3. If there are errors, check the message tab on the Console area, fix the problems, and repeat.

4. If there is no syntactic error, a Synthesis Completed windows appears and asks for the next action. Click the Cancel button since we will invoke the subsequent processes manually.

5. It is recommended to check the warning messages. Many messages relate to design errors, such as unassigned signals and combinational loops. Fix the problems and repeat Step 2.

6. After the synthesis process is completed, a series of analysis and reports are generated. If desired, expand the Synthesized Design subprocess and then select and examine a report.

7. In the Flow Navigator subwindow, expand Implementation and then select Run Implementation.

8. If desired, expand the Open Implemented Design subprocess and select and examine a report.

9. In the Flow Navigator subwindow, expand Program and Debug and then select Generate Bitstream. When the process completes, the Bitstream Generation Completed dialog appears. Click the Cancel button to close the dialog.

Recall that Flow Navigator supports the "auto-make" scheme, which automatically runs the processes necessary to get to the desired step. Another way to complete these tasks is simply starting the bitstream generation process. The synthesis and implementation processes will be invoked automatically.

A.2.6   Program an FPGA device

The last step is to *program* the FPGA device; i.e., to download the configuration file (i.e., the bit file) to the FPGA device. The procedure is as follows:

Figure A.4   Screen capture of Open Target.

1. Connect the USB cable to the micro USB port (labeled PROG UART) on the Nexys 4 DDR board and turn on the power. Make sure that Jumper JP1 is in the JTAG position.
2. In the Flow Navigator subwindow, expand Program and Debug, select Hardware manager, and then select Open Target. A small window appears, as shown in Figure A.4. If the Nexys 4 DDR board was set up before, it appears as localhost:.... Select it and go to Step 4. Otherwise, select Open New Target and an Open New Hardware Target window appears.
3. Click the Next button to go through a series of screens. Select Local server in the second screen and select Digilent board in the third one. Click the Finish button to close the window.
4. In the Flow Navigator subwindow, select Program Device. A small window labeled with xc7a100t appears. Select it and a Program Device dialog appears.
5. Click the Program button to download the bit file.
6. Verify the operation with the switches and the LED.

An alternative way to configure the FPGA is to download the configuration file to a flash device and load the configuration file from it when the power is turned on. More information of this method can be found in Digilent website.

A.3   SHORT TUTORIAL ON VIVADO SIMULATION

Vivado Design Suite integrates the simulator within the framework, which can perform behavioral, post-synthesis, and post-implementation simulations. Recall that the layout of the Task area depends on the process. A window with a typical simulation area is shown in Figure A.5. The area contains three subwindows:

- Scopes subwindow
- Objects subwindow
- Workplace subwindow

The Scopes subwindow shows the hierarchy of *VHDL scope*, which includes entity declaration, architecture body, process, etc. The scope can be expanded or collapsed similar to file folder structure. The Objects subwindow displays the *VHDL objects*, which can be signals, constants, etc., within the highlighted scope. The Workplace subwindow contains multiple documents. The key document is the waveforms of the Waveform subwindow, as shown in Figure A.5.

We use the 2-bit comparator testbench discussed in Chapter 1 for the tutorial. The design codes are shown in Listings A.1 and A.2 and the testbench code is repeated in Listing A.4.

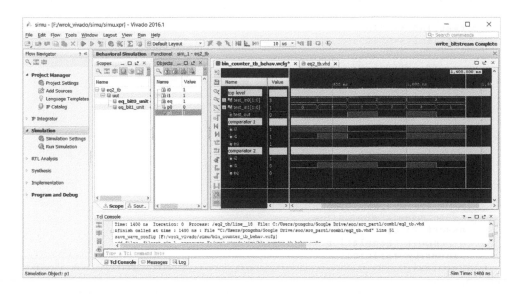

Figure A.5 Complete simulation window.

Listing A.4 Testbench for a 2-bit comparator

```
library ieee;
use ieee.std_logic_1164.all;
entity eq2_tb is
end eq2_tb;

architecture tb_arch of eq2_tb is
    signal test_in0 : std_logic_vector(1 downto 0);
    signal test_in1 : std_logic_vector(1 downto 0);
    signal test_out : std_logic;
begin
    -- instantiate the circuit under test
    uut: entity work.eq2(struc_arch)
        port map(
            a    => test_in0,
            b    => test_in1,
            aeqb => test_out
        );
    -- test vector generator
    process
    begin
        -- test vector 1
        test_in0 <= "00";
        test_in1 <= "00";
        wait for 200 ns;
        -- test vector 2
        test_in0 <= "01";
        test_in1 <= "00";
        wait for 200 ns;
        -- test vector 3
        test_in0 <= "01";
        test_in1 <= "11";
        wait for 200 ns;
        -- test vector 4
```

```
        test_in0 <= "10";
        test_in1 <= "10";
        wait for 200 ns;
        -- test vector 5
        test_in0 <= "10";
        test_in1 <= "00";
        wait for 200 ns;
        -- test vector 6
        test_in0 <= "11";
        test_in1 <= "11";
        wait for 200 ns;
        -- test vector 7
        test_in0 <= "11";
        test_in1 <= "01";
        wait for 200 ns;
        -- terminate simulation
        assert false
           report "Simulation Completed"
        severity failure;
    end process;
end tb_arch;
```

In Vivado Design Suite, the simplified simulation flow consists of the following major steps:

1. Create a design project.
2. Add or create HDL design codes.
3. Add or create an HDL testbench.
4. Perform the initial simulation.
5. Customize the waveform subwindow and re-simulate.

The first and second steps are identical to those in Section A.2. Note that the source subwindow in Figure A.3 includes a Design Sources folder and a Simulation Sources folder. When an HDL design file is added, it is also included in the Simulation Sources folder as well. Thus, the first two steps can be omitted if the project is already set up for synthesis. The remaining steps are discussed in the following subsections.

A.3.1 Add or create an HDL testbench

Adding an existing HDL testbench file is similar to adding a design file and the procedure is as follows:

1. In the Flow Navigator subwindow, expand Project Manager and then select Add Sources.
2. Select the Add or create simulation sources button and click Next to proceed to the next dialog.
3. In this dialog window, press the "+" sign in middle of the dialog window. A small window pops up with three items: Add files.., Add directories.., and Create Files...
4. Click Add files.. item and navigate to the location. Select the testbench file and add it to the list.
5. Click the Finish button. The files will be analyzed and imported to the project and displayed hierarchically in the Simulations folder of the Sources window, as shown in Figure A.3.

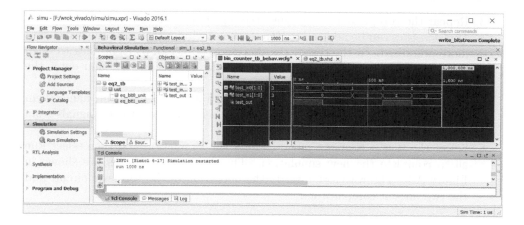

Figure A.6   Initial simulation window.

The procedure is similar to adding a design file except that the Add or create simulation sources button is selected in the second step. Because of this, the testbench file is not included in the Design Sources folder.

A.3.2   Perform initial simulation

After the testbench is set up, the procedure to perform simulation is the following:
1. In the Flow Navigator subwindow, expand Simulation and then select Simulation Settings. A dialog window appears.
2. Select sim_1 in the Simulation set: field and confirm that eq2_tb is in the Simulation top-module name: field. The name mirrors the one in the Simulations folder in Figure A.3.
3. In the Flow Navigator subwindow, select Run Behavioral Simulation.
4. After the codes are successfully compiled, the simulator GUI opens in the Workplace subwindow, as shown in Figure A.6.

By default, the HDL objects of the top-level entity (i.e., eq2_tb) are displayed in the Waveform subwindow and the simulation runs for 1000 ns. If desired, select the Run ≻ Run all menu to complete the simulation.

A.3.3   Customize waveform display

It is frequently necessary to examine signals in lower-level modules. Additional signals can added and the Waveform subwindow can be customized accordingly. For example, we want to check the operations of the two one-bit comparators by adding their I/O signals to the subwindow. This can be done as follows:
1. In the Scopes subwindow, expand the eq2_testbench icon and then the uut icon, and then highlight the eq_bit0_unit instance. The Objects subwindow is updated accordingly and lists I/O ports and internal signals of the eq_bit0_unit instance.
2. In the Objects subwindow, select i0, i1, and eq and then drag and drop them to the Waveform subwindow.

3. Repeat the two previous steps with the `eq_bit1_unit` instance.
4. *Division bars* can be add to the Waveform subwindow to make the waveforms more organized. In the top of the Waveform subwindow, right-click to open the pop-up menu, select New Divider to add a division bar, and enter the `top-level` for its name.
5. Repeat the previous step two more times to add two bars and name them `comparator 1` and `comparator 2`. Drag them to separate the signals from the two one-bit comparator instances.
6. Select the Run ≻ Restart.. menu to reset the simulation and then select Run all to complete the new simulation. The finished waveforms are shown in Figure A.5. The waveforms verify the operation of the top-level circuit and the two one-bit comparators.

A.4 TUTORIAL ON IP INSTANTIATION

Xilinx provides a comprehensive collection of IP cores. Since the focus of the book is on digital hardware design, we develop our own cores in HDL from scratch. However, certain FPGA's macro cells, such as XADC, can only be incorporated as a Xilinx IP core.

There are three general approaches to instantiate a Xilinx IP core in Vivado Design Suite:

- Copy and modify the pre-designed HDL template.
- Instantiate a single IP instance with the IP catalog utility.
- Construct an IP-based system with the IP Integrator utility.

In the first and second approaches, an IP core is instantiated as an isolated HDL module and manually incorporated into the upper level HDL code. In the third approach, IP cores are the only building blocks and used to construct the entire system. We only use the first two approaches in the book.

The following Xilinx IP cores are used in the book:

- MicroBlaze MCS core in Chapter 9
- XADC core in Chapter 12
- Dual-clock FIFO core in Chapter 21
- Clock management core in Chapter 21

The procedures to create and configure these IPs are covered in the following subsections.

The HDL files and core script files (the .xci/.xcix files) can be found in the companion website. They are stored in the xilinx_ip folder. If an Artix-7 device is used, they can be used directly without going through the following tutorials. However, these files are obtained with Vivado version 2016.2 and may need to be regenerated for later versions. The file structure is:

- ip_xci folder
  - cpu.xci
  - xadc_fpro.xci
  - mmcm_fpro.xci
- ip_hdl folder
  - xadc_fpro.vhd
  - mmcm_fpro.v

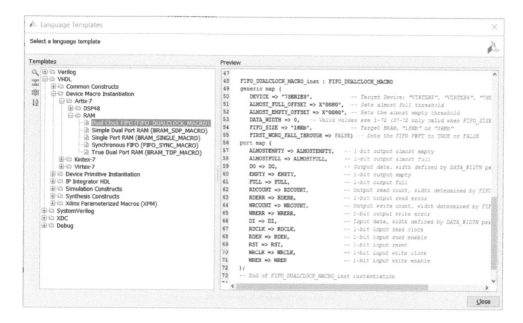

Figure A.7   BRAM-based dual-clock FIFO template window.

The HDL files can be imported as regular HDL source files, as discussed in Section A.2.3. An xci file can be imported by selecting Add existing IP in the Adding Sources.. step. Only one type of file should be imported for a project (e.g., either xadc_fpro.vhd or xadc_fpro.xci).

The dual-clock FIFO is instantiated directly as an HDL primitive and there is no separate file.

A.4.1   Dual-clock FIFO core via HDL templates

Xilinx provides a collection of HDL templates. These templates include codes to instantiate Xilinx's proprietary IP cores. We use this approach to obtain the dual-clock FIFO core. The Artix-7 BRAM itself can be configured as a FIFO buffer without additional logic. A single 16Kb BRAM module can support a 1024-by-13 dual-clock FIFO required by the line buffer of Chapter 21. The procedure to obtain the template is as follows:

1. Select the Tools ≻ Language Templates menu and a subwindow appears.
2. Click VHDL, Device Macro instantiation, Artix-7, RAM, and Dual Clock FIFO. The instantiation code segment is shown in the right panel, as shown in Figure A.7.
3. Copy the code segment to the top-level HDL file and modify the code as needed.

A.4.2   IP catalog utility

A sophisticated IP core has many configurable parameters and optional features. It is too tedious to set up the core with an HDL template. Various "wizard" programs

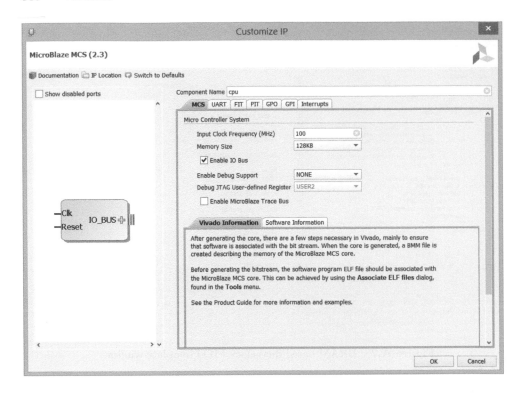

Figure A.8 MicroBlaze MCS configuration window.

guide users to configure the core and generate the HDL file. This is done with the IP catalog utility. Instantiating a single IP instance consists of the following steps:

1. Invoke a specific "wizard" to configure the desired IP core.
2. Generate the HDL file.
3. Instantiate the IP as a VHDL component in the top-level HDL file.

After the first step, a high-level description file with an extension of .xci is created. It is then used as a "blueprint" to generate HDL files. Once created, the .xci file can be copied and reused. It can be incorporated into a project by invoking the Add existing IP procedure, similar to that of adding an HDL source or constraint file.

A.4.3   Generate a MicroBlaze MCS component

The procedure to create and configure a MicroBlaze MCS instance is as follows:

1. In the Flow Navigator subwindow, expand Project Manager and then select IP Catalog. The IP Catalog subwindow appears.
2. Select Embedded Processing, then Processor, and then MicroBlaze MCS. A MicroBlaze MCS dialog window appears, as shown in Figure A.8.
3. In MCS page, configure the core as follows:
    - Enter a name, say cpu, in the Component Name field.
    - Select 128KB in the Memory Size field to obtain maximum memory.

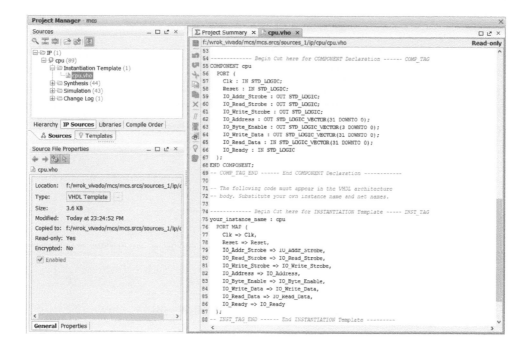

Figure A.9   MicroBlaze .vho file.

- Enter 100 in the Input Clock Frequency (MHz) field since the 100M-Hz system clock is used.
- Check the Enable IO Bus box to include the bus interface. No component in other tabs should be included since we construct the FPro system's I/O subsystem from scratch. The completed configuration should be similar to that in Figure A.8.

4. Click the OK button and a Generate Output Products dialog appears.
5. Click the Global button and then the Generate button. A Xilinx IP description file cpu.xci is created and a collection of HDL and constraint files is generated. These files are automatically added to the Design Sources folder of the Sources subwindow.

After creation, the IP instance's relevant files can be found in the IP Sources tab of the Sources subwindow, as shown in the left panel of Figure A.9. There are several folders. The Synthesis and Simulation folders contain the HDL files for synthesis and simulation, respectively. The Instantiation Template folder contains the HDL templates to instantiate the IP core. The cpu.vho file is for VHDL and its content is shown in the Workplace subwindow of Figure A.9. It contains a segment for component declaration and a segment for component instantiation. These segments can be copied and pasted in an VHDL program, as in Listing 10.8.

A.4.4   XADC IP core

The procedure to create an XADC instance is shown below. It is configured to meet the design requirement specified in Section 12.2.

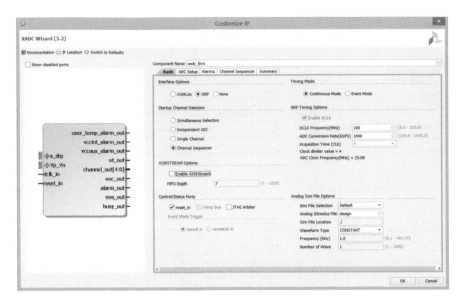

Figure A.10   Basic page of XADC Wizard.

1. In the Flow Navigator subwindow, expand Project Manager and then select IP Catalog. The IP Catalog subwindow appears.
2. Select FPGA Features and Design, then XADC, and then XADC Wizard. An XADC Wizard dialog window appears.
3. In the Basic page, configure the core as follows:
   - Enter `xadc_fpro` in the Component Name field.
   - Select DRP in the Interface Options field.
   - Select Continuous Mode in the Timing mode field.
   - Select Channel Sequencer in the Startup Channel Selection field.

   The completed page is shown in Figure A.10.
4. In the ADC Setup page, confirm that continuous option is selected in the Sequencer Mode field.
5. In the Alarms page, deselect all alarms. Confirm that continuous option is selected in the Sequencer Mode field.
6. In the Channel Sequencer page, uncheck the boxes in Bipolar column and check the following boxes in Channel Enable column: TEMPERATURE (for die temperature), VCCINT (for die core voltage), vauxp2/vauxn2, vauxp3/vauxn3, vauxp10/vauxn10, and vauxp11/vauxn11. Confirm that continuous option is selected in the Sequencer Mode field. The completed page is shown in Figure A.11.
7. Click the OK button and then the Generate button to generate the xadc_fpro.xci file and HDL files.

As in the MicroBlaze MCS IP discussed in Section A.4.3, the HDL template files can be found in the Instantiation Template folder of the xadc_fpro instance in the IP Sources page. The component declaration and component instantiation segments can be copied and pasted, as in Listing 12.1.

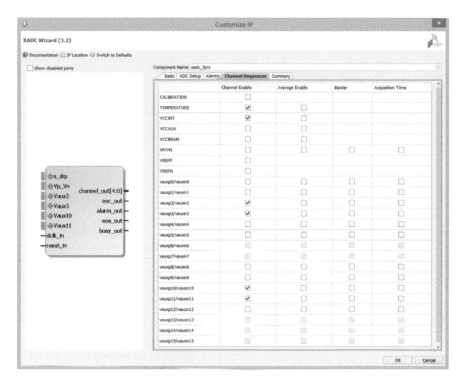

Figure A.11   Channel Sequencer page of XADC Wizard

A.4.5   Clock management IP core

The procedure to create a clock management IP instance is shown below. It is configured to generate the 25M-Hz, 40M-Hz, and 65M-Hz clocks required for the video synchronization. The 25M-Hz clock is used for the VGA synchronization core of Chapter 21 and the other two are for the SVGA and XGA resolutions in homework experiments.

1. In the Flow Navigator subwindow, expand Project Manager and then select IP Catalog. The IP Catalog window appears.
2. Select FPGA Features and Design, then Clocking, and then Clocking Wizard. A Clocking Wizard dialog window appears.
3. In the Clocking Options page, verify the following:
   - MMCM is selected in primitive field.
   - 100 is specified in the Input Frequency (MHz) column of the Input Clock Information field.
   - In the Component Name field, enter mmcm_fpro.
4. In the Output Clocks page,
   - enter 100 in the Output Frequency (MHz) / Requested column of the clk_out1 row.
   - enter 25 in the Output Frequency (MHz) / Requested column of the clk_out2 row.
   - enter 40 in the Output Frequency (MHz) / Requested column of the clk_out3 row.

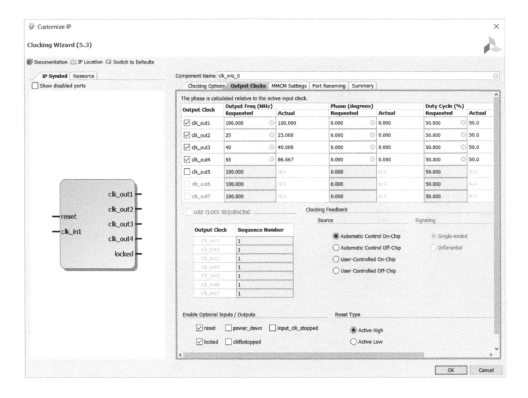

Figure A.12 Output Clocks page of MMCM Wizard

- enter 65 in the Output Frequency (MHz) / Requested column of the clk_out4 row.

The completed page is shown in Figure A.12 Note that the exact 65M-Hz clock cannot be obtained and a 66.667M-Hz clock is generated.

5. In the Port Renaming page, rename the input port and four output ports clk_in_100M, clk_100M, clk_25M, clk_40M, and clk_67M, respectively.

6. Click the OK button and then Generate button to generate the mmcm_fpro.xci file and HDL files.

Similar to the MicroBlaze MCS IP core discussed in Section A.4.3, the HDL template files can be found in Instantiation Template folder of the mmcm_fpro instance in the IP Sources page. The component declaration and component instantiation segments can be copied and pasted, as in Listing 21.12.

Note that in Clocking Wizard version v5.3, a Verilog file, mmcm_fpro.v, is generated. However, since Vivado supports mixed language synthesis, the file can be incorporated into a VHDL project.

A.5 SHORT TUTORIAL ON FPRO SYSTEM DEVELOPMENT

The FPro system development involves the derivation of hardware and software. The procedure consists of following steps:

1. Create a design project.

2. Add or create a MicroBlaze MCS instance.
3. Add or create HDL codes with an MCS instance.
4. Add a constraint file.
5. Perform synthesis, implementation, and bitstream generation.
6. Export hardware configuration.
7. Derive software and generate the executable file (elf file).
8. Embed the elf file into FPGA's memory module and regenerate bitstream.
9. Set up a terminal emulator program.
10. Program an FPGA device.

The procedure expands the previous hardware development procedure in Section A.2 and incorporates three additional steps (Steps 6, 7, and 8) to accommodate the software development. The tutorials of the three steps are provided in the following subsections.

Note that Vivado Design Suite can serve as a platform for SoC development. The IP Integrator process of Flow Navigator is for this purpose. However, the platform is intended for full-featured MicroBlaze and AXI-based IP cores. Support for MicroBlaze MCS is limited and its development does not follow Vivado's general IP-based flow. The FPro system in the book is constructed from scratch and does not use any Vivado's built-in IP integration facilities.

A.5.1 Derive FPro system hardware

We use the vanilla FPro system discussed in Section 10.7 for the tutorial. The cpu.xci IP file and HDL files can be found in the companion website. The FPro system hardware can be derived as follows:

1. Create a design project: same as in Section A.2.1.
2. Add a MicroBlaze MCS instance: add cpu.xci to project. The .xci may need to be updated for the new Vivado version. Follow the procedure in Section A.4.3 to recreate the instance if needed.
3. Add HDL codes: add the following HDL files:
   - Top-level design and slot definition: mcs_top_vanilla and chu_io_map
   - MMIO subsystem and bridge: mmio_sys_vanilla, chu_mmio_controller, and chu_mcs_bridge
   - GPI, GPO, and timer MMIO cores: chu_gpi, chu_gpo, and chu_timer
   - UART core: chu_uart, uart, uart_rx, uart_tx, baud_gen, fifo, fifo_ctrl, and reg_file
4. Add a constraint file: same as in Section A.2.4.
5. Perform synthesis, implementation, and bitstream generation: same as in Section A.2.5.

A.5.2 Export hardware configuration

Many features of MicroBlaze MCS core can be customized and thus the configuration of each instance can be different. This configuration information can be obtained and encapsulated in a *hardware description file*. In Vivado version 2016, the file can be obtained as follows:

1. Select the File ≻ Export ≻ Export Hardware... menu and a subwindow appears.

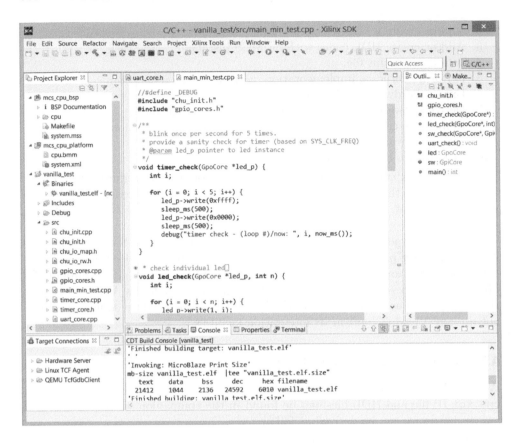

Figure A.13 Xilinx SDK .

2. In the Export to field, navigate to destination folder and then click the Ok button.
3. The hardware description file (with the extension of .hdf) is generated in the designated folder.

Since an FPro system is designed manually from scratch and not from IP Integrator, the hardware description file only contains the information about the MicroBlaze MCS configuration, not the entire FPro system. Thus, this step does not need to be repeated if the same MicroBlaze MCS instance is used.

A.5.3 Derive software

We use Xilinx SDK (software development kit) as the platform for software development. It is based on Eclipse IDE (integrated development environment) with a custom Xilinx plug-in module. A typical Xilinx SDK window is shown in Figure A.13. The basic software model constitutes a three-layer hierarchy:

- Hardware platform specifications
- BSP (board support package)
- Application program

An embedded system is built around a specific application and its configuration is tailored to support the application. *Hardware platform specification* is the bottom layer that captures the relevant hardware information required for software development and deployment. *BSP* is the middle layer. It is a software library that contains drivers and start-up routines based on the information from a specific hardware platform specification file. The term is borrowed from traditional embedded system development, in which the system is usually realized by a custom printed circuit board. An *application program* is the top-layer that uses the routines in the BSP library to access hardware.

This software model is automated for a system derived from Vivado's IP Integrator. Since the FPro system is designed from scratch, the first two layers are just for MicroBlaze MCS and the BSP only contains a start-up routine. We need to manually include the software I/O drivers in an application program.

The software development constructs the three layers in sequence. The basic steps are the following:

1. Select Xilinx SDK from the Windows start menu or click on the SDK icon. Do not launch it from Vivado.
2. Create or select a workspace.
3. Select File ≻ New ≻ Others and a window appears. Expand the Xilinx folder and then select Hardware Platform Specification. The New Hardware Project dialog appears.
4. Enter a name, say mcs_cpu_platform, for the hardware project. In Target Hardware Specification field, navigate to the destination folder specified in Section A.5.2 and select the .hdf file. Click Finish to generate the hardware platform specifications. The new specifications folder appears on the Project Explorer subwindow.
5. Select File ≻ New ≻ Board Support Package and a dialog appears. Enter a name, say mcs_cpu_bsp, for the BSP project, select the previously created mcs_cpu_platform in Target Hardware, and then select standalone for the OS field, as shown in Figure A.14. Click finish to generate BSP. The new BSP folder appears the Project Explorer subwindow.
6. Select File ≻ New ≻ Application Project and a dialog appears. Enter a name, say vanilla_test, for the application project, select the previously created mcs_cpu_platform and mcs_cpu_bsp, and then click on C++ button, as shown in Figure A.15. Click Finish to create a new application project. The new application project folder appears in the Project Explorer subwindow.
7. Import the main program file and driver files discussed in Chapter 9. An easy way to do this is to open Windows File Explorer and drag these files into the src folder. The Project Explorer subwindow of the completed project is shown in Figure A.16.
8. By default, Xilinx SDK is configured to build a project automatically and thus the vanilla_test.elf file is compiled and linked automatically after the files are dragged into the project. Note that the file size is displayed in the console tab, shown in Figure A.17.

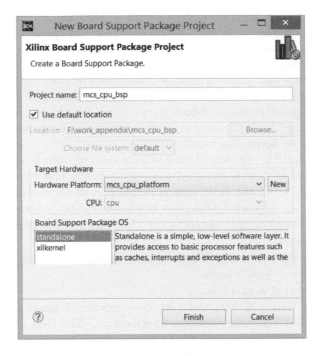

Figure A.14   BSP dialog.

Figure A.15   New project dialog.

Figure A.16 Project Explorer subwindow.

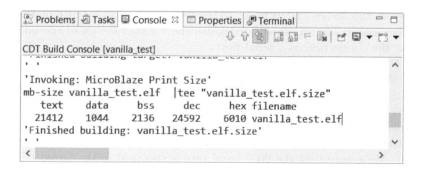

Figure A.17 File size information.

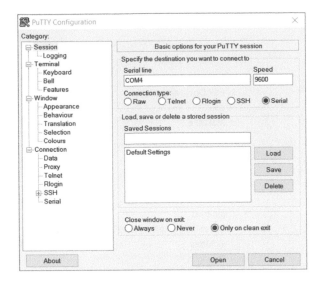

Figure A.18 PuTTY screen.

A.5.4 Embed elf file and regenerate bitstream

After the elf file is generated, it can be used as the *initial values* of FPGA's BRAMs and incorporated into the module definitions. When the bitstream is generated, these values are embedded into the bit file. The steps are as follows:

1. In Vivado Design Suite, select the Tools ≻ Associate ELF Files... menu and a dialog appears. Navigate the folder and select vanilla_test.elf and then click OK. The file will be shown under the ELF subfolder of the Design Sources folder in the Sources subwindow.
2. Follow the procedure in Section A.2 to regenerate the bit file.

A.5.5 Set up a terminal emulator program

To display the UART output character stream, a terminal emulator program is needed. We use a program, PuTTY, for this purpose. PuTTY is a telnet client and can be downloaded for free. The procedure to set up PuTTY is as follows:

1. Connect the Nexys 4 DDR board to the PC's USB port and turn on the power of the board. The PC should recognize the FT2232 device of the board and treat the connection as a serial port.
2. In Windows, open Control Panel, select Device Manager, and expand Ports (COM & LPT). The board should be listed as one of the serial ports, labeled as USB Serial Port (COMn), where n is the designated serial port (i.e., COM port) number. Record this number.
3. Invoke the PuTTY program. In its application window, select the Serial button for serial port and enter the previous recorded COMn in the Serial line field. Make sure that the 9600 baud rate is specified in the Speed field. The completed configuration screen is shown in Figure A.18.
4. Click the Open button and the terminal window appears.

A.5.6   Program an FPGA device

Follow the procedure in Section A.2 to download the bit file to an FPGA device. The program starts upon completion and the UART character data stream from the FPGA board will be displayed in PuTTY window.

Note that that the bitstream needs to be regenerated and downloaded after each software revision; i.e., the Steps 7, 8, and 10 have to be repeated for each revision.

A.6   BIBLIOGRAPHIC NOTES

Vivado is a complex software package. Four Vivado user guides, *UG888 Vivado Design Suite Tutorial: Design Flows Overview, UG986 Vivado Design Suite Tutorial: Implementation, UG936 Vivado Design Suite Tutorial: Programming and Debugging,* and *UG937 Vivado Design Suite Tutorial: Logic Simulation* provide more in-depth tutorials. Additional information can be found in various references, user guides, and application notes. An easy way to search for information is via Xilinx *DocNav (Document Navigator)*, a software utility downloaded along with the Vivado package.

Lots of useful information for Xilinx SDK can be found in its Help menu.

INDEX

Printed and bound by CPI Group (UK) Ltd, Croydon, CR0 4YY